# PROSPECTS OF FRACTURE MECHANICS

Proceedings of an international conference on

# PROSPECTS OF FRACTURE MECHANICS

Held at Delft University of Technology, The Netherlands
June 24-28, 1974

*Editors:*

G. C. SIH
*Lehigh University, Bethlehem, Pennsylvania, USA*

H. C. van ELST
*Metal Research Institute TNO, Apeldoorn, The Netherlands*

D. BROEK
*Delft University of Technology, Delft, The Netherlands*

NOORDHOFF INTERNATIONAL PUBLISHING LEYDEN

ISBN: 90 286 0474 x



Printed in The Netherlands by Drukkerij Vonk Zeist B.V.

# CONTENTS

# PREFACE

An international conference on the Prospects of Fracture Mechanics was held from June 24 to June 28, 1974, at Delft University of Technology, The Netherlands. It was sponsored jointly by the Metal Research Institute TNO, Delft University of Technology and Lehigh University, USA. The conference was organized to encourage interaction among workers in fracture mechanics from industry, government and the universities. Needless to say, the ultimate goal of safe and reliable design of structures requires the close cooperation of all three groups. It was felt that this kind of interaction can be best achieved among a small group of active participants.

The conference was held at an opportune time when Europe, the United States and other parts of the world were faced with the problem of energy shortages. The demand for large numbers of refineries, power plants, air transports, and super-mammoth tankers has made more urgent the need for reliable construction of a variety of structural components such as pressure vessels, pipelines, storage tanks, aircraft frames, turbines, generators, inland and off-shore drilling equipment, etc. A rational procedure for selecting the appropriate material in a given design is no doubt needed. To this end, the concepts of fracture mechanics become indispensable as they deal directly with the size of the tolerable subcritical defect from which adequate life-time predictions of the structural components can be made. This point was well emphasized in an opening address by Dr. F. Lagerwey, Chairman of the Industrial Branch, Dutch Organization for Applied Scientific Research. He also expressed the possibility that fracture mechanics technology could be used to establish future non-destructive testing and quality control procedures in the Netherlands.

A distinctive feature of this conference was that the organizing committee consisting of:

D. Broek (The Netherlands)

H. C. van Elst (The Netherlands)

G. C. Sih (USA)

worked hard to attract participants who are leading experts in fracture mechanics from different countries: France, Germany, Italy, Japan, Sweden, Switzerland, UK, USA, USSR and The Netherlands. The international flavor of this conference clearly demonstrates that the subject of fracture mechanics cannot be contained within national boundaries. By exchanging information from different parts of the world in a conference such as this one, the members of the fracture mechanics community can move more rapidly and more effectively toward the common goal of safe structural design.

The committee also made an attempt to design a technical program covering the various aspects of fracture mechanics problems. This includes void growth, microfracture, marcofracture and the failure of structures. Among the topics discussed were:

Inclusion and void growth

Energy criteria

Path independent integrals

Fracture experiments and testing

Fatigue crack propagation

Dynamics and arrest

Stress analysis of crack problems

Structural analysis

The participants were encouraged to review existing ideas of the field and to explore new research areas. Each session was followed by a panel discussion in which the present and future status of the field in each paper were reviewed in order to provide a perspective view of the subject.

It was hoped that the attendance in one conference of material engineers, continuum mechanists and structural designers concerned with the application of fracture mechanics would be at least a partial step toward bridging the gaps existing between the groups.

The members of the organizing committee gratefully acknowledge the cooperation received from authors, session chairmen, discussers, and panel members. A grant from the Department of Education and Science, Ministry of Cultural Affairs of the Netherlands, provided partial support of the travel expenses of participants from the United States and Japan. Appreciation is also extended to Dr. F. van Wely, Director of Management, Metal Research Institute TNO, for his personal encouragement and official support of this conference. The committee is particularly grateful to Mrs. B. DeLazaro for reading the complete manuscript, making many helpful suggestions, and preparing the format of this book.

Delft, The Netherlands
June 29, 1974

H. C. van Elst
D. Broek
G. C. Sih

The organizing committee and speaker of the opening address. Left to right: D. Broek, H.C. van Elst, G.C. Sih and F. Lagerwey.

Opening session in the Aula Building, Delft University of Technology.

# REFLECTIONS ON THE CONFERENCE

At the request of numerous participants, this effort is made to give a brief critical review of the conference. It is not our purpose here to list the various fracture phenomena in different materials or to summarize the mathematical and experimental results, for such a compilation has already been made in the subject index. On the other hand, it must be recognized that a review of technical discussions is by nature dependent on the viewpoints of the reviewers. Despite its shortcomings this document is offered as a focus for future discussion and a means of clarifying certain basic concepts. It may serve as a useful reference in the years to come.

The present day conception of fracture involves three stages: atomic, microscopic and macroscopic. The relative importance of each stage obviously depends on the subject under investigation. According to the resolution available in electron fractography, fracture begins with the nucleation, growth and coalescence* of microvoids. The material then separates when a sheet of voids creates a fracture surface. On a microscopic scale, the surface contains a series of concave depressions, referred to as "dimples". The size, shape and spacing of the dimples are observed to be strongly influenced by the prevailing stress state and microstructure of the material in a given environment. This characteristic feature of fracture occurs in many metallic materials and has been referred to as "ductile fracture**".

---

* The word coalescence should not be interpreted to describe the mechanism of voids growing or fusing into one another. Technically speaking, there always exists a finite ligament between voids at the time of failure.

** The features of microvoid growth and coalescence have also been observed in predominantly brittle fracture. Hence, ductility on a microscopic scale does not, in general, explain or imply ductility at the macroscopic or continuum level.

Study of microvoid growth has been based strictly on qualitative observations and has not led to a reliable quantitative theory. The question of whether the actual mechanism of fracture is one in which ligaments between the voids break by necking as if they were miniature tensile specimens or one in which the voids actually grow into one another is a matter of definition*. Depending on the temperature and rate of loading, these tensile specimens of the same material may either fail by cup-cone fracture or by chisel-point fracture, leaving a very small, but always finite, ligament at the point of material separation. One of the essential features of this problem is that the stability or instability of the growth of the microvoids determines the ultimate ratio of void diameter to ligament size. Moreover, one should be aware that the material elements in the microsize ligaments experience a multiaxial state of stress while experimental data are collected under uniaxial loading in terms of macrostrain or macrostress on a continuum specimen. The region within which calculations are made differs from that on which measurements are made by several orders of magnitude. The task of relating microquantities to macroquantities should be dealt with carefully. In particular, microstress and microstrain have to be distinguished from macrostress and macrostrain.

When the microvoids and dimples are extremely small, the unmagnified fracture surface appears to be flat and featureless and is said to be of the "brittle" type. In polycrystalline aggregates, the local directions of cracking are often irregular and vary from grain to grain. They do not follow any recognizable pattern with reference to the macroscopic direction of crack propagation. The "brittle fracture" characteristics observed in the grains may not necessarily imply brittleness in a continuum or global sense, i.e., in the behavior of the whole specimen. The brittle-ductile transition behavior of metals is still not well understood or defined.

At the continuum level, the phenomenon of subcritical crack growth observed in metal plates under steadily increasing load remains a mystery to analysts. Attempts made to quantify this growth using two-dimensional plasticity models are of questionable validity because stable crack growth in metal plates is known to be intimately associated with necking in the thickness direction that is not included in a two-dimensional analysis. The basic requirements of a successful analysis is to have an effective criterion of fracture. In addition, it is advisable to choose the simplest permissible idealization in the stress analysis that con-

---

*When the ligaments are much smaller, say by one order of magnitude or more, than the voids, it is justified, as an approximation, to assume that the voids grow into each other.

forms with the physics of the problem. The path independent integral is a useful tool for calculating the stress intensity factor or energy release rate expressions of certain specialized specimen geometries with cracks that extend in a self-similar fashion (Mode I crack extension). Its applicability in characterizing materials with plastic deformation* is not clear. The very concept of path independency stresses the identity of local and global fracture criteria. Some aspects of metal fracture cannot be described by a global criterion, and thus the associated local criterion is inadequate. For example, it cannot be used to explain the transition of stable to unstable crack extension. In addition, correct analytical expressions of path independent integrals have not yet been found for mixed mode crack problems involving $k_1$ and $k_2$.

The state of affairs local to the crack tip region can differ significantly from that at a distance, particularly in the response of the material to different types of loading. This local inhomogeneity can strongly affect the experimental data. A crack always tends to arrest with its tip at an obstacle, an unfavorably oriented grain or inclusion. Failure of a material element usually takes place at a small, but finite, distance away from the crack tip. This distance is an indication of the local inhomogeneity of the system and is introduced in the strain energy density criterion as the radius of a core region around the crack tip. The material elements outside the core region are assumed to possess the gross mechanical properties of the specimen. In the case of a blunt notch, the critical element would be located at some surface layer distance away from the notch surface. Such an approach avoids the difficulty of analyzing branch crack problems.

The need for a single fracture criterion that can simultaneously account for the volume change (dilatation) and shape change (distortion) experienced by material elements ahead of the crack is apparent. Under plane strain fracture**, the material along the path of expected crack extension experiences more volume change while the material in the regions away from the crack plane are highly distorted. The regions of high distortion are known

---

* Path independency applies to a conservative system which releases no energy of the dissipative type as in the plastic deformation of metals.

** This occurs in the stretching of a large cylindrical metal bar containing a small penny-shaped crack which has no thickness effect. Necking becomes important only when the crack approaches the free surface.

to contain plasticity in metals, secondary cracks in glass and crazing in polymers. The proportion of energy absorbed in dilatation and distortion can be calculated for various material properties at the locations of the stationary values of the strain energy density factor.

One of the objectives of macroscopic fracture mechanics is to find a parameter that is not so sensitive to changes in loading, geometry, etc., and that represents a material constant such as fracture toughness. The critical value of the Mode I stress intensity factor, $k_{1c} = \sigma\sqrt{a}$, is one such example, where σ and a can change individually but the combination at instability should remain essentially constant. Any experimentally measured parameter that changes with load and geometry is of limited applicability to material characterization and results in even less usefulness in structural design. The validity of the $k_1$ approach lies in its association with $G_1$, which can be measured from a compliance test. Fracture mechanics relates changes in the stiffness (or compliance) of a specimen to the amount of crack extension. For ductile materials, the increments of crack growth can no longer be accurately monitored by changes of the specimen gage length. The lack of understanding of the basic problem results in the obvious alternative of measuring changes in length over a much shortened gage length, a crack opening displacement (COD). For an edge cracked beam under three-point bending, COD, measured at the edge, is a function of the crack length until net section yielding is reached. The yielded material then acts like a plastic hinge. The two halves of the bending specimen rotate as rigid bodies. In such a case, COD is obviously not directly related to the growth of the continuum crack and its association with fracture toughness defined at the continuum level is even more remote. Test data involving measured COD should be interpreted with extreme care when compared to measurements on specimens of different crack geometries. It is meaningless to compare calculated values of COD with measured values for various specimens when the measurements were made at different locations on the crack surface unless the proper elastic-plastic stress analysis, including the thickness effect, has been made.

Fracture analysis makes use of a stress solution of the crack problem. Near the crack front, the singular part of the stress state is in a state of plane strain except where the crack intersects a free surface. On theoretical grounds, the coefficient of the $1/\sqrt{r}$ term, part of which represents the customary definition of the stress intensity factor, vanishes at the point of intersection of a crack front and a free surface which are at right angles; r is the radial distance measured from the crack front. Such a situation applies to both partial and complete penetration of cracks in a material with finite dimensions. Numerical and experimental results referring to surface stress intensity factors are not meaningful in themselves. They simply represent the inability of the analysis to calculate or measure a surface stress intensity factor which should have vanished in the strict sense. It is more important to report the variations in the stress intensity factors along the crack border as the free surface is approached. Furthermore, the stress intensity factor is an incomplete criterion for analyzing the three-dimensional fracture of fully or partially embedded cracks.

The necessity for having an accurate stress solution can be illustrated by another example dealing with the crack tip stress fields in shells. The classical solution of a crack in thin shells is inadequate because the free surface condition on the crack is not completely satisfied. This approximation leads to a discrepancy between the angular distribution of the extensional and bending stress fields. Therefore, a combined stress intensity factor (i.e., the addition of extensional and bending stress intensity factors) can be defined only by adding the extensional and bending stresses at a fixed position ahead of the crack. This invokes the criterion of maximum local stress and is in conflict with the concept of linear fracture mechanics which defines the stress intensity factor to be independent of the local coordinate variables. This difficulty does not arise in a theory which includes the effect of transverse shear deformation*.

Although fracture produced under cyclic loading of a fatigue crack exhibits a characteristically different surface topography from that of other types of fracture, it should not be considered necessarily as a basic fracture mechanism. The fatigue fracture surface usually consists of a series of parallel, regularly spaced, miniature, wave-like fracture markings oriented normally to the the direction of local crack propagation. These features, loosely called "striations", are affected by loading, specimen geometry, etc. What causes the formation of striations is still not known. The application of fracture mechanics to fatigue-crack growth has been centered on identifying cracks at the earliest possible stage and following their growth to the point of unstable crack extension. Fatigue failure is now known to initiate with the nucleation and growth of localized microcracks. There is no doubt that dislocations, vacancies, inclusions, second-phase precipitates,

---

*Although the theories of plates and shells which include the effect of transverse shear permit a more complete satisfaction of free surface conditions, there remains the fact that the stresses may not be distributed through the thickness as assumed by the theory. The boundary terms are not stresses, but rather resultants and moments of stresses.

environmental conditions, nature of loading, crack geometries, etc., all affect the fatigue crack growth rate. A rational synthesis of how these variables could alter resistance to crack initiation and growth in metal alloys is needed. Experimental fatigue data from a single test containing the combined effects of loading, specimen geometry and material properties are of limited value unless the interaction of these effects can be better understood.

There are two ways of arresting a running crack in a structure. One way is to place stiffeners along the path of expected crack propagation so as to absorb the excess energy that drives the crack. Another possibility is to replace some portions of the structure by a tough material that has sufficient resistance to arrest the crack. In either case, the dynamic fracture toughness should depend only on the material properties and the crack velocity. A continuum crack may be arrested without exhausting all the available energy in the system. The excess energy may be dissipated by the creation of free surfaces at the microscopic scale level. This is a reversal of the sequence of events in crack initiation where microscopic damage in the plastic enclaves takes place prior to crack propagation.

A few additional remarks to conclude this review are in order. The general problem of understanding the physics of a fracture phenomenon and then quantifying it analytically or experimentally presents the greatest conceptual difficulties. Attempts to include too many variables into an analysis or experiment without knowing the physics can lead quickly to confusion. Common fracture terminology such as ductile, brittle cleavage, shear, etc., should be carefully defined when used to describe fracture at the microscopic and/or macroscopic level. Cleavage and shear fracture on a microscopic scale may or may not correspond to brittle and ductile fracture on a macroscopic scale. Too often, fractures of steel structures are reported as "brittle" failures that occurred at conditions beyond the elastic limit of the material. Ambiguities of this type are common and cannot be overcome by the effort of one conference. The various questions raised here are indicative of the need for more research.

G. C. Sih
H. C. van Elst
D. Broek

# Section I

# INCLUSIONS AND VOID GROWTH

# A CRITICAL ASSESSMENT OF PLASTIC HOLE-GROWTH MECHANISMS OF DUCTILE FRACTURE

P. F. Thomason

University of Salford, Salford M5 4WT, England

## ABSTRACT

A number of hole-growth models, for normal modes of ductile fracture, are critically assessed in relation to experimental results for fracture in uniaxial tension tests, notched bend tests and metalworking processes. It is shown that, in most practical situations, transverse hole-growth is likely to have a relatively minor effect, in comparison to longitudinal or extensional growth, in bringing about the conditions where holes can coalesce by internal microscopic necking.

## INTRODUCTION

Several types of hole-growth models of ductile fracture have been presented, and these range from the dislocation models of Ashby {1} and Broek {2} to the continuum-plasticity models of McClintock {3}, Rice and Tracey {4}, Tracey {5} and Thomason {6}{7}. In the present paper, however, we consider only the continuum-plasticity models.

The continuum models of ductile fracture have two basic features in common. Firstly, they contain an expression for the growth-rate of holes under the remote stress and strain-rate fields. Secondly, they include a criterion for the coalescence of adjacent holes to form a fracture surface; thus establishing an upper limit to the integral of the growth-rate equation and a corresponding value to the fracture strain. In the following work the basic features of the ductile-fracture models are briefly described and this is followed by an assessment of the validity of the models, in relation to experimental results for ductile fracture under widely different histories of plastic flow.

## HOLE-GROWTH MODELS OF DUCTILE FRACTURE

### Elliptical Holes in a Strain-Hardening Solid

McClintock {3} presented an approximate model for the transverse growth of cylindrical holes, of elliptical cross-section, where the major and minor elliptical axes and the cylindrical axes are aligned with the principal stress directions. In developing the model, McClintock used the solution for the transverse growth of a single circular-cylindrical hole, in an infinite perfectly-plastic body, along with a solution for the growth of elliptical holes in a viscous body, to obtain (by extrapolation) a growth-rate equation for elliptical holes in a strain-hardening plastic solid. The growth-rate equation neglects interactions between neighbouring holes and thus each hole is assumed to grow as though it existed alone in the infinite body.

The hole-coalescence or fracture criterion employed by McClintock {3}, was a simple 'transverse-contact' condition in which the transverse hole dimension increases uniformly, under the applied stress and strain field, until adjacent hole surfaces achieve intimate contact. The magnitude of transverse hole-growth at fracture is thus directly related to the initial volume fraction of holes. With the aid of this fracture criterion, and by neglecting the transient nature of the stress history, McClintock was able to integrate the growth-rate equation to obtain an expression for the equivalent strain at fracture. The resulting equation was found to be in qualitative agreement with experiment since it predicts an increase in ductility with increase in strain-hardening exponent and a strong inverse relation between fracture strain and transverse stress. However, the equation is not in quantitative agreement with experiment since it considerably overestimates the fracture strains in simple tension tests, when comparisons are made with the experimental data of Edelson and Baldwin {8}, Fig. (1). (In obtaining the theoretical results in (Fig.(1), Bridgman's formula {9} was used to estimate the transverse stresses at the centre of a tension specimen 'neck' and a volume fraction $V_f$ of holes was assumed to be equivalent to the same volume fraction of second-phase particles). In an attempt to reduce the discrepancy between the theoretical and experimental results, McClintock {10} introduced a criterion for hole coalescence by localised shear between adjacent holes; this proved to be unsuccessful, however, and the relatively low values of the experimental fracture strains could not be accounted for on this basis.

## Circular Holes in a Strain-Hardening Solid

A model for the transverse growth of a circular hole lying along the axis of a rigid-plastic strain-hardening cylinder of finite diameter, was recently presented by Tracey {5}. The hole growth-rate was determined for the case where the cylinder was subject to a prescribed axial principal strain-rate and a uniform transverse stress on the outer wall. Unlike the McClintock model {3}, which was derived for a single hole in an infinite body, the Tracey model, in which the finite cylinder diameter can be set equal to the hole spacing in the fracture model, can provide a rough indication of the interaction effects of adjacent holes on the transverse growth-rate. Tracey obtained upper and lower bounds to the hole growth-rate, in terms of the hardening exponent and the stress triaxiality. His results show that an increase in strain-hardening exponent can cause a marked reduction in the initial growth-rate, but at large strains the growth-rate converges rapidly on the perfectly plastic (non-hardening) solution. The extent of the reduction in growth-rate, with strain-hardening, was found to be large for high triaxial stress levels, but only small at low stress levels.

Tracey used his transverse growth-rate model to estimate the fracture strain in the neck of a uniaxial tension specimen, on the assumption that the cylindrical cells were distributed uniformly throughout the central regions of the neck, with axes parallel to the specimen centreline. The cells were also assumed to have hole diameter to spacing ratios equal to the cube root of the volume fraction of holes in the neck region. The growth-rate equation was integrated with the aid of Bridgman's formula {9} for the stress distribution in a necked tension specimen, and the hole-coalescence criterion adopted was the transverse-contact condition of McClintock {3}. Hence, Tracey's fracture strain results, in common with those of McClintock, neglect the effects of a possible unstable coalescence of holes at strains where the hole surfaces are still widely separated. The theoretical tensile fracture strains, in Tracey's work, were obtained for material properties similar to those in the experiments of Edelson and Baldwin {8}, and the two sets of results are compared in Fig.(1). It is clear from Fig.(1) that the theoretical results greatly overestimate the experimental fracture strains, with the possible exception of the very low volume-fraction data, and in an attempt to determine whether this large overestimate was due to strain-hardening effects, Tracey obtained the theoretical fracture strains for the same model with a zero hardening exponent. The non-hardening results were found, however, to lie only slightly lower than the strain-hardening curve in Fig.(1). This result is in marked contrast to the strong hardening effect predicted by McClintock's approximate strain-hardening model {3}. Tracey's results {5} suggest,

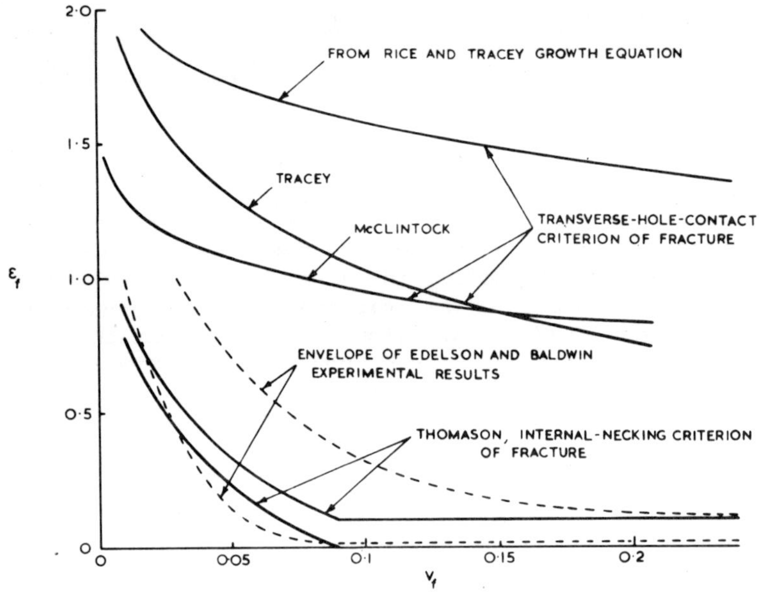


Figure (1) - Theoretical and experimental relationships between ductile fracture strain in uniaxial tension test and volume fraction of holes or particles.

however, that the strong hardening effect, in McClintock's fracture strain equation {3} is due to the failure of the approximate theory to account for the transient nature of hole growth-rate under strain-hardening conditions.

## Spherical Holes in a Non-Hardening Solid

Rice and Tracey {4} presented a model for the growth of an isolated spherical hole in an infinite rigid-plastic solid, subject to remotely uniform stress and strain-rate fields. However, they did not introduce a hole-coalescence criterion, for estimating fracture strains, and were concerned primarily with the problem of obtaining expressions for hole growth-rates in the directions of the principal axes of the remote plastic field.

The solution to the hole-growth problem was obtained with the aid of a variational principle and numerical results were obtained for a non-hardening Mises solid, subject to a remotely uniaxial tensile strain-rate field and a remotely uniform stress field. The velocity field for the variational solution was written in terms of three basic components consisting of :-
(i) the uniform incompressible strain-rate component;

(ii) the shape changing component $E$, giving shape change at constant volume; (iii) the dilational component $D$, giving volume change at constant shape. The term $(1 + E)$ thus represents an incompressible shape-changing factor which, together with $D$ completely determines the change in hole geometry. Rice and Tracey found that the D factor has an exponential dependence on the mean normal stress $\sigma_m$ and, at high triaxial stress levels, it completely overwhelms the $(1 + E)$ factor. The latter result enabled the following approximate expression to be developed for the D factor in a general remote strain-rate field (characterised by the Lode variable $\nu$) and this was shown to be a good approximation to $D$ at all triaxial stress levels {4}.

$$D = 0.558 \sinh\left(\frac{\sqrt{3}}{2}\frac{\sigma_m}{k}\right) + 0.008\,\nu \cosh\left(\frac{\sqrt{3}}{2}\frac{\sigma_m}{k}\right) \qquad (1)$$

where $k$ is the non-hardening yield stress in shear.

Rice and Tracey then went on to develop an approximate expression for the growth-rate of the spherical hole in an arbitrary remote strain-rate field and obtained the following equation for the radial velocities on the hole surface in the directions of the remote principal strain-rates $(\dot{\varepsilon}_I, \dot{\varepsilon}_{II}, \dot{\varepsilon}_{III})$.

$$\dot{R}_{oK} = \{(1 + E)\,\dot{\varepsilon}_K + \dot{\bar{\varepsilon}}\,D\}\,R_o\,, \qquad K = I, II, III \qquad (2)$$

where $\dot{\bar{\varepsilon}}$ is the remote equivalent strain-rate, $R_o$ is the spherical-hole radius, and the incompressible shape-changing factor $(1 + E)$ takes the value $^5/_3$ for low stress-triaxiality and the value 2 for high triaxiality. Equation (2) can be integrated in an approximate manner to give the finite growth of a cavity {4}, by replacing $R_o$ with the mean of the ellipsoidal principal radii. Although Rice and Tracey did not use Eqs.(1) - (2) to predict the fracture strains in a uniaxial tension test, we can perform this task by following the methods of McClintock{3} and Tracey {5} and assume that neighbouring holes will grow, as though they existed alone in the body, until a transverse hole-contact criterion of fracture {3} is satisfied. The resulting tensile fracture strains, from Rice and Tracey's non-hardening hole-growth model, are compared with the experimental results of Edelson and Baldwin {8} in Fig.(1), where it is clear that the theoretical curve greatly overestimates the experimental results and this discrepancy would be even greater if strain-hardening effects had been included in the growth-rate model. In presenting these results it must, however, be acknowledged that the method of replacing $R_o$ by the mean of the principal radii, in Eq.(2), will lead to appreciable errors at the high strains involved in the results.

The fact that the three hole-growth models of ductile fracture {3}{4}{5} overestimate the tensile fracture strains by an order of magnitude, Fig.(1), suggests that ductile fracture is occurring by an unstable-coalescence process, at a sheet of holes, well in advance of the transverse-hole-contact criterion of fracture, proposed by McClintock {3}. A hole-growth model for ductile fracture {6}, which takes account of the experimental observation {11} that ductile-fracture surfaces are created by the unstable process of internal-necking of the matrix between adjacent holes, is described in the following section. A particularly important feature of this model is the significance attached to longitudinal or extensional hole-growth and its effect on the conditions leading to the unstable coalescence of holes. In the previous fracture models {3}{4}{5} no importance was attached to extensional growth and attention was focused on the effects of uniform transverse growth.

## Square Prismatic Holes in a Non-Hardening Solid

The hole-growth model of ductile fracture, presented by Thomason {6}, is based on a simple incompressible shape-changing plastic field, in which the growth-rate amplification effects of the stress field are neglected. In the context of the Rice and Tracey analysis of spherical-hole-growth {4}, this plastic field is the part remaining when the D and E factors are set equal to zero. An inspection of Eqs.(1) - (2) shows that this simple growth-rate model will lead to a serious underestimate of hole growth, at high triaxial stress levels. However, at low and moderate stress levels there will not be a serious growth discrepancy, because D is then small and the (1 + E) factor of $(^5/_3)$ compares to a value of unity in the simple hole-growth model. Thomason's model of ductile fracture {6} was developed for a rigid-plastic non-hardening solid, deforming in plane-strain and containing a uniform distribution of square prismatic cavities, Fig.(2); each hole deforming as though it existed alone in the body. Under normal conditions of plastic flow the length to width ratio $a/b$ of the inter-cavity matrix, Fig.(2), will be such that, $a/b \ll 1$, and an unstable internal-necking of the matrix, along a sheet of holes, will be prevented by the high constraint factor $\sigma_n/2k$ needed to initiate this mode of localised deformation {6}, Fig.(2). When this condition is satisfied, plastic deformation continues uniformly throughout the body, thus elongating the holes in the direction of the tensile principal strain $\varepsilon_z$ and causing a transverse contraction, Fig.(2). This mode of deformation has the effect of reducing the 'neck' geometry ratio $a/b$, and with it the constraint factor for localised flow in the inter-cavity matrix, thus leading to a condition where ductile fracture can occur due to the unstable coalescence of holes by internal-necking along the weakest sheet of matrix 'necks'. The fracture strain will therefore correspond to the uniform strain achieved just at the onset of unstable hole coalescence, since the displacement associated with internal-necking at a single sheet of holes corresponds to a minute strain-increment on the macroscopic level, Fig.(2).

An expression for the current 'neck' geometry ratio $a/b$, in terms of the volume fraction of holes $V_f$ and the tensile principal strain $\varepsilon_z$, can be obtained by integrating the growth-rate equation for the simple incompressible shape-changing field, and this takes the form {6},

$$\frac{a}{b} = \frac{\sqrt{V_f}}{1 - \sqrt{V_f}} \exp(2\,\varepsilon_z) \; . \tag{3}$$

A criterion for ductile fracture, by the onset of internal microscopic necking, was obtained in {6} by considering the flow-mode bifurcation problem for a small element of the hole-growth model, Fig.(2). Neglecting the effect of holes on the macroscopic yield stress of the solid, the condition for the load at the onset of internal necking $L_n$, to become just less than that for continuing uniform flow on the macroscopic scale $L_u$, was shown {6} to be given by,

$$\frac{\sigma_n}{2k}\left(1 - \sqrt{V_f}\right) < \left(\frac{1}{2} + \frac{\sigma_m}{2k}\right) \; . \tag{4}$$

It is with this emphasis on a stress-dependent criterion for the unstable coalescence of holes, that Thomason's model for ductile fracture {6} differs fundamentally from the McClintock-Rice-Tracey models {3}{4}{5}. It is proposed in Thomason's model that, right up to the very onset of internal necking, large transverse hole expansions are not an essential preliminary to ductile fracture, Fig.(2), since the large transverse growth needed for hole coalescence occurs rapidly once internal necking has begun at the sheet of holes associated with fracture surface formation. This aspect of the model is consistent with experimental observations from microsections of regions immediately adjacent to ductile fracture surfaces {12}, which show the holes to be still widely separated.

Estimates of the ductile fracture strains in uniaxial tension tests, obtained from Eqs.(3) - (4) with the aid of Bridgman's formulae {9} for the stresses in an external neck, are compared with the experimental results {8} in Fig.(1). The results from the internal-necking model are seen to be in good agreement with experiment, but tend to give a slight underestimate of the fracture strains. A primary factor contributing to this small discrepancy between theory and experiment is undoubtedly the non-hardening nature of the ductile-fracture model. When second-phase particles are weakly-bonded to the strain-hardening matrix of a real metal, the conditions for the onset of internal necking are likely to be delayed until the matrix approaches a saturation level in the strain-hardening effect; this would be expected to occur at a strain approximately equal to the strain for tensile plastic instability $\varepsilon_I$ (external necking). In addition to this, when the particles are strongly-bonded to the strain-hardening matrix, experiments show {13} that a strain of the order $\varepsilon_I$ is needed to nucleate appreciable volume fractions of holes. It seems reasonable, therefore, to correct the non-hardening fracture strains, for the effects of strain-hardening, by simply adding the strain at tensile plastic instability $\varepsilon_I$ to the non-hardening results. This corrected internal-necking fracture curve is given in Fig.(1), assuming a typical instability strain $\varepsilon_I \simeq 0.1$ , and this shows a remarkable agreement between theory and experiment, particularly at high volume fractions of holes where the McClintock-Rice-Tracey results are in serious error.

## COMPARISON OF HOLE-GROWTH FRACTURE MODELS WITH EXPERIMENTAL RESULTS

In the previous discussion of ductile fracture models, comparisons were made between theoretical and experimental fracture strains in uniaxial tension tests, where the mean normal stresses achieve relatively low tensile magnitudes. We now consider the relative merits of the hole-growth fracture models, in relation to experimental results for ductile fracture at the root of notched specimens, where large triaxial tensile stresses exist, and in metalworking processes where ductile fracture often occurs in regions subject to highly compressive transverse stresses.

### Ductile Fracture in Notched Specimens

Slip-line field analyses of tension specimens with circular-root notches {14} and initially-sharp cracks subject to large-geometry-change effects {15} show that, on the axis of symmetry, the tensile principal strain achieves its greatest value at the notch surface, where the mean normal stress is small ($\sigma_m/2k = 0.5$ at $x/r = 0$). Also, with increasing distance from the root into the logarithmic spiral field, the tensile principal strain decreases rapidly to become an order of magnitude smaller than the surface strain, in the region of high stress triaxiality ($\sigma_m/2k = (1 + \pi)/2$ at $x/r = 3.82$ , for parallel-sided notch). These results suggest that, for the case of a notched specimen in which the root radius is large compared to the size and spacing of second-phase particles or holes, extensional hole-growth will greatly exceed transverse growth close to the notch root; since Eq.(1) shows that D is then small compared to the value of (1 + E) in Eq.(2). Away from the notch root, however, in the region of high stress triaxiality, transverse hole-growth will be virtually equal to extensional growth, since the D factor then overwhelms the (1 + E) factor {4}. Now in the region of low stress-triaxiality, where the tensile strain is an order of magnitude greater than that in the high triaxial stress region, it is clearly possible for the large extensional growth of holes to bring about the conditions for hole coalescence by internal-necking {16}, at a stage where the holes in the high triaxial stress region are still in the early stages of the growth process and too small for coalescence to occur. This suggests, therefore, that extensional hole-growth could be a far more important factor than transverse growth, in controlling the onset of ductile fracture at the root of notches or blunted cracks.

An indication of the relative importance of extensional and transverse hole-growth, in the ductile fracture process, can be gained from an experimental determination of the location of the initial ductile fracture-surface at the root of a notched test specimen. Experiments of this type have been performed {17} on a high-sulphur mild steel in the annealed condition; in the form of V-notched Charpy test specimens of $45^{\circ}$ notch angle and 0.25 mm. root radius. The notched specimens were deformed in three-point bending until the initial ductile fracture damage had occurred, and a micropolished section of a notch root in this condition is shown in Fig.(3(a)). This microsection clearly shows that the fracture surface has initiated in the high-strain low triaxial-stress region of large extensional hole-growth and small transverse growth. Direct confirmation that this mode of hole-growth has in fact occurred in the low triaxial-stress region, is given by the etched microsection of the area adjacent to the initial fracture surface of a similar specimen, Fig.(3(b)), where the transverse dimensions of the holes, away from the fracture surface, are very nearly equal to the size of the sulphide particles. These experimental results clearly support the suggestion that extensional hole-growth can have a predominant influence on the conditions leading to ductile fracture by unstable hole coalescence.

Since ductile fracture can initiate preferentially in regions of low triaxial stress, where transverse hole-growth is small, Thomason's incompressible hole-growth model and internal-necking fracture criterion {6} can be applied directly when fracture occurs under these conditions at a notch root {16}. However, if a situation existed where fracture initiated away from the root, in a region of high triaxial stress, Thomason's growth-rate model would have to be modified to account for the growth-rate amplification effects of the stress field {4}. The hole-growth models of ductile fracture {3}{4}{5}, which concentrate only on the effects of uniform transverse growth, are clearly inadequate models of the fracture process shown in Fig.(3).

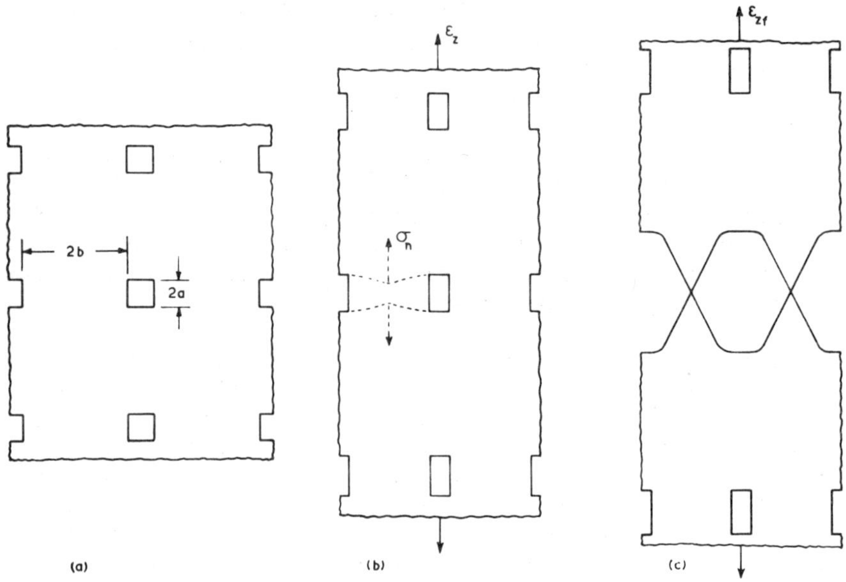


Figure (2) - (a) Initial geometry of plane-strain element with square holes. (b) Change in hole shape due to simple incompressible hole-growth law; large constraint factor $\sigma_n/2k$ prevents hole coalescence by internal necking ($L_n > L_u$). (c) Onset of ductile fracture by coalescence of single sheet of holes, when extensional growth of holes has reduced constraint factor sufficiently to give $L_n < L_u$ .

## Ductile Fracture in Metalworking Processes

In the present section we consider the hole-growth models of ductile fracture in relation to results from fracture experiments on metalworking processes, of the type where ductile fracture initiates at a free surface. The advantage of this type of process, in fracture studies, is that the complete stress-history at a free surface can be determined accurately from experimental measurements of the plastic strain history {18}.

Ductile fracture studies have been performed on cylindrical compression specimens {19}{20} to simulate conditions in the upsetting or heading type of metalworking process, where fracture initiates at the 'equator' of the barrelled free-surface, Fig.(4(a)). This type of test is particularly useful in fracture experiments since the plastic stress and strain histories, on the equatorial surface, can be varied widely by testing cylinders of different height to diameter ratio and varying the compression platen friction conditions {18}. Experimental results for a

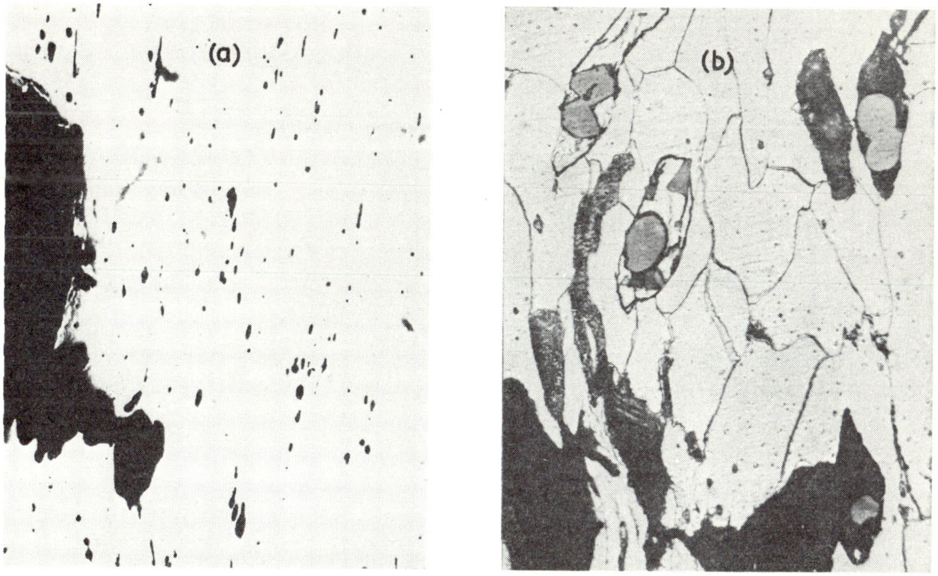


Figure (3) - (a) Location of initial ductile fracture surface at root of notched bend specimen; polished and unetched microsection at magnification: X 50. (b) Etched microsection adjacent to fracture surface at notch root; magnification: X 450. Material: high-sulphur mild steel. Notch plane horizontal.

spheroidised 1% carbon steel compression specimen, in the annealed condition and compressed between lubricated platens, are shown in Fig.(4(a)) where the circumferential $\sigma_\theta$ , axial $\sigma_z$ and mean normal $\sigma_m$ stresses at the equatorial surface are plotted in normalised form against the tensile circumferential strain $\varepsilon_\theta$ . These results show that the free-surface barrelling effect introduces an increasingly tensile $\sigma_\theta$ stress and a decreasingly compressive $\sigma_z$ stress, as the compression test proceeds; the $\sigma_z$ stress is, however, still highly compressive when ductile fracture intervenes.

Similar experiments have been performed for the rolling of a medium-carbon steel strip {20}, where ductile fracture initiates at the mid-thickness region of the free edges, Fig.(4(b)). The plastic strain-history data, from the rolling experiments of Kuhn et al {20}, can be used to determine the longitudinal $\sigma_\ell$ , thickness $\sigma_t$ and mean normal stresses at the middle region of the bulged strip edges, and the results are plotted in normalised form against the tensile longitudinal strain $\varepsilon_\ell$ in Fig.(4(b)). These results show similar trends to the compression tests, with the $\sigma_\ell$ stress becoming increasingly tensile and the $\sigma_t$ stress becoming decreasingly compressive; the $\sigma_t$ stress, however,

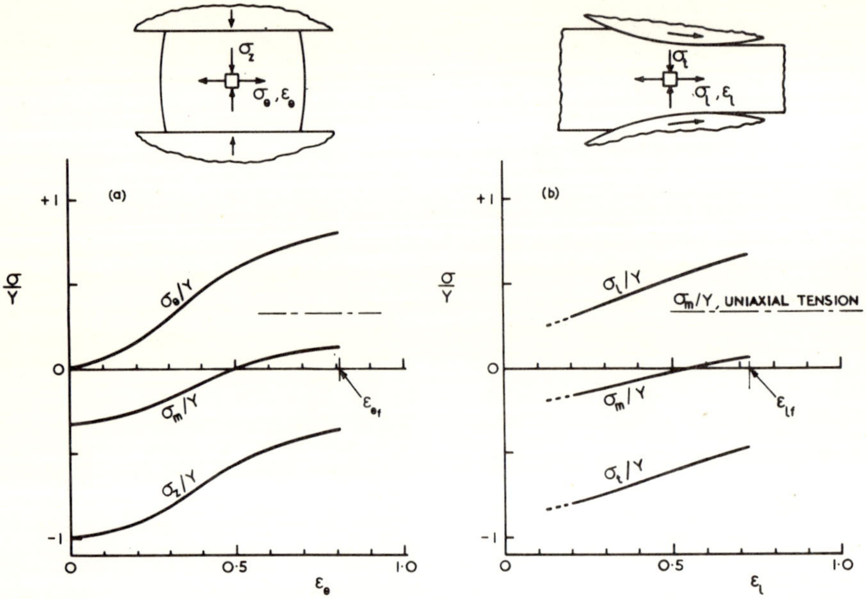


Figure (4) - (a) Stress history to fracture at equatorial free-surface of 1% carbon steel compression specimen. Initial height to diameter ratio of cylindrical specimen: 1.5. Platens lubricated with graphite and tallow; results from Thomason {19}. (b) Stress history to fracture at mid-thickness free surface of medium-carbon steel strip in rolling process. Stresses estimated from results of Kuhn et al {20}. All stresses normalised by the uniaxial yield stress Y .

remains highly compressive up to the point of ductile fracture.

In assessing the relative merits of the various hole-growth models of ductile fracture, the experimental results in Fig.(4) are of critical importance. These results show that ductile fracture can occur when the $\sigma_z$ or $\sigma_t$ stress, acting transversely to the direction of the tensile principal strain ($\varepsilon_\theta$ or $\varepsilon_\ell$), is compressive throughout the deformation process (or alternatively, when the mean normal stress $\sigma_m$ remains less than the value Y/3 , corresponding to a simple uniaxial-tension field). For this type of stress history, the McClintock-Rice-Tracey models {3}{4}{5} predict transverse hole closure and thus an unlimited ductility if McClintock's {3} transverse-hole-contact criterion of ductile fracture is employed. The ductile-fracture models {3}{4}{5} which regard uniform transverse hole-growth as the primary feature of the fracture process are therefore in complete disagreement with the results in Fig.(4). The failure of these models is not in any way connected with the respective hole growth-rate equations; the failure results from the fact that the effects of extensional hole-growth, on the conditions leading to hole coalescence by internal-necking, are neglected.

The important difference between the McClintock-Rice-Tracey models {3}{4}{5} of hole-growth and coalescence and that of Thomason {6}, lies in the fact that the latter model places great emphasis on the effects of extensional growth on hole coalescence. Under a compressive transverse-stress history, of the type shown in Fig.(4), the simple incompressible hole-growth model employed by Thomason, gives extensional growth with transverse hole contraction but not closure. In this respect the simple hole-growth model may be providing a reasonable approximation to the situation in real metals, since the results of Rice and Tracey {4} suggest that extensional growth will not be seriously restricted by moderate compressive mean-normal-stresses and in addition strain-hardening effects in real metals should inhibit transverse hole closure {4}{5}. Due to the fact that extensional hole-growth continues, under transverse compressive stresses, Thomason's ductile-fracture model {6} predicts that hole-coalescence can occur under compressive transverse stresses, when the neck geometry ratio has been increased sufficiently to satisfy the criterion for the onset of internal necking at a sheet of holes, Eq.(4). Thus Thomason's internal-necking model of ductile fracture is consistent with the experimental results in Fig.(4), showing ductile fracture under a completely compressive history of transverse stress.

## CONCLUDING REMARKS

Experimental studies of the ductile fracture process, at the root of notches and the free surfaces of metalworking specimens, show that fracture often initiates in regions of low-tensile or compressive transverse stress, where uniform transverse hole-growth can have little or no influence on the conditions leading to the onset of ductile fracture. In such situations extensional hole-growth is the predominant factor controlling the onset of hole-coalescence by internal necking. Extensional hole-growth is also likely to have a primary influence on the ductile fracture process, when fracture occurs under high triaxial tensile stresses, however, under these conditions transverse hole-growth will have a significant effect on the geometry of the inter-cavity matrix 'necks'; thus influencing the conditions for the onset of internal necking. It can be concluded, therefore, that a simple incompressible hole-growth equation, coupled with an internal-necking criterion of hole coalescence {6}, will form a completely satisfactory model of ductile fracture under low and intermediate values of tensile or compressive mean normal stresses. However, a valid model for ductile fracture under large tensile or compressive mean normal stresses would require a hole-growth equation which accounted for the growth-amplification effects of the remote stress field {4}. Thomason's model of ductile fracture could be modified to account for the effects of high stress triaxiality by developing a plane-strain growth-rate equation for cylindrical holes instead of the square holes in Fig.(2); this could be achieved with the aid of the variational principle of Rice and Tracey {4}. This approach would also require slight modifications to the form of the internal necking criterion in Eq.(4).

## REFERENCES


{1} Ashby, M.F., Work hardening of dispersion-hardened crystals, Phil.Mag., 14, 1157, 1966.

{2} Broek, D., A study on ductile fracture, National Aerospace Laboratory, the Netherlands, Publication NLR TR 71021 U, 1971.

{3} McClintock, F.A., A criterion for ductile fracture by the growth of holes, J.Appl.Mech., 35, Trans.A.S.M.E., 90, 363, 1968.

{4} Rice, J.R. and Tracey, D.M., On the ductile enlargement of voids in triaxial stress fields, J.Mech.Phys.Solids, 17, 201, 1969.

{5} Tracey, D.M., Strain-hardening and interaction effects on the growth of voids in ductile fracture, Eng.Fract.Mech., 3, 301, 1971.

{6} Thomason, P.F., A theory for ductile fracture by internal necking of cavities, J.Inst.Metals, 96, 360, 1968.

{7} Thomason, P.F., A theory for the effects of strain-rate-sensitivity on ductile fracture, Met.Sci.Journ., 3, 139, 1969.

{8} Edelson, B.I. and Baldwin, W.M.Jr., The effect of second phases on the mechanical properties of alloys, Trans.Q.A.S.M., 55, 230, 1962.

{9} Bridgman, P.W., Studies in Large Plastic Flow and Fracture, McGraw-Hill, New York, 1952.

10} McClintock, F.A., On the mechanics of fracture from inclusions, in Ductility, Amer.Soc.for Metals, 1968, 255.

{11} Cottrell, A.H., Theoretical aspects of fracture, in Fracture, Averbach, B.L. et al, Eds., Technology Press - Wiley, New York, 1959, 20.

{12} Liu, C.T. and Gurland, J., The fracture behaviour of spheroidized carbon steels, Trans.Q.A.S.M., 61, 156, 1968.

{13} Brindley, B.J., The effect of dynamic strain-ageing on the ductile fracture process in mild steel, Acta Met., 18, 325, 1970.

{14} McClintock, F.A., Crack growth in fully plastic grooved tensile specimens, in Physics of Strength and Plasticity, Argon, A.S., Ed., M.I.T. Press, Cambridge, Mass., 1969, 307.

{15} Rice, J.R. and Johnson, M.A., The role of large crack tip geometry changes in plane strain fracture, in Inelastic Behaviour of Solids, Kanninen, M.F. et al, Eds., McGraw-Hill, New York, 1970, 641.

{16} Thomason, P.F., A theoretical relation between $K_{IC}$ and basic material properties in ductile metals, Int.J.Fract. Mech., 7, 409, 1971.

{17} Nixon, W.K., On the mechanism of ductile fracture at the root of a notch, Undergraduate Student Project Report, University of Salford, June 1973.

{18} Thomason, P.F., The use of pure aluminium as an analogue for the history of plastic flow in studies of ductile fracture, Int.J.Mech.Sci., 10, 501, 1968.

{19} Thomason, P.F., Tensile plastic instability and ductile fracture criteria in uniaxial compression tests, Int.J. Mech.Sci., 11, 187, 1969.

{20} Kuhn, H.A., Lee, P.W. and Erturk, T., A fracture criterion for cold forming, J.Eng.Mat. and Techn., 1, Trans.A.S.M.E., 95, 213, 1973.

## DISCUSSION by J. N. Robinson, Imperial College, UK

I would first like to congratulate Dr. Thomason on his paper which helps clarify an important aspect of fracture. The theory presented throws light on some puzzling results obtained by Professor Tetelman and myself [1]. We measured critical crack tip opening displacement at fracture initiation, $(COD)_c$, on a low alloy pressure vessel steel, A533B. We also printed grid lines on the notch root so that we could measure strain in the transverse direction (parallel to the axis of the notch). The results are given in Table 1.

TABLE 1

$(COD)_c$ MEASUREMENTS ON CHARPY GEOMETRY SLOW BEND SPECIMENS OF A533 STEEL, 0.05 mm ROOT RADIUS

| Specimen Thickness (mm) | Transverse Strain in Central Third of Specimen | $(COD)_c$ at 24°C (mm) |
|---|---|---|
| 10 | 0 | 0.290 |
| 2.5 | 8% | 0.279 |
| 1.25 | - | 0.292 |

Theoretical treatments by Rice and Tracey [2] and McClintock [3] indicate that the magnitude of the transverse stresses has a marked effect on the longitudinal strain to fracture. Since transverse stress will be lower in the thinner specimens where transverse strain occurs, the $(COD)_c$ might be expected to increase with decrease in specimen thickness. The experimental values of $(COD)_c$ are not significantly different in the three thicknesses however. The theory presented by Dr. Thomason, where coalescence is not dependent on transverse void growth, implies a lesser dependence of longitudinal fracture strain on transverse stress. This theory thus fits the above experimental results much better than do those of McClintock or Rice and Tracey.

REFERENCES

[1] J. N. Robinson and A. S. Tetelman, "The relationship between crack tip opening displacement, local strain and specimen geometry", to be published in Int. J. of Fracture.

[2] J. R. Rice and D. M. Tracey, "On the ductile enlargement of voids in triaxial stress fields", J. Mech. Phys. Solids, 17, pp. 201-217, 1969.

[3] F. A. McClintock, "A criterion for ductile fracture by the growth of holes", J. Appl. Mech., 35, pp. 353-371, 1968.

## AUTHOR'S CLOSURE

The critical COD results quoted by Dr. Robinson are very interesting and give strong support to the suggestion in my paper that the onset of ductile fracture is controlled by internal microscopic necking at a sheet of holes, rather than by uniform hole-growth to the point of transverse hole contact.

The COD results were obtained with specimens of similar geometry to that in Fig. (3), and it appears likely that the insensitivity of the results to transverse stress is due to ductile fracture initiating in the immediate vicinity of the notch root, where extensional hole-growth greatly exceeds transverse growth. Under these conditions the low mean-normal stress at the notch root has only a minor influence on the criterion for internal necking, Eq. (4), the onset of ductile fracture being primarily dependent on the reduction in constraint factor $\sigma_n/2k$, which is brought about by the extensional hole-growth.

# PARTICLES AND CRACK GROWTH IN ALUMINIUM ALLOYS

D. Broek

Delft University, Metallurgy Dept., Delft, Netherlands

## ABSTRACT

A brief survey is given of progress made in research to the role of particles on the cracking behaviour of aluminium alloys. Two types of particles are considered: large size intermetallic constituents failing early in the fracture process, and small particles responsible for the final separation. Experimental evidence is presented to show that large particles fracture by cleavage when the largest principal stress exceeds a critical value. This takes place already outside the plastic zone. Final separation occurs by initiation and coalescence of small voids formed in bands of intense shear between the cracked large particles.

A brief analysis is made of the prospects of the development of materials with low particle content for improved toughness.

## INTRODUCTION

Fracture of structural alloys is the result of the initiation and growth of voids at second phase particles. The mechanisms and the conditions by which these voids develop are not yet well understood. The criteria for void initiation may be different for different types of materials. The discussions in this paper are limited to the role of particles in the fracture of aluminium alloys.

High strength aluminium alloys contain two types of particles that are of direct influence to the fracture behaviour.

Particles of the first type are large intermetallic constituents (1-20 microns) which start cracking already after little straining when fracture of the matrix is still remote [1-4]. Cracking of these particles results in a large number of small voids around the tip of an existing crack. In several fracture models [e.g. 5, 6], the voids are considered to constitute the ligaments for plastic instability, the average distance between the nearest voids and the crack tip being the ligament width. Fracture is supposed to occur when the ligaments fail by plastic instability.

The other type of particles of interest for the fracture process are of sub-micron size (0,1 - 0,5 micron). In many alloys these particles are near to spherical. Final separation of the material takes place upon decohesion of these particles, because the voids show almost spontaneous growth and coalescence [2, 3, 4].

Mathematical evaluation of models representing the sequence of events described above, may yield the parameters that govern ductility and toughness. The main problem in designing such models is in the fact that the criteria for fracture and decohesion of particles are not well established. In this paper some evidence is presented concerning the fracture criterion of the large particles.

The role of particles in the fracture process is of scientific interest because it is essential to the explanation of fracture mechanisms. On the other hand, knowledge of the effect of particles is of practical interest, because it may lead the way to the development of better materials. The manufacturing of particle-free materials requires a large expenditure of time and money. Then it is a pertinent question how much success can ultimately be achieved. In other words: what is the absolute maximum toughness a material might exhibit. A second question is how much the structural safety would be improved and at which costs, if materials of absolute maximum toughness could be made available. A brief discussion is devoted to these problems as the last part of this paper.

## LARGE PARTICLES

### Experiments and results

There exist many types of intermetallic constituents occurring as irregularly shaped particles of a size in the order of 1 - 20 microns. The particles are readily visible by means of an optical microscope.

Tanaka, Pampillo and Low [1] studied an Al-Cu-Mg-alloy (2014-T6) and found many particles cracking at low strains. They observed two distinct types of particles, one being very likely $Al_{12}$ (FeMn)$_3$ Si, the other containing Al, Si, Cu and possibly Mg.

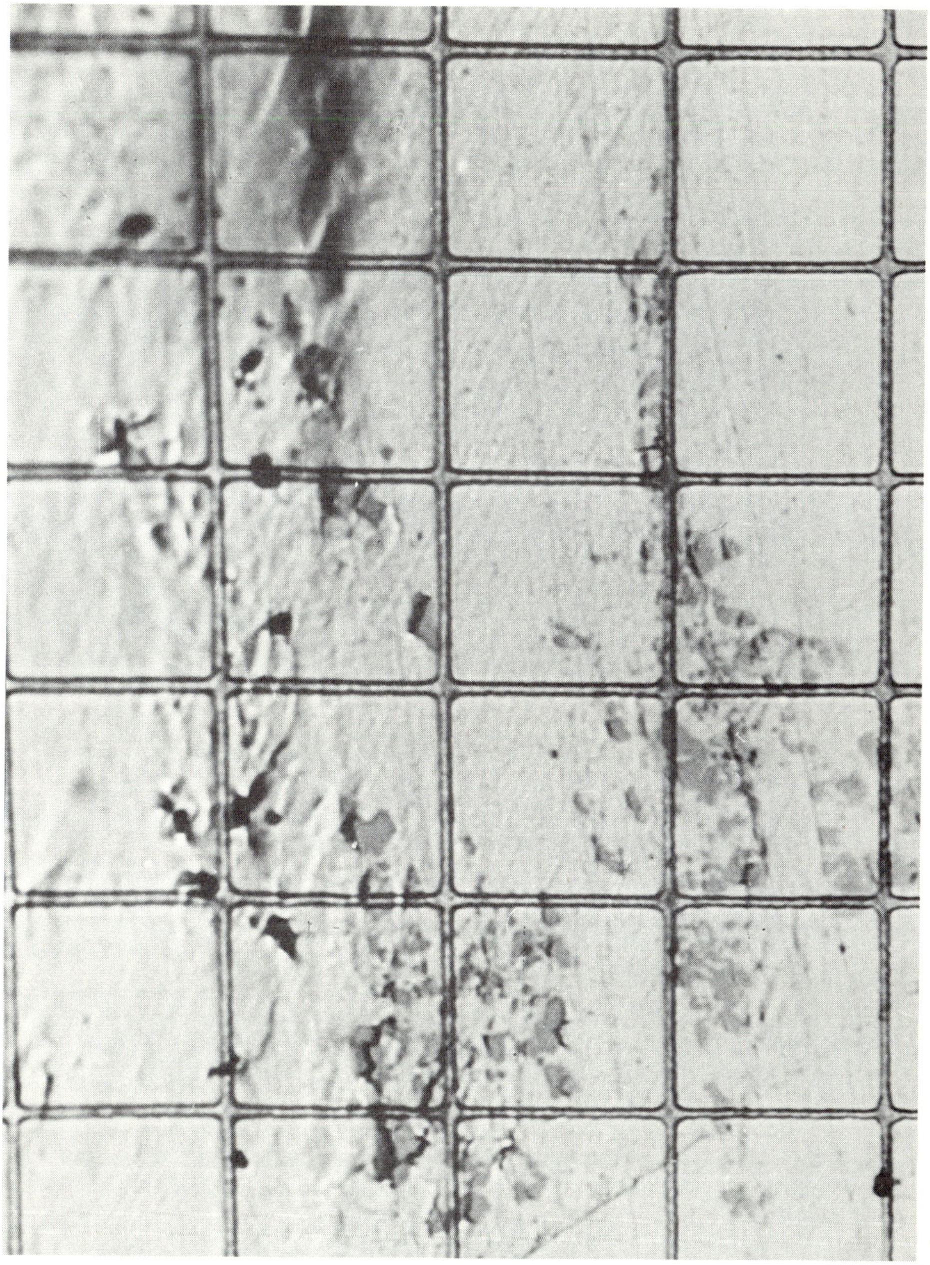

Figure 1. Cracked particles in front of crack tip in Al-Zn-Mg alloy 7075-T651. Small specimen under test in straining device operated under optical microscope. Grid size is 50 micrometers.

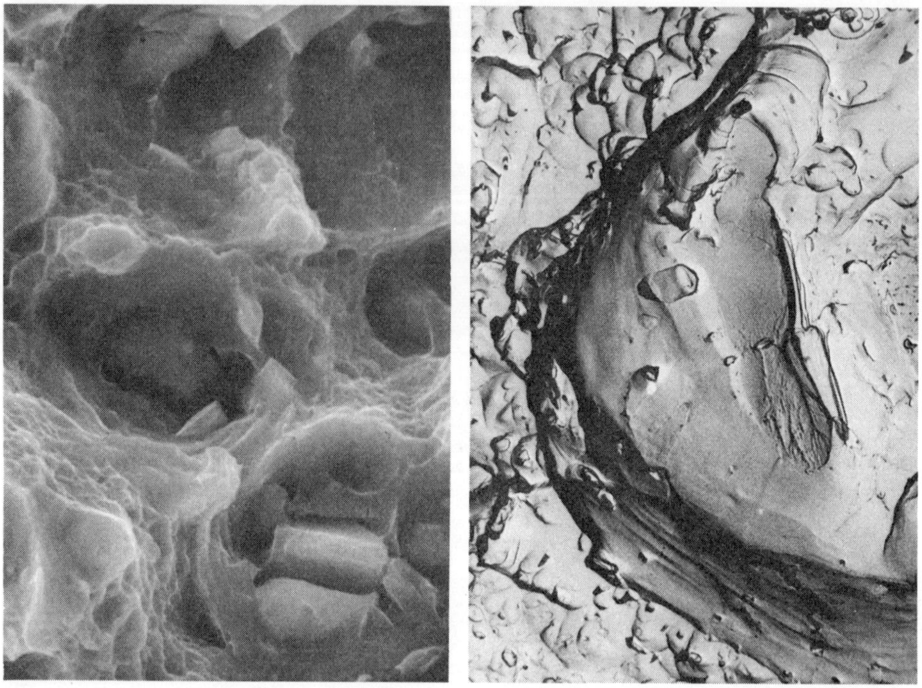

Figure 2. Electron fractographs of Al-Zn-Mg alloy, showing large particles and areas with dimples. Left: scanning fractograph at 2700 X; Right: transmission fractograph at 9000 X.

Similar observations were made by Broek [2-4] on a variety of aluminium alloys.

Thus, it must be expected that the majority of these particles present in the region of high stretch in front of a crack tip are cracked long before the crack is ready to propagate. A number of failed particles around the tip of a crack in an Al-Zn-Mg alloy is shown in Fig. (1). The straightness of the cracks in the particles suggests that failure occurs by cleavage. This is confirmed [1-4, 7, 8] by electron fractography, as illustrated in Fig. (2). The ductile fracture surface consists of dimples resulting from voids initiated at the small particles, and of large areas of shear around prematurely cracked large particles. The river markings on the fractured particles show that particle fracture was by cleavage, which must be dictated by tensile stresses.

In order to obtain information about the fracture criterion for these particles an investigation was made to cracked particles around a crack tip. The material used was 7075-T651 plate with

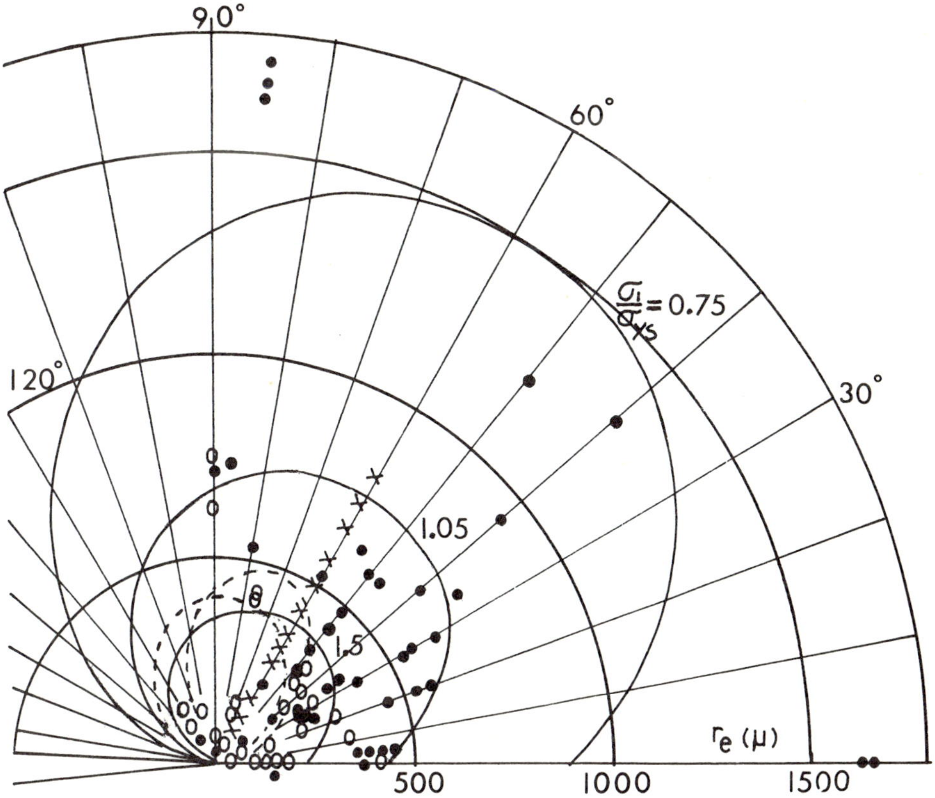


Figure 3. Particles found cracked around crack tip in 7075-T651 aluminium alloy at $K = K_{Ic}$. Continuous lines are lines of constant maximum principal stress. Dashed lines represent extent of plastic zone.

$\sigma_{ys}$ = 57.6 kg/mm$^2$, $\sigma_v$ = 63.2 kg/mm$^2$ and $\delta$ = 9.5 per cent. Optical microscopy revealed two types of large particles, one type appearing as light grey on the micrographs, the other as dark grey. Qualitative X-ray micro-analysis showed the light particles to be rich in Fe and Cu and low in Mg and Zn. The dark particles were high in Mg and Si and low in Zn.

Fracture toughness values were determined by means of standard bend specimens at 100 kg/mm$^{3/2}$ in longitudinal directin and 90 kg/mm$^{3/2}$ in transverse direction. A number of bend specimens was loaded to different percentages of $K_{Ic}$ and then unloaded. The crack tip regions were sectioned and polished but left unetched, and the samples were searched for cracked particles with an optical microscope.

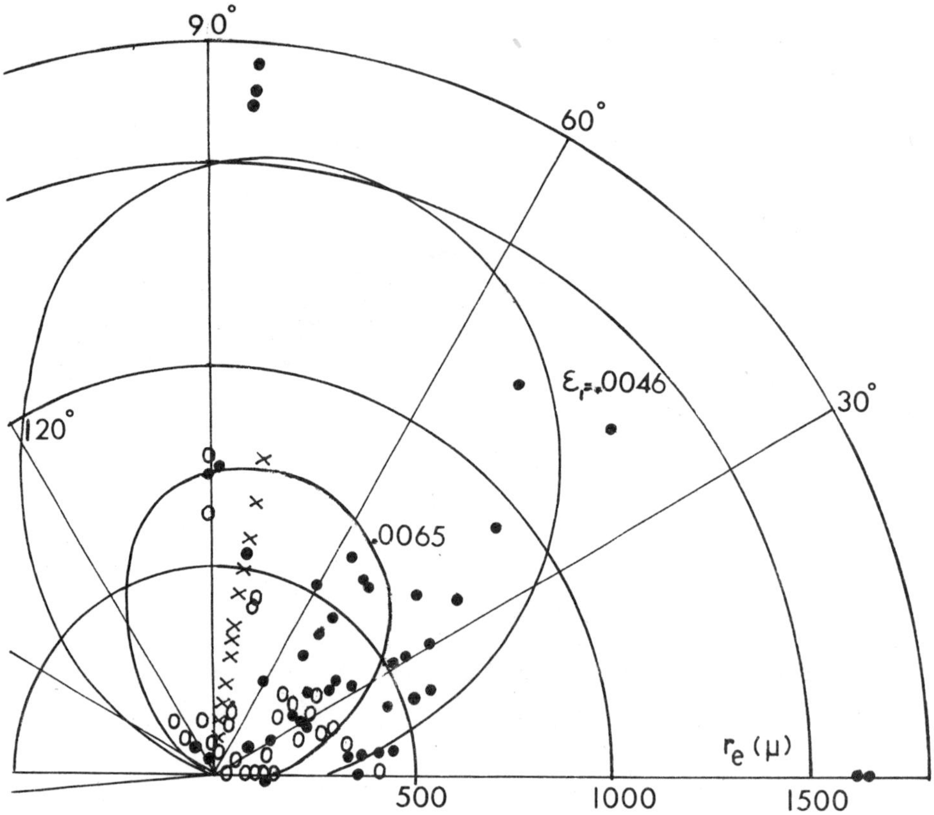


Figure 4. Particles found cracked around crack tip in 7075-T651 aluminium alloy at $K = K_{Ic}$. Lines of constant strain, $\varepsilon_1$, indicated.

The coordinates r and θ of the cracked particles were determined as well as the apparent particle size. The data were normalized by calculating $r_e = (K_{Ic}/K)^2 r$. Thus, the distance $r_e$ gives the location where the stresses and strains under the action of $K = K_{Ic}$ are the same as at the distance r at K = K. The value of $K_{Ic}$ was taken arbitrarily at 100 $kg/mm^{3/2}$ for all data. The data were not made dimensionless in order to preserve the possibility of a direct appreciation of the magnitude of r.

Data plots are presented in Figs. (3) and (4). Light particles are presented by open data points, dark particles by solid data points. The other data represent cracked particles along the fatigue crack. These are discussed later. Figure (3) shows lines of constant principal stress $\sigma_1$. Also, the magnitude of the plastic zone is given (dashed lines), the inner one according to the Von Mises criterion, the outer one according to calculations by Levy et al [9]. Figure (4) is a replot of the data and the lines are for constant value of $\varepsilon_1 = \sigma_1/E - \nu(\sigma_2 + \sigma_3)/E$.

Discussion of data

First of all, it should be noted that the majority of the data points are outside the plastic zone. Thus, the elastic stress field solution can be used (the plastic zone is only small). Particle failure due to cleavage must be the result of exceedance of a critical tensile stress. If the particles have the same modulus as the matrix there is no stress concentration (in the absence of plasticity) and the maximum tensile stress on the particle would be given by $\sigma_1 = (1 + \sin\theta/2)\ K\ \sqrt{2\pi r}$.

According to Goodier's solution [10] the tensile stress across spherical particles would only be about twice the stress $\sigma_1$ (even in the case of a hydrostatic tension $\sigma_1$), if the particles had infinite stiffness. In view of the alloying elements involved the ratio between the modulus of the particle and the modulus of the matrix is likely to be between 1 and 2. This means that the stress concentration is between 1 and 1.5 [10]. Added to these are the stresses set up during cooling by a difference in thermal expansion of the particles and the matrix. In view of the compatibility requirement, the stress in the inclusion may be determined by the strain $\varepsilon_1$, rather than by stress.

Consider the data in Figs. (3) and (4). The stresses and strains decrease with increasing distance from the crack tip. The furthest cracked particles must indicate where stress or strain were just sufficient to cause cracking, i.e., the particle stress at that location is the minimum required fracture stress. Particles closer to the crack must have fractured at lower K-values. Hence, the fracture criterion must be given by the envelope of the data.

The curve for $\sigma_1 = 1.05\ \sigma_{ys}$ in Fig. (3) very nearly envelopes 80 per cent of the solid data points. The curve $\sigma_1 = 0.75\ \sigma_{ys}$ includes more data points, but the area enclosed by this curve is four times as large. This means that the total amount of particles (cracked and uncracked) within this curve is four times as large as the amount located within the curve $1.05\ \sigma_{ys}$. It implies that the relative amount of cracked particles within the line $1.05\ \sigma_{ys}$ is twelve times as large as in the area between the two lines. Therefore the data points outside r = 800 microns were discarded. These particles may have cracked because they were particularly weak (Also, the gradient of $\sigma_1$ in the outer area is only small, so the stress at some of them was not too much below $1.05\ \sigma_{ys}$). This leads to the preliminary conclusion that the particles fail if

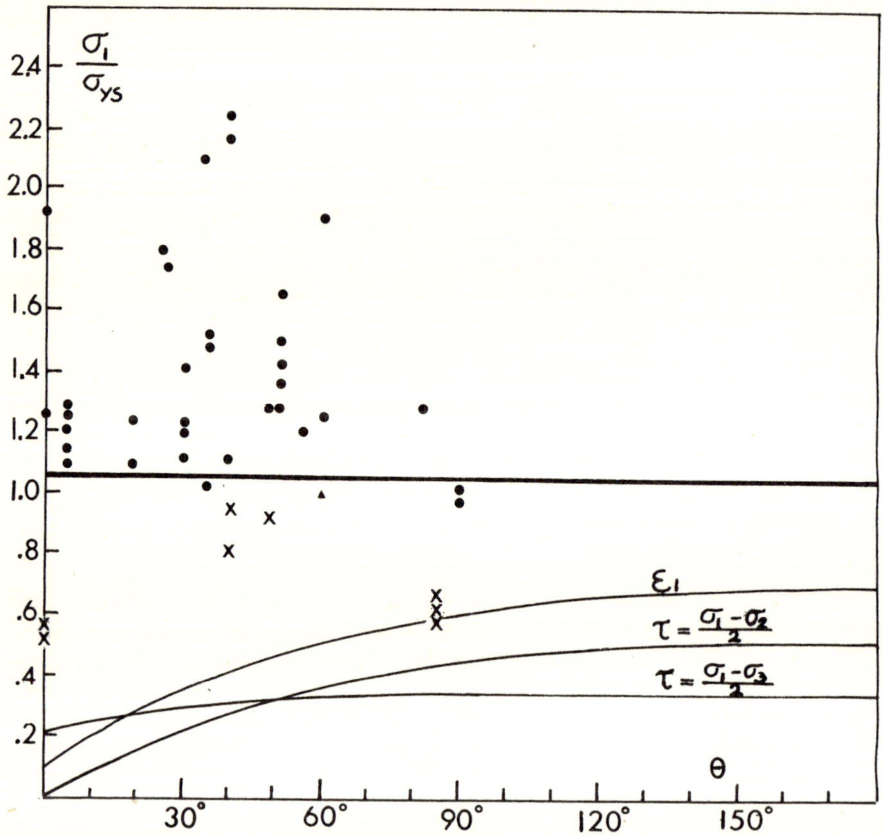


Figure 5. Maximum principal stress at location of cracked particles. Lower lines give magnitude of shear stresses along contour $\sigma_1/\sigma_{ys}$ = 1.05. See text for explanation of symbols.

the matrix stress exceeds 1.05 $\sigma_{ys}$. Similar reasoning leads to a failure stress of 1.5 $\sigma_{ys}$ for the light particles.

It appears from Fig. (4) that lines of constant $\varepsilon_1$ do not fit the data points as well as lines of constant $\sigma_1$. It is possible to think of other fracture criteria, such as maximum shear stress or Von Mises equivalent stress. Such criteria would show a significantly different shape of the envelope (see plastic zone in Fig. (3)), which would not include as many data points. Therefore, it is likely that the particles fail upon exceedance of a critical value of $\sigma_1$. This implies that the modulus of the particles is not too different from the modulus of the matrix, otherwise the strain would be the governing parameter. In view of the fact that the particles are intermetallics containing much aluminium [1, 7], indeed, only small differences in modulus are to be expected.

The majority of the cracked particles was found at $\theta < 60$ degrees. Therefore, it has to be checked whether some other parameter contributes to cracking. In Fig. (5) the value of $\sigma_1$ at the location of the cracked particles is plotted as a function of $\theta$. The x-data are the ones that were discarded on the basis of Fig. (3). The lower boundary is given by $\sigma_1 = 1.05\ \sigma_{ys}$. Also shown are the magnitude of the maximum shear stress and the magnitude of $\varepsilon_1$ along the line $\sigma_1 = 1.05\ \sigma_{ys}$. These are rising lines. If one of these parameters were of importance more data points should be found at large $\theta$ then at small $\theta$. This is not the case.

The lack of data points at $\theta > 60$ degrees is independent of r. The shape of the plastic zone is such that cracked particles at large $\theta$ should have been mostly inside the plastic zone. After unloading, the material inside the plastic zone is subject to residual compressive stresses. The compressive stresses may have closed the cracks, thus preventing the detection of cracked particles.

Along the path of the fatigue crack several particles were found cracked. Their distance from the fatigue crack could be measured, but it is impossible to determine $\theta$ after passage of the crack. The lines of constant $\sigma_1$ have their largest distance from the crack tip at $\theta \simeq 60$ degrees. This means that the most remote particles must have cracked when they were at $\theta = 60$ degrees, since further progression of the fatigue crack would have left them in an area with lower stress. Therefore, all fatigue crack data were corrected by a factor $1/\sin \pi/3$ and plotted at $\theta = 60$ in Fig. (3). It appears that the furthest cracked particle is near the boundary $1.05\ \sigma_{ys}$.

Gurland and Plateau [11] established a particle fracture criterion showing a dependence of fracture stress on particle size. Since sections may not cut the particles at their largest cross section it is not possible to determine the absolute particle size, unless sufficient data are available for statistical analysis. An analysis of the data with respect to their apparent size did not yield additional information.

## FINAL SEPARATION; THE ROLE OF SMALL PARTICLES

It must be concluded that all large particles within a limited distance from the crack tip are fractured long before the crack is ready to advance. The conditions for fracture then depend upon the behaviour of the ligaments. Continuum analyses by

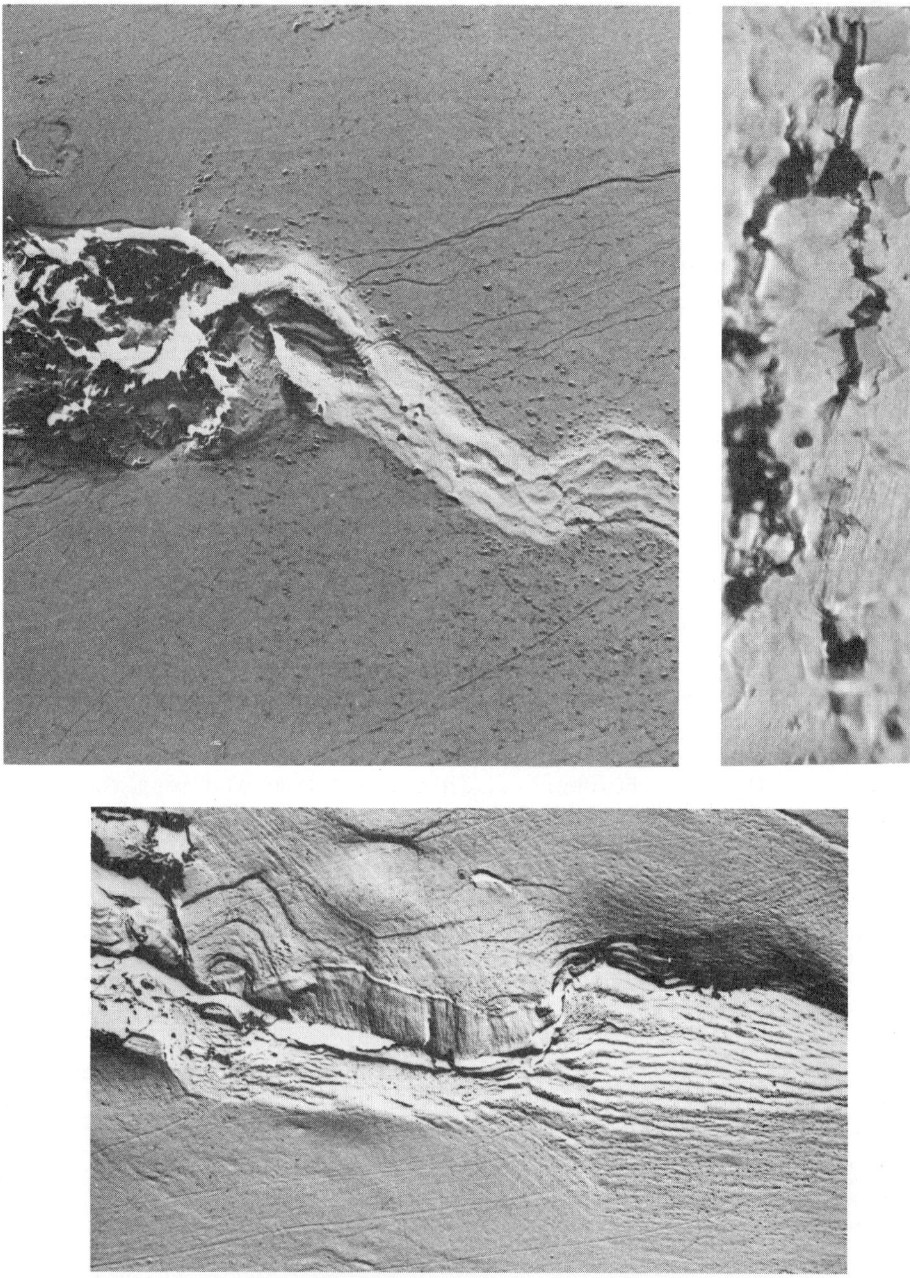

Figure 6. Linking of cracked particles by bands of intense slip, followed by cracking. Top right: Optical micrograph at 450 X. Top left and bottom: electron micrographs of surface replica's at 7000 X.

Thomason [6, 12] and by McClintock et al [13] lead to a dependence of fracture strain on the volume fraction of cracked particles and on the state of triaxiality. Failure is due to plastic instability of the ligaments. This is supported by the experimental observation that shear is confined to narrow bands between cracked particles [14]. Similar evidence is presented in Fig. (6) showing electron micrographs of surface replicas which reveal narrow bands of intense slip connecting the cracked particles.

Although plastic instability at constant load leads to fracture, it is not a fracture mechanism in itself. In the absence of small particles, final separation after plastic instability is a geometrical consequence of slip. If small particles are present, voids are initiated which may grow and coalesce, also as a geometrical consequence of slip [2]. In a tensile test these voids are initiated just before fracture, which is far beyond the onset of plastic instability. The question then arises whether failure of the ligaments takes place upon reaching plastic instability, or upon reaching the critical condition for micro-void initiation.

The shortcoming of the instability models is that they predict an increasing toughness with increasing yield stress, contrary to the practical behaviour. This can be avoided by assuming that fracture is determined by a void initiation criterion, given approximately [2] by:

$$\overline{\sigma\varepsilon} = c_1 \sigma_d f^{-\alpha} \tag{1}$$

In this equation c is a constant, f is the volume fraction of the small particles, and $\sigma_d$ is the interface stress required for their decohesion. The latter is directly related to the true fracture stress, $\sigma_f$, of the material, since decohesion -followed by immediate coalescence- occurs if the true fracture stress is reached. Initiation is governed by a critical combination of stress and strain. If the stress during plastic deformation is higher (i.e., higher yield stress and higher strain hardening coefficient) a lower strain (less dislocations) are required for void initiation. Hence, the equation can be written as:

$$g(n)\, \sigma_{ys}\, \varepsilon_d = c_2\, \sigma_f\, f^{-\alpha} \quad \text{or} \quad \varepsilon_d = \frac{c_2}{g(n)} \frac{\sigma_f}{\sigma_{ys}} f^{-\alpha} \tag{2}$$

Void initiation is supposed to occur if $\varepsilon_d$ is reached throughout the ligament $\lambda$, the average spacing of the (cracked) large particles. According to the numerical results presented by Rice and Johnson [5] the true strain at the crack tip is given approximately by

$$\varepsilon \approx \frac{c_3}{r/d_t} \quad \text{or} \quad \varepsilon = c_3 \frac{d_t}{r} \quad \text{with} \quad d_t = 0.5\, K^2/\sigma_{ys}E \tag{3}$$

In this equation $d_t$ is the size of the heavily stretched region in front of the crack tip, which has the order of magnitude of the crack tip opening displacement. Under the assumptions made, fracture occurs if $\varepsilon = \varepsilon_d$ at $r = \lambda$, hence:

$$K_{Ic}^2 = \frac{c}{g(n)} \sigma_f E \lambda f^{-\alpha} \tag{4}$$

Assuming a regular distribution of spherical coarse particles, the equation can be rewritten as:

$$K_{Ic} = \sqrt{\frac{c}{g(n)} \sigma_f E\, F^{-1/3}\, f^{-\alpha}\, d} \tag{5}$$

where F is the volume fracture of the coarse particles and d is their average size. This equation only differs from the one presented by Hahn and Rosenfield [14] by the replacement of $\sigma_{ys}$ by $\sigma_f$, and by the introduction of the volume fraction of the small particles. It shows $K_{Ic}$ to increase with true fracture strain, implying that more ductile materials have higher toughness. Hahn and Rosenfield have shown that the predicted dependence on volume fraction of coarse particles is confirmed fairly well by test data. They show two sets of data, one for $\sigma_{ys}/E \approx 7$ and one for $\sigma_{ys}/E \approx 8$. The latter set is for materials of low ductility, i.e., $\sigma_y/\sigma_f \approx 1$. For a given volume fraction their toughness is about half that of the materials with $\sigma_{ys}/E = 7$. If these materials have a ratio $\sigma_{ys}/\sigma_f \approx 0.7$ (which is a reasonable value) the two sets of data would coincide if plotted on the basis of Eq. (5). The dependence of $K_{Ic}$ upon n as predicted by Eq. (5) is questionable.

## PARTICLES AND FATIGUE

It has been shown by Grosskreutz and Shaw [15], by Bowles and Schijve, and by several others that fatigue cracks in unnotched samples are initiated at the large particles. The fatigue life until crack initiation can be improved considerably if materials are made particle free. However, this beneficial effect is largely reduced if the samples are notched: the stress concentration of the notch is sufficient to initiate cracks irrespective of the presence of particles. Since notches in real structures are unavoidable (and often associated with fretting) the practical benefit of clean materials is limited.

Particles appear to have little influence on fatigue crack propagation, because a fatigue crack is a far more severe stress raiser than the particles. This was shown by means of electron fractography by Pelloux [17] and Broek [18] and more recently in a systematic study by El-Soudani and Pelloux [19].

Once fatigue growth rates become very high the large particles may fracture in front of the crack tip, and the voids may induce small amounts of ductile fracture [17, 18, 19]. This enhances the rate of fatigue cracking considerably. However, in this regime the growth rates are already high anyway. This last part of the crack propagation is only in the order of 10 per cent of the crack growth life, and hence, it is only a few per cent of the total fatigue life. Consequently, the effect of particles on fatigue of aluminium alloys, in general, is of limited technical importance.

## PRACTICALITY OF ALLOY DEVELOPMENT

Particles are of interest for our understanding of fracture mechanisms. There is evidence that they reduce toughness but have little effect on fatigue. The question now arises whether it is rewarding to develop alloys with low particle content. Then it is interesting to know how much can be gained ultimately in toughness, but more important is how much can be gained in safety. An increase of the critical crack size in itself does not contribute much to safety if the crack growth rates are so high that the time gained for crack detection is still small. The argument that longer cracks can be detected more easily is not of great significance, because cracks are small during the greater part of their existence. Once they are large there is little time left for detection and therefore the chance of discovery is still limited.

The formulae relating toughness to volume fraction of particles all lead to the conclusion that $K_{Ic}$ becomes infinite for zero volume fraction. Of course, this is irrelevant. The highest critical stress that can be reached in a cracked structure is the yield stress (assuming no work-hardening). Hence, the practical limit of $K_{Ic}$ is reached if the critical stress is equal to the yield stress, $\sigma_{ys}$. For a structure of given width W this would be the case if the curve $\sigma_c = K_{Ic}/\sqrt{\pi a}$ is tangent to the line of net section yield. Feddersen [20] has shown that this tangency point is at $\sigma_c = 2/3\ \sigma_{ys}$ and $a = W/6$.

Taking a heavy section aluminium alloy structure in which plane strain prevails, the maximum W found in practice may be in the order of 300 mm. Hence, if $K_{Ic} = 2/3\ \sigma_{ys}\ \sqrt{\pi .50}$ all structures of this size and smaller would fail at net section yield. For a high strength aluminium alloy with $\sigma_{ys} = 50$ kg/mm$^2$ this gives $K_{Ic} \simeq 400$ kg/mm$^{3/2}$. It should be noted that a structure of smaller size of e.g. W = 150 mm would fail at an apparent $K_{Ic} = 2/3.50\ \sqrt{\pi .25} \simeq 300$ kg/mm$^{3/2}$, even if the toughness of the material was at 400 kg/mm$^{3/2}$.

Practical toughness values presently obtained are in the order of 100 kg/mm$^{3/2}$. Hence, the toughness might be improved ultimately by a factor 3 or 4. According to the data collected by Hahn and Rosenfield [14], materials with low particle contents now available show about 60 per cent increase in toughness.

Improvement of $K_{Ic}$ by a factor 4 would increase the critical crack size by a factor 16. What would this mean to structural safety, i.e. to the period available for crack detection. The crack growth life may be calculated from Forman's equation, which includes the effect of $K_{Ic}$ on crack growth. However, this discussion is concerned with the effect of particles, which are known not to affect fatigue cracking until in the last stage. (Of course, both fatigue and toughness depend upon yield strength and strain hardening; hence, if toughness is improved by affecting these parameters also fatigue behaviour may improve).

Hence, in an approximate evaluation the lines of log da/dn versus log $\Delta K$ can be assumed straight until $K_{Ic}$. Then crack growth is simply given by (assuming m = 4 and R = 0, i.e. $\Delta K = K_{max}$):

$$\frac{da}{dn} = C\,K_{max}^4 \qquad (6)$$

Also for constant amplitude loading with amplitude $\sigma_a$:

$$\frac{da}{dn} = \frac{d}{dn}\left(\frac{K_{max}}{4\pi\sigma_a}\right)^2 = \frac{K_{max}}{2\pi\sigma_a}\,\frac{d\,K_{max}}{dn} \qquad (7)$$

Integration gives the crack propagation life as:

$$n = \frac{C}{2\pi\sigma_a}\int_{K_d}^{K_{Ic}} \frac{d\,K_{max}}{K_{max}^3} = \frac{C}{2\pi\sigma_a\,K_d^2}\left[1 - \left(\frac{K_d}{K_{Ic}}\right)^2\right] \qquad (8)$$

where $K_d$ is the stress intensity for the minimum detectable crack. From this equation it can be derived which gain in life is obtained if toughness is improved to $\bar{K}_{Ic}$. This is shown as a function of $\bar{K}_{Ic}/K_{Ic}$ in Fig. (7). The figure is valid for the particular case that the minimum detectable crack size is 5 mm and the critical crack size at $K_{Ic}$ is 10 mm, i.e., $(K_d/K_{Ic})^2 = 2$. The maximum obtainable gain in life is 100 per cent, most of which is realized already if $K_{Ic}$ is improved by a factor 2. Obviously, a much larger gain in life is obtained by decreasing the detection limit. Reduction of the minimum detectable crack by a factor 2, gives already an increase in life of 100 per cent.

Figure (7) is a little deceiving. If the critical crack is equal to the minimum detectable then n = 0; any increase of $K_{Ic}$ would give a life improvement by an infinite factor. However, this life would still be short, and again, most of the gain would be attained already if $K_{Ic}$ improved by a factor 2. Apparently, it becomes less and less effective to produce cleaner materials; it soon becomes more efficient to the users of materials to put more effort into reducing detection limits, than to buy expensive clean materials.

The research into the role of particles should certainly be continued until satisfactory quantitative models are established, but the effort put into the development of cleaner materials may

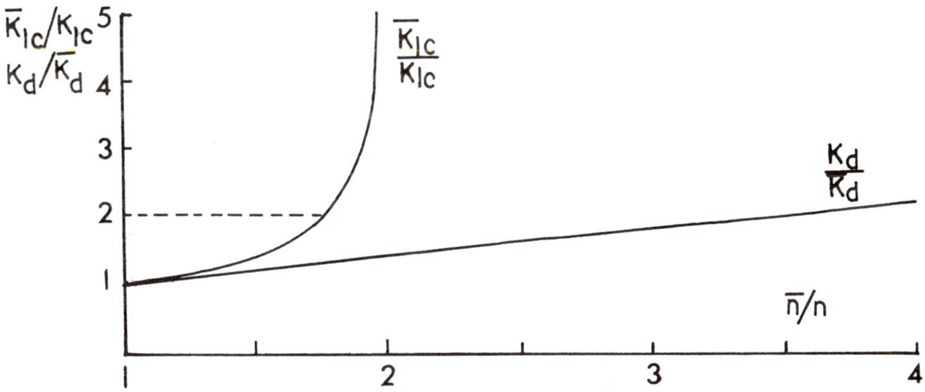


Figure 7. Increase of time available for crack detection as a result of improved $\bar{K}_{Ic}$, and as a result of decreasing detection limit $\bar{K}_d$.

be of low cost-effectiveness. The elimination of large particles should not be continued if it affects material costs too much, unless the strain hardening rate and fracture strain can be affected by dissolving the particles. The latter has been reported by Singh and Flemings [7]. It should be noted that elimination of large particles always reduces the bulk fracture strain, but this may not be of importance [2] since it is the fracture strain of the matrix that counts.

The effect of strain hardening and yield strength on toughness needs more exploration. Improvement of materials along these lines is promising, since it may also lead to reduced fatigue crack growth rates, which contributes more to safety (detection period). A plea is made that experimental investigations on these subjects include both standard fracture toughness tests and fatigue crack growth tests. Composition, volume fraction of particles, and true stress-strain curves should be reported. This would facilitate analysis and comparison of data to make the work more worthwhile. Publication of work not giving all relevant data should be discouraged.

## REFERENCES

[1] Tanaka, J.P., Pampillo, C.A., Low, J.R., Fractographic analysis of the low energy fracture of an aluminium alloy, ASTM STP 463, 1970, 191.

[2] Broek, D., A study on ductile fracture, NLR rept. TR 71021, 1971.

[3] Broek, D., Some contributions of electron fractography to

the theory of fracture, NLR rept. TR 72029, 1972.

[4] Broek, D., The role of inclusions in ductile fracture and fracture toughness, Eng. Fract. Mech. 5, 55, 1973.

[5] Rice, J.R., Johnson, M.A., The role of large crack tip geometry changes in plane strain fracture, in Inelastic behaviour of solids, M.F. Kanninen et al eds., McGraw Hill, 1970, 641.

[6] Thomason, P.F., A theoretical relation between $K_{Ic}$ and basic material properties in ductile metals, Int. J. Fract. Mech. 7, 409, 1971.

[7] Singh, S.N., Flemings, M.C., Influence of ingot structure and processing on mechanical properties and fracture of a high strength wrought aluminium alloy. AIME Trans. 245, 1811, 1969.

[8] Peel, C.J., Forsyth, P.J.E., The fracture toughness of Al-Zn-Mg-Cu-Mn alloys to DTD 5024 - Effect of iron content. RAE. TR 70162, 1970.

[9] Levy, N., Marcal, P.V., Ostergren, W.J., Rice, J.R., Small scale yielding near a crack in plane strain: a finite element analysis, Inst. J. Fract. Mech. 7, 143, 1971.

[10] Goodier, J.N., Concentration of stress around spherical and cylindrical inclusions and flaws, J. Appl. Mech. 1, 39, 1933.

[11] Gurland, J., Plateau, J., The mechanism of ductile rupture of metals containing inclusions, ASM Trans 56, 442, 1963.

[12] Thomason, P.F., A theory for ductile fracture by internal necking of cavities, J. Inst. Met. 96, 360, 1968.

[13] McClintock, M.F. et al, Ductile fracture by hole growth in shear bands, Int. J. Fract. Mech. 2, 614, 1966.

[14] Hahn, G.T., Rosenfield, A.R., Relations between microstructure and the fracture toughness of metals, 3rd ICF Conf., Munich, III-211, 1973.

[15] Grosskreutz, J.C., Shaw, G.C., Critical mechanisms in the development of fatigue cracks in 2024-T4 aluminium, Fracture 1969, 620, Chapman and Hall, 1969.

[16] Bowles, C.Q., Schijve, J., The role of inclusions in fatigue crack initiation in an aluminium alloy. Int. J. Fract. Mech. 9, 171, 1972.

[17] Pelloux, R.M.N., Fractographic analysis of the influence of constituent particles on fatigue crack propagation in aluminium alloys, ASM Trans. 57, 511, 1964.

[18] Broek, D., The effect of intermetallic particles on fatigue crack propagation in aluminium alloys, Fracture 1969, 754, Chapman and Hall, 1969.

[19] El-Soudani, S.M., Pelloux, R.M.N., Influence of inclusion content on fatigue crack propagation in aluminium alloys. Met. Trans. 4, 519, 1973.

[20] Feddersen, C.E., Evaluation and prediction of the residual strength of center cracked tension panels, ASTM STP 486, 50, 1971.

Acknowledgement: The experimental work was performed by J.C.P. van de Kasteele.

# OBSERVATIONS OF CAVITY NUCLEATION, GROWTH AND COALESCENCE IN DIMPLED RUPTURE

John R. Low, Jr.

Carnegie-Mellon University
Pittsburgh, Pennsylvania 15213 USA

## INTRODUCTION

This report summarizes the more important findings on the stages of void nucleation, growth and coalescence involved in the plastic fracture of a number of high strength aluminum alloys and two high strength steels. The work to be described is published in more detail in [References 1, 2, 3].

## MATERIALS

The aluminum alloys examined were 2014, 2024, 2124, 7075, and 7079 which showed a variation in plane-strain fracture toughness ($K_{Ic}$). The two steels were AISI 4340, quenched and tempered to approximately 200,000 psi yield strength, and an 18 Ni maraging steel, aged to approximately the same yield strength. Two heats of each steel were examined. One was a standard commercial purity heat and the other was a high-purity laboratory heat. The reason for choosing these two steels for comparision was the fact that, at the same yield strength, the maraging steel exhibits a considerably higher fracture toughness.

The pertinent mechanical properties of these materials are listed in Table 1. All mechanical tests were made at room temperature (20-25° C).

TABLE 1

Mechanical Test Results of Alloys Examined

| Alloy | Yield Strength (ksi) | $K_{Ic}$ (ksi-$\sqrt{in.}$) | Fracture Strain |
|---|---|---|---|
| 2014-T6 | 63.5 | 17.7 | - |
| 2024-T851 | 64.4 | 18.8 | - |
| 2124-T851 | 63.3 | 23.8 | - |
| 7074-T651 | 73.0 | 24.9 | - |
| 7075-T7351 | 56.8 | 28.1 | - |
| Commercial 4340 | 204.8 | 67.9 | 0.29 |
| High Purity 4340 | 204.6 | 97.2 | 0.52 |
| Commercial Maraging | 193.5 | 113.0 | 0.75 |
| High-Purity Maraging | 189.9 | 149.1* | 1.01 |

* Invalid test. True value probably higher.

## EXPERIMENTAL PROCEDURES

The general experimental scheme used was to make standard $K_{Ic}$ tests, using WOL specimens [4] and prepare transmission electron microscope (TEM) replicas of the fracture surfaces produced, using the two-step method. Attention was concentrated on the region at the beginning of the fracture, i.e., adjacent to region of transition from the initial fatigue crack to the region of unstable fracture. In some cases, these TEM fractographs were supplemented with scanning electron microscope (SEM) fractographs. In addition, tensile specimens were prepared with the axis of the specimens parallel to the direction of the maximum tensile stress in the fracture toughness test specimens. Separate tensile specimens were then plastically strained in successive steps up to and including fracture. The individual tensile specimens were then metallographically sectioned at a midplane parallel to the tension axis, polished and examined for voids. Observations were made on the nucleation sites of voids and measurements made on the growth of these voids with increasing plastic strain until final coalescence and fracture occurred. In the case of aluminum alloys, sheet tensile specimens were used; and in the case of the steels, the specimens were both smooth, and mildly notched, round tensile specimens. Matrix structures, and in one case (the 4340 steel) void nucleation, were observed using thin-film transmission electron microscopy.

## OBSERVATIONS - ALUMINUM ALLOYS

A typical aluminum alloy fractograph from a $K_{Ic}$ specimen is shown in Fig. (1). These aluminum alloy fractographs were generally made up of two populations of dimples differing greatly in size. The larger dimples were nucleated by the cracking of large (~ 10μ) second phase particles in the alloys. The fracture surface of these particles are commonly seen within the boundaries of the large dimples and presumably were the sites of initial void nucleation at the beginning of plastic straining at the advancing crack-tip front of the fracture toughness specimen. Statistical data from quantitative metallographic examination of fractographs for a number of the alloys are shown in Table 2. It will be noted that 45 to 65% of the fracture areas of the aluminum alloys are made up of large dimples. The ratio of the area of cracked particles to that of large dimples is of the order of 0.05 to 0.1 indicating substantial void growth between nucleation and final coalescence at complete fracture.

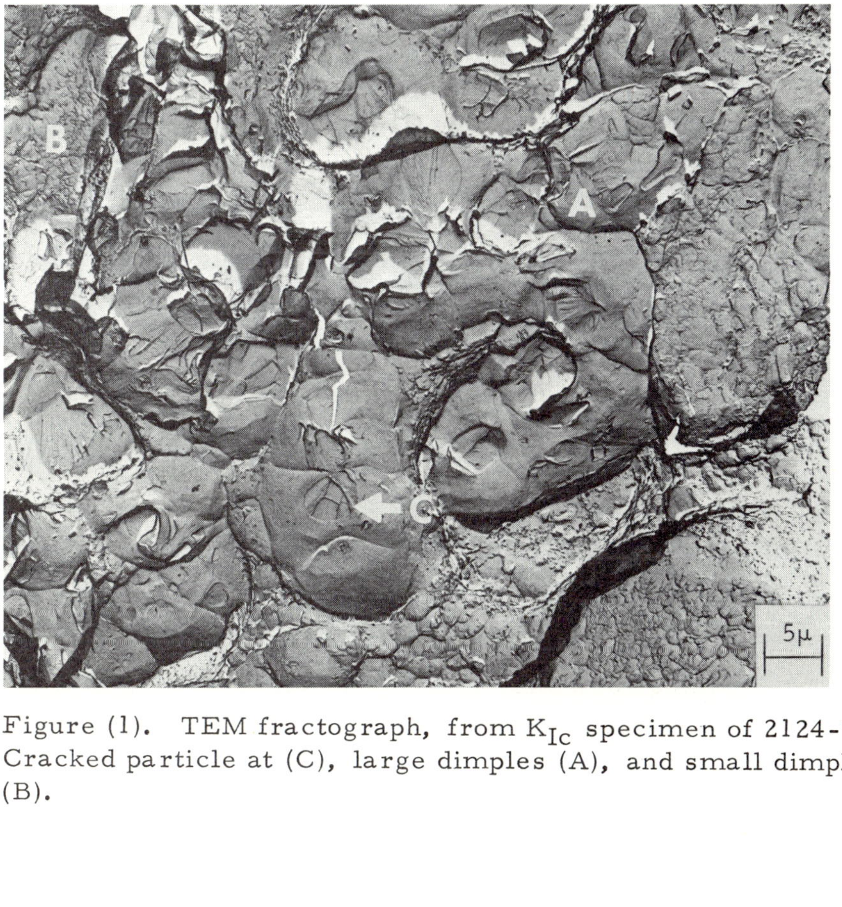


Figure (1). TEM fractograph, from $K_{Ic}$ specimen of 2124-T851. Cracked particle at (C), large dimples (A), and small dimples (B).

TABLE 2

Some Fractographic Data for the Alloys
($K_{Ic}$ Specimens)

| Alloy | Large Dimples | Particle Fracture Surfaces |
|---|---|---|
| | (% of Area | Examined) |
| 2014-T6 | 57.5 | 5.4 |
| 2024-T851 | 46.9 | 2.7 |
| 2124-T851 | 64.1 | 3.4 |
| 7079-T651 | 43.6 | 2.1 |
| Commercial 4340 | 8.8* | - |
| High Purity 4340 | 3.0* | - |
| Commercial Maraging | 100* | - |
| High Purity Maraging | 100* | - |

* Values were similar for notched and un-notched tensile tests.

In the aluminum alloys, the larger second-phase particles tend to cluster so that the large dimples are also clustered. Within these clusters of large dimples, coalescence frequently occurs by simple impingement of growing voids. The coalescence of these groups of large voids at fracture appears to occur by the formation of approximately coplanar arrays of smaller voids at an angle to the tension axis, connecting the groups of large voids. This is suggested by the SEM fractograph shown in Fig. (2) where level differences are more clearly shown than in the TEM fractographs. This view is supported by the micrograph in Fig. (3) of a region near the fracture in a tension specimen of a 2124 alloy strained to fracture and sectioned. In this figure, a row of small voids connects a pair of large voids, presumably caused by the localization of plastic deformation between large voids and the nucleation of voids at small, non-deformable particles in the matrix. This type of coalescence will be discussed in more detail when the behavior of 4340 steel is described later.

For the aluminum alloys, the presumed nucleating particles for these small dimples are the so-called high-temperature precipitate particles commonly present in commercial high-strength alloys. One example of this type of dispersoid is shown in Fig. (4). The large, dark particles in this alloy are

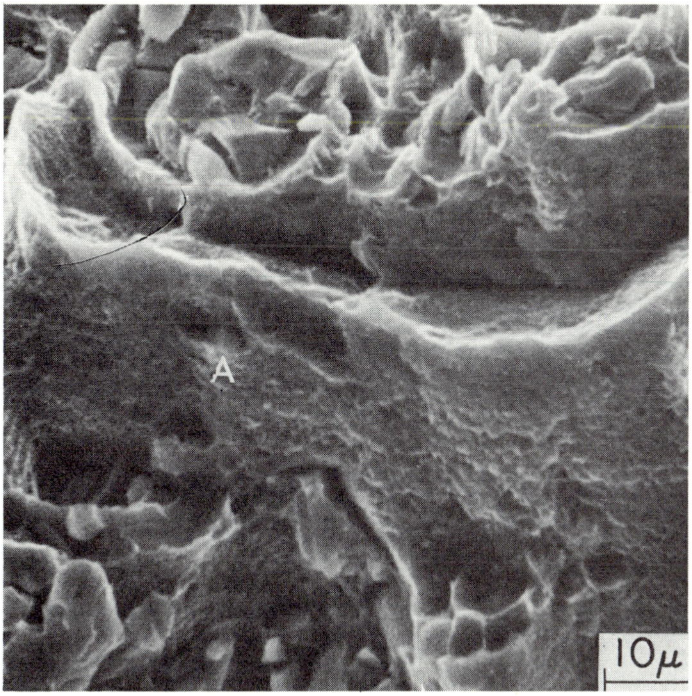


Figure (2). SEM fractograph of 2014-T6. Region of small dimples at, A, connecting two regions of large dimples.

$Al_{12}(Mn, Fe)_3Si_6$ precipitated during high temperature processing of the material; the acicular particles in the background are $CuAl_2(\theta')$ precipitated during the precipitation hardening treatment of the material. The assignment of these particles to the role of nucleation sites for the small dimples is based wholly, for the aluminum alloys, on the similarity between the average diameters of the small dimples and of the spacing of the "high-temperature precipitate" particles. The average linear intercepts of the small dimples, in the region of small dimples ranged from 0.5 to 0.8μm, while the center-to-center spacings of the dispersoid ranged from 0.2 to 0.4μm for the alloys examined. Presumably not all dispersoid particles nucleated voids.

From the metallographic examination of variously strained and sectioned tensile specimens, it was observed that the fracture process began, at small strains, by the cracking of certain

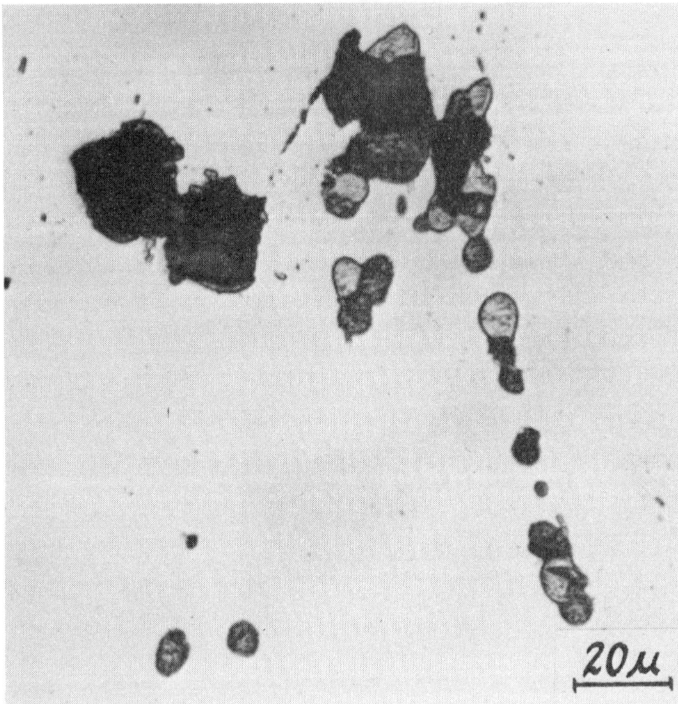


Figure (3). Void sheet of small voids forming between two large, inclusion nucleated, voids. 2124-T851. Tension axis vertical.

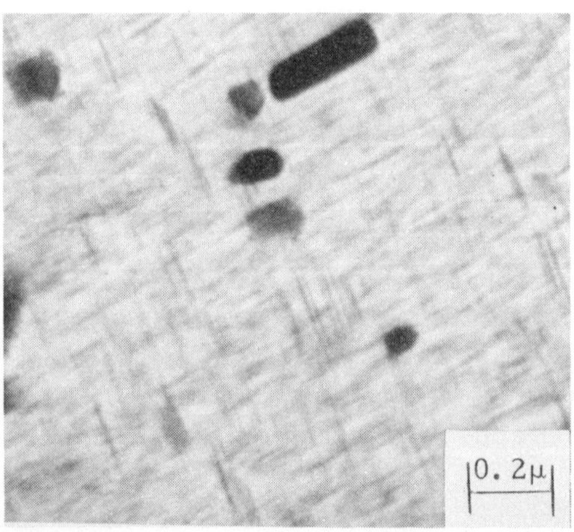


Figure (4). Thin-film electron transmission structure of 2024-T851 showing two types of precipitate particles (see text).

types of large second-phase particles, these cracks in particles then grew into voids with further plastic straining and finally coalesced to produce fracture. An example of the cracking of particles of this type are shown in Fig. (5). The increase in the fraction of the population of large (> 2μm) particles cracked as a function of plastic strain for the tension specimens of the 2000-series alloys is shown in Fig. (6).

Not all of these larger (2-20μm) second-phase particles exhibited cracking and void nucleation during tensile plastic straining, the particles containing Fe or Si appeared to be more susceptible to cracking while $Mg_2Si$ when it appeared, resisted cracking and therefore void nucleation. In addition, the qualitative observation was made that larger particles in this population always cracked first and the smaller particles did not crack until higher strains and therefore higher applied stresses were reached. This effect is shown in another way in Fig. (7) when the strain to produce cracking (and therefore void nucleation) is plotted as a function of the mean size of the particles which did crack.

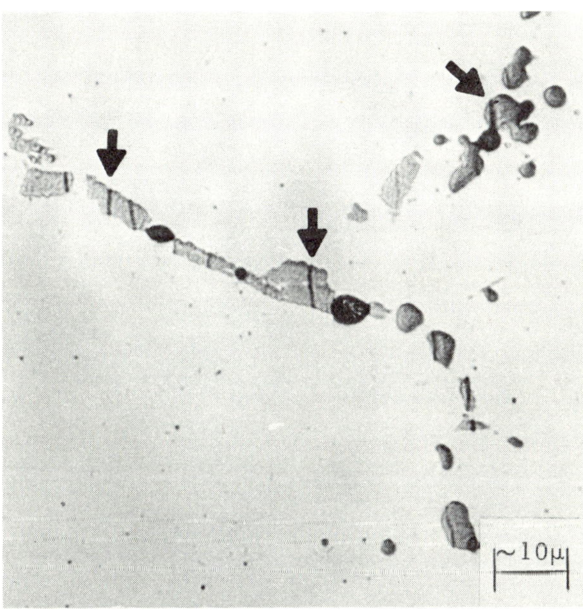


Figure (5). Optical metallography section of 2124-T851 after 2.5% elongation. Note cracks in second phase particles. Tension axis horizontal.

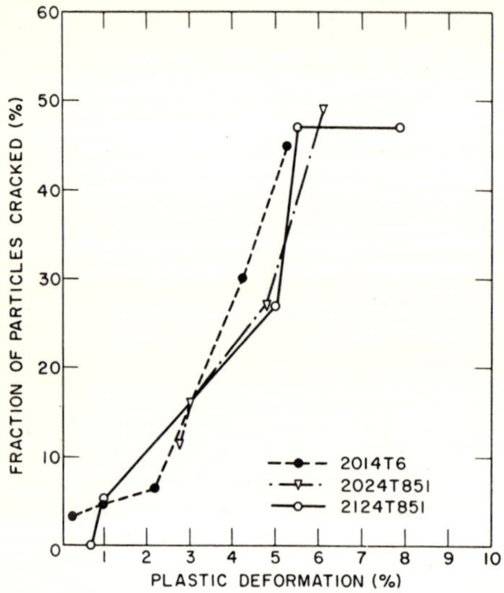


Figure (6). Fraction of large (> 2μm) second phase particles cracked vs. plastic strain.

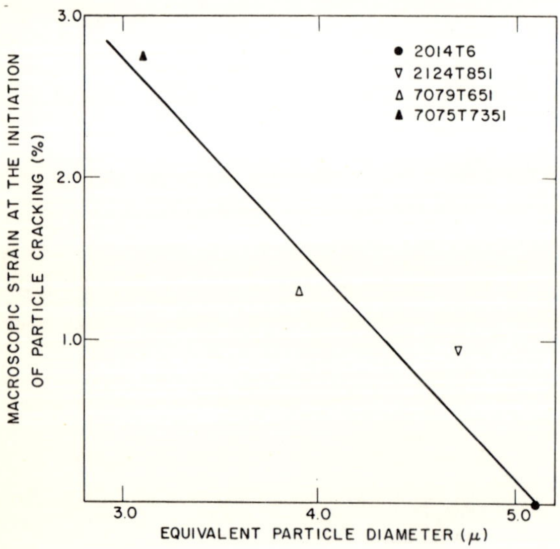


Figure (7). Plastic elongation to begin particle cracking plotted against $\overline{d}$, the "equivalent particle diameter". $\overline{d} = \sqrt{d_1 d_2}$ ; $d_1$ and $d_2$ minimum and maximum dimensions in plane normal to tension axis.

The influence of the void-nucleating particle size is also shown in the two steels. The data for the two maraging steels are shown in Table 3.

TABLE 3

Effect of Void-Nucleating Particle Size on Strain to Fracture and Toughness of Maraging Steels

| | Commercial Purity | High Purity |
|---|---|---|
| Number of Particles per $cm^3$. | $3 \times 10^6$ | $5 \times 10^6$ |
| Average Size (μm) | 8.6 | 3.0 |
| Volume Fraction (%) | 0.21 | 0.09 |
| Strain to Fracture in Tension Test | 0.75 | 1.01 |
| $K_{Ic}$ (ksi-√in.) | 113 | >149 |

A second correlation is shown in Fig. (8) where $K_{Ic}$ is seen to increase as the volume fraction of the void nucleating phase is decreased. Volume fractions were approximated by measuring

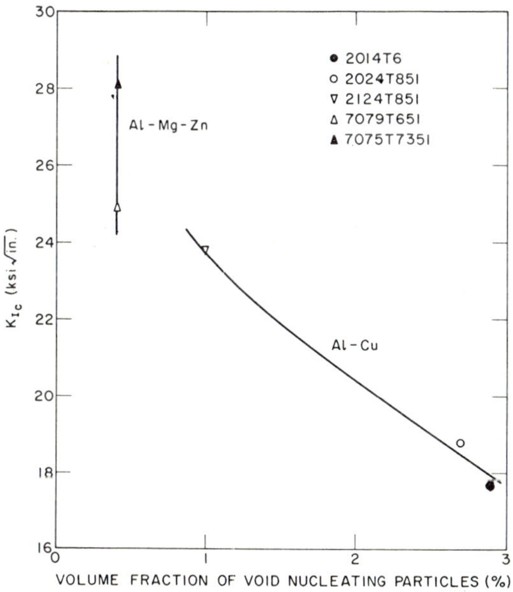


Figure (8). Fracture toughness of aluminum alloys as a function of the volume fraction of void-nucleating second-phase particles.

average diameters on three cross sections taken normal to each other and treating the particles as ellipsoids having these dimensions as principal axes.

SUMMARY - ALUMINUM ALLOYS

From these studies we would expect an improvement in the fracture toughness to result, without sacrifice of strength, if:

1. The iron and silicon impurity levels of the alloys could be substantially reduced.

2. A better distribution of the larger second-phase particles could be accomplished, particularly a reduction in the size of individual particles and the elimination of the clustering of these particles into closely spaced groups.

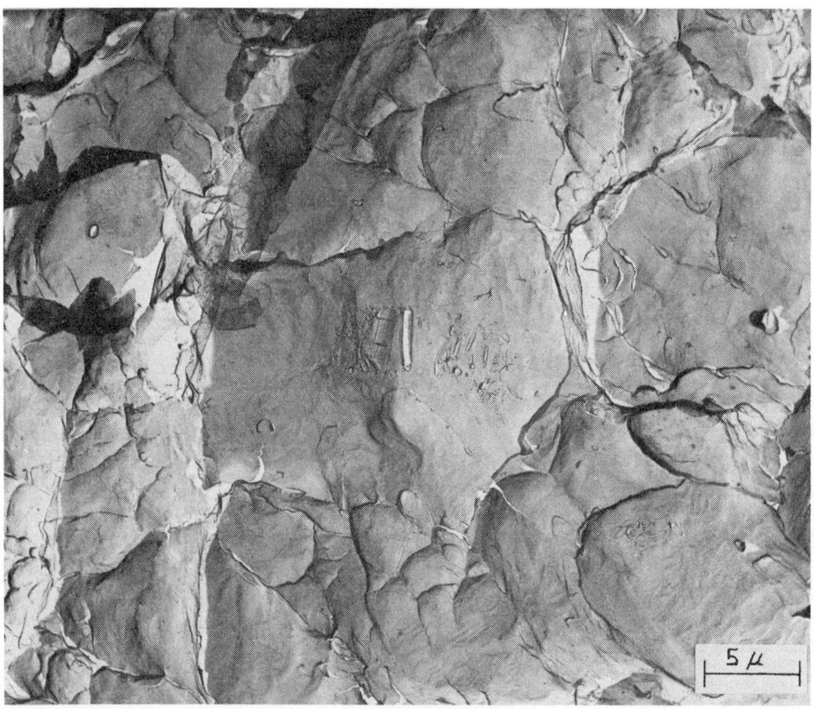


Figure (9). Electron fractograph from a notched-tension specimen of high-purity maraging steel.

## OBSERVATIONS - AISI 4340 AND MARAGING STEELS

An experimental procedure, similar to that used for the aluminum alloys, was used for the steels, with the difference that round tensile specimens were used and notched, round tensile specimens were included. The notch dimensions were chosen to give fully plastic behavior in the notched cross-section before fracture. In addition, a careful comparison was made of fractographs from the $K_{Ic}$ specimens and both tensile specimens to examine their similarity. In both steels, the principal fractographic characteristics were found to be independent of the type of specimen in which the fracture had been produced.

Each steel contained only a single inclusion type: cuboidal TiCN for the maraging steel and MnS for the 4340. Fractographs taken from notched-tension specimens are shown in Fig. (9) for the maraging steel and Fig. (10) for the 4340.

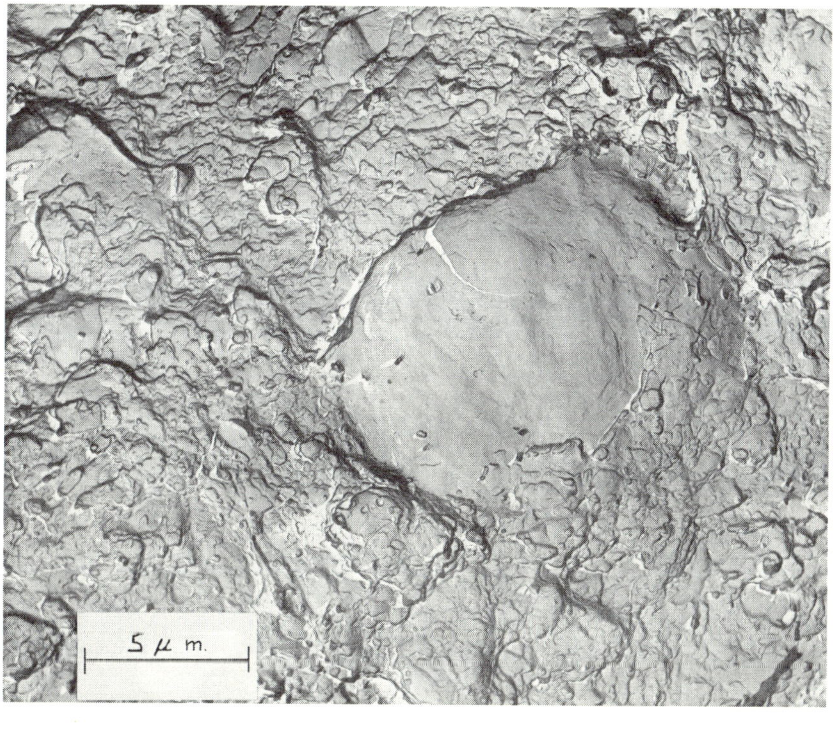


Figure (10). Electron fractograph from a notched-tension specimen of commercial purity AISI 4340.

There is a qualitative difference in these two fractographs: that for the maraging steel is made up almost entirely of a single population of large dimples, whereas the 4340 fracture surface is composed of a few large dimples and a very large fraction (> 90%) of the surface is covered with small dimples.

This fracture topography is explained by differing void coalescence processes in the two steels. If one follows the fracture process metallographically, for the two steels, what is observed is as follows:

Voids are first nucleated at small plastic strains at the inclusions, TiCN for the maraging steels by cracking and MnS in the 4340 by interface separation. With further plastic deformation, and concentrating at the center line of the specimen in the necked, or notched region where large scale separation begins, these voids grow in three dimensions until they coalesce to produce fracture.

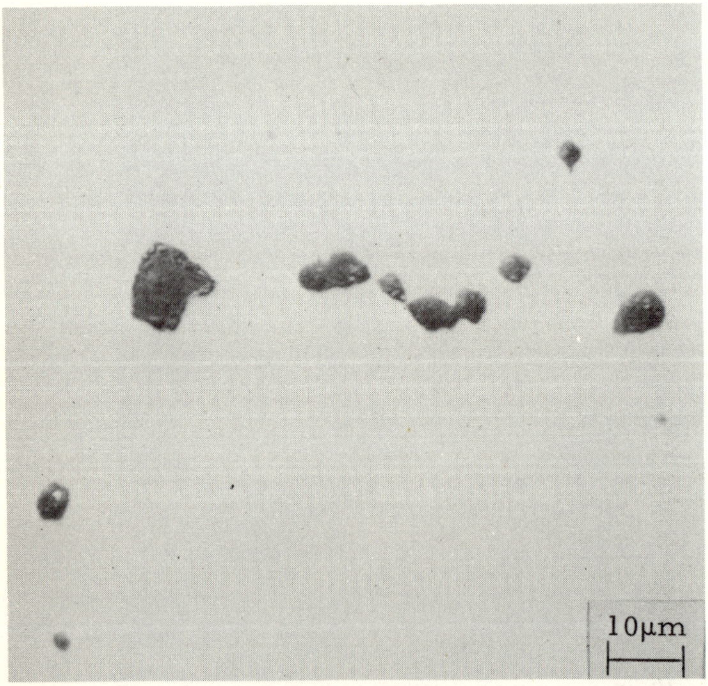


Figure (11). Void coalescence by impingement in smooth tension specimen of high purity maraging steel. Tension axis vertical. $\epsilon = 0.81$.

In the case of the maraging steel, the coalescence is by simple impingement of growing voids, as illustrated in Fig. (11). This type of coalescence does not occur until large plastic strains, e.g., $\epsilon_p = 0.81$ in Fig. (11). In the case of the AISI 4340, spherical growth of the voids takes place for a time, but as the void surfaces approach each other, this growth is terminated at much smaller plastic strains than those for the maraging steels, (e.g., $\epsilon_p = 0.25$) by the formation of sheets of very small voids forming on planes at approximately 45° to the tension axis and connecting the growing inclusion-nucleated voids. An example of this process is shown in Fig. (12). These small voids continue to grow and eventually connect together to form a surface of separation covered with small dimples. Thin-film transmission examination of the region of a void sheet has shown that these small voids are being nucleated at the $Fe_3C$ particles of the tempered martensite, by interface separation. The $Fe_3C$ particles for the tempering treatment (one hour at 430° C) were about 0.1μm in diameter, judging from extraction replicas and thin film transmission metallography. In the maraging steel,

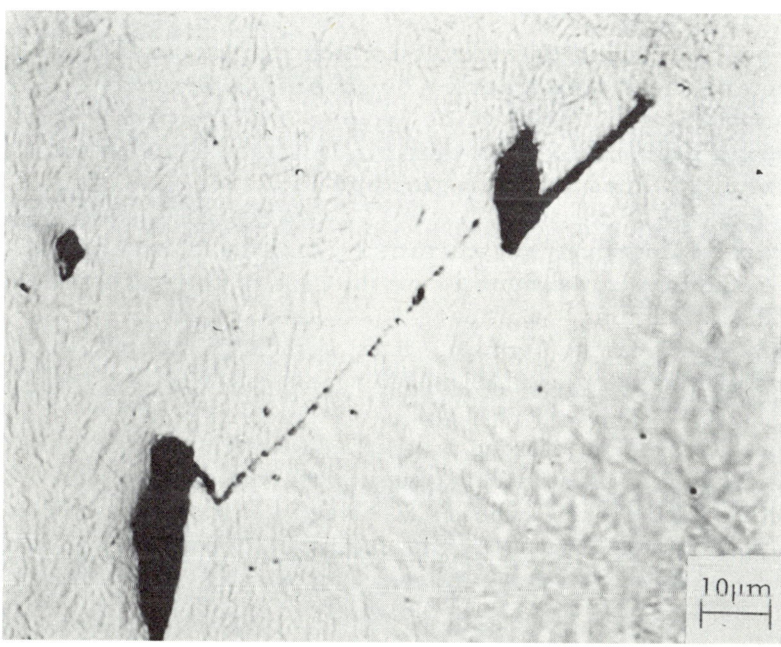


Figure (12). Void coalescence by void sheet formation in smooth tensile specimen of AISI 4340. Tension axis vertical. $\epsilon = 0.25$.

the $Ni_3Mo$ precipitate is much smaller (estimated at 0.01μm from other work) and does not nucleate voids between the large inclusion nucleated voids. This difference in the coalescence process thus accounts for the higher $K_{Ic}$ values of the maraging steels compared to the quenched and tempered 4340 steel.

## SUMMARY

One may summarize the findings of these various studies as follows (not all of what follows has been illustrated in this brief paper):

1. The first voids are initiated at the larger second phase (mainly impurity) particles by cracking or by interface separation almost from the beginning of plastic deformation.

2. In the presence of a hydrostatic tension component in the stress field during plastic deformation, these voids grow in three dimensions until they impinge on each other or the growth is arrested by the formation of sheets of small voids between the growing large voids.

3. The amount of plastic deformation which may occur before fracture and, also, the fracture toughness, is strongly affected by whether or not voids grow to impingement or the void growth process is interrupted by void sheet formation (compare $K_{Ic}$ values for maraging and AISI 4340 steels).

4. A substantial improvement in fracture toughness results from a reduction in the number or the size of the initial void nucleating inclusions. Compare 2024 and 2124 aluminum alloys (number) or the "commercial" and "high purity" versions of maraging and AISI 4340 steels (size).

5. Void nucleation is unaffected by the hydrostatic tension component during plastic flow but, comparing notched and un-notched specimens, appears to be dependent only on the maximum principal stress acting during deformation.

## ACKNOWLEDGEMENTS

The investigations which have been briefly described are the work of a number of graduate students at Carnegie-Mellon University: Thomas B. Cox, Rajen H. Merchant, Joseph A.

Psioda, Junichi P. Tanaka, Robert H. Van Stone, and a post-doctoral fellow, Carlos A. Pampillo. The work has been made possible by a grant from the National Aeronautics and Space Administration and the advice and assistance of Dr. William F. Brown, Jr. of that agency is gratefully acknowledged.

REFERENCES

[1] Tanaka, J. P., Pampillo, C. A., and Low, J. R., Jr., Fractographic Analysis of the Low-Energy Fracture of an Aluminum Alloy, Am. Soc. Testing and Materials, Spec. Tech. Publ. No. 463, 191, 1970.

[2] Van Stone, R. H., Merchant, R. H., and Low, J. R., Jr., Investigation of the Plastic Fracture of High-Strength Aluminum Alloys, Am. Soc. Testing and Materials, Spec. Tech. Publ. No. 556, to be published in 1974.

[3] Cox, T. B. and Low, J. R., Jr., An Investigation of the Plastic Fracture of AISI 4340 and 18 Nickel--200 Grade Maraging Steels, to be published in Metallurgical Transactions in 1974.

[4] ASTM Standard Test Method E399-72, Vol. 31 of ASTM Standards, 1973.

# PROPAGATION OF CLEAVAGE MICROCRACKS IN TRIAXIAL STRESS FIELDS

George J. Dvorak

Duke University, Durham, North Carolina 27706, USA

## ABSTRACT

Selective propagation of elastic cleavage microcracks along a continuous path on the {100} planes in individual grains of a b.c.c. polycrystal is analyzed. It is shown that the cleavage path is a nearly planar, multiply connected surface consisting of cleavage facets which are joined at intercept points or lines at the grain boundaries. The extent of the cleavage fracture process is evaluated in terms of the probability of successful grain boundary crossings by the microcracks for a given macroscopic state of stress. Results are presented for the plane stress and plane strain states, and for a combination of uniaxial tension with a transverse hydrostatic stress. The influence of the stress state on the initial material weakening by elastic cleavage, and on the resulting fracture toughness of the material is discussed.

## INTRODUCTION

It is now generally recognized that the microscopic fracture mechanisms associated with brittle crack propagation in steel are dominated by the process of elastic or easy cleavage. Specifically, the material at the moving crack tip is first weakened by cleavage microcracks which propagate in a selective but continuous fashion through grain boundaries and individual grains with favorably oriented systems of cleavage planes. The cleavage front spreads well ahead of the main crack tip, and leaves behind unfractured grains and other disconnected links of material. As a result, there are two distinct modes of local material separation at the tip of a propagating crack, confined within a discrete weakened zone which extends over many grains: Initial weakening by the propagating cleavage microcracks, and failure of the remaining ligaments. Of course, the latter failure mode which is probably dominated by ductile rupture, provides the major contribution to fracture toughness of the material. However, since the density of the remaining ligaments depends upon the extent of the initial material weakening, one should expect the magnitude of fracture toughness to be affected by the factors which control the microscopic elastic cleavage.

It has been suggested by Dvorak [1]-[3] that the microcrack propagation process takes place in an elastically strained region ahead of the main crack, and, therefore, is of the Griffith type, controlled by the local state of stress. An experimental confirmation of both the continuity of the cleavage path, and of the elastic nature of the initial cleavage process was provided by the work of Hoagland, Rosenfield, and Hahn [4] on crack propagation and arrest in silicon iron at -75 to +100 deg. C. Although this material is more brittle than mild steel, it has similar fracture characteristics and permits the detection of microscopic plastic deformation by etching.

The present fracture propagation model differs appreciably from the earlier ones which assumed that the cleavage microcracks would form in a discontinuous manner, by crack nucleation mechanisms involving plastic straining. In fact, except at very low temperatures, cleavage fracture propagation was considered in terms of plastically induced hole formation and coalescence [5], which is the established mechanism of ductile fracture [6]. However, as discussed by Smith [7], the more recent results essentially confirm the dominant role of continuous elastic cleavage in initial material weakening. A comparison of the present cleavage fracture model with that of ductile fracture reveals interesting duality relationships between the two fracture mechanisms. The cleavage microcracks form a continuous, multiply connected path through the polycrystal and the remaining material consists of disconnected ligaments. The opposite is true in ductile fracture where the initial material weakening consists of formation of disconnected holes in a continuous matrix. Furthermore, the cleavage front propagates in an elastic material, whereas the formation of holes is contingent upon the existence of local plastic straining. However, it seems reasonable to assume that these distinctions will become gradually less evident as the final material separation is approached.

The purpose of the present paper is to continue the eariler work [3] on a statistical analysis of the influence of the local state of stress on the probability of successful grain boundary crossing by cleavage microcracks in a randomly oriented b.c.c. polycrystal. Specifically, the analysis is extended to the case of an arbitrary three-dimensional state of stress. Practically interesting results, e.g., for the case of plane stress and plane strain will be presented.

THE MODEL OF ELASTIC CLEAVAGE FRACTURE

The complexity of the statistical analysis requires the introduction of several simplifying assumptions. The polycrystal will be regarded as an elastic, isotropic solid, divided by a continuous network of grain boundaries into grains of similar size with an average diameter d. In each grain there exists a system of the {100} cleavage planes which are available for microcrack propagation, and which do not have a preferential orientation. Each microcrack in a grain will be regarded as a planar penny-shaped crack of diameter d. The crossing of a microcrack through a grain boundary into the next grain will be assumed to take place at a single intercept point or line between the existing cleavage crack plane and one of the {100} planes in the next grain. It is recognized that many such intercepts may exist at an actual grain boundary, especially between systems of cleavage planes with a considerable twist misorientation, and that several microcracks may be initiated in the next grain. These nucleated microcracks may eventually join into a single one, and create a number of large cleavage steps. One would expect, therefore, that the effective surface energy for microcrack propagation will fluctuate for individual grains, depending on the orientation of the incoming microcrack. However, these and other microstructural details of the propagation process cannot be specifically considered in quantitative terms at this time. They can be accounted for only in an average sense, by assuming that the surface energy for the elastic cleavage process in a polycrystal will be considerably greater than that observed in single crystals. Based on earlier considerations [2], this average surface energy for steels should be of the order of $10^4$ergs/cm$^2$.

The existence of the cleavage steps at the microcrack planes will not be completely disregarded, since their height is likely to be comparable to the maximum crack opening displacement [3]. Therefore, the shear fracture modes should be very limited in the elastic cleavage process, and it is appropriate to assume that all microcracks will open only in the direction normal to the cleavage plane, i.e., in mode I.

Furthermore, the inertia terms will be neglected on the microscale, and it will be assumed that the remotely applied stresses, resolved at any system of cleavage planes in a grain are not influenced by the presence of other microcracks in the polycrystal. Finally, we shall require that in all grains which may cleave during the fracture process, the microcracks can propagate only on those cleavage planes which are most nearly normal to the direction of the largest principal stress and, as will be shown later, are also most nearly parallel to each other. Accordingly, the cleavage path will be a nearly planar, multiply connected continuous surface, consisting of the favorably oriented cleavage facets joined at grain boundary intercept points or lines.

## MICROCRACK PROPAGATION THROUGH A GRAIN BOUNDARY

The criterion for propagation of a single microcrack through a grain boundary was derived in [3]. The results will be restated here for completeness, and the criterion will be compared with experimental data. Figure 1 shows the geometry of the two microcracks. The front c of an existing microcrack in the $x_1x_3$ cleavage plane intersects the grain boundary at the origin 0. The axis $x_2$ is a [100] normal to the microcrack plane, and $x_3$ is the tangent to c at 0. Assume that the {100} cleavage plane system in the next grain is associated with the coordinates $x_i'$, $i = 1,2,3$. The axis $x_2'$ is a normal to the cleavage plane system which is most nearly perpendicular to the direction of the maximum principal stress; the $x_1'x_3'$-plane intersects the existing $x_1x_3$ crack plane at 0 and, therefore, it is the assumed cleavage plane available for microcrack propagation in the next grain. If a new microcrack is nucleated in the next grain at the intercept 0, the new elementary crack front can be indicated by the contour c' such that $x_3'$ is the tangent of this contour at 0. The transformation from the $x_i$ to the $x_i$ system can be described either by the angles $\phi_1$ and $\phi_2$, or more conveniently by the angles $\beta$ and $\gamma$, Fig. 1. Since we assume that microcrack propagation will take place on those cleavage planes which are most nearly parallel to each other, we must require in a b.c.c. polycrystal that:

$$0 \leq \beta \leq \cos^{-1}(1/\sqrt{3}) \doteq 54.735 \text{ deg.}$$
$$\pi/4 \leq \gamma \leq \frac{\pi}{2}, \quad \beta + \gamma \geq \frac{\pi}{2} \tag{1}$$

These relations, of course, reflect the symmetry of the crack nucleation problem about the $x_1x_2$-plane in Fig. 1. The nucleation criterion for the new microcrack was based on the requirement that the resolved normal stress on the $x_1'x_3'$-plane at 0 must be equal in magnitude to that which would be needed for extension of the exiting microcrack in the $x_1x_3$-plane, i.e., at $\beta = 0$. Assuming that the Griffith condition is valid in this case, one finds for $\beta = 0$:

$$\sigma_R \geq \left[\frac{\pi\Gamma E}{(1-\nu^2)d}\right]^{1/2} = \sigma_0 \tag{2}$$

and for $\beta \neq 0$:

$$\sigma_R \geq 4(1+3\cos\beta)^{-1}\sigma_0 , \tag{3}$$

where $\sigma_R$ is the resolved normal stress at the $x_1x_3$-plane caused by remotely applied stresses; E and $\nu$ are the elastic constants of the polycrystal; d is the grain diameter; and $\Gamma$ is the effective surface energy for elastic cleavage in a polycrystal, discussed in the previous section. It is assumed that $\sigma_R$ is always uniform within each grain of diameter d. The derivation of Eqn. (3) was described in Ref. [3]. It is readily seen that the required magnitude of $\sigma_R$ increases with $\beta$, which suggests that the microcracks will indeed propagate in a selective manner, and more easily on planes which are most nearly parallel to each other.

Equation (3) is valid for the case of plane strain deformation at the existing microcrack, which is likely to be satisfied along the entire periphery of a cleavage front, with the exception of the close vicinity of its intercepts with a free surface. It is important to distinguish between the macroscopic deformation states caused by the remotely applied stresses, and the microscopic ones which pertain only to the relatively close vicinity of existing microcracks. For example, the local stress field at the tip of a microcrack which is located at the depth of 10 d from a free surface can be expected to satisfy the plane strain assumptions of Eqn. (3), even in the presence of a macroscopic plane stress state.

We shall assume that the nucleation criterion (3) is also sufficient for propagation of the nucleated microcrack across the next grain. Specifically, Eqs. (1) and (3) suggest that all microcracks which may initiate new microcracks and thus successfully participate in the elastic cleavage fracture process may exist only on cleavage planes where the resolved normal stress exceeds $\sigma_0$. Accordingly, the assumed condition for the propagation of a nucleated microcrack in the next grain is

$$\sigma'_R \geq \sigma_0 , \tag{4}$$

where $\sigma'_R$ is the resolved normal stress on the $x'_1x'_3$-plane, Fig. 1. The verification and a possible refinement of (4) would have to be based upon the solution of an elasticity problem for two cracks intersecting at an arbitrary angle. This is a very difficult problem, the chances for its solution appear to be rather remote at this time.

It is of interest to compare the criterion (3) with experimental data on propagation of cracks through grain boundaries in polycrystaline 3% silicon iron, which were reported by Gell and Smith [8]. The results were obtained on 1/8 in. × 1/4 in. × 1/2 in. specimens with coarse grain sizes of 1 mm, and 2 mm. The specimens were cathodically charged with hydrogen at -75 deg. C, to form cleavage microcracks on {100} planes by the precipitation of molecular hydrogen in the metal lattice at the charging temperature. After ageing, the microcracks were studied by successive repolishing and etching in stages, and direct observations were made of microcracks which had been stopped at grain boundaries, and also of cracks that had propagated across the boundaries. These observations were correlated with the misorientation of the two adjacent grains. Specifically, the misorientation between the initial cleavage crack in one grain and the cleavage crack or potential cleavage plane in the neighboring grain was determined. The ability of the microcracks to cross a grain boundary was found to depend on two misorientation components: The component B corresponding to a tilt about an axis in the grain boundary, i.e., to a rotation by an angle $\Phi$ about the axis $x_3$ in Fig. 1, and the component C corresponding to a twist about the grain boundary normal, i.e., to a rotation by an angle $\psi$ about the $x_1$ axis in Fig. 1. Note that $\Phi = \phi_2$ for $\phi_1 = 0$, and $\psi = \phi_1$ for $\phi_2 = 0$ in Fig. 1. The results of the experiments are shown in Fig. 2, in terms of the misorientation components B, and C. The dots denote that the crack crossed a grain boundary, the crosses indicate that the initial cracks were stopped by a grain boundary.

The plane strain and plane stress curves shown in the drawing were constructed as follows. In plane strain, using Eqn. (3), and assuming that the microcracks are acted upon by a constant internal pressure p, one finds that

$$p = p_c \geq (\frac{1}{4} + \frac{3}{4} \cos \beta)^{-1} \sigma_0 \qquad (5)$$

for a microcrack to be able to cross the grain boundary; $p_c$ is a critical magnitude of p needed for microcrack propagation, which was not determined in the experiments. Furthermore, it can be shown that

$$\cos \beta = \cos \Phi \cos \psi \quad . \qquad (6)$$

Hence, assuming that $\sigma_0$ and $p_c$ are constants, the microcrack propagation criterion can be written as

$$\cos \Phi \cos \psi = \frac{4}{3} \left[ \frac{\sigma_0}{p_c} - \frac{1}{4} \right] = \text{const.} \qquad (7)$$

Equation (7) represents a nearly circular contour in the BC-plane (or $\Phi\psi$-plane) of Fig. 2. However, since the actual experiments were performed on specimens the thickness of which was equal to about 2 - 3 d, a comparison of experiments with the present criterion requires a plane stress analysis of the microcrack

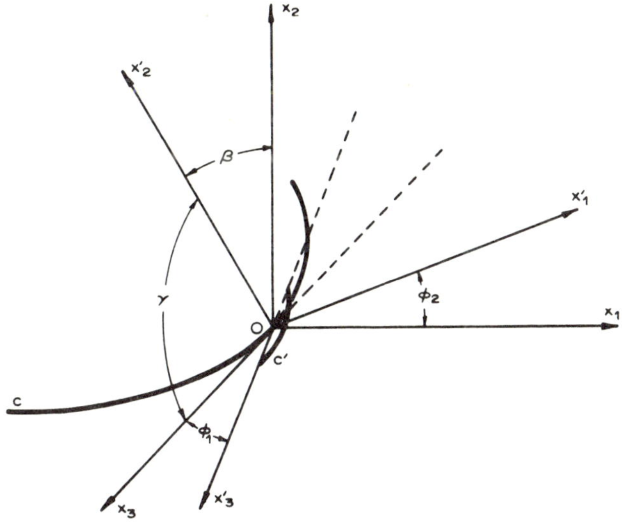


Fig. (1) Nucleation of a new elementary microcrack front c' at the intercept O of an existing crack point c with a grain boundary.

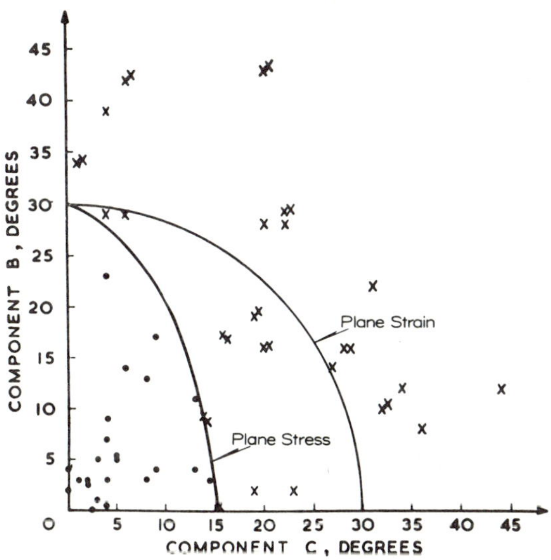


Fig. (2) The effect of misorientation components $B = \Phi$ and $C = \psi$ on the ability of cleavage microcracks to nucleate new cleavage cracks in an adjacent grain, Ref. [8].

propagation problem. Following the same approach which was outlined in Ref. [3] for the plane strain case, one obtains the plane stress curve of Fig. 2. This curve was drawn to such size which approximated closely the boundary between the data points in Fig. 2, and the plane strain curve (7) was matched with the plane stress one for the case of pure tilt (component B), where the theoretical results should coincide, if one neglects the minor difference between the plane stress and plane strain stress intensity factors. As expected, the microcrack propagation through grain boundaries is more difficult in the case of plane stress, especially in the presence of a large twist misorientation component C. On the other hand, the lateral constraint in the plane strain case causes a relatively large transverse normal tensile stress component (in the $x_3$ direction in Fig. 1) which facilitates microcrack propagation in the case of a large misorientation component C.

The validity of Eqn. (3) is supported by these experiments in the following sense. If one uses the fracture criterion leading to Eqn. (3) in the plane stress case, one obtains a good agreement between the theoretical prediction and plane stress data. Therefore, it is expected that an equally good agreement between theory and experiment would be observed in the plane strain case, if appropriate experimental results were available.

Finally, we note that if the two curves in Fig. 2 are matched for component B equal to 30°, one obtains from (7) that $p_C = 1.11\ \sigma_0$. This suggests that only small values of $p_C$ at the existing microcrack permit to overcome relatively large misorientations between the cleavage planes at a grain boundary.

PROPAGATION OF MANY MICROCRACKS

The first step towards the evaluation of the relative portion of the grains on the fracture surface which can fail by cleavage is the determination of the probability of successful grain boundary crossings by the microcracks. Assuming that the polycrystal remains elastic and that the stresses acting at an existing microcrack are not significantly affected by other microcracks, the propagation process will be controlled only by the local macroscopic stress state and by the orientation of the cleavage planes.

The state of stress can be described in terms of the principal stresses $\sigma_i$, $i = 1,2,3$. It is advantageous to introduce a new dimensionless stress system $T_i = \sigma_i/\sigma_0$, such that

$$T_1 \geq T_2 \geq T_3 \ , \quad T_2 \geq 0 \ , \tag{8}$$

where $\sigma_0$ is defined by Eqn. (2). The orientation of the cleavage planes at a grain boundary can be related to the principal stress system $T_i$, as shown in Fig. 3. The angles $\beta_j$ and $\theta_j$ define the normals $n_j (j=1,2)$ to the two cleavage planes intersecting at a grain boundary point O. The subscript $j=1$ refers to the cleavage plane in the next grain. The microcrack propagation criterion, Eqn. (3), for a microcrack with the normal $n_1$ is:

$$T_1\left\{\left(\frac{T_1-T_3}{T_1}\right)\cos^2\beta_1 + \left(\frac{T_2-T_3}{T_1}\right)\sin^2\beta_1\cos^2\theta_1 + \frac{T_3}{T_1}\right\} \geq 4(1+3\cos\beta)^{-1}, \tag{9}$$

where the angle $\beta$ can be expressed as

$$\cos\beta = \sin\beta_1 \sin\beta_2 \cos(\theta_1-\theta_2) + \cos\beta_1 \cos\beta_2 \quad . \tag{10}$$

The requirement that the cleavage planes must be most nearly perpendicular to the direction of $T_1$ imposes the following constraints upon the orientation parameters in a b.c.c. system:

$$\left.\begin{array}{l} 0 \leq \beta_j \leq \cos^{-1}(1/\sqrt{3}) \\ 0 \leq \theta_j \leq 2\pi \end{array}\right\} j=1,2 \quad .$$

$$0 \leq \beta \leq \cos^{-1}(1/\sqrt{3}) \quad \text{for } \beta_1+\beta_2 \geq \cos^{-1}(1/\sqrt{3}) \tag{11}$$

$$0 \leq \beta \leq \beta_1+\beta_2 \quad \text{for } \beta_1+\beta_2 \leq \cos^{-1}(1/\sqrt{3}) \quad .$$

These inequalities, in conjunction with Eqs. (8) and (9) suggest the following limits on the applied stresses $T_i$:

$$T_1 + 2T_3 \leq \frac{12}{1+\sqrt{3}} \doteq 4.4 \tag{12}$$

$$T_1 \geq 1, \quad T_1 \geq T_2, \quad T_2 \geq T_3 \quad . \tag{13}$$

The equality sign in Eqn. (12) represents a stress state which ensures crack propagation through all grain boundaries; $T_1=1$ is, of course, the lowest permissible value at which crack extension can proceed only in the plane of an existing microcrack when $\beta_1=\beta_2=0$ in Fig. 3. The inequalities (12) and (13) define a semi-infinite stress space of all unique stress states that satisfy Eqs. (8) to (11).

The probability S of successful crossings of grain boundaries by cleavage microcracks which propagate in a b.c.c. polycrystal under an applied stress state $T_i$ can be defined as follows. Let us specify the stress ratios $T_3/T_1$, and $T_2/T_1$ in Eqn. (9). Then for each pair of these ratios, $T_1$ is the only unknown stress in the crack propagation criterion (9), and it is a function of the angles $\beta_j$ and $\theta_j$. If these angles are defined as random variables, $T_1$ is also a random variable, and the probability S can be found

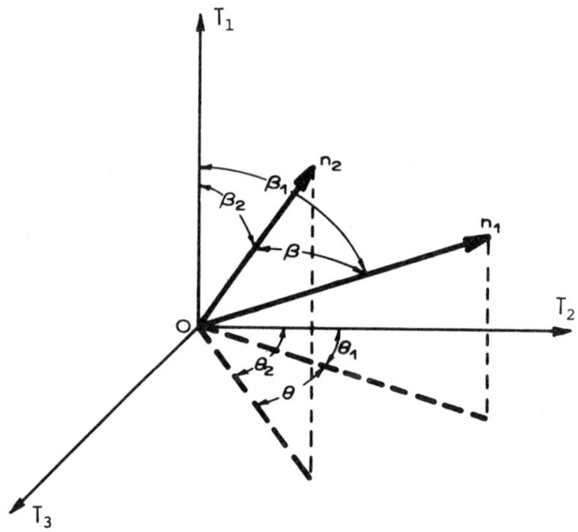


Fig. (3) The normals $n_j$ and orientation angles $\beta_j$, $\theta_j$ of the existing $(j = 1)$ and prospective $(j = 2)$ crack planes in the principal stress system $T_i$.

as

$$S = P\{T_1 > \overline{T}_1\} = 1 - F_T(T_1) \quad , \tag{14}$$

where $F_T(T_1)$ is the probability distribution function of the random variable $T_1$. It follows that S can be expressed in terms of $T_1$ for any combination of the stress ratios $T_3/T_1$, and $T_2/T_1$, i.e., for any selected stress state $T_i$ in the stress space defined by Eqs. (12) and (13).

The probability density functions of the random variables $\beta_j (j = 1,2)$ were found in Ref. [9] as:

$$\begin{aligned} f_\beta(\beta_j) &= 3 \sin \beta_j \quad \text{for } 0 \leq \beta_j \leq \pi/4 \\ f_\beta(\beta_j) &= 3 \sin \beta_j [1 - \tfrac{4}{\pi} \cos^{-1}(\cot \beta_j)] \\ &\qquad \text{for } \tfrac{4}{\pi} \leq \beta_j \leq \cos^{-1}(1/\sqrt{3}) \quad . \end{aligned} \tag{15}$$

The variables $\theta_j$ are defined by:

$$f_\theta(\theta_j) = \frac{1}{2\pi} \quad \text{for } 0 \leq \theta_j \leq 2\pi \text{ , } j=1,2 \quad . \tag{16}$$

A brief outline of the evaluation of S will be now presented. Let us introduce a new auxiliary variable x such that

$$T_1 x \geq 1 \text{ ,} \tag{17}$$

where x is a solution of Eqs. (9) and (17). Then, $x=1/T_1$ and the probability S can be found in terms of the probability distribution function $F_x(\overline{x})$. In order to determine this new distribution function, one must find the joint probability density function $f(\beta_1,\beta_2,\theta_1,x)$. The variable x can be expressed from Eqs. (9) and (10), in the form:

$$x = x(\beta_1,\beta_2,\theta_1,\theta_2) \quad . \tag{18}$$

The new joint density function is:

$$f(\beta_1,\beta_2,\theta_1,x) = \frac{f(\beta_1,\beta_2,\theta_1,\theta_2)}{\left|\frac{\partial x}{\partial \theta_2}\right|} \tag{19}$$

Since the random variables $\beta_j$, and $\theta_j$ are independent, their joint density function is:

$$f(\beta_1,\beta_2,\theta_1,\theta_2) = f_\beta(\beta_1)f_\theta(\beta_2)f_\theta(\theta_1)f_\theta(\theta_2) \quad , \tag{20}$$

and can be easily evaluated from Eqs. (15) and (16).

The probability distribution function of the variable x can be written as:

$$F_x(\overline{x}) = \int_0^{\overline{x}} \left[ \int_{\beta_1}\int_{\beta_2}\int_{\theta_1} f(\beta_1,\beta_2,\theta_1,x)\, d\beta_1 d\beta_2 d\theta_1 \right] dx \tag{21}$$

The evaluation of this integral is simplified by the following substitution:

$$x - \frac{3}{4}(\frac{1}{3} + \cos\beta) \left[ \cos^2\beta_1 + \frac{T_3}{T_1} + (\frac{T_2-T_3}{T_1}) \sin^2\beta_1 \cos^2\theta_1 \right] \text{ ,} \tag{22}$$

which follows from Eqs. (9) and (17).

Then, one obtains $F_x(\overline{x})$ for $0 \leq \beta_j \leq \pi/4$ in the form:

$$F_x(\bar{x}) = c\int_0^{\bar{\beta}}\left[\int_{\beta_1}\int_{\beta_2}\int_{\theta_1} \frac{\sin\beta_1 \sin\beta_2 \sin\beta d\beta_1 d\beta_2 d\theta_1}{\left[\sin^2\beta_2 - \cos^2\beta_1 - \cos^2\beta + 2\cos\beta_1 \cos\beta_2 \cos\beta\right]}\right] d\beta, \tag{23}$$

where $\bar{\beta} = \bar{\beta}(\beta_1, \theta_1, \bar{x})$ can be found from (22), and c is a constant. The integral (23) exists as an improper integral within the integration limits imposed by Eqs. (11). One integration step (e.g., in $\beta_1$) can be completed in closed form, and the singularity thus removed. The remaining steps must be performed by numerical integration. A somewhat similar procedure can be applied to find $F_x(\bar{x})$ for $\pi/4 \leq \beta_j \leq \cos^{-1}(1/\sqrt{3})$, but the functions $f_\beta(\beta_j)$ in Eqs. (15) must be approximated by a finite trigonometric series.

A more detailed exposition of the statistical analysis is beyond the scope of this paper. The theory of functions of many random variables is described in several textbooks, e.g., in Feller [10], and Papoulis [11].

## RESULTS

The probability S of successful grain boundary crossings was evaluated for a number of stress states within the limits specified by Eqs. (12) and (13). The results of particular interest are those for the plane stress and plane strain states of stress, and for the combination of simple tension with lateral hydrostatic stress. These will be now presented.

### Plane Stress

The principal stresses can be specified as $T_1 \geq T_2 \geq 0$, $T_3 = 0$, or as $T_1 \geq 0 \geq T_3$, $T_2 = 0$. Figure 4 shows these two cases, in the right and left quarter planes, respectively. For the first stress combination one finds from Eqs. (12) that $S \leq 1$ for $T_1 \leq 4.4$. For the second case, $S \leq 1$ for $T_1 \leq 4.4 - 2T_3$. These limits, together with those of Eqs. (13) define the solution space of Fig. 4. The calculations of S were made at the points denoted by circles.

### Plane Strain

The principal stresses can be described by one of the following relations: $T_3 = \nu(T_1 + T_2)$, or $T_2 = \nu(T_1 + T_3)$, $\nu$ is the Poisson's ratio. $T_1 = T_2$ at $T_1 = T_2(1-\nu)/\nu$, as shown in Fig. 5. For $T_2 \leq T_1 \leq T_2(1-\nu)/\nu$, $T_1$ and $T_2$ are considered as the given stress components. Then, $S \leq 1$ for $T_1 \leq (4.4 - 2\nu T_2)/(1+2\nu)$. For $T_1 \geq T_3(1-\nu)/\nu$, $T_1$ and $T_3$ are the given principal stresses, and $S \leq 1$ at $T_1 \leq 4.4 - 2T_3$. These rules indicate the appropriate assignment of the symbols $T_2$ and $T_3$ to the principal stresses. The

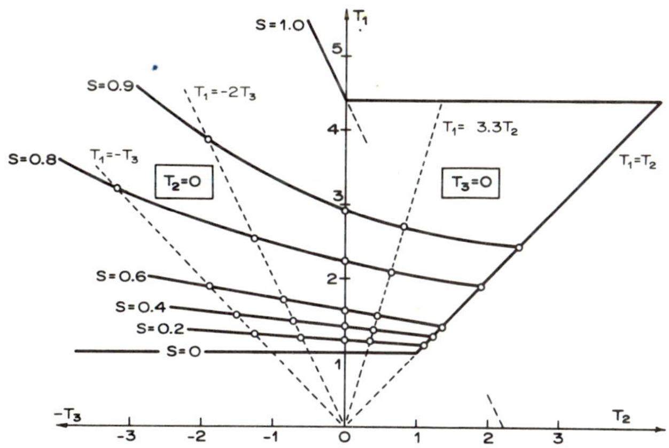


Fig. (4) The probability S of successful grain boundary crossings by cleavage microcracks. The case of plane stress.

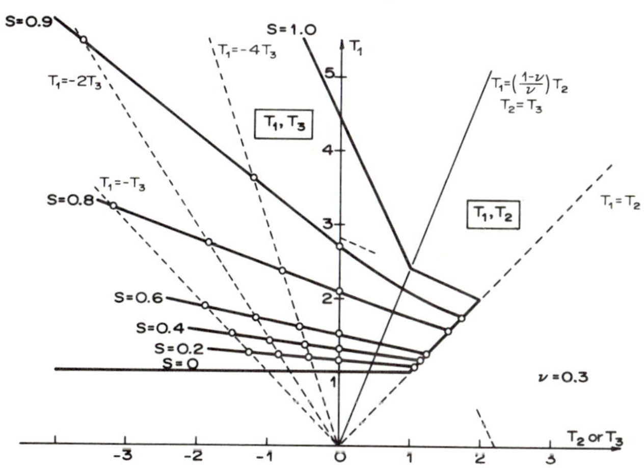


Fig. (5) The probability S of successful grain boundary crossings by cleavage microcracks. The case of plane strain.

limits suggested by Eqs. (12) and (13), and the calculated S values for the plane strain case are shown in Fig. 5.

Simple Tension and Transverse Hydrostatic Stress

This stress state is defined as $T_1 > 1$, $T_2 = T_3$. The limits were determined in the same way as in the previous two cases, and are presented with the calculated values of S in Fig. 6. Note that the horizontal scale has been reduced by the factor of $1/\sqrt{2}$. These results include the state of pure hydrostatic tension stress. As one would expect, the lowest value of $T_1 = 1.4\overline{6}$ at $S \to 1$ is found on the hydrostatic stress axis.

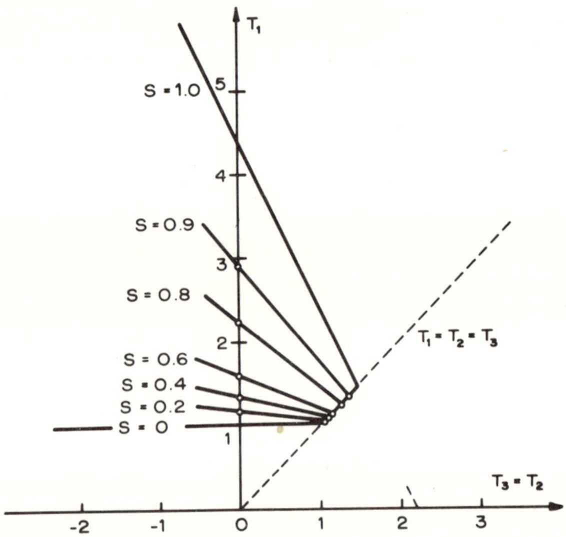


Fig. (6) The probability S of successful grain boundary crossings by cleavage microcracks. The combination of uniaxial tension $T_1$ and a lateral hydrostatic stress $T_2 = T_3$.

DISCUSSION AND CONCLUSIONS

The elastic cleavage cannot affect all grains, the stresses required would be much too high, so that plastic deformation and ductile rupture of the ligaments will be preferred. However, as the present results suggest, the density of the ligaments at a given stress level should depend considerably on the local state of stress at the crack tip. Specifically, a comparison of the plane strain and plane stress results, at $T_1 \to T_2$, or $T_1 \to T_3$, shows that much lower stress levels are needed for complete cleavage fracture in plane strain. Since the in-plane stresses on the prospective fracture plane ahead of a crack are essentially equal, these results have a direct bearing on the expected levels of fracture toughness under these respective states of stress.

The determination of the density and size of the ligaments is the most important result which is expected to be obtained later in the course of this research.

REFERENCES

[1] Dvorak, G.J., Motion of cracks in brittle fracture, in Fracture 1969, ed. by P.L. Pratt, Chapman & Hall, London, 1969, p. 338.

[2] Dvorak, G.J., A model of brittle fracture propagation, Part I — Continuum aspects, Engineering Fracture Mechanics, 3, 351, 1971.

[3] Dvorak, G.J., Statistical criteria for microcrack propagation in b.c.c. polycrystals, in Dynamic Crack Propagation, ed. by G.C. Sih, Noordhoff International Publishing, Leyden, 1973, 49.

[4] Hoagland, R.G., Rosenfield, A.R., and Hahn, G.T., Mechanisms of fast fracture and arrest in steels, Metallurgical Transactions, 3, 123, 1972.

[5] Tetelman, A.S. and McEvily, A.J., Fracture of Structural Materials, Wiley, New York, 1967, p. 210.

[6] McClintock, F.A., A criterion for ductile fracture by growth of holes. Journal of Applied Mechanics, 35, 363, 1968.

[7] Smith, E., Flat fracture propagation in mild steel: some comments on the role of plastic deformation processes, in Dynamic Crack Propagation, ed. by G.C. Sih, Noordhoff International Publishing, Leyden, 1973, 63.

[8] Gell, M., and Smith, E., Propagation of cracks through grain boundaries in polycrystalline 3% silicon-iron, Acta Metallurgica, 15, 253, 1967.

[9] Kaechele, L.E., and Tetelman, A.S., A statistical investigation of microcrack formation, Acta Metallurgica, 17, 463, 1969.

[10] Feller, W., An Introduction to Probability Theory and Its Applications, Wiley, New York, 1957.

[11] Papoulis, A., Probability, Random Variables, and Stochastic Processes, McGraw-Hill, New York, 1965.

# SECTION II

# ENERGY CRITERIA AND PATH INDEPENDENT INTEGRALS

# THE CALCULATION OF ENERGY RELEASE RATES

J. D. Eshelby

Department of the Theory of Materials, University of Sheffield, U.K.

## ABSTRACT

The connection between the elastic energy-momentum tensor and the surface or path-independent integral for energy release rates and kindred concepts is recalled and extended to grade 2 materials. For a completely general elastic field there are apparently only the three types of path-independent integrals derived by Günther from Noether's theorem, but if the field is anti-plane (mode III) there is an indefinite number of them. In plane strain the energy-momentum tensor can be re-interpreted as an ordinary compatible stress. This observation, coupled with the elastic reciprocal theorem, leads to an indefinite number of path-independent integrals for this case also. Some other possibilities are also touched on. The paper contains some specimen calculations for specific cases.

## NOTATION

The following symbols are used throughout:

| | |
|---|---|
| $F_\ell$ | force on a singularity |
| C | energy release rate |
| $J, L_i, M$ | path-independent integrals |
| K | stress intensity factor |
| $n_i$ | unit normal vector |
| $p_{ij}$ | nominal stress tensor |

| | |
|---|---|
| $P_{\ell j}$ | energy-momentum tensor |
| $dS, dS_j$ | surface element, $n_j\, dS$ |
| $u_i$ | displacement vector |
| $W$ | energy density |
| $X_i$ | initial coordinates |
| $x_i$ | final coordinates |
| $\delta_{ij}$ | Kronecker delta |
| $\varepsilon_{ijk}$ | is completely antisymmetric with $\varepsilon_{123} = 1$ |
| $\xi_\ell$ | position of a singularity |
| $\tilde{\omega}$ | rotation |

Symbols used in a single context are defined in their place. Repeated suffixes are summed over 1,2,3 in three or 1,2 in two dimensions. Suffixes following a comma denote differentiation with respect to the corresponding $X_i$ or $x_i$. In the linear theory, where $X_i$ and $x_i$ are identified, $x_i$ is used, with $x, y$ replacing $x_1, x_2$ when convenient.

## INTRODUCTION - THE BASIC PATH-INDEPENDENT INTEGRAL

The matters to be discussed here go back to a famous paper by Emmy Noether [1]. (There is now an English translation.) From it there follows, among other things, the result that if we have a set of quantities $u_i(X_m)$ which depend on the variables $X_m$ (in our case Cartesian coordinates) and which satisfy the equations

$$\partial W/\partial u_i - (\partial W/\partial u_{i,j})_{,j} = 0 , \tag{1}$$

where $W = W(u_i, u_{i,j})$ does not depend explicitly on the $X_m$, that is, it is the same function of the $u_i$ and $u_{i,j}$ for any $X_m$, then the tensor

$$P_{\ell j} = W\delta_{\ell j} - (\partial W/\partial u_{\ell,j})\, u_{i,\ell} \tag{2}$$

has zero divergence, that is, $P_{\ell j,j} = 0$. This is easily checked. It then follows from Gauss's theorem that the integral

$$F_\ell = \int_S P_{\ell j}\, dS_j \tag{3}$$

is zero if there are no singularities of the integrand within the surface S, and that, if there are such singularities, $F_\ell$ has the same value for all surfaces S which can be deformed into one another without passing through these or any other singularities of the integrand. In two dimensions S is a plane circuit and we describe the above property by saying that (3) is path-independent. We shall use this familiar term also in three dimensions in place of the more appropriate 'surface-independent'.

If W is the energy density of an electrostatic field, $P_{\ell j}$ is Maxwell's tensor and $F_1$, $F_2$, $F_3$ give the components of the force on any charge there may be inside S. Similarly, if W is the energy density of an elastic field (3) gives the force on any elastic singularity within S. 'Force' now has the meaning it has in the phrases crack extension force in fracture mechanics, or force on a dislocation, point defect etc. in solid state physics [2]. It is defined so that if the defect is displaced by $\delta\xi_\ell$ then $F_\ell \delta\xi_\ell$ is the decrease in the total energy of the system, made up of the elastic energy of the material containing the singularity and the potential energy of any external loading mechanism. This aspect is emphasized in the synonym energy release rate for crack extension force.

A succession of imaginary physical operations establishes what has just been said. As the writer has presented it several times already [3,4,5] the argument is here cut to the bone, thus begging several questions, but, for variety, it is done for a grade 2 material.

Take Cartesian coordinates $X_m$ in the undeformed material. On deformation a particle of material originally at $X_i$ goes to $x_i(X_m)$ and $u_i(X_m) = x_i(X_m) - X_i$ is the displacement. So far as possible everything is referred to the undeformed state (Lagrangian point of view). In particular, in what follows W is the energy (in the deformed state, of course) per original unit volume and, in the vector surface element $dS_j = n_j dS$, $n_j$ is the normal to and dS the area of an element of a surface S drawn in the undeformed medium.

In a grade 2 material [6] the energy density

$$W = W(u_{i,j}\ u_{i,jk})$$

depends on the first and second derivatives of the displacement with respect to the $X_i$. The requirement that the volume integral of W be stationary with respect to small variations of the $u_i$ leads to the Euler equation

$$(\partial W/\partial u_{i,j})_{,j} - (\partial W/\partial u_{i,jk})_{,jk} = 0 . \qquad (4)$$

We shall suppose that (4) is in fact the equilibrium equation governing the $u_i$.

As a preliminary we work out the energy change produced by a small change of displacement, equation (5) below. Suppose that the displacement field alters slightly from $u_i$ to $u_i + \delta u_i$ as a result, say, of a change in the external loads or the re-arrangement of sources of internal stress; there is no need to introduce the word 'virtual'. The energy density changes by

$$\delta W = (\partial W/\partial u_{i,j})\,\delta u_{i,j} + (\partial W/\partial u_{i,jk})\,\delta u_{i,jk}\,.$$

The total change of elastic energy inside a closed surface S is this quantity integrated over its interior. With the help of (4), 'differentiation by parts' and Gauss's theorem it may be converted into the integral

$$\delta E = \int\{[\partial W/\partial u_{i,j} - (\partial W/\partial u_{i,jk})_{,k}]\,\delta u_i + (\partial W/\partial u_{i,jk})\,\delta u_{i,k}\}\,dS_j \tag{5}$$

To find the force on an elastic singularity we draw a surface S enclosing it and cut out the material inside S from the surrounding material and apply suitable distributions of force and couple over both the freshly formed surfaces so as to prevent relaxation. We next displace the whole elastic field inside S by $\delta\xi_\ell$. This means that at a given material element $u_i$ changes to $u_i + \delta u_i$ with

$$\delta u_i = -\,\delta\xi_\ell\, u_{i,\ell}\,. \tag{6}$$

After this change the energy inside S is evidently equal to the energy which was previously inside a surface S' derived from S by giving it a rigid translation $-\delta\xi_\ell$. The volume inside S can be derived from the volume inside S' by adding a crescent-shaped volume to one side of S' and subtracting another from the other side. The change of energy inside S is thus the integral of W over the first crescent-shaped area minus its integral over the second, or in all

$$\delta E(\text{in}) = -\,\delta\xi_\ell \int W\,dS_\ell\,. \tag{7}$$

We next change the displacement on the inner surface of the hole from $u_i$ to $u_i + \delta u_i$ (with the $\delta u_i$ of (6)) so that it matches that of the material inside S. The work required to do this is the expression (5) with the $\delta u_i$ of (6) and the sign reversed since the outward normal of the piece removed is the inward normal of the hole left. We do not include a similar term for the outer boundary of the material because any energy inflow through it is accompanied by an equal decrease of the potential energy of the loading mechanism, and we are only interested in changes in the <u>total</u> energy of the system. Next weld the piece back into the hole. Although the displacements now match, the surface tractions do not quite do so: there is a layer of body force and couple with surface density proportional to $\delta\xi_\ell$ on the interface which induces a displacement of order $\delta\xi_\ell$. Thus when these unwanted layers are relaxed an amount of work only of order $(\delta\xi_\ell)^2$ is extracted from the system. Hence the total change of energy is (7) minus (5) with the $\delta u_i$ of (6) so that the force on the singularity is given by the integral (3) with

$$P_{\ell j} = W\delta_{\ell j} - \{\partial W/\partial u_{i,j} - (\partial W/\partial u_{i,jk})_{,k}\} u_{i,\ell} - (\partial W/\partial u_{i,jk}) u_{i,k\ell} . \qquad (8)$$

This agrees with the formal field theory expression for the energy-momentum tensor when the Lagrangian depends on the first and second derivatives of the field variables [7,8,9]. A material of grade N can be treated similarly; the appropriate $P_{\ell j}$ agree with the formal results of [8] and [9] for a Lagrangian containing derivatives up to the N th.

The elastic field near a crack tip (or, indeed, some other elastic singularity) in a grade 2 (or N) material may be notably different from what it would be in an ordinary elastic (grade 1) material, but in the body of the material the difference will be negligible. Hence if there is nothing to stop a small surface S surrounding the crack tip from being expanded to macroscopic size, the energy release rate will be the same as it would be for a suitably chosen grade 1 material.

From now on we shall only consider ordinary (grade 1) elastic materials. Then in (3) we may put

$$P_{\ell j} = W\delta_{\ell j} - p_{ij} u_{i,\ell} \qquad (9)$$

where

$$p_{ij} = \partial W(u_{m,n})/\partial u_{i,j}$$

is the nominal, Boussinesq or first Piola-Kirchhoff stress tensor which gives the component parallel to the (rectangular) $X_i$ coordinate axis of the force on a surface element which was of unit area and perpendicular to the $X_j$ axis before deformation.

It is not hard to see that in a two-dimensional context a crack tip qualifies as an elastic singularity within the framework of the above argument provided $\delta\xi_\ell$ is parallel to the crack. Equation (3) can also be applied locally to the edge of a three-dimensional crack [5] and to plane stress [10].

When the path is prevented from closing by a mathematical cut representing, for example, a crack or a dislocation, one must distinguish between two kinds of path-independent integrals: in type (a) the path may be deformed provided the points where it meets the cut remain fixed, and in type (b) these points may, in addition, be slid along the cut, not necessarily remaining opposite one another. For a crack parallel to the $X_1$ axis with stress-free surfaces $F_1$ is of type (b). However, even in this case, if, moving away from the tip, the crack is at first parallel to the $X_1$ axis and then deviates from it beyond a certain point, $F_1$ ceases to be path-independent when its ends are slid along the crack beyond that point. (There is a similar distinction in three dimensions [11].)

In the interval between the appearance of Griffith's energy criterion and the rise of modern fracture mechanics quantities equivalent to the energy release rate had already appeared in the physical and technical literature.

In his 1930 paper on the surface energy of mica Obreimoff [12] gave what is in effect the first formula for the energy release rate for a specific crack configuration. As shown in Fig. 1 a flake, of length $\ell$ and thickness h say, is made to peel off by

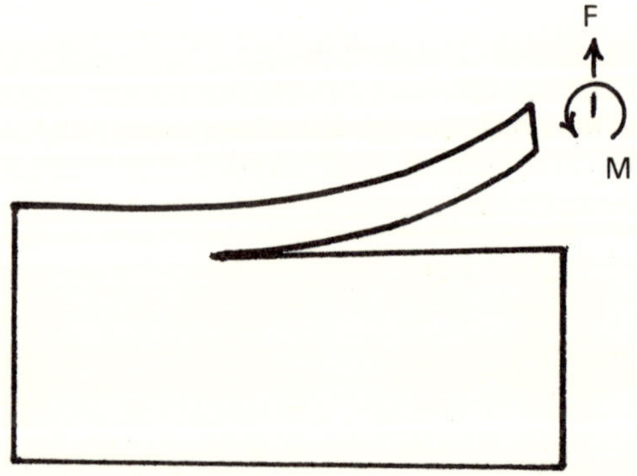


Fig. 1. Obreimoff's method of measuring surface energy.

applying a force F and couple M at the end. Its deformation $y = y(x)$ is calculated by treating it as a beam or plate clamped to a block at the position of the crack tip, say $x = 0$.

As is well known, of the work done by the loading mechanism when the crack extends only half contributes to the energy release rate, the other half being needed to increase the elastic energy. Consequently we have

$$G = 2\gamma = \tfrac{1}{2}F\partial y(\ell)/\partial\ell - \tfrac{1}{2}M\partial y'(\ell)/\partial\ell \qquad (10)$$

where the two terms are respectively, half the work done per unit length perpendicular to the paper by the force and by the couple per unit crack extension; the minus sign follows from the usual sign convention for M. Beam theory gives

$$G = \tfrac{1}{2}(M - \ell F)^2/E'I \qquad (11)$$

where $E'$ is $E$ or $E/(1-\nu^2)$ according to whether we treat the flake as a beam or a plate, and $I = h^3/12$. If we cut through the beam at $x$ and provide the appropriate local bending moment $M(x)$ and shear force $F(x)$ to the cut end to prevent relaxation the (unchanged) G will be given by (11) with $M - \ell F = M(\ell) - \ell F(\ell)$ replaced by $M(x) - xF(x)$. Beam theory shows that, as expected, this is independent of $x$. In particular it is equal to $M(0)$, whence

$$G = E'h^3[y''(0)]^2/24 ,$$

the form given by Obreimoff. (Actually by a trivial mistake he got a factor 4 wrong. On the other hand Landau and Lifshitz's [13] rather cryptic derivation gives a value twice too big because of an error of principle; in effect they calculate the work done by the load but forget the fifty percent rake-off exacted by the elastic energy.)

To find G from (3) we surround the tip by a small circuit and then deform it into a circuit which is made up of a large semicircle in the block, centered on the tip, and any cross-section of the flake. (These two segments are actually joined by straight lines along the crack faces and the upper surface of the flake, but they contribute nothing.) On the semicircle the stress is essentially the same as that of a point force applied to a semi-infinite solid so that $p_{ij} \sim 1/r$, $P_{\ell j} \sim 1/r^2$ and it contributes nothing. Hence $G = -F_1$ is the integral of

$$-P_{11} = -W + p_{11}u_{1,1} + p_{12}u_{2,1} \qquad (12)$$

across the flake. If we force in the assumptions of elementary beam theory, $p_{11}$ proportional to $x_2$, $u_{2,1} = y'$ independent of $x_2$, equation (12) easily gives

$$G = -\tfrac{1}{2}My'' + Fy' \tag{13}$$

evaluated at any cross-section. This is equivalent to (10) or (11). More elegantly, having reduced the integration to one across the beam we may recognize that the beam is a one-dimensional grade 2 medium (Cosserat continuum with constrained rotation) with energy density $W = \tfrac{1}{2}E'I(y'')^2$. Variation in the body of the integral $\int W dx_1$ gives correctly $y'''' = 0$, the one-dimensional version of (4), and in addition variation at the end points shows that the quantities $M = -E'Iy''$ and $F = -E'Iy'''$ must be interpreted as couple and shearing force, while the one-dimensional version of (8) gives

$$G = -F_1 = \tfrac{1}{2}E'I(y'')^2 - E'Iy'''$$

which is the same as (13).

In 1953 Rivlin and Thomas [14] gave the theory of several tests used on rubber and plastics. An interesting one is the 'trouser test' shown in Fig. 2. The following treatment illustrates how the formalism with nominal stresses and Lagrangian

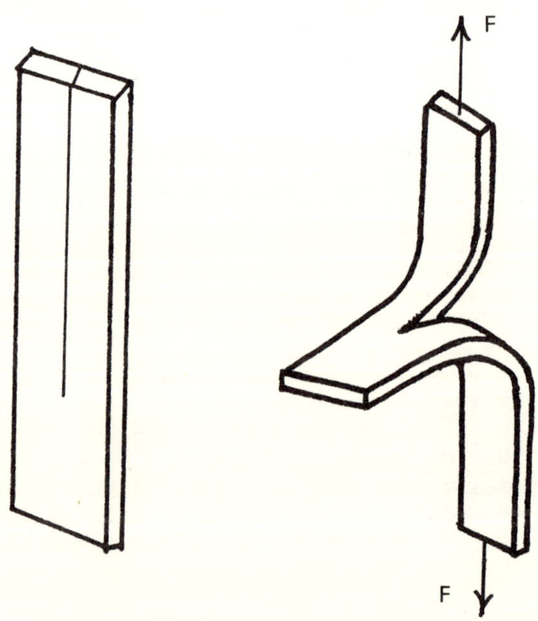


Fig. 2. The trouser test.

coordinates copes neatly with finite deformation. The two legs are extended by forces F and we want to find the force on the tip of the crack (or tear), equal to the total energy release rate Gt (t is the specimen thickness), or in the language of rubber testing Tt, where T is the tearing energy. To get it we must work out the integral (3) where S is a surface embracing the tip of the tear. It can be expanded until it becomes a surface made up of parts of the specimen surface and a cross-section of the unstressed tail, all of which contribute nothing, together with cross-sections of the two legs, one of which contributes

$$\tfrac{1}{2}F = \int (W - p_{i1}\, u_{i,1})\, dS$$

$$= WA - F\partial u_2/\partial X_1 \qquad (14)$$

where A is the unstrained area of one leg. A point on one leg with initial coordinates $X_1$, $X_2 = 0$ is first swung round to $x_1 = 0$, $x_2 = X_1$ and then stretched vertically by F to the final position $x_1 = 0$, $x_2 = X_1\ell/\ell_o$ where $\ell$, $\ell_o$ are the unstrained and strained lengths of a leg. Hence $u_2 = x_2(X) - X_2 = X_1\ell/\ell_o - X_2$ and so $\partial u_2/\partial X_1 = \ell/\ell_o$ and we have

$$Gt = -F_1 = 2F\ell/\ell_o - 2WA .$$

The minus sign before $F_1$ merely says that the tear will run to the left. The expression on the right can, of course, be interpreted as the work done by the forces minus the increase in elastic energy when the tear extends by unit length of unstrained material. (Note that this lengthens each leg by $\ell/\ell_o$ and that W is the final energy per unit initial volume.) When the forces are small enough for Hooke's law to be valid the second term on the right of (14) is minus one half of the first, just as in the Obreimoff example, and generally on the linear infinitesimal theory. This example shows that a large geometrical deformation alone need not upset the principle.

At the other extreme, far beyond the Hooke's law region, some rubbers reach a final extension $\ell'/\ell_o$ which scarcely increases with further loading until the material breaks. In that case the first term can be neglected, $|F_1| = 2F\ell'/\ell_o$, and the force on the cut tip is no longer proportional to $F^2$ but is, so to speak, equal to the total applied force 2F magnified by the mechanical advantage factor $\ell'/\ell_o$ which takes into account the fact that one unit of unstressed tail becomes $\ell'/\ell_o$ units of stressed leg.

The formula (14) does not concern itself with the hardness or softness of the loading mechanism and so it applies also when the legs are extended and clamped before the tear is allowed to propagate. $|F_1|$ is then the decrease in internal energy alone per unit increase in tear length since no external work is done. In fact

Rivlin and Thomas treat this as the basic case, defining Tt to be $-(\partial U/\partial c)_\ell$ where U is the strain energy of the specimen and c the (unstrained) tear length. The equation

$$dU = (\partial U/\partial c)_\ell \, dc + (\partial U/\partial \ell)_c \, d\ell$$

combined with the thermodynamic relation $F = (\partial(\frac{1}{2}U)/\partial \ell)_c$ and the geometrical one $(\partial \ell/\partial c)_F = \ell/\ell_0$ then takes us back to (14).

## GÜNTHER'S PATH-INDEPENDENT INTEGRALS

It was Günther [15] who first applied Noether's theorem systematically to elastostatics. As well as the $F_\ell$ of equation (3) he obtained the path-independent integrals

$$L_i = \varepsilon_{ik\ell} \int_S (x_k P_{\ell j} + u_k p_{\ell j}) \, dS_j \tag{15}$$

and

$$M = \int_S (x_\ell P_{\ell j} - \tfrac{1}{2} u_\ell p_{\ell j}) \, dS_j \; . \tag{16}$$

(We follow the notation of Budiansky and Rice [16]). Suppose that we make a picture of an elastic field by drawing the appropriate displacement vector at each point. Then the results that $F_\ell$, $L_i$, M are path-independent depend on the fact that the picture still represents a possible elastic state after it has been, respectively, translated, rotated or enlarged. From this one can read off under what conditions each result is valid. For $F_\ell$ the deformation may be finite and the material may be non-linear, but it must be homogeneous. For $L_i$ the conditions are the same, except that the material must, in addition, be isotropic. For M the stresses must be linear in the displacement gradients, so we are limited to the usual linear infinitesimal theory, though the material can be anisotropic. The term 'homogeneous' implies that the <u>form</u> of the energy density W is independent of $X_1$, $X_2$, $X_3$. A few relaxations are possible. If the form of W depends on $X_2$, $X_3$ but not $X_1$ then $F_1$ is still path-independent. (This takes care of the application of (3) to cracks at interfaces.) Similarly $L_3$ is path-independent if the form of W depends on $r = (x_1^2 + x_2^2)^{\frac{1}{2}}$ and $x_3$ but not on $\theta = \arctan(x_2/x_1)$ (Curvilinear aeolotropy).

It is hard to think of any operations other than the above three which will convert one picture of a general equilibrium elastic field into another possible one, and in fact Knowles and Sternberg [17] have given an argument to show that $F_\ell$, $L_i$ and M are the only path-independent integrals derivable from Noether's theorem for a general field. However, we shall see later that when certain limitations are imposed on the form of the displacement (anti-plane strain, plane strain) many more path-independent integrals appear, some of the form envisaged in Noether's theorem, some not.

In a two-dimensional state of plane strain (16) reduces to the simpler-looking integral

$$M = \int_S x_\ell P_{\ell j} dS_j \tag{17}$$

taken along a plane curve S with normal $(n_1, n_2)$. Summation over $\ell$ and j is now, of course, only over 1,2. There is no real difference between (16) and (17). To apply the three-dimensional (16) to the two-dimensional situation the surface of integration can be taken to be the surface of the disk generated by letting the plane area bounded by the curve S sweep through, say, unit distance along the $x_3$-axis. The two flat faces of the disk make a contribution to (16) which can, with the help of Gauss's theorem, be transformed into an integral over the rim of the disk, where it cancels the second term of (16), leaving (17). The following is a simple application of (17). The G for the right-hand tip of the crack of Fig. 3 is given by the integral $F_1$ of (3) evaluated over any loop which embraces the right tip but not the left one. It would be nice if we were able to slip the loop over the left-hand tip, because then we could expand the loop to infinity and make use of the simple asymptotic form of the crack field. However, if we do this we actually get G-G', where G' is the energy release rate for the left-hand tip. To see this, contract the large loop back onto the crack to form the keyhole-shaped contour in the figure; the right-hand circle gives G and the

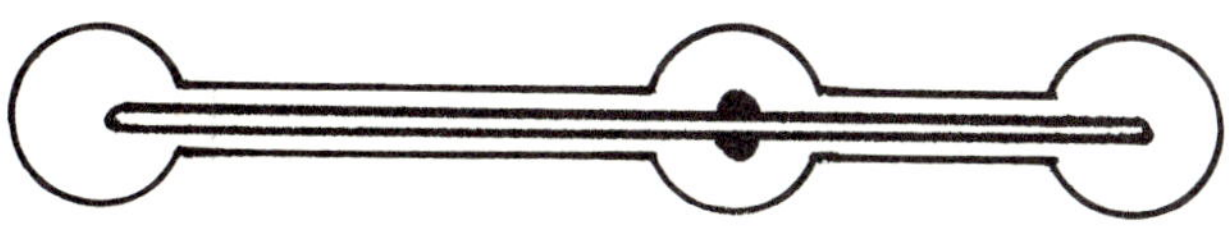


Fig. 3. Contour embracing a crack.

left hand one -G'. (The middle indentations relate to another problem.) To get over this difficulty we can use M instead of $F_1$, and site the origin of $x_\ell$ at the left-hand tip. On the keyhole-shaped contour the factor $x_\ell$ suppresses the contribution from the left-hand circle and on the right-hand circle it can be taken outside the integral to give G x length of crack, and this quantity is also the value of M over a large contour completely enclosing the crack. This argument shows at once, for example, that there is a respectable energy release rate for a crack at an interface between different media, despite the odd behaviour of the elastic field at the tips.

Fig. 4 shows a couple of situations which can be handled with the help of (17). On the left an edge crack of length a is opened

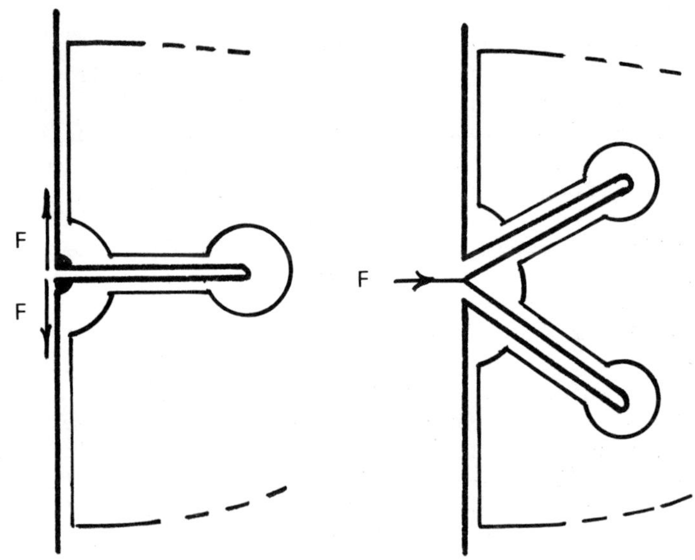


Fig. 4. Corner-loaded edge crack and plane version of conical crack.

in mode I by transverse concentrated forces $\pm$ F applied (perhaps by forcing in a wedge) at the points where the crack faces intersect the vertical free surface. For the circuit shown the straight parts contribute nothing to M. In particular the presence of $x_\ell$ in (17) gives a factor $x_1 = 0$ on the vertical surface which takes care of the non-zero $P_{11}$ there. The contribution from the tip is a G and from the large semicircle, zero. To evaluate the contribution from one of the small quarter-circles surrounding the forces we can use standard expressions for the field at the tip of an infinite wedge loaded by a point force [18]. The result is

$$G = 2(1-\nu)\ \pi F^2/\mu(\pi^2-4)\ a\ , \quad K = \alpha.2F/(\pi a)^{\frac{1}{2}}$$

with $\alpha = 1.2967...$; for the analogous mode III situation $\alpha = 1$. Howard was able to use this result in discussing edge cracks under large-scale yielding. The left of the figure shows the two-dimensional analogue of the intractable problem of a conical crack; a knife-edge produces two cracks of length a inclined at an angle $\beta$.

In this case 2Ga is equal to the difference of the M-integrals over the throat of the loaded wedge and over an infinite semi-circle. This gives

$$G = (1-\nu)(F^2/2\mu a)\{(\beta + \sin\beta)^{-1} - \pi^{-1}\} \tag{18}$$

## OTHER PATH-INDEPENDENT INTEGRALS

For linear anti-plane strain we have

$$P_{11,1} + P_{12,2} = 0 \quad , \quad P_{21,1} + P_{22,2} = 0 \tag{19}$$

and also

$$P_{11} + P_{22} = 0 \quad , \quad P_{12} = P_{21} \tag{20}$$

whence, by Gauss's theorem,

$$I = \int \phi_\ell \, P_{\ell j} \, dS_j \tag{21}$$

is path independent if $\phi_{1,1} = \phi_{2,2}$ and $\phi_{1,2} = -\phi_{2,1}$ i.e. if $f(z) = \phi_1 + i\phi_2$ is any analytic function of $z = x_1 + ix_2$. In plane strain the first, but not the second, of (20) holds and the nearest analogue of (21) is that, with the same $\phi_1$, $\phi_2$,

$$I = \int\{\phi_\ell P_{\ell j} + \tfrac{1}{2}(x_1\phi_2)_{,j}(P_{12}-P_{21}) - \tfrac{1}{2}x_1\phi_2(P_{12}-P_{21})_{,j}\} \, dS_j \tag{22}$$

is path independent. This can be checked with the help of the easily-verified relation

$$\nabla^2(P_{12}-P_{21}) = 0 \,. \tag{23}$$

Equation (21) or (22) can be used to reproduce various known results quite simply. For example, suppose that the crack of Fig. 3 has point forces $\pm F$ applied to the upper (lower) faces at $x_1 = x'$. On putting $f(z)$ equal successively to $1/(z-x')$ and $z-x'$ we get, in mode III,

$$(a-x')\,G + (a+x')\,G' = F^2/\pi\mu$$

and

$$(a-x')^2 G - (a+x')^2 G' = 0$$

whence

$$G = F^2(a+x')/2\pi a\mu(a-x')$$

or

$$K = F\{(a+x')/\pi a(a-x')\}^{\frac{1}{2}} \,, \tag{24}$$

the latter valid on all three modes, and, by superposition, leading at once to the well-known result

$$K = \int_{-a}^{a} \{(a+x')/a-x')\}^{\frac{1}{2}} P(x')\, dx'/(\pi a)^{\frac{1}{2}}$$

for general internal or external loading.

The fact that $P_{12}-P_{21}$ is harmonic, eqn. (23), leads to a more general class of plane-strain path-independent integrals. Write

$$p'_{ij} = \varepsilon_{i\ell 3} P_{\ell j} , \quad P_{ij} = -\varepsilon_{i\ell 3} p'_{ij} , \quad i,j,\ell = 1,2 .$$

Then (19), the first of (20) and (23) show that the $p'_{ij} = p'_{ji}$ satisfy the equilibrium equations without body force and the compatibility equation $\nabla^2(p'_{11} + p'_{22}) = 0$ and are thus ordinary elastic stresses derivable from a displacement $u'_i$. Choose any other elastic field $p^0_{ij}$, $u^0_i$. Then the Rayleigh-Betti reciprocal theorem states that

$$\int(u^0_i\, p'_{ij} - u'_i\, p^0_{ij})\, dS_j \qquad (25)$$

is path-independent. It can be verified that, if, with the traditional Muskhelishvili notation we write

$$2\mu(u_1 + iu_2) = (3-4\nu)\,\phi(z) - z\overline{\phi'(z)} - \overline{\psi(z)}$$

together with the related expressions for the $p_{ij}$, then $u'_i$ and the $p'_{ij}$ are given by similar expressions with $\phi'$ and $\psi'$ replaced by

$$\Phi' = (1-\nu)\,\phi'^2/i\mu , \quad \Psi' = 4(1-\nu)\,\phi'\psi'/i\mu$$

respectively.

The $P_{\ell j}$ of a rightward-pointing mode I crack along the $x_1$-axis with its tip at the origin translates into a field $p'_{ij}$, $u'_i$ which near the tip is summed up in the Airy stress function

$$\chi' = G x\theta/2\pi$$

and thus represents a force (0,G) plus an edge dislocation with Burgers vector

$$b' = (1-2\nu)\; G/2\mu$$

coexisting at the origin. In order to use (25) to find G for some particular geometry and loading we have to try and choose a contour which embraces the crack tip and elsewhere as far as possible runs where $p'_{ij}n_j$ or $u'_i$ are known, and arrange that $u^0_i$ or $p^0_{ij}\, n_j$ vanishes where they are not. The writer has not had much success with (25) so far. An example of what can go wrong: an attempt to find G for an edge crack under uniform tension T gave for it a multiple of $T^2$ plus a multiple of the square of the relative rotation of the faces of the crack at the corners where they meet the free surface, a quantity which can, presumably, only be found by solving the problem in detail.

REFERENCES


1. Noether, E., Nachr. Ges. Wiss. Göttingen, math.-phys. Klasse, p.235, 1918. (English translation by Tavel, M., Transport Theory and Statistical Physics, 1, 183, 1971.)

2. Nabarro, F. R. N., Theory of Crystal Dislocations, Clarendon Press, Oxford, 1967, 562.

3. Eshelby, J. D., in Solid State Physics, Seitz, F. and Turnbull, D., Eds., Academic Press, New York, 1956, 79.

4. Eshelby, J. D., in Internal Stresses and Fatigue, Rassweiler, G. M., and Grube, W. L., Eds., Elsevier, Amsterdam, 1959, 41.

5. Eshelby, J. D., in Inelastic Behavior of Solids, Kanninen, M. F. et al. Eds., McGraw-Hill, New York, 1970, 77.

6. Toupin, R., Arch. rat. Mech. Anal., 11, 385, 1962.

7. Podolsky, B., and Kikuchi, C., Phys. Rev., 65, 228, 1944.

8. Chang, T. S., Proc. Cant. Phil. Soc., 42, 132, 1946.

9. Thielheim, K. O., Proc. Phys. Soc. Lond., 91, 798, 1967.

10. Eshelby, J. D., Sci. Prog., Oxf., 59, 161, 1971.

11. Eshelby, J. D., Phil. Trans., A244, 87, 1951.

12. Obreimoff, J. W., Proc. Roy. Soc., A127, 290, 1930.

13. Landau, L. D., and Lifshitz, E. M., Theory of Elasticity, Pergamon, Oxford, 1970, 52.

14. Rivlin, R. S., and Thomas, A. G., J. Polym. Sci., 10, 291, 1953.

15. Günther, W., Abh. braunschw. wiss. Ges., 14, 54, 1962.

16. Budiansky, B., and Rice, J. R., J. appl. Mech., 40, 201, 1973.

17. Knowles, J. K. and Sternberg, Eli, Arch. rat. Mech. Anal., 44, 187, 1972.

18. Timoshenko, S. and Goodier, J. N., Theory of Elasticity, McGraw-Hill, New York, 1951, 96.

19. Eshelby, J. D., in Fracture Toughness, I.S.I. Publication 121 The Iron and Steel Institute, London, 1968, 13.

# SURFACE LAYER ENERGY AND STRAIN ENERGY DENSITY FOR A BLUNTED CRACK OR NOTCH

G. C. Sih

Institute of Fracture and Solid Mechanics, Lehigh University, Bethlehem, Pennsylvania 18015 USA

## ABSTRACT

The problems of failure by fracture of sharp cracks and blunt notches have been correlated into a coherent theory where a sharp crack is simply the limiting case of the elliptical notch. The theory is based on the hypothesis that fracture occurs when a small element of material near the crack or notch front has absorbed a critical amount of energy, and then releases it to allow crack extension. This strain energy density theory [1,2] is also extended to notch boundaries where in addition the energy in a surface layer is calculated and the location of failure initiation is determined.

The concept of a boundary layer near the surface of a blunted notch is discussed. This stems from a concern that at the notch surface, variations in material behavior, such as localized inhomogeneity or anisotropy, preclude an accurate local solution, but an analysis presuming a valid elastic solution external to this region, provides a more realistic measure of the failure behavior of the actual material. The thickness of this surface layer is left unspecified at present, but subsequent research should provide means to establish its size, perhaps as a parameter depending on the microstructure of the material.

The example treated is that of an elliptical notch loaded uniformly at an angle to the major axis of the ellipse; the results are shown to approach that of the crack solution for narrow ellipses, and to display satisfactory agreement with recently published experimental data. Results also indicate that in globally unstable configurations, the original loading and notch geometry are sufficient to predict the subsequent crack trajectory.

## INTRODUCTION

The approach that has been applied in engineering fracture mechanics is by in large based on the concept of a critical stress intensity factor. Griffith [3] assumed that this factor is a material constant and is related to the strain energy release of a mathematically sharp crack undergoing a small extension. An extensive literature on stress intensity factor solutions has amassed for a wide variety of crack geometries and loading conditions [4]. Although the stress intensity factor concept [5-7] were originally proposed in terms of three independent modes - normal, in-plane shear, and out-of-plane shear - it is now known that the energy release rate formulation is correct only in the normal case and that the shear modes may result in non-planar crack extension. Attempts to extend this energy release concept to three dimensions appear possible only for the case of a penny-shaped crack under symmetric loading, and suffer maladies similar to those for the two-dimensional cases when non-symmetric loading occurs. This is because crack extension will no longer be planar and the extended crack configuration will not be known as an apriori.

An additional consideration that has rarely been included in fracture analysis is that of local inhomogeneity such as inclusions, strong grains, etc., in the region near the notch surface. The incorporation of these effects into the analysis requires the concept that the criterion of failure of a layer of material next to the notch surface is different from that of an element of material inside the solid. The surface layer provides a means of absorbing the local inhomogeneities into the fracture theory. The size of this layer may be experimentally determined for a given material. The proposed theory also offers a common basis for evaluating the behavior of cracks and notches under all loading conditions.

## SURFACE LAYER ENERGY - BLUNTED CRACK OR NOTCH

It is well known that a physical boundary is not a sharp mathematically defined curve, but a rough surface composed of microscopic pits, holes, ridges, cracks, etc. Materially, the surface is also unlike the bulk of the solid. In general, thin layers of material foreign to the interior form on the surface and will behave differently from that of the interior. Additionally, forming new surfaces or boundaries, whether by rolling, drawing, or machining, results in layers of material that may be harder, less porous, etc., properties that may significantly differ mechanically from those further inside these regions.

At the continuum level, modeling the characteristics mentioned above is largely inhibited by lack of physical understanding, and by the very nature of the continuum approximations. The smooth curve of a continuum cannot describe the microscopic features of a surface as it physically occurs, so the approximate averaged curve is employed to predict behavior in regions away from local disturbances. However, as noted, it is these very surface features that could affect the failure load of a solid, and to this end a possible means of incorporating the effects of the solid surface into a continuum model is proposed.

In order to include any surface phenomenon into a continuum model, it will be necessary to do so in an averaged way. As an initial analysis, it will be assumed that a thin layer, of approximately continuum dimensions in thickness, extends into the solid from the boundary surface. In this boundary layer will be included any surface inhomogeneities, faults, etc., that characterize the physical surface (Fig. (1)). To a first approximation, it will be further assumed that the gross elastic properties of the layer are similar enough to those of the interior to permit treatment as an isotropic homogeneous medium. The criterion for failure will be that at some location or region along the boundary layer, the local energy caused by loading will exceed some materially determined constant, and failure will initiate. Necessarily, this criterion provides but one location point of failure at the surface only, and the behavior interior to the solid is not expected to emerge from consideration of the surface layer. It may also be noted that failure is expected to occur in a region where the surface layer is in tension, subject to local separation, although local inhomogeneities may alter slightly the actual location of failure as predicted by the model.

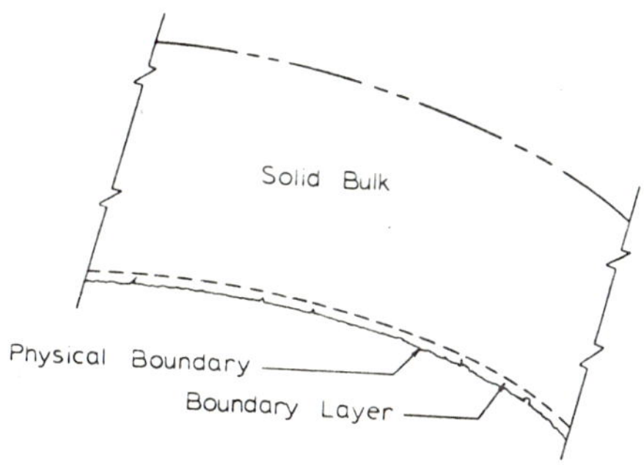

Fig. 1 - Surface boundary layer

Sih [8] has suggested that the form of this surface layer energy, represented by $\gamma_e$, be given as the product of the local radius of curvature, $\rho$, the tangential strain at the continuum surface, $\varepsilon_s$, and the normal stress, $\sigma_n$, on the interior surface of the boundary layer, designated to be of thickness $\delta$; i.e.,

$$\gamma_e = \rho\varepsilon_s\sigma_n \tag{1}$$

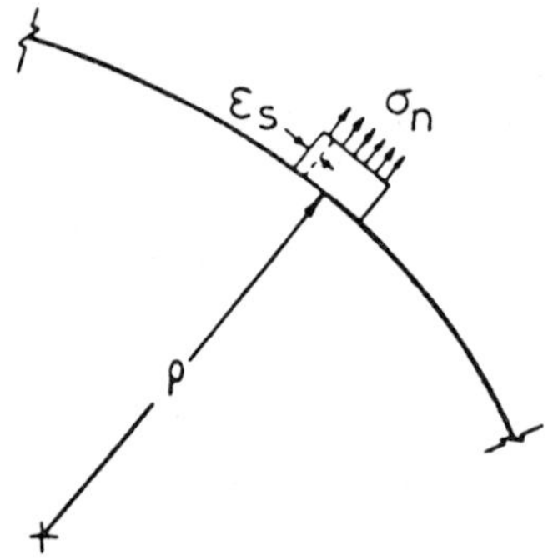


Fig. (2) - Local parameters of the surface layer energy

where $\gamma_e$ is the energy contained by a small element at the surface of the notch, for unit thickness of material. The quantities involved in $\gamma_e$ are shown in Fig. (2). The limitations on geometry are left unspecified.

For an elastic isotropic material, the tangential strain along the boundary is given by the ratio of the tangential stress, $\sigma_t$, to Young's modulus, E,

$$\varepsilon_s = \sigma_t/E \tag{2}$$

The normal stress acting on the interior of the boundary layer, of thickness $\delta$, can be shown to be related to the tangential stress, local radius of boundary curvature, and boundary thickness (to first order in $\delta$),

$$\sigma_n = \sigma_t\delta/\rho \tag{3}$$

Combining expressions (2) and (3) into (1), the surface layer energy now takes the form

$$\gamma_e = \delta\sigma_t^2/E \tag{4}$$

Since neither $\gamma_e$ nor $\delta$ has explicitly been given, it is convenient to form the quantity $\gamma_e E/\delta$ and treat it as a material parameter. A notched specimen may be loaded to failure and from Eq. (4), $\gamma_e E/\delta$ computed; then for any notch geometry of this material, the failure load may be obtained.

For the purposes of illustrating numerically the behavior of the surface layer tension, consider an elliptical notch cut into a large solid loaded far from the hole at an angle $\beta$ to the major axis of the ellipse as shown in Fig. (3). The boundary surface of the notch is assumed rough. Around the notch is depicted a boundary layer of uniform thickness, $\delta$. At some point along the layer, represented by $(x,y)$, or $(a\cos\eta, b\sin\eta)$, where $\eta$ is the eccentric angle for the ellipse, the magnitude of the surface layer energy, $\gamma_e$, may be calculated from Eq. (4) when the local tangential stress is known. In the case of the elliptical hole, the known tangential stress expression may be substituted into Eq. (4) to give

$$\gamma_e = \frac{\delta p^2}{E}\left[\frac{(a+b)^2(\sin\eta\cos\beta-\cos\eta\sin\beta)^2-a^2\sin^2\beta-b^2\cos^2\beta}{a^2\sin^2\eta+b^2\cos^2\eta}\right]^2 \quad (5)$$

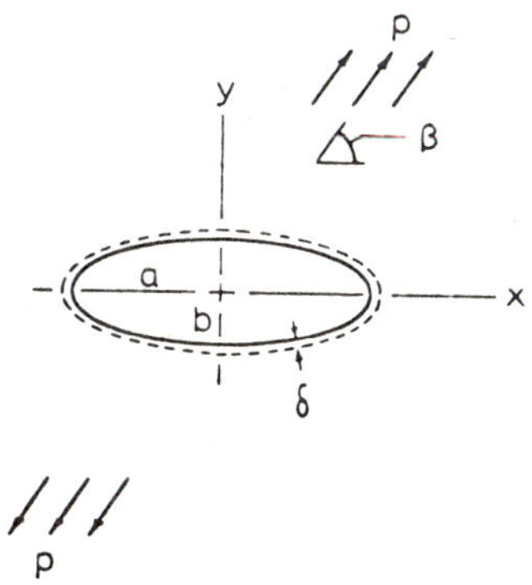


Fig. (3) - Loading and geometry of an elliptical notch

In its present form, Eq. (5) obscures the difference between tensile and compressive loading. Along the boundary of an elliptical cavity, $\sigma_t$ may be tensile or compressive, as indicated in Fig. (4). However, $\gamma_e$ remains positive at all points along the

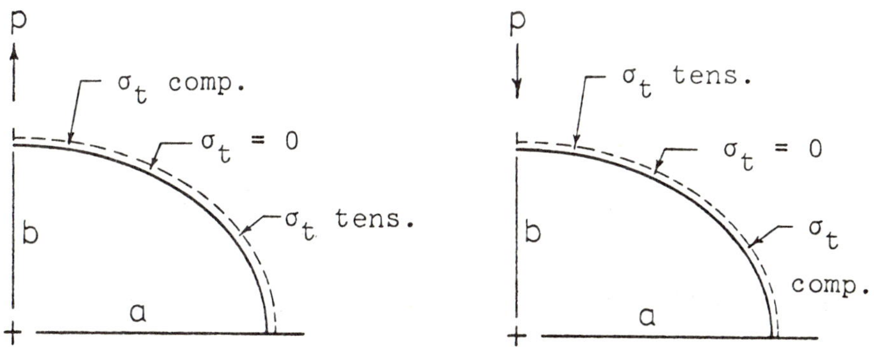


Fig. (4) - Tangential stress under tensile and compressive loading

boundary, and hence, requires that the nature of the loading be dictated in order to select the region in which the surface layer is in tension. As the angle of loading varies, the regions of tensile and compressive surface stresses also shift positions. The location of failure on the elliptical notch boundary corresponds to the maximum value of $\gamma_e$ and may be determined from the condition, $\partial\gamma_e/\partial\eta = 0$.

In general, $\gamma_e$ in Eq. (5) may be normalized to the form $\gamma_e E/\delta p^2$, where p is the applied uniform stress, then this quantity may be represented as a function of the loading angle $\beta$, semi-minor to semi-major axes ratio b/a, and the eccentric angle $\eta$. Figure (5) gives a plot of the surface layer energy as a function of the eccentric angle $\eta$ for $\beta = 90^\circ$ and several values of b/a from 0.04 to 0.10 inclusive. It is seen that for small values of b/a, the surface layer energy displays a sharp peak near $\eta = 0$, and this peak rapidly disappears as b/a increases. This effect is more pronounced for ellipses that are very slender, i.e., $0.04 \leq b/a \leq 0.01$ as illustrated in Fig. (6).

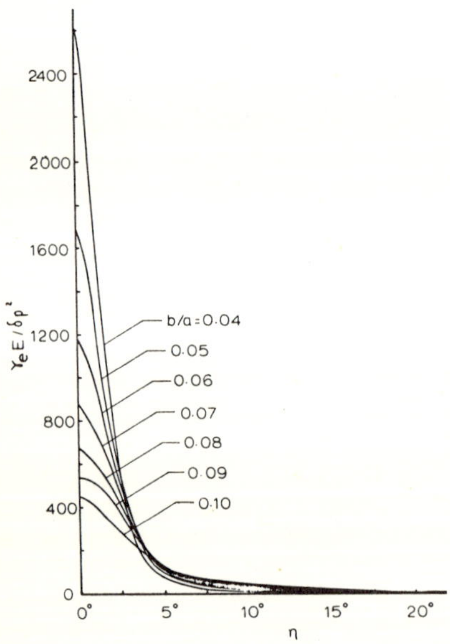


Fig. (5) - Variation of surface layer energy with eccentric angle for $0.04 \leq b/a \leq 0.1$

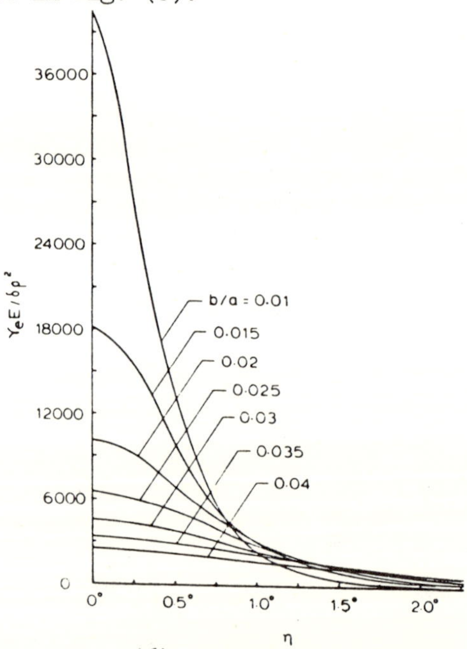


Fig. (6) - Variation of surface layer energy with eccentric angle for $0.01 \leq b/a \leq 0.04$.

It has been assumed that failure will occur at the location of maximum surface layer energy (in tension), and both the magnitude and location along the surface vary with angle of loading. The results may be best illustrated by plotting $p\sqrt{\delta/\gamma_e E}$ as a function of β for several values of b/a, where p represents the permissible load for a given material with $\sqrt{\delta/\gamma_e E}$ being a constant. As shown in Fig. (7), the minimum load to failure under tension is at β = 90°, while under compression (Fig. (8)), the minimum load to failure occurs in the region β = 45°. The elasticity solution has assumed no surface contact under compression, and the notch sizes entered are free of this problem. One of the values of the skew-symmetric loading geometry is manifested by the behavior of the elliptical cavity under compression.

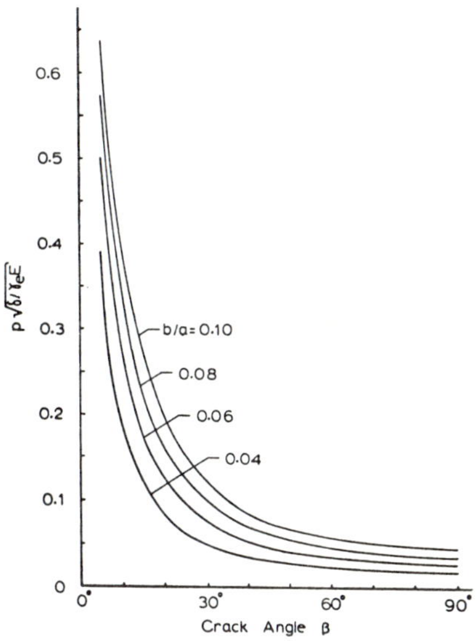


Fig. (7) - Tensile load versus angle of loading

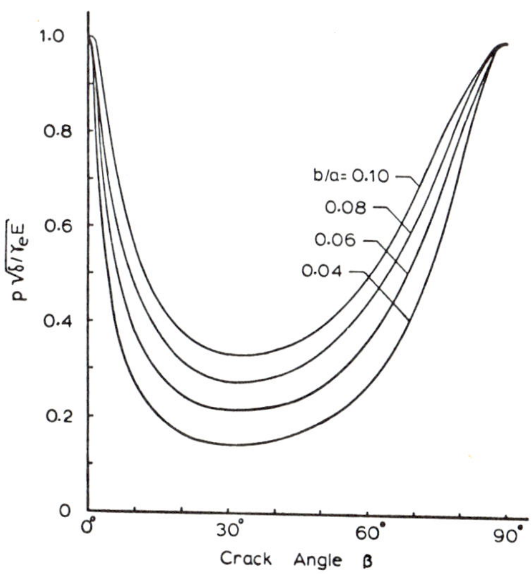


Fig. (8) - Compressive load versus angle of loading

For an elliptical notch with β = 90° and η = 0, Eq. (5) takes the form

$$\gamma_e = \frac{\delta p^2}{E}\left(1 + \frac{2a}{b}\right)^2 = \frac{\delta p^2}{E}\left(1 + 2\sqrt{\frac{a}{\rho}}\right)^2 \qquad (6)$$

If a/ρ is large in comparison with unity as in the case of a sharp crack, Eq. (6) further simplifies to

$$\gamma_e = \frac{4p^2 a}{E}\left(\frac{\delta}{\rho}\right), \quad \rho \ll a \qquad (7)$$

It should be emphasized that $\gamma_e$ is not the Griffith surface energy term and is derived from an entirely different physical view point. Equation (7) will not give an accurate measure of crack initiation should the material next to the crack tip vary substantially from that nearby.

## RELATIONSHIP OF SURFACE LAYER TENSION TO MAXIMUM STRESS CRITERION

It may be questioned that the results on the surface layer energy are similar to those of the maximum surface tangential stress failure criterion. Although quantitatively identical in normalized form, Eq. (4), there is a major conceptual difference that must be emphasized: the maximum stress criterion is basically applied in a layer of zero thickness, at surface points, whereas the surface layer energy has been applied over an element of non-zero thickness, and resulted in an additional unknown parameter, $\delta$, in Eq. (4). As outlined earlier, the surface of the notch contains many discontinuities, and the use of the surface tangential stress alone is unable to include such local phenomena. The form of $\gamma_e$ is such that, although also including the tangential stress implicitly, may include these discontinuities on an averaged basis.

The relationship of the internal normal stress to the tangential stress expressed in Eq. (3) provides a means to determine the layer thickness. Photoelastic experiments indicate the existence of such a surface layer, and measurements may be made to determine the normal and tangential stresses. In this way, a region of validity of the relationship may be set, and the layer thickness acquired. Then $\gamma_e$ may be experimentally determined, completing the information required to predict surface failure phenomena.

Hence, although the limitations on the load-to-failure prediction capabilities of the surface layer energy formulation appear to be the same as those of the maximum tangential stress theory, the presence of $\delta$ provides a means of absorbing the local surface inhomogeneities into the energy theory; that is, different surface roughnesses would require different values of the constants, although the bulk material properties remain unaffected.

A correction to the maximum surface layer energy theory that could possibly be made would be to use a continuum solution where the thin boundary layer is actually assigned different material parameters than those of the bulk material. This step would perhaps remove some of the arbitrariness associated with assigning $\gamma_e E/\delta$ as a new parameter, when it appears at first glance merely to bear the image of the maximum tangential stress criterion.

## STRAIN ENERGY DENSITY THEORY - FAILURE OF INTERIOR ELEMENTS

As discussed earlier, the location on the (free) surface of a notch of the maximum surface layer tension may be postulated to be the region of initial failure of the notch. Additional behavior, such as the immediate post-failure crack direction, may be described by the local features of the strain energy density function as proposed by Sih [1,2]. Referring to Fig. (9), a radius vector, $\underset{\sim}{r}$, is attached to the point of maximum surface layer tension. The line normal to the elliptical boundary at this point forms an angle $\phi$ with the major axis of the ellipse. The vector is of length r, and its position relative to a fixed axis is given by the angle $\theta$, measured positive counterclockwise. The origin of the vector is a function of both loading and notch geometry. Further, for each notch and loading geometry, there is usually a separate origin for the tension and compression cases. In the degenerate case, when the notch becomes a line crack, the origin moves to the crack tip, and the fracture path will always be assumed connected to the crack tip.

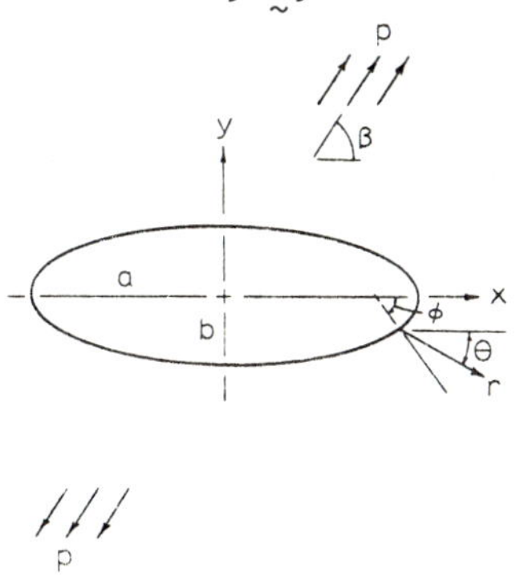


Fig. (9) - Off-axis loading of an elliptical notch and radius vector

In general, the failure criterion of an interior element will not be the same as that of the material immediately adjacent to the crack tip or notch surface.

The concept of some critical value of the strain energy density [1,2] being a material constant may be used to predict fracture of a material, whether near a crack tip, notch tip, re-entrant corner, or in an unflawed structure. In its original form, the theory takes the singular terms of the stresses near a line crack tip, and substitutes them into the strain energy density expression. This results in a quadratic form for the strain energy density function,

$$\frac{dW}{dV} = \frac{1}{r}\left(a_{11}k_1^2 + 2a_{12}k_1k_2 + a_{22}k_2^2 + a_{33}k_3^2\right) + \ldots \qquad (8)$$

Here the quadratic

$$S = a_{11}k_1^2 + 2a_{12}k_1k_2 + a_{22}k_2^2 + a_{33}k_3^2 \qquad (9)$$

represents the amplitude of the energy density field, where the coefficients $a_{ij}$ (i,j = 1,2,3) vary with a polar angle, $\theta$, measured from the crack tip, and the stress intensity factors $k_1$, $k_2$, $k_3$ are dependent only upon loading and geometric conditions. It has been shown, based on potential energy considerations, that crack initiation will start on a radial direction along which the strain energy density is a minimum. In general, for a notch, the form of the strain energy density will not be as concise as that of Eq. (8), but the evaluation of the energy field will be the same as that for the crack, and the resulting failure loads and angles also determined on the basis of the strain energy density assuming a critical constant value in a given material. It should be noted that in the considerations here, the exact strain energy density function is evaluated, at some finite distance from the notch boundary. This has been done since, as alluded, there is no simple expansion of the energy available for a blunt notch. It is for this reason that as the notch degenerates to a line crack, the evaluation is unaffected, since the locations of the points of interest do not lie on the surface of the notch itself.

GLOBAL INSTABILITY

Prior to becoming involved in the details of an example, a few additional observations and conjectures on failure at notches will be advanced for consideration. Having already noted that while in principle the strain energy density theory may be extended with suitable constitutive and kinematic relationships to account accurately for behavior near the notch front, the primary limitation placed on the present discussion of the theory is that the material undergoes brittle fracture. Within such a framework, it does not seem unreasonable to assert that the initial notch and loading geometries (and material properties) determine not only the load required to precipitate fracture and the initial fracture angle, but also the subsequent path the fracture will follow. That is, upon being loaded, the notch must seek a path of release, through knowledge of the local properties of the strained material, seen from its *present configuration*. That is, once loading is initiated, it should be immediately clear from the geometry on exactly what path the fracture will proceed. There are, however, many configurations for which crack growth is only locally unstable while globally stable: one in particular involves the loading of a punch resting on a glass block [9]. As the load is increased, the crack grows a corresponding amount, stops, and remains stationary until a further load increase occurs. Under such conditions, long range path predictions would not be possible. In the configurations where local and global instability occur simultaneously, such as notches loaded uniformly at distances far from the notch, crack growth is quite sudden and unstable. Then the material has no time to readjust to a new configuration, and as suggested above, quite possibly this path is predetermined. A primary argument in favor of such behavior comes from an understanding of the global energy field. In brittle materials, for blunt notches, or line cracks at low angles of loading, the loads required for initial fracture are relatively high (when compared to that of the line crack in symmetric loading) so that there already exists a high density energy field in the regions at both near and far distance from the point of incipient failure. Such a field will do much to contribute to rapid fracture. The path of fracture will be determined from the family of curves generated for a wide range of radius vectors which provide the locations of the sequence of points that indicated stationary values of the strain energy function. It is here suggested that this trajectory should be in close agreement with the actual fracture path taken from a notch on a brittle material.

## CRACK TRAJECTORIES AND FAILURE LOADS

The problem of an elliptical notch under uniform loading at an angle β with the x-axis as shown in Fig. (9) will again be used as an example. The evaluation of the local strain energy density field requires a knowledge of the position of maximum surface layer energy. For the elliptical notch, this position may be analytically determined: setting the first derivative of $\gamma_e$ in Eq. (5) with respect to eccentric angle η to zero, the resulting equation, quadratic in tanη, may be solved to give

$$\tan\eta = \frac{2b}{a(a+b)\sin 2\beta}\left[a\sin^2\beta - b\cos^2\beta \pm \sqrt{a^2\sin^2\beta + b^2\cos^2\beta}\right] \quad (10)$$

The positive root provides the location for failure initiation under compressive loading, and the negative root for failure initiation under tensile loading (Fig. (10)). The exact expression for the strain energy field surrounding the elliptical notch is first computed from the known stress distribution. All stationary values of the strain energy are then obtained numerically as the contours followed do not lend themselves to analytic evaluation. Since the behavior of the strain energy field is load dependent, there are fundamental differences in the fracture behavior of notches loaded in tension and compression.

Figure (11) gives the predicted crack trajectories emanating from the front of a blunt notch with b/a = 0.1 for several angles of loading. The theoretical prediction suggests that the fracture path is initially normal to the notch surface and then bends sharply towards the major axis of the notch. It extends in this way for a short distance and then again, a sharp turn occurs, now into a direction whose tangent matches that for the initial

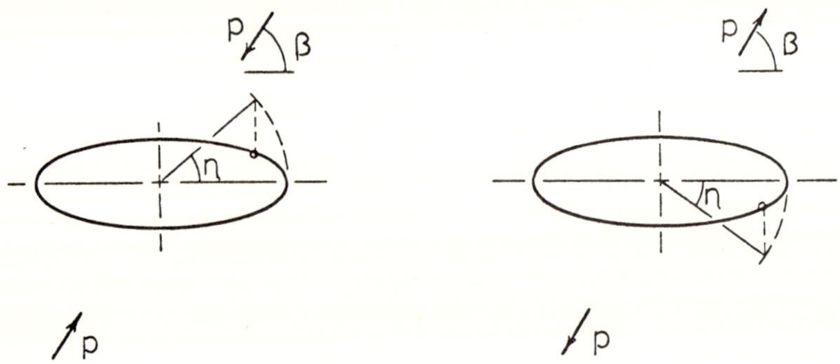


Fig. (10) - Location of points of initial failure on elliptical notch

fracture angle of the line crack [10]. The short extension distance is approximately equal to the notch tip radius, in accordance with the experimental observations [11]. Localized inhomogeneity or unaccountable grain size effects would likely obscure much of this local behavior in a real material. Moreover, the theory assumes that the material has no time to react to the new geometry at each instant of time as the crack propagates.

The locus of stationary values of strain energy may also be plotted to predict the fracture path of a notch under compression, Fig. (12). As in the tension case, the path is initially normal to the notch surface before turning away; then there is a quick return into the general trend leading the path towards the load. The same behavior prevails for $\beta$ greater than 60°.

The variation of the tensile load with notch or crack angle is plotted in Fig. (13) for $b/a = 0.1$ for several ratios of $r/a$. For a given material, $r(=r_o)$ represents a characteristic dimension of the core region and may be determined experimentally. For instance, an elliptical notch may be loaded to failure at several notch or crack angles, then from the corresponding family of curves for the same size notch, the curve that matches the data may be extracted, and that choice will provide the characteristic dimension of the core region, assuming that the elastic strain energy density constant is known. A knowledge of these two parameters has been postulated to be sufficient to describe the failure of this material in the presence of any geometric configuration, whether line crack or notch. In the case at hand, given the core dimension, a family of curves may now be established that indicates the change in loading patterns as the geometry changes. For example, should $r_o/a = 0.01$, then for various

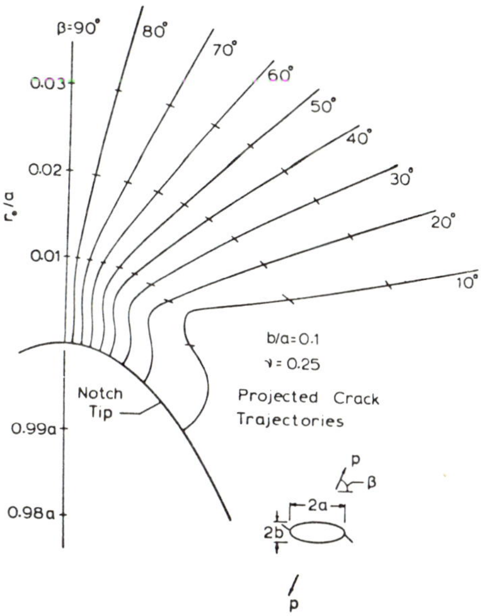


Fig. (11) - Projected fracture trajectories from a notch in tension

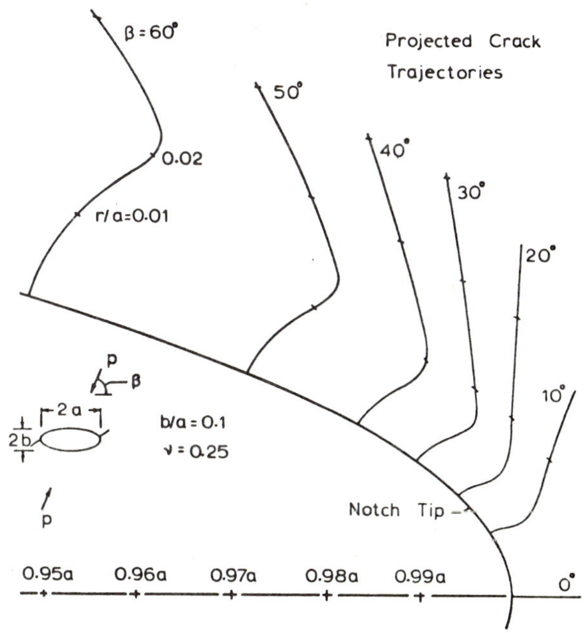


Fig. (12) - Projected fracture trajectories from a notch in compression

values of b/a, the variation in loading appears as in Fig. (14). The value of the angled crack problem should now be more clearly seen: the failure criterion is based not on a single load to failure, but a function, and while more difficult to obtain, in principle, the prediction reliability should be considerably greater than presently is the case.

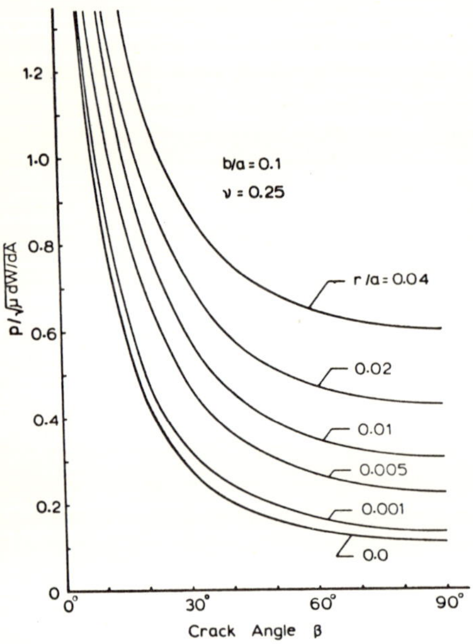


Fig. (13) - Tensile load versus notch angle for b/a = 0.1

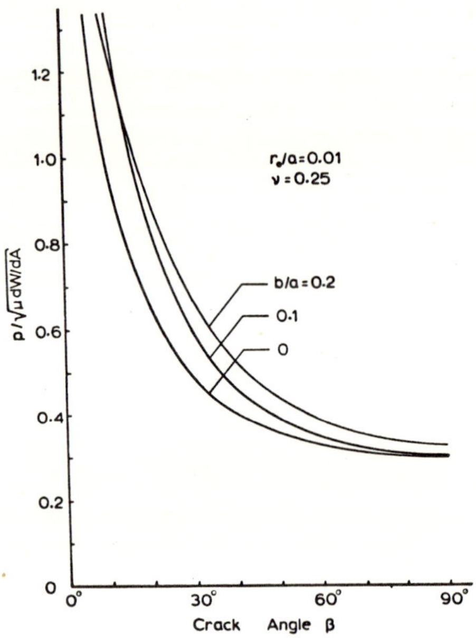


Fig. (14) - Tensile load versus notch angle for $r_o/a = 0.01$

Consider now the case of notch fracture under compression. As before, assuming a small applied uniform stress, the surface position of maximum surface layer energy may be located, and an origin established to evaluate the load behavior as a function of radial distance from the surface. Figure (15) describes the load behavior as a function of notch angle, for various values of r/a.

Once a core dimension has been established, say $r_o/a = 0.01$, the variation in load with geometry may be described, as in Fig. (16). For a given material, and the present analysis, it is expected that a new core dimension, $r_o/a$, must be determined, since in general, the behavior of a material in compression differs considerably from that in tension. That is to say, another sequence of tests would be required to obtain the core dimension.

A characteristic of compressive loading is the minimum load to fracture occurrence in the non-symmetric case. The families of curves plotted indicate a wide range of failure loads over the various values of β from zero the normal, but experimentally, there exists much less variation. Cotterell [12] has published the experimental data that appear in Fig. (17) for an elliptical notch, b/a = 0.1, in glass. The curves in this figure are those of Fig. (15), normalized to the load at β = 90°. As the position of failure criterion is moved into the material, there appears quite satisfactory agreement between r/a = 0.005 and 0.008. Hence, the energy criterion may be expected to provide quite reasonable failure predictions. On the basis of the data, one could choose a critical core dimension of, say, $r_o/a = 0.005$, and then use this dimension as set forth in previous sections for load predictions for other geometries in this material.

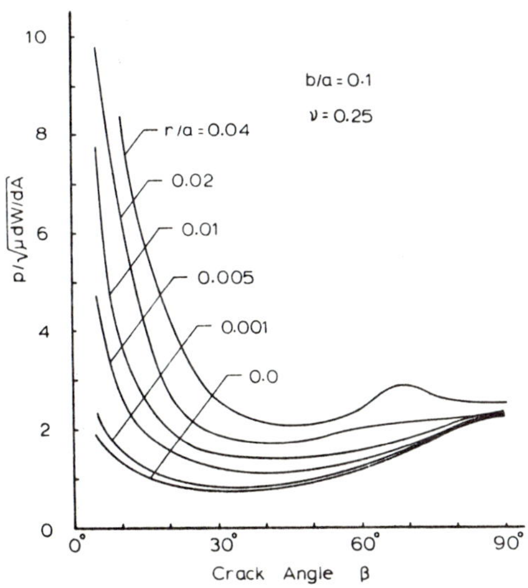


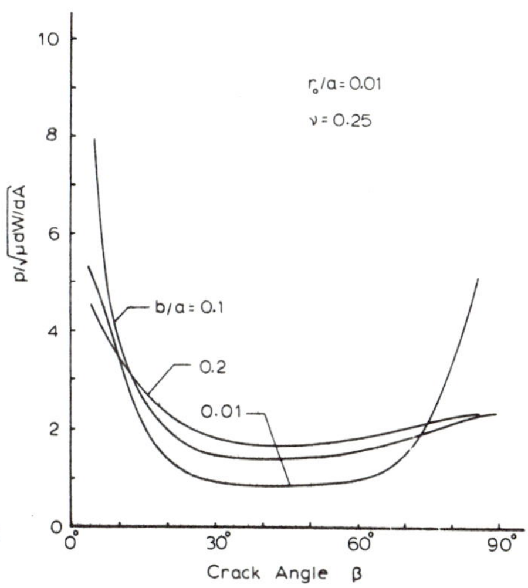


Fig. (15) - Compressive load versus notch angle for b/a = 0.1

Fig. (16) - Compressive load versus notch angle for $r_o/a = 0.01$

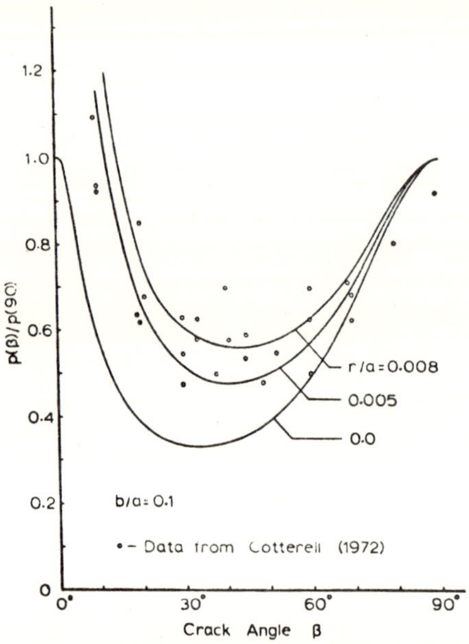


Fig. (17) - Compressive failure load versus notch angle for b/a = 0.1; data from [12]

CONCLUDING REMARKS

One of the most important purposes of the application of the theory of surface layer energy and strain energy density was to join the behavior of cracks to that of notches. This has essentially been accomplished through the addition of another material constant - a length parameter. Neuber [13] has also used a length parameter in his work on notches though the argument differs from the present one. By specifying the failure to be dependent upon internal conditions being met before it can occur, the local singular behavior of the line crack is avoided, hence disposing of a barrier that has separated notch and crack failure theories. Another engaging result was the description of the phenomenon of post-fracture behavior. The local behavior has been shown to agree well with observable details of crack paths emanating from a notch front. The theory upon which all the results were based is that of total strain energy density. Under tensile loading conditions, it is required to locate the stationary values of the strain energy density as described, and determine whether the energy of volume change exceeds that of distortion to give the direction of fracture. Sih [10,14] has made a detailed study of two- and three-dimensional crack problems and argued that the crack propagates in a direction other than that of yielding since the strain energy density theory distinguishes the path of fracture from that of yielding. Moreover, fracture surface created at the continuum level is essentially an elastic process. The plasticity formed in the region away from the direction of crack propagation simply reduces the amount of available energy to drive the macroscopic crack in the elastic portion of the material. It must also be said that the present discussion has avoided any effect such as necking due to yielding of the material in the thickness direction of metal sheet specimens. Current studies on the interplay between the shear lips and crack tunnelling as analyzed by the strain energy density theory are underway.

REFERENCES

[1] Sih, G. C., Methods of analysis and solutions to crack problems, Mechanics of Fracture, Volume I, Ed. G. C. Sih, Noordhoff International Publishing, Leyden, 1973, 21.

[2] Kipp, M. E. and Sih, G. C., The strain energy density failure criterion applied to notch elastic solids, Int. J. of Solids and Structures (in press).

[3] Griffith, A. A., The phenomena of rupture and flow in solids, Phil. Trans. Roy. Soc., London, 163, 1921.

[4] Sih, G. C., Handbook of stress intensity factors, Institute of Fracture and Solid Mechanics, Lehigh University, Bethlehem, Pa., 1973.

[5] Irwin, G. R., Analysis of stresses and strains near the edge of a crack traversing a plate, J. Appl Mech., 24, 361, 1957.

[6] Sih, G. C., Paris, P. C. and Erdogan, F., Crack-tip stress-intensity factors for plane extension and plate bending problems, J. Appl. Mech., 29, 306, 1962.

[7] Sih, G. C., Stress distribution near internal crack tips for longitudinal shear cracks, J. Appl. Mech., 32, 51, 1965.

[8] Sih, G. C., Private Notes, 1972.

[9] Roesler, F. C., Brittle fractures near equilibrium, Proc. Phys. Soc., B69, 981, 1956.

[10] Sih, G. C., Strain energy density criterion applied to fracture problems, Proc. of Symposium on Mech. Behavior of Materials (in press).

[11] Cotterell, B., The paradox between the theories for tensile and compressive fracture, Int. J. Frac. Mech., 5, 251, 1969.

[12] Cotterell, B., Brittle fracture in compression, Int. J. Frac. Mech., 8, 195, 1972.

[13] Neuber, H., Theory of notch stresses, Edwards, Michigan, 1946.

[14] Sih, G. C., Three dimensional crack problems, Mechanics of Fracture, Volume II, Ed. G. C. Sih, Noordhoff International Publishing, Leyden, (in press).

ACKNOWLEDGEMENT

This work was partially supported by the Office of Scientific Research under Contract No. AFOSR-74-2586 with the Institute of Fracture and Solid Mechanics, Lehigh University.

# DEVELOPMENT OF THE NONLINEAR ENERGY METHOD FOR DETERMINATION OF FRACTURE TOUGHNESS VALUES

H. Liebowitz, D. L. Jones and P. K. Poulose

School of Engineering and Applied Science, The George Washington University, Washington, D.C. 20006 USA

## ABSTRACT

The nonlinear energy method, which was previously introduced by Liebowitz and Eftis, has been evaluated for different specimen geometries and for cyclic as well as static loading. Considerable testing has been conducted on thin, center-cracked sheets of 7075-T6 and 2024-T3 aluminum for the purpose of comparing the nonlinear energy toughness parameter, $\tilde{G}_c$, with values obtained by the R-curve method. The dependence of these two methods on specimen geometry has been established and comparisons were made between them.

The nonlinear energy method has also been applied to the load-displacement record obtained from compact tension tests on several aluminum and steel alloys for the purpose of comparing the toughness parameter, $\tilde{G}_{Ic}$, with other measures of fracture toughness. As for the $\tilde{G}_c$ evaluations, the point of maximum load was selected for comparison purposes even though this procedure did not satisfy for all tests the guidelines of the ASTM, COD and J-integral methods. For tests on an A533B steel, for which the ASTM thickness requirement was not satisfied, the J-integral values were highest with the nonlinear energy, COD and ASTM results 10, 15 and 70 percent below $J_{Ic}$. For the aluminum alloys the COD toughness was considerably higher than the other three results, possibly due to an uncertain amount of subcritical crack growth.

Cyclic loading tests were conducted on thin, center-cracked sheets of 7075-T6 and 2024-T3 aluminum and the envelope of the cyclic load-displacement curve was used to evaluate $\tilde{G}_c$.

Although only eight tests were conducted, $\tilde{G}_c$ exhibited a definite tendency to decrease with increasing cyclic life. Further testing is needed in order to better understand the possibilities of this application of the nonlinear energy method.

## INTRODUCTION

In recent years two significant developments have underscored the need for a straightforward and reliable method for including nonlinear effects in fracture toughness testing. One is the acceptance of linear fracture mechanics concepts by the metals industry which has resulted in the development of high toughness alloys often displaying significant nonlinearity prior to unstable fracture. The second is the development of very large engineering structures, i.e., the wide-body aircraft, supertankers, and nuclear reactor containment vessels, whose unexpected structural failures result in tragedies on a much greater scale than heretofore possible. Thus several methods for including nonlinear effects in fracture toughness testing have been proposed, none of which have yet gained universal acceptance.

Several methods which possess the potential for encompassing nonlinear behavior will be discussed and comparison will be made between them. In thin-sheet fracture mechanics testing of many medium and high strength materials, considerable crack growth occurs prior to unstable fracture. The crack growth resistance curve (R-curve) method [1] has been employed under these circumstances to determine a nonlinear fracture toughness parameter which includes the nonlinearity due to subcritical crack growth. The nonlinearity due to crack-tip plasticity may also contribute to the R-curve toughness in an indirect manner, since crack-tip plasticity may allow the specimen to sustain greater loads than would otherwise be possible. For fracture testing under plane strain conditions, the contribution of subcritical crack growth to the nonlinear material response is usually much less significant than the contribution due to crack-tip plasticity. Several methods have been proposed for incorporating the nonlinearity due to crack-tip plasticity into the plane strain fracture toughness. The concept of a path-independent energy line integral (J-integral) has been introduced [2,3] and evaluated as a failure criterion [4]. Another approach, based on the crack opening displacement (COD) at the crack tip has been proposed [5] to account for the considerable amounts of plastic flow observed in the fracture of mild steels. The amount of crack opening is identified as a measure of the work done in extending the crack and has been related to the G and K concepts of fracture mechanics.

An additional nonlinear fracture method advanced by Liebowitz and Eftis [6,7] defines a fracture toughness parameter $\tilde{G}_c = \tilde{C}\,\overline{G}_c$ based on the total (linear and nonlinear) energy applied to the test specimen or structural component. The $\tilde{C}$ term is determined solely by the curvature of the load-displacement record through the use of a three-parameter Ramberg-Osgood relation, although other curve fitting techniques could also be used. The quantity $\overline{G}_c$ represents the critical value of the strain energy release rate or crack extension force for completely brittle fracture. This method is applicable to any configuration for which linear G or K values exist or can be established. Certain refinements to the nonlinear energy method have been discussed from an experimental viewpoint by Liebowitz and Jones [8] and from a theoretical orientation by Eftis and Liebowitz [9].

Because of the current interest in establishing a "standard" nonlinear fracture test method, many theoretical and experimental studies of the various methods have been published. It is the purpose of this paper to present experimental comparisons between the nonlinear energy method and other test methods. For tests of thin sheet, center-cracked panels, fracture toughness comparisons were made between the nonlinear energy and R-curve methods. Several test series were conducted on two aluminum alloys, 7075-T6 and 2024-T3, in which the specimen lengths, widths and $\frac{a}{w}$ ratios were each independently varied so that the effect of specimen geometry on each method could be ascertained. Further, comparisons were made between toughness values obtained by the nonlinear energy method and the ASTM, COD and J-integral methods for compact tension tests on several aluminum alloys.

Additional tests have been conducted for the purpose of ascertaining the validity of the nonlinear energy method when applied to cyclic loading situations. Several different variable amplitude, cyclic loading programs have been applied to center-cracked sheet specimens of 7075-T6 and 2024-T3 aluminum alloys. The amplitude was varied in such a manner that failure was obtained in less than 50 cycles. The envelope of the loading curves was employed for determination of the nonlinear toughness which was then compared with the results of static tests.

## THE NONLINEAR ENERGY METHOD

The nonlinear energy method, as proposed by Liebowitz and Eftis [6], was originally applied only to the tensile fracture of center-cracked panels under fixed-grip conditions. Subsequently [7], it was shown that the method could be applied to all of the common fracture mechanics test geometries. The nonlinear fracture toughness was defined as

$$-\frac{\partial U_t}{\partial c}\bigg|_{crit} = \tilde{G}_c . \quad (1)$$

where $U_t$ represents the total (linear plus nonlinear) strain energy in the system. The partial derivative was evaluated at the critical point which was taken to be the point of onset of unstable crack propagation. It was seen that the nonlinear fracture toughness could be evaluated for all geometries considered by the relation

$$\tilde{G}_c = \tilde{C}\,\overline{G}_c. \tag{2}$$

In Eq.(2) $\tilde{C}$ is a measure of the nonlinearity of the load-displacement record in a fracture mechanics test while $\overline{G}_c$ represents the critical value of the strain energy rate or crack extension force for completely brittle fracture.

The principal output of a fracture mechanics test is a load-displacement record which consists of a linear portion and usually a nonlinear portion resulting from crack-tip plasticity and subcritical crack growth. For tensile tests of center-cracked, double edge-cracked, or single edge-cracked specimens of unit thickness, the three-parameter (Ramberg-Osgood) relation

$$v = \frac{F}{M} + k\left(\frac{F}{M}\right)^n \tag{3}$$

usually can be employed quite satisfactorily to represent the actual load-displacement record. Then by the use of Eqs.(1&3) it was shown [6] that

$$\tilde{G}_c = \left\{\left[1 + \frac{2nk}{n+1}\left(\frac{F}{M}\right)^{n-1}\right]\frac{F^2}{2}\frac{\partial(1/M)}{\partial c}\right\}_{crit} \tag{4}$$

where, from Eq.(2)

$$\overline{G}_c = \left[\frac{F^2}{2}\frac{\partial(1/M)}{\partial c}\right]_{F=F_c} \tag{5}$$

and

$$\tilde{C} = \left[1 + \frac{2nk}{n+1}\left(\frac{F}{M}\right)^{n-1}\right]_{F=F_c} \tag{6}$$

It can easily be seen from Eqs.(3&6) that as the load-displacement curve becomes a straight line, $\tilde{C}$ approaches unity since $k \to 0$ and $n \to 1$. Thus $\overline{G}_c$ represents the limiting case of $\tilde{G}_c$ as $\tilde{C} \to 1$.

The original procedure [6] for evaluating Eq.(5) for center-cracked panels employed the relation

$$G_c = \frac{K_c^2}{E} = \frac{F_c^2}{E(WB)^2} \left[\pi(a_c + r_y) \sec \frac{\pi}{w} (a_c + r_y)\right] \tag{7}$$

where $r_y$ is a measure of the crack tip plastic zone size. Jones, Liebowitz and Eftis [10] later proposed that the plastic zone correction should not be included in Eq.(7) since the effect of crack-tip plasticity is included in $\tilde{C}$. They also employed a more direct compliance calculation for evaluating Eq.(5) but concluded that the method consistently underestimated $G_c$.

The procedure introduced by Liebowitz and Eftis [6] for evaluating $\tilde{C}$ depended upon determining the three parameters M, k and n by drawing the tangent and two reduced modulus (secant) lines to the load-displacement curve, Fig. 1. At the two points of intersection of the secant lines with the load-displacement

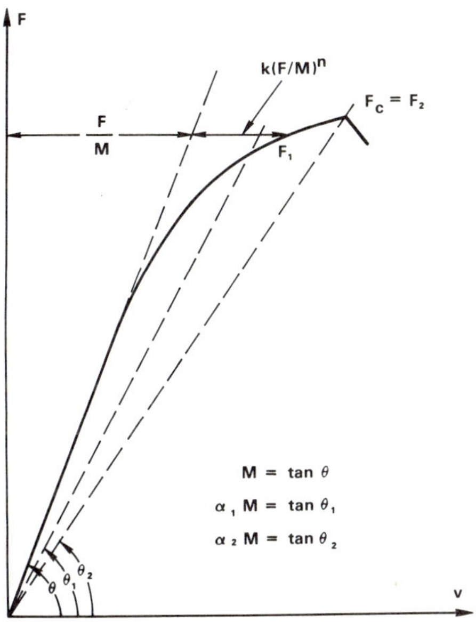


FIG. 1. TYPICAL NONLINEAR LOAD-DISPLACEMENT CURVE SHOWING THREE-PARAMETER REPRESENTATION AND REDUCED MODULUS LINES

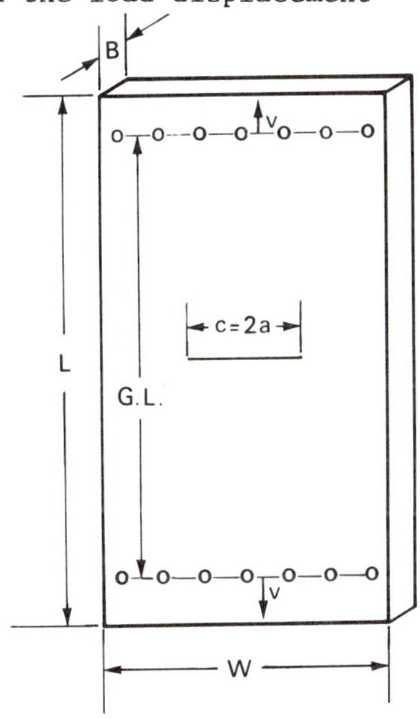


Fig.2. General specimen geometry

record, it is possible to write

$$v_1 = \frac{F_1}{\alpha_1 M} = \frac{F_1}{M} + k\left(\frac{F_1}{M}\right)^n, \quad v_2 = \frac{F_2}{\alpha_2 M} = \frac{F_2}{M} + k\left(\frac{F_2}{M}\right)^n \quad . \tag{8}$$

from which the following two equations are obtained

$$\frac{1-\alpha_1}{\alpha_1} = k\left(\frac{F_1}{M}\right)^{n-1}, \quad \frac{1-\alpha_2}{\alpha_2} = k\left(\frac{F_2}{M}\right)^{n-1} \quad . \tag{9}$$

The parameter k can be eliminated from Eqs.(9) which then allows n to be obtained as

$$n = 1 + \frac{\ell n[\alpha_2(1-\alpha_1)/\alpha_1(1-\alpha_2)]}{\ell n(F_1/F_2)} \, . \tag{10}$$

Also, by use of Eq.(9), k can be eliminated from Eq.(6) so that

$$\tilde{C} = 1 + \frac{2n(1-\alpha_1)}{\alpha_1(n+1)} \left(\frac{F_c}{F_1}\right)^{n-1} = 1 + \frac{2n(1-\alpha_2)}{\alpha_2(n+1)} \left(\frac{F_c}{F_2}\right)^{n-1} \tag{11}$$

which is independent of M. No definitive statements were made in the original papers about the selection of points of intersection of the secant curves with the load-displacement record.

Additional studies of the nonlinear energy method have led to certain refinements in both the analytical foundation and the experimental procedure for determining $\tilde{G}_c$. For fracture mechanics tests conducted in load control (approximating fixed-load conditions at failure), the definition of $\tilde{G}_c$ given in Eq.(1) is not adequate because of the absence of the work rate term. However, Eftis and Liebowitz [9] have shown that the nonlinear energy rate can be represented in this case by

$$\tilde{G} = \frac{\partial W}{\partial c} - \frac{\partial U_t}{\partial c} \tag{12}$$

where W is the work done by the applied forces. This result was established in the context of a general energy balance on a body containing a traction-free crack and subjected to a state of tensile stress normal to the crack surface. It was then shown that, for the case in which the nonlinearity of the load-displacement record is due entirely to crack-tip plasticity, Eq.(12) can be evaluated by the relation

$$\tilde{G}_c = \left[1 + \frac{2nk}{n+1} \left(\frac{F_c}{M}\right)^{n-1}\right] \frac{F_c^2}{2} \frac{\partial(1/M)}{\partial c}\bigg|_{c=c_0} = \tilde{C}\,\overline{G}_c \, . \tag{13}$$

It is seen that Eq.(13) has the same form as Eq.(4) so that the results correspond to the elastic case in that the same expression is obtained for $G_c$ regardless of whether the boundary work term is included.

For the case of fracture after a significant amount of crack-tip plasticity and subcritical crack growth have occurred, a different situation exists. In this case both crack tip plasticity and subcritical crack growth contribute to the nonlinearity in the load-displacement record. Under these conditions it has been shown [9] that

$$\tilde{G}_c = \left[1 + \frac{2nk}{n+1}\left(\frac{F_c}{M_c}\right)^{n-1}\right]\left(\frac{M_o}{M_c}\right)^2 \frac{F_c^{\,2}}{2}\left.\frac{\partial(1/M)}{\partial c}\right|_{c=c_o} \tag{14}$$

In Eq.(14) $M_o$ and $M_c$ are the elastic compliances corresponding to the original and critical crack sizes, respectively.

The experimental procedure originally proposed by Liebowitz and Eftis [6] for evaluating $\tilde{G}_c$ has been followed in subsequent work with only slight modifications intended to achieve greater accuracy or simplicity. It was shown by Jones, Liebowitz and Eftis [10] that both the accuracy and simplicity of the experimental procedure could be enhanced by requiring that the second reduced modulus line intersect the load-displacement record at the critical point. In this case $F_2 = F_c$ and the second portion of Eq.(11) reduces to

$$C = 1 + \frac{2n(1-\alpha_2)}{\alpha_2(n+1)} \; . \tag{15}$$

It was originally suggested only that the second reduced modulus line intersect the load-displacement curve near the critical point, since it was felt that $\tilde{C}$ was rather insensitive to variations in $\alpha_2$. It has since been established that $\tilde{C}$ is significantly dependent upon $\alpha_2$ and that the only consistent characterization is to require that the lines intersect at the critical point. However, $\tilde{C}$ does appear to be rather insensitive to variations in $\alpha_1$ so it has been suggested that the first reduced modulus line approximately bisect the angle between the tangent and second reduced modulus lines.

Several different formulas and procedures have been utilized in evaluating $\overline{G}_c$. The original procedure and a modification of it has already been discussed. A later modification [8] recognized that, since $\overline{G}_c$ includes only linear effects, the most appropriate K formulas and the relationship $G = K^2/E$ should provide the most accurate representation for $\overline{G}_c$. For most of the common fracture mechanics geometries, accurate K formulas have been published in polynomial form by ASTM [11] and these were used for the results presented in this paper.

Some consideration has also been given to the use of the nonlinear energy method for cyclic loading applications. One procedure for incorporating the effect of cyclic loading on G is to utilize the envelope of the cyclic load-displacement record for the determination of $\tilde{G}_c$. A variety of loading programs could be employed with the only restriction being that a usable load-displacement envelope be generated. Another procedure that could be employed for cyclic loading applications would be to precycle each specimen and then conduct the pull test in the same manner as for the static case. It would thus be possible to establish the magnitude of the reduction in $\tilde{G}_c$ due to a specified amount of precycling. In this case the reduction in $\tilde{G}_c$ would represent a measure of the damage accumulation in the specimen. For each of the two types of loading programs discussed above, the value of $\tilde{G}_c$ obtained from cyclic loading could be compared directly with the static $\tilde{G}_c$ results.

## EXPERIMENTAL PROCEDURE

Since the results of three different testing programs are to be reported in this paper, each experimental procedure will be discussed in the same order that the results are to be presented. The first test program involved the static testing of thin, center-cracked sheets of 7075-T6 and 2024-T3 aluminum alloys. Comparisons were made between $\tilde{G}_c$ and the fracture toughness obtained from the R-curve method for a number of tests in which the length, width and a/w ratio of the specimens were independently varied. A number of tests were then conducted on compact tension specimens of several aluminum and steel alloys. It was possible to obtain direct comparisons between $\tilde{G}_{Ic}$, $J_{Ic}$, $G_{Ic} = K_{Ic}^2/E$ and $G_{COD}$ from these tests. Finally, cyclic loading tests were conducted on thin center-cracked sheet specimens of 7075-T6 and 2024-T3 aluminum alloys.

### $G_c$ and R Curve Comparisons

A considerable number of tests have been conducted on 0.063 in., center-cracked sheets of 7075-T6 and 2024-T3 aluminum alloys for comparison between $\tilde{G}_c$ and R curve toughness estimations. The general specimen geometry is shown in Fig.(2) where it is emphasized that the displacements were measured at the loading holes. These displacements were obtained by measuring the deformation of the specimen plus the grips and then subtracting the deformation of the grips. Tests were conducted in load control and subcritical crack growth was monitored visually with a hand-held magnifier (5-20x). Three test series were conducted on each alloy: (a) one series having constant w and a, with L = 10,15, 20,25,30, and 35 in., (b) one series having constant L and a/w, with w = 4,6,8,10,12,14 in., and (c) one series having constant L and w, with 2a/w = 0.3, 0.4, 0.5, 0.6. Duplicate tests were run for each geometry so that some appreciation of the repeatibility of the results could be obtained. The gage length was nominally 3 inches less than the overall specimen length and all tests were conducted in the LT orientation. The notch tip radii were filed to $0.003<r<0.005$ in. One variation from the usual thin-sheet test procedures employed for these tests was that stiffeners were not used to prevent tension buckling. Utilization of stiffeners has the advantage of eliminating out-of-plane behavior while it possesses the disadvantage of not representing failures as they would actually occur in engineering applications.

Values of $\tilde{G}_c$ were obtained for each test by the use of Eq.(13) and compared with the toughness determined by the R curve method. In order to evaluate $\tilde{G}_c$ it is necessary to determine $\tilde{C}$ from Eq.(15) and $\overline{G}_c$ from the ASTM [11] formula

$$\overline{G}_c = \frac{K_c^2}{E} = \frac{\sigma_c^2 \pi a}{E} \left[1-0.1\left(\frac{2a}{w}\right) + \left(\frac{2a}{w}\right)^2\right]^2 \tag{16}$$

where $\sigma_c$ is the gross critical stress and a is the original crack size. According to recent discussions on the R curve method [1,9], the fracture toughness parameter, $G_R$, for these specimens can be obtained by substituting the critical crack size and critical stress into Eq.(16). This was the procedure followed in obtaining the R curve toughness values reported in this paper. Equation (14) was also evaluated for some of these tests so that a comparison could be made between the values obtained by Eq.(13) and Eq.(14). For these tests it was found that toughness values calculated from these two equations generally differed by less than five percent. Also, since the procedure for evaluating Eq.(14) is different and somewhat more involved, detailed discussion of this method will not be presented here.

## $G_{Ic}$, $\tilde{G}_{Ic}$, $G_{COD}$ and $J_{Ic}$ Comparisons

Fracture mechanics tests have been conducted on several different aluminum alloys for the purpose of obtaining conventional $K_{Ic}$ toughness data and then making comparisons between the various nonlinear fracture mechanics methods. Compact tension tests were conducted, by Reynolds Metals Company on 2124-T851, x2048-T851, 7075-T651 and 7175-T7651 aluminum alloys and the data, including the load-displacement records were made available to the authors for subsequent comparisons. The test specimens were manufactured from 3 in. rolled plate and tests were conducted in the TL, LT and ST orientations in accordance with the ASTM standard method E399. Some additional test data from compact tension tests on an A533B steel were made available by Westinghouse Research Laboratories. These tests were conducted primarily for the purpose of J-integral evaluations so all of the requirements of E399-72 were not satisfied. Although the evaluation procedure was different for each method, all of the methods were evaluated at the same point on the load-displacement curve (the point of maximum load). This procedure was followed because of the desirability of ascertaining the behavior of each method up to the point of maximum load, which in most cases coincided with the instability point.

The nonlinear energy, COD, and J-integral methods will be discussed in some detail; however, the standard ASTM method will not be discussed since it is adequately treated in E399-72. The ASTM method was followed except for the utilization of the maximum load and the lack of satisfaction of the thickness requirement for the A533B steel. The procedure for calculating $\tilde{G}_{Ic}$ is quite straightforward since the relation

$$\tilde{G}_{Ic} = \tilde{C}\,\overline{G}_{Ic} \tag{17}$$

can be employed for plane strain conditions. Equation (15) can still be used to evaluate $\tilde{C}$ and $\overline{G}_{Ic}$ is obtained from

$$\overline{G}_{Ic} = \frac{K_{Ic}^{\,2}(1-\nu^2)}{E}\ , \tag{18}$$

where $K_{Ic}$ is the fracture toughness obtained by the use of the ASTM procedure and the maximum load.

The concept involved in the determination of a toughness parameter by the COD method is relatively simple; however, its application is not necessarily a straightforward matter. It has been shown by a number of researchers that a relation exists between the crack-tip opening and the linear fracture mechanics parameter G. For example, Egan [12] has shown the relation to be

$$\delta = \alpha \frac{G}{\sigma_Y} \tag{19}$$

$$\text{where} \quad \alpha = \begin{cases} \dfrac{4}{\pi} & \text{for generalized plane stress conditions} \\ \dfrac{4}{\pi\sqrt{3}} & \text{for plane strain conditions,} \end{cases}$$

$\delta$ is the crack-opening displacement, and $\sigma_Y$ is the yield strength. It has been concluded more recently [13] that, because of the approximations involved in the development of Eq.(19), the result is equally satisfactory if $\alpha$ is arbitrarily equated to unity. This leads to the result

$$G_{COD} = \sigma_Y\,\delta_c \tag{20}$$

which is the expression utilized in this paper. Since $\sigma_Y$ is readily determined, effective application of this method is dependent upon obtaining a reasonably accurate determination of $\delta_c$, the critical crack-opening displacement. Although other methods were employed originally, the current procedure for evaluating $\delta_c$ has been to modify the displacement obtained from a standard clip gage to approximate the opening at the crack front. Dover [14] assumed that the ratio between the clip gage reading, $V_g$, and $\delta$ can be given by

$$\frac{V_g}{\delta} = 1 + \frac{a+z}{r(w-a)} \tag{21}$$

where the terms in Eq.(21) are defined in Fig.(3). The derivation of Eq.(21) is based on the assumption that the specimen deformation can be represented by a rotation about some point r(w-a) ahead of the crack.

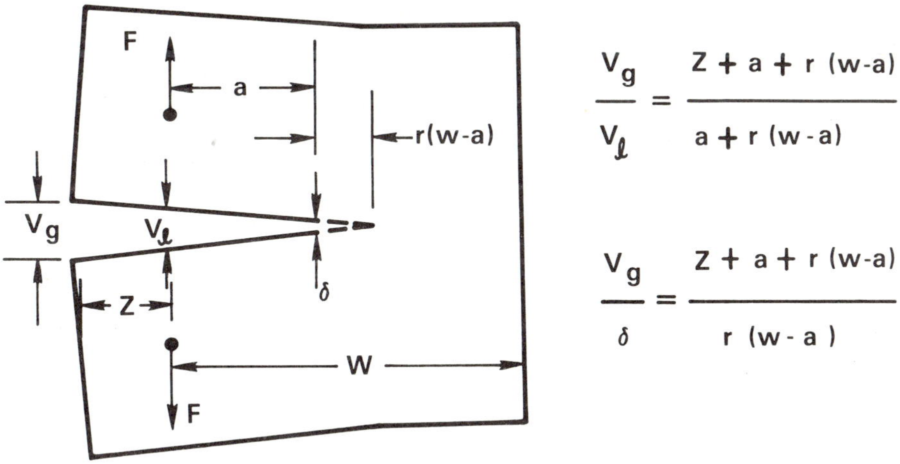


FIGURE 3. COMPUTATIONAL RESULTS FOR $K_q/K_{Ic} = 2.0$ AND $a_0/h = 2.0$ USING A CRACK SPEED DEPENDENT FRACTURE ENERGY GIVEN BY EQ. (11) WITH $K_{Im}/K_{Ic} = 0.7$ AND $V_m/C_o = .05$

The rotation factor, r, is generally believed to be constant after the displacement has attained a certain, rather large value. The magnitude of $G_{COD}$ is strongly dependent upon the value selected for r and, unfortunately, general agreement upon one value of r has not been attained. Usually researchers have determined that the rotation factor should be selected within the range $2 \leq \frac{1}{r} \leq 3.1$. Another procedure has been reported by Egan [12] in which an elastic-plastic, finite element analysis of crack-tip profiles was employed to establish a relationship between gage point and crack-tip displacements. For the A533B data provided by Westinghouse, it was found that Eq.(21) with $\frac{1}{r} = 3.0$ gave satisfactory results. However, for the aluminum specimens $\frac{1}{r} = 3.0$ resulted in $G_{COD}$ values which were unreasonably large. Thus the relationship between crack opening and clip gage displacements established by Egan was employed for determining the critical COD from clip gage measurements.

The J-integral was defined originally as an energy line integral for which path independence had been established for an elastic medium. However, the principal interest in the J-integral has been related to the possibility of its assuming a constant value, $J_{Ic}$, for a given material. This concept was proposed and developed by Begley and Landes [15] although their experimental procedure was rather cumbersome. It has since been shown [16] that $J_{Ic}$ can be evaluated by the relation

$$J_{Ic} = \frac{2A}{Bb} \tag{22}$$

where A is the area under the load displacement curve up to the critical point and b is the remaining ligament length. Although other authors have used different methods, it was determined that the Ramberg-Osgood characterization of the load-displacement record could be integrated directly to provide an expression for the area under the curve in the form

$$A = \frac{\tilde{C}F^2}{2M} . \tag{23}$$

The critical point for all tests was taken to be the point of maximum load which coincided with the instability point for the aluminum alloys but not for the A533B steel. In evaluating $J_{Ic}$ it is necessary that the displacements be measured at the loading points. Therefore, the clip gage data were modified by an equation similar to Eq.(21) to provide displacements corresponding to the loading points. A method of evaluating $J_{Ic}$ by the use of an R curve procedure has been suggested by Landes and Begley [17] but this method also requires significant variation from the standard ASTM procedure and so could not be employed on these data.

### Cyclic Loading Tests

Cyclic loading programs were applied to several thin sheet specimens of 7075-T6 and 2024-T3 aluminum alloys for the purpose of ascertaining the suitability of applying the nonlinear energy method to cyclic loading situations. For the tests reported here the loading programs were limited to successively increasing load levels until the specimen fractured. The envelope of the load-displacement curve was then analyzed in the same manner as the static load-displacement curves for the establishment of a critical value of $\tilde{G}$. The specimen dimensions were not varied in these tests but different load increments were used, with the same starting point, which effectively caused a variation in the number of cycles to failure.

## RESULTS AND DISCUSSION

A number of tests have been conducted under plane stress (static and cyclic ) and plane strain (static) conditions for the purpose of evaluating the nonlinear energy toughness parameter and verifying regimes of validity of the test method. The appropriate comparisons were made for the static tests and certain trends were identified in the toughness values for the different cyclic loading programs.

Comparisons between $\tilde{G}_c$ and the R curve toughness values for the center-cracked sheet tests are shown in Figs.(4-6). The results of the test series on both the 7075-T6 and 2024-T3 alloys in which the specimen length was varied while other dimensions were held constant are seen in Fig.(4). Each data point represents the average results of two duplicate tests. It is seen that both materials gave high fracture toughness values at short specimen lengths but, as the length increased, the toughness stabilized to a relatively constant minimum value. The minimum values for the two alloys were $\tilde{G}_c$ = 335 lb./in. for 7075-T6 and $\tilde{G}_c$ = 530 lb./in. for 2024-T3. The toughness values obtained from the R curve method were $G_R$ = 315 lb./in. for 7075-T6 and $G_R$ = 460 lb./in. for 2024-T3. For the 7075-T6 test series the specimen width was 10 in. and for 2024-T3 the width was 12 in. The 2a/w ratio was maintained at 0.5 for all test results reported in Fig. 4. It is noted that the minimum gage length requirement of four times the crack length was satisfied for the longer specimens of both alloys.

The results of the test series in which the specimen width was varied are shown in Fig.(5) in the same manner as for the

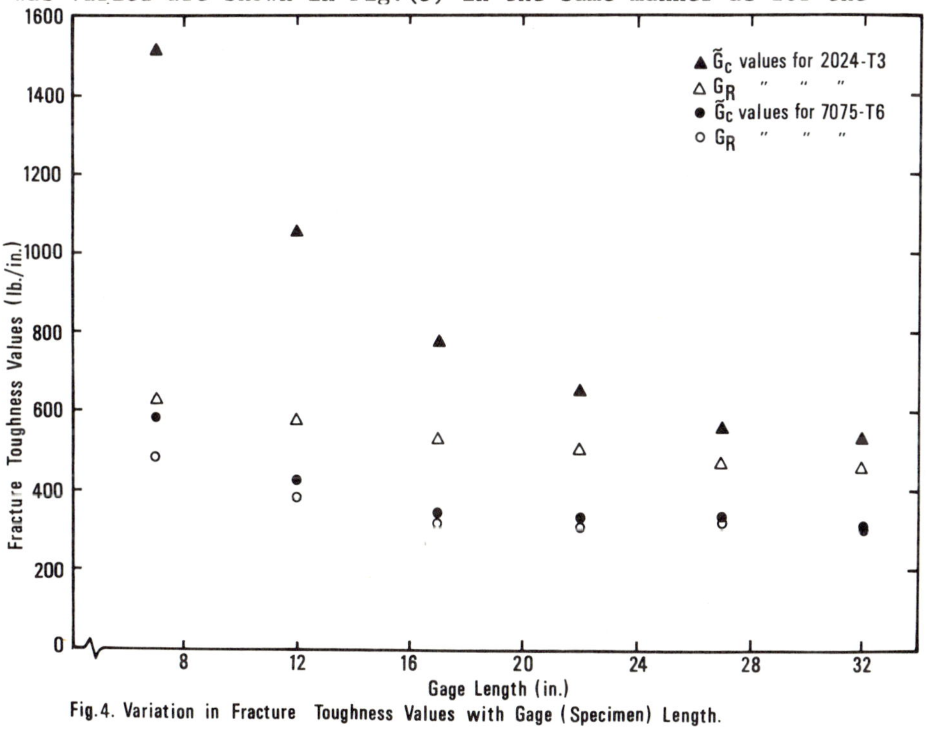


Fig.4. Variation in Fracture Toughness Values with Gage (Specimen) Length.

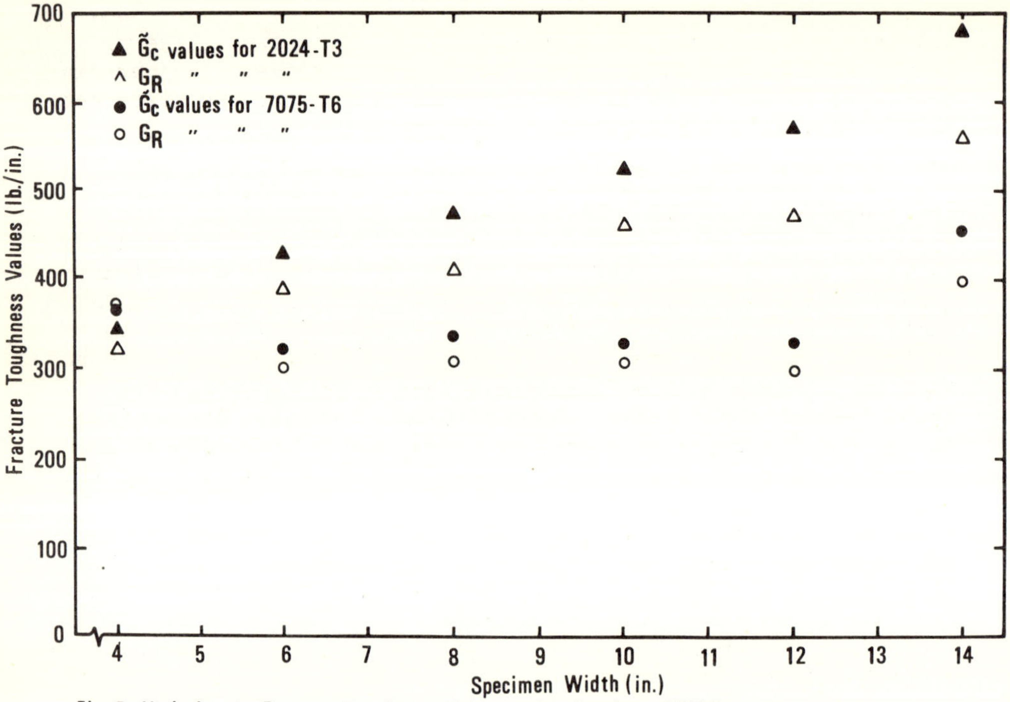


Fig. 5. Variation in Fracture Toughness Values with Specimen Width.

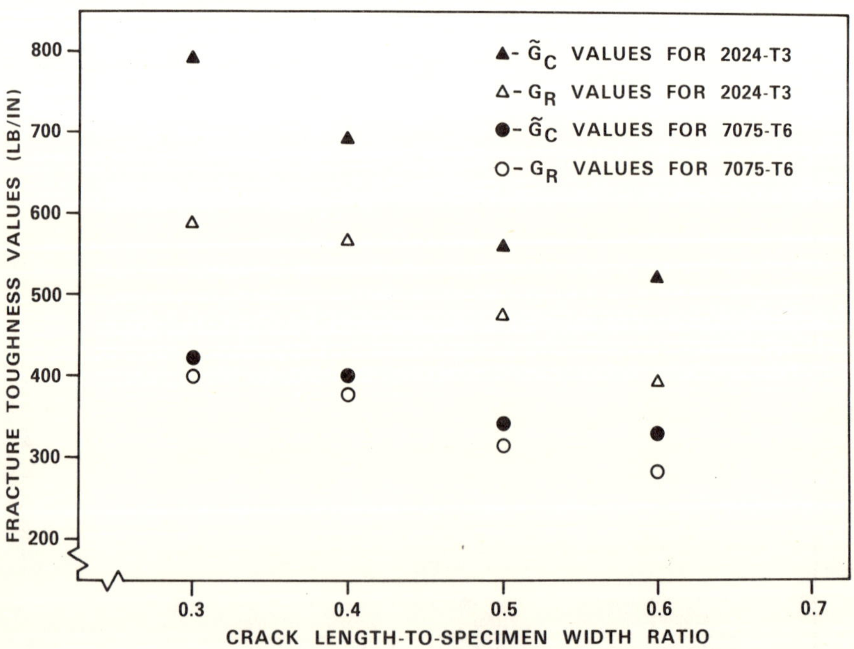


FIG. 6. VARIATION IN FRACTURE TOUGHNESS VALUES AS A FUNCTION OF THE CRACK LENGTH-TO-SPECIMEN WIDTH RATIO

length series. Contrary to the situation for the length series, the toughness did not approach a lower limiting value as the width increased. For 7075-T6 the toughness was reasonably constant at $\tilde{G}_c$ = 320-340 lb./in. and $G_R$ = 295-310 lb./in. for variation of the specimen width between 6 and 12 in., although the results for w = 4 and 14 in. were significantly higher. However, the 2024-T3 alloy exhibited no tendency to approach a constant value with increasing specimen width. Both the nonlinear energy and R curve toughness exhibited gradually increasing values with increasing specimen width. The nonlinear toughness ranged from 350 to 680 lb./in. while $G_R$ varied from 325 to 560 lb./in. For these tests the 2a/w ratio was maintained at 0.5 and the specimen lengths were 20 in. for 7075-T6 and 30 in. for 2024-T3.

The effect of the variation of 2a/w on the fracture toughness parameters is seen in Fig.(6). For this test series the toughness values exhibited the tendency to stabilize at larger 2a/w ratios to a lower limiting value for both alloys. For 7075-T6 the minimum toughness values were about $\tilde{G}_c$ = 330 lb./in. and $G_R$ = 285 lb./in., which also agreed rather well with the minimum values observed in Fig.(4). For 2024-T3 the minimum toughness values were $\tilde{G}_c$ = 520 lb./in. and $G_R$ = 395 lb./in. which also agreed with the minimum values observed in Fig.(4). For all of the tests in the 7075-T6 series the length and width of the specimens were 20 and 10 in. respectively while, for the 2024-T3 series the length and width were 30 and 12 in.

The load-displacement records of the compact tension tests were used to compare toughness values obtained from the ASTM, nonlinear energy, COD and J-integral test methods. The comparative results differ markedly depending on whether the thickness requirement for plane strain fracture toughness testing was satisfied. The results obtained from data provided by Westinghouse Research Laboratories on an A533B steel are shown in Table I. For these tests the ASTM thickness requirements were not satisfied and large amounts of plastic deformation were recorded prior

Table I. Plane strain toughness comparisons for A533B steel.

| Test | Source | $G_{Ic}$ lb/in | $\tilde{G}_{Ic}$ lb/in | $G_{COD}$ lb/in | $J_{Ic}$ lb/in |
|---|---|---|---|---|---|
| 18-8-2 | Westinghouse | 271 | 834 | 781 | 925 |
| 18-8-3 | Research | 264 | 826 | 743 | 898 |
| 18-8-4 | Laboratories | 242 | 699 | 629 | 766 |

to the onset of unstable fracture. It is noted that the $J_{Ic}$ values were the largest of any of the results and were about 10 percent below the values obtained by the original graphical procedure suggested by Begley and Landes [15]. The values of $\tilde{G}_{Ic}$ and $G_{COD}$ were approximately 90 and 85 percent of the $J_{Ic}$ results for these tests. However, all three nonlinear methods provided results much higher than the ASTM standard method which was apparently due to the lack of satisfaction of the ASTM requirements.

Plane strain toughness comparisons of several aluminum alloys are seen in Fig.(7) in which a very different behavior is evident.

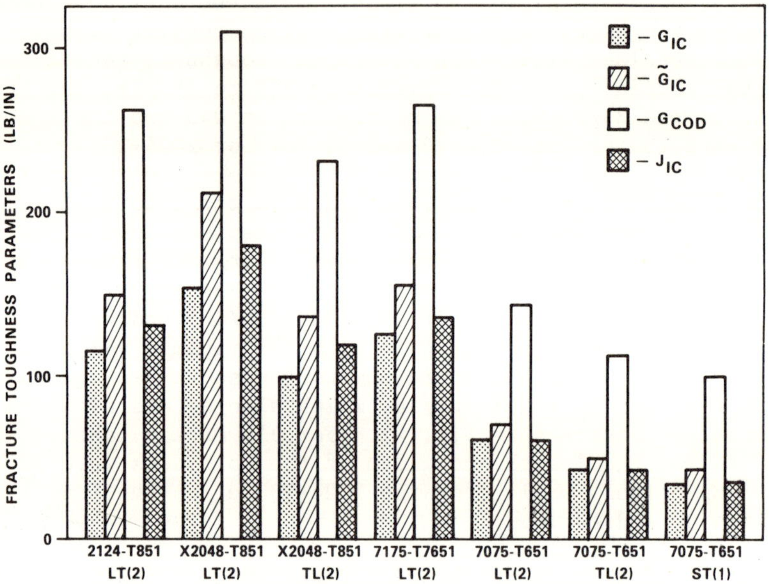


FIG. 7. COMPARISON OF TOUGHNESS FOR SEVERAL ALUMINUM ALLOYS BY THE ASTM, NONLINEAR ENERGY, COD, AND J INTEGRAL METHODS. LOAD-DISPLACEMENT RECORDS PROVIDED BY REYNOLDS METALS COMPANY.

The toughness values for these alloys were determined from data provided by Reynolds Metals Co. from compact tension specimens which satisfied the ASTM E399-72 specifications. There were two variations from the standard method in that the maximum load was used and only one or two duplicate tests, as indicated in Fig.(7), were conducted rather than the required three. It is seen from these results that the toughness values determined by the COD method were much higher than for the other three methods. It is surmised that subcritical crack growth was a major contributing factor to the high $G_{COD}$ values but it was not possible to determine either the magnitude of the subcritical crack growth or its relative effect on $G_{COD}$. The consistency of this method is also very dependent upon the value selected for the rotation factor whereas the other three methods are considerably less sensitive to changes in r. It is further noted that the other three methods give results in rather close agreement with each other. The nonlinear energy method always yielded toughness values higher than the other two methods for tests on the aluminum alloys. For the tougher alloys $J_{Ic}$ was intermediate between $G_{Ic}$ and $\tilde{G}_{Ic}$ but for

7075-T651 the $J_{Ic}$ values were the same as or slightly lower than $G_{Ic}$. The improvement in fracture toughness of the newer alloys x2048, 2124 and 7175 over 7075 is quite impressive.

The results of the application of three different cyclic loading programs to thin center-cracked sheets of 7075-T6 and 2024-T3 aluminum alloys are summarized in Table II. The first cycle in each of the programs consisted of loading to approximately 50 percent of the estimated fracture strength and unloading. The load applied at each subsequent cycle was then increased a constant increment above the preceeding cycle. The three

Table II. Nonlinear energy toughness of 7075-T6 and 2024-T3 aluminum alloys under static and cyclic loading

| | Static | 5% inc. | 2% inc. | 1% inc. |
|---|---|---|---|---|
| 2024-T3 | | | | |
| Nonlinear energy toughness (lb/in) | 562* | 593 | 584 | 547 |
| Cycles to failure | 1 | 8 | 16 | 28 |
| 7075-T6 | | | | |
| Nonlinear energy toughness (lb/in) | 342* | 334 | 326* | 317 |
| Cycles to failure | 1 | 10 | 22 | 36 |

* Average of two tests.

loading programs utilized for these tests consisted of the first cycle as discussed above and then load increments of 1, 2, and 5 percent of the estimated fracture strength. For each of the tests the envelope of the cyclic load-displacement curve was drawn and the nonlinear energy toughness established in the usual manner. It is seen that for both alloys there is a decrease in the toughness as the loading increment is decreased and the number of cycles increased. It is reasonable to expect $\tilde{G}_c$ to decrease with an increasing number of cycles since damage accumulation is generally considered to be more localized for cyclic loadings. It thus appears possible to establish a $\tilde{G}_c$-N curve analogous to the S-N curve for unnotched specimens. The $\tilde{G}_c$-N curve could then be used to determine a reduction factor to be applied to the static $\tilde{G}_c$ for estimating the critical value for a given cyclic life. However, it should be emphasized that considerably more testing is necessary so that the experimental scatter can be reduced and the nature of the $\tilde{G}_c$ versus N behavior better understood. Further research is currently being focused on this problem. It is also noted that the static value of 2024-T3 is somewhat lower than two of the cyclic tests. This is also felt to be caused by scatter in the experimental results and should be reduced with further testing. Other loading programs and procedures should also be employed so that their effect on the toughness can be properly ascertained.

CONCLUSIONS

1. The nonlinear energy method has been developed to allow the determination of the nonlinear energy toughness for both plane stress and plane strain applications and for cyclic as well as static loading.

2. Comparisons have been made between the nonlinear energy toughness parameter $\tilde{G}_c$ and the toughness established from the R curve method for thin, center-cracked sheets of 7075-T6 and 2024-T3 aluminum. The effect of specimen geometry on the toughness values determined from these two methods has been examined and comparisons were made between them. The toughness values of both methods showed similar variation with changes in specimen geometry; however, $\tilde{G}_c$ was consistently 5-20 percent higher than $G_R$.

3. For plane strain testing of compact tension specimens, comparisons were made between the nonlinear energy method and the J-integral, ASTM and COD methods for one steel and several aluminum alloys. Tests on the aluminum alloys, for which the ASTM E399-72 specifications were satisfied, resulted in very high toughness values for the COD method with the other three methods showing considerably better agreement. The nonlinear energy method maintained the most consistent correlation with $G_{Ic}$ from the ASTM method and thus appears to represent the best measure of nonlinear effects up to the point of maximum load. For the A533B steel the behavior was quite different which would be expected since the E399-72 thickness requirement was not nearly satisfied.

4. The nonlinear energy method has been applied to thin sheet testing for which certain cyclic loading programs were employed. The envelope of the cyclic load-displacement curves was utilized for the determination of a nonlinear energy toughness parameter. It was found that the values of $\tilde{G}_c$ decreased with an increasing number of cycles in a manner similar to the classical S-N diagrams. This suggests that a procedure of this type might be employed to determine toughness values for a specified cyclic life.

5. The nonlinear energy method has been applied to a variety of materials, specimen geometries and loading procedures and has been found to be well-behaved in every case. This

method appears promising for quantifying nonlinear effects for a wide variety of potential applications.

ACKNOWLEDGEMENTS

The authors would like to express appreciation to J. D. Landes of Westinghouse Research Laboratories and R. E. Zinkham of Reynolds Metals Company who provided test data for some of the results presented in this paper. The authors also wish to acknowledge support for this work by the NASA Langley Research Center through NASA Grant NGR 09-010-053.

LIST OF REFERENCES

[1] Fracture Toughness Evaluation by R-Curve Methods, ASTM STP 527 Am. Soc. Test. Mat'ls, Philadelphia, 1973.

[2] Rice, J. R., Mathematical analysis in the mechanics of fracture, Fracture, ed. by H. Liebowitz, Vol. 2, Academic Press, New York, 1968.

[3] Cherepanov, G. P., Crack propagation for continuous media, J. Appl. Math Mech (PMM), Vol. 31, No. 3, 1967.

[4] Fracture Toughness, Proceedings of the 1971 National Symposium on Fracture Mechanics, Part II, ASTM STP 514, Am. Soc. Test. Mat'ls., Philadelphia (1972).

[5] Wells, A. A., Unstable crack propagation in metals, Proc. Conf. Crack Propagation, Cranfield, U.K., 1962.

[6] Liebowitz, H. and J. Eftis, On nonlinear effects in fracture mechanics, Engrg. Fract. Mech., 3, 267, 1971.

[7] Liebowitz, H. and J. Eftis, Correcting for nonlinear effects in fracture toughness testing. Nucl. Engrg. Design, 18, 457, 1972.

[8] Liebowitz, H. and D. L. Jones, On test methods for nonlinear fracture mechanics, Tenth Anniv. Mtg. Soc. Engrg. Sci., Raleigh, N.C., 1973.

[9] Eftis, J. and H. Liebowitz, On fracture toughness evaluation for semibrittle fracture, Second Int'l Conf. on Struct. Mech. in Reactor Tech., Berlin, 1973.

[10] Jones, D. L., H. Liebowitz and J. Eftis, Fracture toughness characterization of several aluminum alloys in semibrittle fracture, to appear in Engrg. Fract. Mech.

[11] Brown, Jr., W. F. and J. E. Srawley, Plane Strain Crack Toughness Testing of High Strength Metallic Materials, ASTM STP 410, Am. Soc. Test. Mat'ls, Philadelphia, 1966.

[12] Egan, G. R., Compatibility of linear elastic ($K_{Ic}$) and general yielding (COD) fracture mechanics, Engrg. Fract. Mech., 5, 167, 1973.

[13] Sumpter, J. D. G. and C. E. Turner, Fracture analysis in areas of high normal strain, Second Int. Conf. on Pres. Vessel Tech., San Antonio, Texas, 1973.

[14] Dover, W. D., Fatigue crack growth under COD cycling, Engrg. Fract. Mech., 5, 11, 1973.

[15] Begley, J. A. and J. D. Landes, The J-integral as a fracture criterion, Fracture Toughness, Proceedings of the 1971 National Symposium on Fracture Mechanics, Part II, ASTM STP 514, Am. Soc. Test. Mat'ls, Philadelphia, 1972.

[16] Rice, J. R., P. C. Paris and J. G. Merkle, Some further results on J-integral and estimates, Progress in Flaw Growth and Fracture Toughness Testing, ASTM STP 536, Am. Soc. Test. Mat'ls, Philadelphia, 1973.

[17] Landes, J. D. and J. A. Begley, Tests results from J-integral studies - an attempt to establish a $J_{Ic}$ testing procedure, Scientific Paper 73-1E7-FMPWR-P3, Westinghouse Research Laboratories, 1973.

## DISCUSSION by J. N. Robinson, Imperial College, UK

Professor Liebowitz points out in this paper that calculation of crack tip opening displacement values from clip gauge displacement is sensitive to the position of the centre of rotation. It is not advisable to calculate COD from clip gauge displacement in specimens prior to general yield without having first experimentally determined the rotational constant applicable to that geometry and COD. Neglect of this precaution will result in inaccurate COD values and probably accounts for the poor values of G obtained using COD in this paper.

## AUTHORS' CLOSURE

The authors share the concern expressed by Dr. Robinson about the high values of toughness obtained by the COD method as illustrated in Fig. (7). Further, we recognize the problem of associating clip gage and COD displacements prior to general yield, especially in regard to the constancy of the rotational factor,

r. For example, the variation of r with clip gage displacement has been presented by Dover* for the three-point bend specimen geometry.

For the compact tension geometry, which was employed in our tests, the authors utilized the results of a finite-element analysis by Egan** which established the variation in the crack-tip profile with increasing clip gage displacement. Thus it was not necessary to use the rotational factor for these results. For comparison purposes the critical COD obtained from the finite-element calculations and the critical clip gage displacements were used to calculate applicable values for the rotational factor appropriate to these tests. The rotational factor was found to vary from 4.5 to 6.5 for this series of aluminum alloys.

The authors conclude that, for compact tension tests in which the ASTM E399-72 specifications are satisfied, the COD method is inappropriate because of the considerable variation in values of r and also because subcritical crack growth often occurs prior to significant crack-tip plasticity. For such tests the problems associated with determining a critical value of crack-tip opening displacement preclude the use of the COD method in its present form as a suitable failure criterion.

---

* Refer to reference [14] in the paper.

** Refer to reference [12] in the paper.

# THE IMPORTANCE OF STABLE CRACK EXTENSION IN LINEAR AND NON-LINEAR FRACTURE MECHANICS

K. B. Broberg

Division of Solid Mechanics, Lund Institute of Technology, S-220 07 Lund Sweden

## ABSTRACT

Before catastrophic fracture occurs certain changes in the distribution of the plastic strains near the crack-tip seem to take place. These changes make crack extension possible but contribute to a gradually increasing screening of the energy flow through the plastic region towards the crack-tip. Stable crack extension results. This phenomenon appears to be a common feature to almost all fractures. Therefore a unified theory may be developed to describe fracture at both small scale and large scale yielding as well as stable crack extension which is accompanied by plastic collapse rather than superseded by catastrophic fracture. An attempt is made to show how such a unified theory would work. Examples are given treating cases of plane stress, where substantial stable crack extension occurs, and also cases of plane strain, where the stable crack extension itself is small, but is initiated at a rather early state of the loading. Within the linear fracture mechanics certain properties of the stable crack extension are important. Thus the linear fracture toughness depends on plastic flow properties as well as on microstructural properties of the region of material instability close to the crack-tip and on the elastic constants.

## NOTATIONS

| | | |
|---|---|---|
| $a$ | [m] | Crack length parameter, usually the crack-length or half the crack-length |
| $a_o$ | [m] | Crack length parameter for the virgin crack |
| $a_c$ | [m] | Crack length parameter at the onset of unstable fracture |

| | | |
|---|---|---|
| $E$ | $[N/m^2]$ | Modulus of elasticity |
| $J$ | $[N/m]$ | The Rice integral |
| $J_c$ | $[N/m]$ | Critical value of J, determining incipient crack growth |
| $J_1$ | $[N/m]$ | Equal to J or $J_c$, according to the text |
| $K$ | $[N/m^{3/2}]$ | Stress intensity factor |
| $K_{Ic}$ | $[N/m^{3/2}]$ | Linear fracture toughness in plane strain |
| $K_c$ | $[N/m^{3/2}]$ | Linear fracture toughness in plane stress |
| $Q$ | $[N]$ | Load parameter |
| $Q_o$ | $[N]$ | Load parameter at incipient crack growth |
| $Q_c$ | $[N]$ | Load parameter at the onset of unstable fracture |
| SCE | | Abbreviation of Stable Crack Extension |
| $\alpha$ | | Fraction of the energy flow to the plastic region that reaches the end-region at steady-state motion of the crack-tip |
| $\beta$ | | Numerical constant |
| $\delta$ | $[m]$ | Displacement under the load |
| $\varepsilon_{ij}$ | | Strain tensor |
| $\sigma_{ij}$ | $[N/m^2]$ | Stress tensor |
| $\nu$ | | Poisson´s ratio |

## INTRODUCTION

Stable crack extension, abbreviated SCE in the following text, seems to be a very common phenomenon in fracture processes. In fact, catastrophic fractures which are not preceded by SCE seem to belong to rare exceptions. Some investigations, for instance [1] and [2], indicate that the amount of SCE is comparable to the extension of the plastic region at the start of SCE. Therefore calculations of the stress-strain field near the crack-tip, performed under the assumption that the crack-tip position is the virgin one, do not give the adequate information of the state just before catastrophic fracture. In fact, the stress-strain field near the crack-tip seems to be drastically changed when the crack advances. The situation just before catastrophic fracture should therefore be much more similar to the situation for a steadily moving crack-tip, which is the other extreme situation. The problem of calculating the stress-strain field near a steadily moving crack-tip is unfortunately very difficult.

The energy flow to the end-region [3],[4] per unit of crack

extension seems to be much smaller than the energy release-rate calculated in linear fracture mechanics. The reason is the screening action of the plastic zone [1],[5]. The energy flow to the end-region is, however, not zero as some calculations indicate. In plane stress and the extensional mode or in the anti-plane case, a non-zero energy flow to the end-region can be calculated even under the assumption of a point-size end-region. In plane strain the extension of the end-region should be essential in the calculations, as the energy flow to the end-region should be zero if a point-size end-region is assumed.

The screening action of the plastic zone increases gradually at extension of the crack from the virgin position. This explains the mechanism of SCE.

By making certain assumptions of the energy flow to the end-region an approximate quantitative description of the whole fracture process, from incipient crack extension to catastrophic fracture, can be made [5]. Such a description is quite general and can be applied even to cases where the SCE is superseded by plastic collapse rather than by catastrophic fracture.

## THE FOUR STAGES IN THE FRACTURE PROCESS

The following four stages can be recognized in the fracture process:

1. When loading the body from the virgin state, the crack opens, but the crack-tips do not move in the first stage. Certain theoretical [6],[7] and experimental [8],[9] investigations indicate that the originally sharp crack-tip becomes blunted due to plastic flow.
2. At a certain load the crack starts moving. This is a critical state, very easily recognized in theory, but, in fact, generally very difficult to observe in experiments.
3. During continued loading the crack moves in a stable manner, i.e. the movement is controlled by the load (or the displacement under the load). Certain experimental investigations [8],[9] indicate that the crack-tip is rather sharp during SCE, i.e. SCE is not a result of progressive blunting.
4. At a certain load (or displacement under the load) catastrophic fracture starts. This is a critical state, very easily recognized in experiments, but in several cases difficult to derive theoretically. Among other things the onset of catastrophic fracture depends on the stiffness of the body and the loading system.

Stage 3 (SCE) can proceed either at monotonously increasing load or at a load which passes a maximum.

In some cases the catastrophic fracture in stage 4 is exchanged by a plastic collapse. The difference between these two phenomena is the following one. Catastrophic fracture occurs at fixed grip conditions. The energy necessary to extend the crack is supplied from strain energy in parts of the body which are unloaded. In a plastic collapse crack extension is likely, but the energy needed to extend the crack and to deform the body further is supplied by the outer load.

## THE THREE REGIONS OUTSIDE THE CRACK-TIP

In almost any crack problem three regions are distinguished:

1. The elastic region.
2. The plastic region where $d\sigma_{ij}d\varepsilon_{ij} \geq 0$.
3. The end-region, where the stability criterion $d\sigma_{ij}d\varepsilon_{ij} \geq 0$ is not applicable.

In terms of engineering stresses and strains this region is characterized by decreasing stresses at increasing strains.

Regions 1 and 2 can be regarded as homogeneous, stable continua. Region 3 cannot be regarded as a homogeneous, stable continuum.

Region 1 is non-dissipative. Regions 2 and 3 are dissipative.

Size, shape and stress and strain distributions in region 1 depend on body geometry, load distribution and crack size even at critical loads (e.g. when the crack starts moving).

Size, shape and state in region 3 do not depend on body geometry, load distribution and crack size at critical loads. Region 3 is autonomous [10],[4].

Size, shape and stress and strain distributions in region 2 can be more or less dependent on body geometry, load distribution and crack size at critical loads. At small scale yielding a fair autonomy holds, but not at large scale yielding.

The nature of the end-region is different in different solids. Generally speaking, its size is given by the distance between inhomogeneities which cause separations in the materials. In the purely brittle case these inhomogeneities are the atoms. In many real cases the significant inhomogeneities are inclusions in the material. The separation processes may be cleavage (in the purely brittle case), quasicleavage, opening of holes (dimple fracture) etc. These processes require energy, the surface energy in the purely brittle case and a much larger amount of energy in other cases.

Part of the elastic region can have been previously plasticized. This is an obvious consequence of SCE, but also during loading with the crack-tip at the virgin position some unloading from the plastic region near the end-region may take place.

THE ENERGY FLOW TO THE END-REGION

As was mentioned in the Introduction, the energy flow to the end-region during SCE is smaller than the energy flow to the plastic region even in cases of small-scale yielding. This fact is a natural consequence of the dissipation that takes place in the plastic region outside the end-region. In one theoretical case, however, no such partitioning of the energy dissipation between the end-region and the plastic region outside the end-region takes place. This case is the plane stress case at the extensional mode in a perfectly plastic solid, obeying the Tresca yield condition. In this case, there is no plastic region outside the end-region, but the whole plastic region is confined to an infinitesimally thin layer, the Dugdale zone [11],[12].

Supposing SCE to proceed several plastic zone lengths, but still be small compared to the length of the crack, is approximately equivalent to a steady-state movement of the crack-tip region.

This steady-state case has been studied by Chitaley & McClintock [13] for anti-plane strain in a perfectly plastic solid. Their calculations seem, however, to be in error, as pointed out in [5]. Recent attempts to analyse this problem have been done by Andersson [14]. He finds that about 25% of the energy flow to the plastic region reaches the end-region. Previously Andersson [1] showed that this percentage should increase with the degree of strain-hardening. He also considered the steady-state case of plane stress at the extensional mode in a solid obeying the v.Mises yield criterion. At low degrees of strain hardening the energy flow to the end-region was estimated to be less than 50% of the energy flow to the plastic region. Analysis of experimental results on SCE [5] indicates that in the case of plane strain the energy flow to the end-region can be of the order of 10% of the energy flow to the plastic region in a steady-state case.

At initial crack growth, i.e. when SCE starts, almost 100% of the energy flow to the plastic region reaches the end-region at small scale yielding. That so is the case is understood by the following discussion: Suppose the end-region to be implanted at the tip of a crack in a perfectly elastic solid. When the J-integral reaches a certain critical value, $J = J_c$, the crack starts moving. But in this case it is obvious that J equals the energy flow to the end-region per unit of crack extension. In the original (plastic) surrounding the end-region becomes unstable at the same critical value $J = J_c$ and therefore the energy flow to the end-region per unit of crack extension equals J at initial crack growth. Now, at small-scale yielding, the energy flow to the plastic region is approximately equal to J. Thus, at initial crack-growth almost 100% of the energy flow to the plastic region reaches the end-region at small scale yielding.

When SCE proceeds, the plastic zone gives rise to a gradually increased screening of the energy flow to the end-region. Conditions very similar to those in a steady-state case seem to prevail after a fairly small amount of SCE, viz. about one length of the plastic zone [1]. Thus, during SCE, the fraction of the energy flow to the plastic region (at small-scale yielding) that reaches the end-region decreases from about unity to some fraction, which approximately equals the fraction $\alpha$, valid in the steady-state case. According to the review given earlier one has, roughly, the following relations:

| | |
|---|---|
| Anti-plane strain: | $\alpha \approx 0.25$ |
| Plane stress, extensional mode and Tresca yield condition: | $\alpha \approx 1$ |
| Plane stress, extensional mode and v.Mises yield condition: | $\alpha < 0.5$ |
| Plane strain, extensional mode: | $\alpha \approx 0.1$ |

ENERGY CRITERION FOR SCE

It is supposed that the autonomy of the end-region prevails approximately during SCE. (See, however, discussion in [15]). Then the energy flow to the end-region per unit of crack extension ought to be constant during the whole period of SCE. In terms of the J-integral this means (for the case of small-scale yielding):

$$J \approx J_c \quad \text{at the start of SCE} \tag{1}$$

$$\alpha J \approx J_c \quad \text{at the end of SCE} \tag{2}$$

as J equals the energy flow to the plastic region during the whole period of SCE.

Now the stress-intensity factor

$$K = \sqrt{\frac{EJ}{1-\nu^2}} \quad \text{for plane strain} \tag{3}$$

$$K = \sqrt{EJ} \quad \text{for plane stress} \tag{4}$$

Thus

$$K_{Ic} = \sqrt{\frac{EJ_c}{(1-\nu^2)\alpha}} \qquad \text{for plane strain} \tag{5}$$

$$K_c = \sqrt{\frac{EJ_c}{\alpha}} \qquad \text{for plane stress} \tag{6}$$

These expressions show the structure of the linear fracture toughness: Each one of the three factors $E/(1-\nu^2)$ or $E$,$J_c$ and $\alpha$ refers to each one of the three regions outside the crack-tip, viz. the elastic region, the end-region and the plastic region outside the end-region. $J_c$ is the only critical quantity.

J should be calculated for a path completely outside the plastic region (and regions which have been in a plastic state). However, before the crack starts moving, the path may be chosen rather close to the end-region without changing the J-integral very much, because then the case is approximately one given by proportional loading. Likewise, during SCE, the J-integral calculated for a certain position of the crack-tip ought to be approximately the same as the J-integral calculated for the hypothetical case of a non-moving crack with the same position of the crack-tip and at the same load. Thus we may write

$$J(Q_o,a_o) = J_c \qquad \text{at the start of SCE} \tag{7}$$

$$\alpha J(Q_c,a_c) \approx J_c \qquad \text{at the end of SCE} \tag{8}$$

where $Q_o$,$Q_c$,$a_o$ and $a_c$ is the load at the start of SCE, the load at the end of SCE, the original crack length parameter and the crack length parameter at the end of SCE. $J(Q,a)$ is calculated as for a non-moving crack with length parameter a at the load Q.

Following [5], intermediate stages are described by writing

$$J\left[\alpha+(1-\alpha)e^{-\beta\sigma_Y^2(a-a_o)/EJ_1}\right] = J_c \tag{9}$$

where $J_1 = J$ for cases of plane stress and $J_1 = J_c$ for cases of plane strain. $\beta$ is a numerical factor and $\sigma_Y$ the yield stress. It is further supposed, as in [5], that Eq.(9) can be used even in other cases than small-scale yielding. Then, obviously, the complete problem of SCE can be treated if the relations

$$J = J(Q,a) \quad \text{or} \quad J = J(\delta,a) \tag{10}$$

$$\text{and } \delta = \delta(Q,a) \quad \text{or} \quad Q = Q(\delta,a) \tag{11}$$

are known. There $\delta$ is the displacement under the load. Eqs.(9), (10) and (11) can be used to determine the relations between $\delta$

and Q and between $\delta$ and a. It may be observed that Eqs.(10) and (11) are interrelated (cf [5]).

Onset of catastrophic fracture occurs when

$$\frac{d\delta}{da} = 0 \tag{12}$$

at fixed grip conditions. Unstable fracture occurs at fixed load conditions when

$$\frac{dQ}{da} = 0 \tag{13}$$

The reason why $J_1$ was chosen equal to J for plane stress but equal to $J_c$ for plane strain is that the amount of SCE ought to be related to the size of the plastic zone (at small-scale yielding) in the case of plane stress, but to the size of the end-region in the case of plane strain [5].

The particular choice of the dependence on $(a-a_o)$ in Eq.(9) is rather arbitrary. Existing experimental data are too sparse to motivate a more thourough choice.

The extension of the validity of Eq.(9) to other cases than small-scale yielding is perhaps not obvious. However, it must be borne in mind that, for a given material, SCE occurs at small-scale yielding if the original crack is very long and at large-scale yielding if the original crack is very short. The state near the crack-tip should be approximately the same at corresponding phases in both cases. This state may be fairly accurately described by the J-integral, calculated as previously described.

In Figs. 1-7 some examples of calculations of SCE (from [5]), using Eqs.(9), (19) and (11), are shown.

## THE MECHANISM OF SCE

The mechanism of SCE becomes obvious when studying Eqs.(9), (10) and (11), which, however, may be regarded more as illustrations or approximate descriptions than as accurate relations. In the case of small-scale yielding

$$\frac{EJ_1}{\beta\sigma_Y^2} << a_o$$

Thus J reaches the neighbourhood of $J_c/\alpha$ already for

$$a - a_o << a_o$$

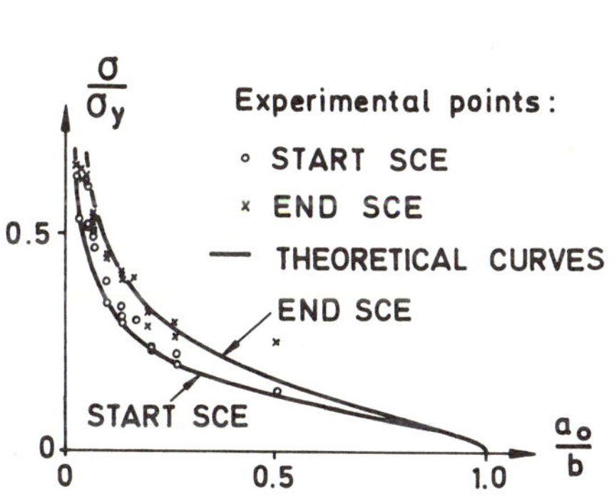


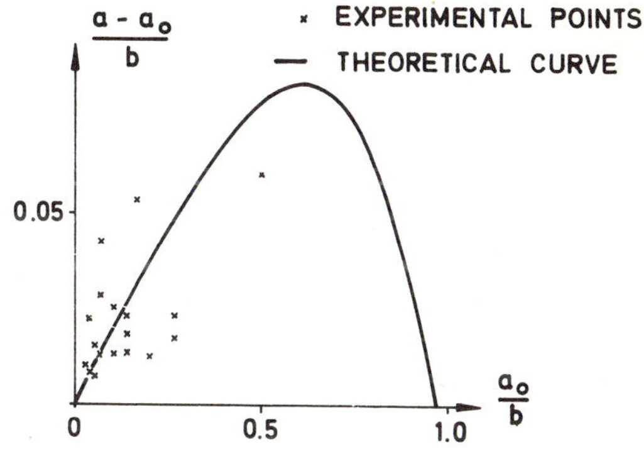


**Fig. 1** Sheet width 600 mm

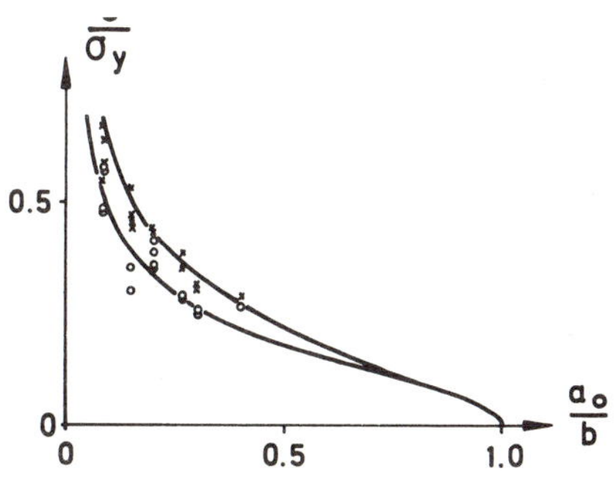


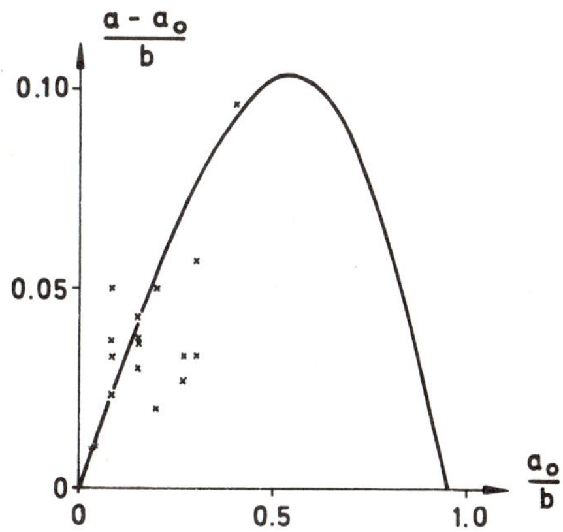


**Fig. 2** Sheet width 300 mm

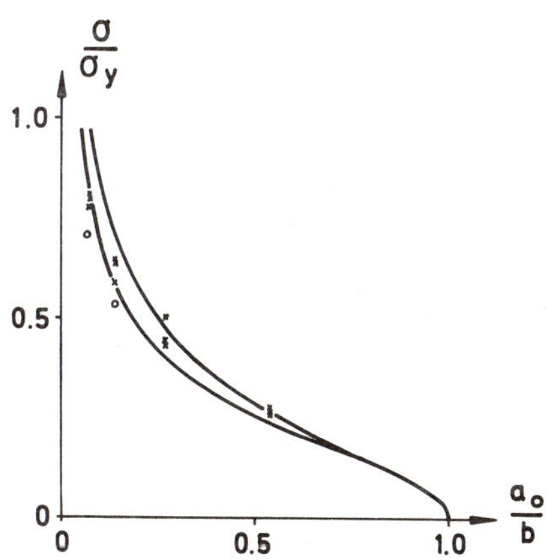


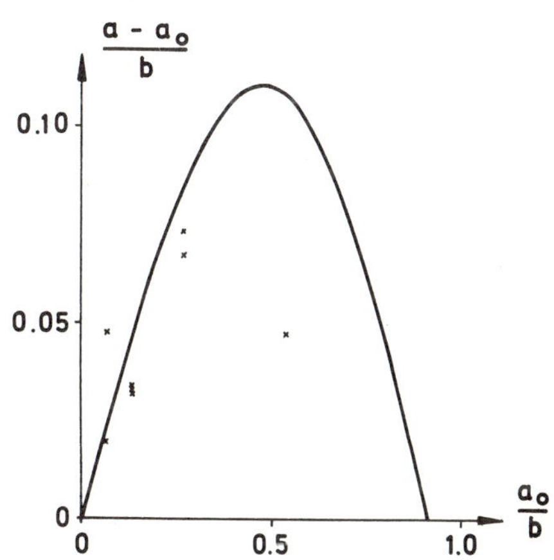


**Fig. 3** Sheet width 150 mm

Fig. 1-3 SCE for aluminium 7075-T6 sheet with widths 2b = 600 mm, 300 mm and 150 mm, containing a central crack with original length $2\,a_o$. Remote stress = σ. Experimental data are given by Broek [2]. Theoretical curves are obtained by choosing
$\alpha = 0.4$, $\beta = 1.5$ and $EJ_c = 58 \times 10^{13}\ N^2/m^3$.

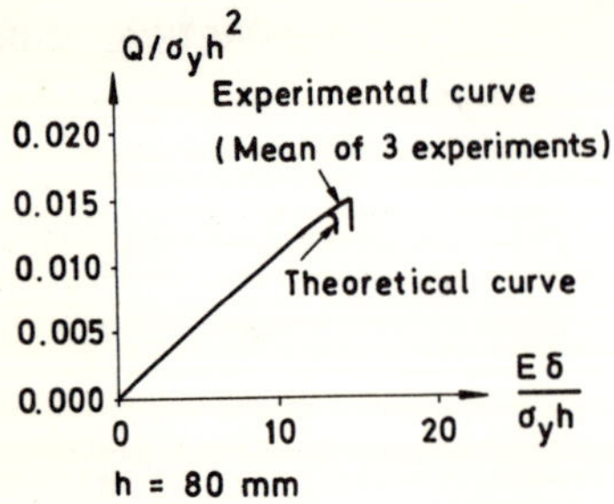


Fig. 4

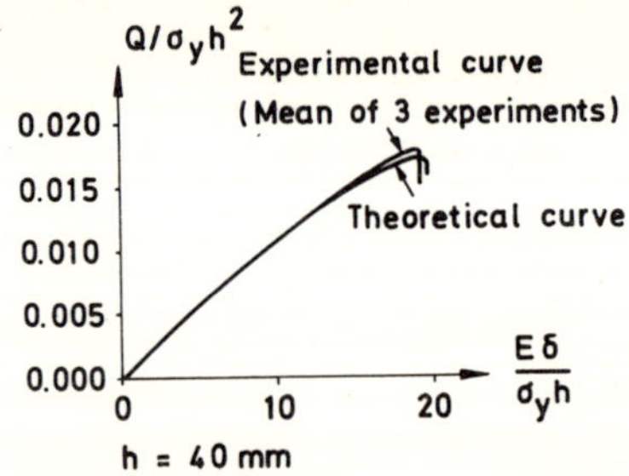


Fig. 5

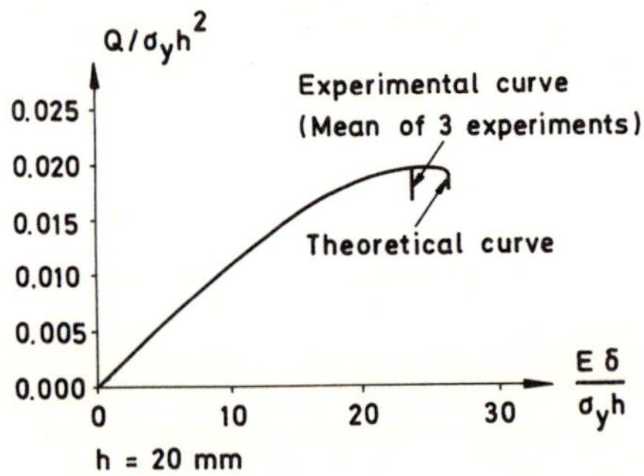


Fig. 6

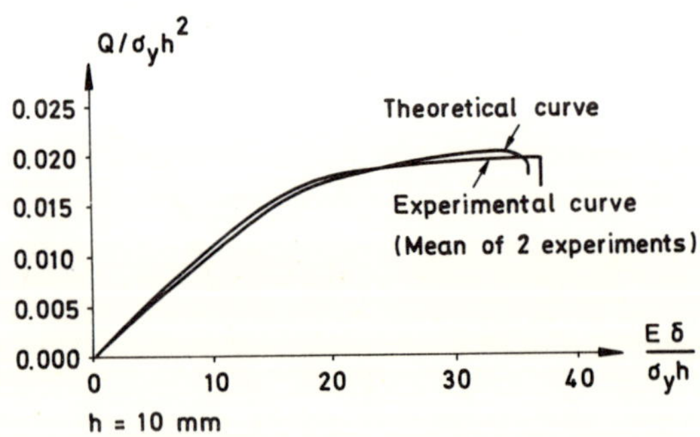


Fig. 7

Fig.4-7. Load displacement curves for three-point bend test bars. Experimental data are obtained for steel DOMEX 450 at -100°C. Displacement $\delta$ is measured under the force. Test-piece heigth is h. Theoretical curves are obtained by choosing $\alpha = 0.1$, $\beta = 25$ and $J_c = 5860$ N/m.

i.e. for a small amount of SCE. Now, in this case

$$J(Q,a) \approx J(Q,a_o) \propto Q^2$$

Thus the load Q is increased by the factor $1/\sqrt{\alpha}$ from the start of SCE to the end of SCE. In other words, in order to meet the gradually increasing screening action of the plastic region outside the end-region, the load must be increased if the crack should be extended. As $\sqrt{\alpha}$ may be of the order of 0.5 initial crack growth may start already at about half the load which is recorded as the critical load in tests within the linear fracture mechanics.

On the other hand, in the case of very large scale yielding

$$\frac{EJ_1}{\beta\sigma_Y^2} >> a_o$$

Then J increases more slowly with $(a-a_o)$. Therefore $J(\delta,a)$ changes both on account of the increase of $\delta$ and on the increase of a. Unstable fracture may therefore result well before J reaches the neighbourhood of $J_c/\alpha$.

## THE ONSET OF CATASTROPHIC FRACTURE

From experiments within the non-linear fracture mechanics it is known that catastrophic fracture occurs suddenly, i.e. without pre-warning in the Q-δ-curve. This fact seems to be difficult to explain otherwise as a result of bifurcation [5]. At a certain moment energy seems to be directed towards the crack-tip by unloading of certain parts of the structure. Therefore it is possible that catastrophic fracture occurs somewhat earlier than expected when using Eqs.(9), (10), (11) and (12).

## THE STRAIN DISTRIBUTION NEAR THE CRACK-TIP

As was shown in [5] and further studied by Andersson [14], substantial changes in the strain distribution near the crack-tip occur after the onset of SCE. Therefore, strain (and stress) distribution at the critical state of unstable fracture will differ rather much from those calculated for a static case [6], [7]. The most obvious difference seems to be a strong tendency towards a concentration of large strains in a narrow zone along the prospective fracture path. This theoretical observation [5], [14] seems to be in agreement with certain experimental observations [8], [9]. In these experiments it is observed that, though the crack-tip becomes blunted before SCE, it is sharp during SCE.

These observations may, perhaps, be remembered when asking for the cause to striations on fatigue fracture surfaces. Fatigue fracture could, perhaps, be regarded as cycles of blunting (at the first stage of loading), SCE (at continued loading) and restoration (at unloading).

## CONCLUSIONS

A quantitative description of stable crack extension (SCE) can be done by using energy considerations. One difficulty is the determination of the energy flow to the end-region, because of the (generally) substantial screening of the energy flow which takes place in the plastic region. One further difficulty, especially at large degrees of non-linearity, is to decide when unstable fracture occurs, as this process possibly starts as a result of bifurcation.

The structure of the linear fracture toughness consists of three parts, each one referring to properties of each one of the three regions outside the crack-tip, viz. the elastic region, the plastic region outside the end-region and the end-region. The only truly critical quantity is the factor referring to the end-region. This factor reflects micro-structural properties. The factor referring to the elastic region is given directly by the elastic constants. The factor referring to the plastic region outside the end-region reflects the screening action of the plastic region.

## REFERENCES

[1] Andersson,H., Finite-element treatment of a uniformly growing elastic-plastic crack tip, Report 73-1, Div.of Solid Mechanics, Lund Institute of Technology,1973. (To appear in J.Mech.Phys.Solids).

[2] Broek,D., The effort of specimen width on the residual strength of light alloy sheet, Report NLR-TR M.2152, National Lucht-en Ruimtevaart Laboratorium, Amsterdam,1968.

[3] Broberg,K.B., Critical review of some theories in fracture mechanics, Int.J.Fracture Mech.,4,11,1968.

[4] Broberg,K.B., Crack-growth criteria and non-linear fracture mechanics, J.Mech.Phys.Solids,19,407,1971.

[5] Broberg,K.B., On stable crack extension, Report 73-4, Div. of Solid Mechanics, Lund Institute of Technology,1973.

[6] Rice,J.R. and Rosengren,G.F., Plane strain deformation near a crack tip in a power-law hardening material, J.Mech.Phys.Solids, 16,1,1968.

[7] Hutchinson,J.W., Singular behaviour at the end of a tensile crack in a hardening material, J.Mech.Phys.Solids,16,13, 1968

[8] Broek,D., Some considerations on slow crack growth, Int.J. Fracture Mech.,4,19,1968.

[9] Tetelman,A-S., Private communication,1973

[10] Barenblatt,G.J., J.Appl.Math.Mech.,23,622

[11] Dugdale,D.S., Yielding of steel sheets containing slits, J.Mech.Phys.Solids,8,100,1960.

[12] Drucker,D.C. and Rice,J.R., Plastic deformation in brittle and ductile fracture, Engng.Fracture Mech.1,577,1970.

[13] Chitaley,A.D. and McClintock,F.A., Elastic-plastic mechanics of steady crack growth under anti-plane shear, J.Mech.Phys.Solids,19,147,1971.

[14] Andersson,H., Private communication,1974.

[15] Andersson,H., A Finite-element representation of stable crack-growth, J.Mech.Phys.Solids,21,337,1973.

# PATH INDEPENDENT INTEGRALS IN FRACTURE MECHANICS AND THEIR RELATION TO VARIATIONAL PRINCIPLES

A.J. Carlsson

The Royal Institute of Technology, S-100 44
Stockholm 70, Sweden

## INTRODUCTION

In the study of crack growth problems in the J-integral developed by Eshelby [1] and Rice [2] has come to extensive use. For linearly elastic materials a fracture criterion based on the J-integral is equivalent to the Griffith [3] energy criterion and to the Irwin [4] crack extension force criterion for crack growth.

The J-integral has also been proposed to have a wider applicability than only to elastic materials.

In this paper the J-integral will be interpreted as a stability criterion for a body containing a crack. This interpretation is arrived at through derivation of the J-integral from the variational principle for the strain energy potential for an hyperelastic material. This approach leads to the introduction of an complementary integral $\bar{J}$ derivable from the complementary strain energy potential.

In an analogous manner the Knowles-Sternberg integrals have their complementary counterparts which will also be given

The J and $\bar{J}$-integrals may be combined to an I-integral which also may be derived from the combined strain energy and complementary strain energy potential.

At least formally the I-integral may be connected to a stress or a strain energy criterion for fracture.

## FORMAL DERIVATION OF CONSERVATION LAWS

Using the Cauchy equation of motion and the Green-Gauss theorem one easily derives the so called Signorini theorem [5]

$$\int_S f\sigma_{ij} n_j dS = \int_V \left( \sigma_{ij} \frac{\partial f}{\partial x_j} + f\rho(\ddot{u}_i - b_i) \right) dV \tag{1}$$

Here $f$ is an arbitrary function of $x_j$, continuous in $V$, $\sigma_{ij}$ is the Cauchy stress tensor, $n_j$ is the normal to $S$, $\ddot{u}_i$ is acceleration, $b_i$ body force and $\rho$ density. A Cartesian coordinate system $x_j$ is used.

From this theorem the Rice integral $J_k$ and the Knowles-Sternberg integrals |6| may be derived by assigning appropriate forms to $f$.

Eq. (1) may be generalized by introducing instead of $\sigma_{ij}$ a function $g(\sigma_{ij}, x_j)$ (no sum on $j$). It then reads

$$\int_S (f \cdot g) n_j dS = \int_V \left( g \frac{\partial f}{\partial x_j} + \rho f \frac{\partial g}{\partial \sigma_{ij}}\Big|_x \cdot (\ddot{u}_i - b_i) + \right.$$

$$\left. + f \cdot \frac{\partial g}{\partial x_j}\Big|_\sigma \right) dV \tag{2}$$

Here $f$ and $g$ may be vector or tensor functions $f_i$, $f_{ik}$, etc. Now a conservation law or in the two-dimensional case a path-independent integral is obtained whenever the functions f and g are given such a form that the volume integral in eq. (2) may be transformed to a surface integral via the Green-Gauss theorem. This requires in general specific constitutive assumptions to be made. From eq. (2) we derive the following for the static case already known conservation laws (body forces absent):

$$J_{\underset{\sim}{1}} = \int_S \left( Wn_k + \left| \frac{\rho c^2}{2} \frac{\partial u_i}{\partial x_m} \frac{\partial u_i}{\partial x_n} l_m l_n n_k \right|_{s.s.} - T_i \frac{\partial u_i}{\partial x_k} \right) l_k dS \tag{3}$$

$$J_{\underset{\sim}{x}} = \int_S \Big[ W x_k n_k - \frac{1}{2} T_i u_i - T_i x_k \frac{\partial u_i}{\partial x_k} + \Big| \rho c^2 \Big( x_k u_i \frac{\partial^2 u_i}{\partial x_m \partial x_n} n_k +$$

$$+ x_k \frac{\partial u_i}{\partial x_k} \frac{\partial u_i}{\partial x_m} n_n - \frac{1}{2} u_i \frac{\partial u_i}{\partial x_n} n_m +$$

$$+ \frac{1}{2} x_k \frac{\partial u_i}{\partial x_m} \frac{\partial u_i}{\partial x_n} n_k - \frac{\partial}{\partial x_n} \left( x_k u_i \frac{\partial u_i}{\partial x_k} \right) n_m \Big) l_m l_n \Big|_{s.s.} \Big] dS \quad (4)$$

$$J_{\underset{\sim}{x} \times \underset{\sim}{l}} = \int_S \varepsilon_{klm} \left( x_m W n_l - T_i x_m \frac{\partial u_i}{\partial x_l} + T_l u_m \right) l_k dS \qquad (5)$$

To arrive at these integrals we have put in eq. (2)

$$f_i = \frac{\partial u_i}{\partial x_k} l_k \,, \quad g_{ij} = \sigma_{ij} \quad \text{for} \quad J_{\underset{\sim}{l}} \,;$$

$$f_i = \frac{1}{2} u_i + x_k \frac{\partial u_i}{\partial x_k} \,, \quad g_{ij} = \sigma_{ij} \quad \text{for} \quad J_{\underset{\sim}{x}}$$

$$\text{and} \quad f_i = \varepsilon_{klm} \left( x_m \frac{\partial u_i}{\partial x_l} - \delta_{il} u_m \right) l_k \,,$$

$$g_{ij} = \sigma_{ij} \quad \text{for} \quad J_{\underset{\sim}{x} \times \underset{\sim}{l}}$$

Further we have assumed the material to be hyper-elastic, i.e. the stress tensor to be derivable from a strain energy potential $W$ :

$$\sigma_{ij} = \frac{\partial W}{\partial \varepsilon_{ij}}$$

In the case of $J_{\underset{\sim}{x}}$ additional restrictions have been put on the material, see [7].

For the volume integral of the inertia term to be transformable we have for the dynamic case in addition assumed steady state conditions, i.e. the time derivative is given through

$$\dot{h} = -\frac{\partial h}{\partial x_k}\frac{\partial x_k}{\partial t} = -\frac{\partial h}{\partial x_k} c\, l_k$$

This implies a steady field propagating in direction $l_k$ with velocity $c$. The corresponding term is put in parentheses denoted s.s. Another new class of conservation integrals complementary to those of eqs. (3)-(5) may be derived in a similar way from eq. (2). These are

$$\bar{J}_{\underset{\sim}{l}} = \int_S \Big[\bar{W} n_k + \Big|\Big(\rho c^2 u_i \frac{\partial^2 u_i}{\partial x_m \partial x_n} - \frac{\rho c^2}{2}\frac{\partial u_i}{\partial x_m}\frac{\partial u_i}{\partial x_n}\Big) l_m l_n n_k\Big|_{s.s.} - u_i \frac{\partial \sigma_{ij}}{\partial x_k} n_j\Big] l_k \, dS \qquad (6)$$

$$\bar{J}_{\underset{\sim}{x}} = \int_S \Big[\bar{W} x_k n_k - \tfrac{3}{2} T_i u_i - u_i x_k \frac{\partial \sigma_{ij}}{\partial x_k} n_j - \Big|\rho c^2 \Big(x_k u_i \frac{\partial^2 u_i}{\partial x_m \partial x_n} n_k - x_k \frac{\partial u_i}{\partial x_k}\frac{\partial u_i}{\partial x_m} n_n + \frac{1}{2} x_k \frac{\partial u_i}{\partial x_n}\frac{\partial u_i}{\partial x_m} n_k + \frac{1}{2} u_i \frac{\partial u_i}{\partial x_n} n_m\Big) l_m l_n\Big|_{s.s.}\Big] dS \qquad (7)$$

$$\bar{J}_{\underset{\sim}{x} * \underset{\sim}{l}} = \int_S \varepsilon_{klm}\Big(\bar{W} x_m n_l - u_i x_m \frac{\partial \sigma_{ij}}{\partial x_l} n_j\Big) l_k \, dS \qquad (8)$$

These integrals are obtained from eq. (2) by putting

$$f_i = u_i \quad , \quad g_{ij} = \frac{\partial \sigma_{ij}}{\partial x_k} l_k \quad \text{for} \quad \bar{J}_{\underset{\sim}{l}} \; ;$$

$$f_i = u_i \quad , \quad g_{ij} = x_k \frac{\partial \sigma_{ij}}{\partial x_k} \quad \text{for} \quad J_{\underset{\sim}{x}} \quad \text{and}$$

$$f_i = u_i \quad , \quad g_{ij} = \varepsilon_{klm} x_m \frac{\partial \sigma_{ij}}{\partial x_l} l_k \quad \text{for} \quad \bar{J}_{\underset{\sim}{x} \times \underset{\sim}{l}} \; ,$$

respectively.

Here $l_k$ is a constant vector, $\bar{W}$ is the complementary strain energy and thus we have for a hyperelastic material allowing a

Legendre transformation: $\varepsilon_{ij} = \dfrac{\partial \bar{W}}{\partial \sigma_{ij}}$

The dynamic terms, denoted s.s., relate to the steady state case mentioned in connection with $J_{\underset{\sim}{l}}$ and $J_{\underset{\sim}{x}}$ . This term is omitted for $J_{\underset{\sim}{x} \times \underset{\sim}{l}}$ and $\bar{J}_{\underset{\sim}{x} \times \underset{\sim}{l}}$ .

All these surface integrals are zero for fields continuous in $V$ if the surface $S$ satisfies certain regularity conditions. By combining $J_{\underset{\sim}{l}}$ and $\bar{J}_{\underset{\sim}{l}}$ , $J_{\underset{\sim}{x}}$ and $\bar{J}_{\underset{\sim}{x}}$ , etc. we obtain a new class of conservation laws which do not rest on any specific constitutive assumption. These new conservation laws are

$$\left\{ \begin{aligned} I_{\underset{\sim}{l}} &= \int_S \left( \sigma_{ij} \varepsilon_{ij} n_k + \rho u_i \ddot{u}_i n_k - \frac{\partial}{\partial x_k} (\sigma_{ij} u_i) n_j \right) l_k dS \\ I_{\underset{\sim}{x}} &= \int_S \left( \sigma_{ij} \varepsilon_{ij} n_k x_k + \rho u_i \ddot{u}_i x_k n_k - 2 \sigma_{ij} u_i n_j - \right. \\ &\qquad \left. - x_k \frac{\partial}{\partial x_k} (\sigma_{ij} u_i) n_j \right) dS \end{aligned} \right. \qquad (9)$$

$$I_{\underset{\sim}{x}\times\underset{\sim}{l}} = \int_S \varepsilon_{klm}\Big(x_m\sigma_{ij}\varepsilon_{ij}n_l + \rho u_i\ddot{u}_i x_m n_l + T_l u_m -$$

$$-x_m \frac{\partial}{\partial x_l}(\sigma_{ij}u_i)n_j\Big)l_k dS$$

Eq. (9)[1], e.g., is obtained by adding the two integral expressions obtained from eq. (2) with

$$f_i = \frac{\partial u_i}{\partial x_k} lk \quad , \quad g_{ij} = \sigma_{ij} \quad \text{and}$$

$$f_i = u_i \quad , \quad g_{ij} = \frac{\partial \sigma_{ij}}{\partial x_k} lk \quad \text{respectively.}$$

The transformation of the resulting volume integral does then not require any constitutive assumption.

Above we have assumed absence of discontinuity surfaces in the considered volume. If there are such surfaces an additional term would be obtained from the volume to surface integral transformation.

## ALTERNATIVE, FORMAL DERIVATION OF CONSERVATION LAWS

A more general approach may be used to derive the integrals of eq. (9). This approach also generalizes the form of these integrals and makes the derivation of any number of $I_l$-integrals a straight-forward procedure. To this end we note that for any vector field $\underset{\sim}{a}$

$$\text{div curl } \underset{\sim}{a} = 0$$

Thus for an arbitrary volume the following relation holds

$$\int_V \text{div curl } (\underset{\sim}{h}\times\underset{\sim}{p})dV = 0 \tag{10}$$

We may e.g. choose $\underset{\sim}{p} = \sigma_{ij}u_i$ and allow $\underset{\sim}{h}$ to be an arbitrary vector function.

We now apply the Green-Gauss theorem to eq. (10). For a volume V bounded by the surface S the theorem reads

$$\int_V \operatorname{div} \underset{\sim}{a}\, dV = \oint_S \underset{\sim}{a}\cdot\underset{\sim}{n}\, dS \qquad (11)$$

Thus eq. (10) may be transformed to a surface integral:

$$\int_V \operatorname{div} \operatorname{curl} (\underset{\sim}{h}\times\underset{\sim}{p})\, dV = \oint_S \underset{\sim}{n} \operatorname{curl} (\underset{\sim}{h}\times\underset{\sim}{p}) dS = I_{\underset{\sim}{h}} \qquad (12)$$

We may also write in a Cartesian coordinate system

$$I_{\underset{\sim}{h}} = \oint_S \left(h_k \frac{\partial p_l}{\partial x_l} - h_l \frac{\partial p_k}{\partial x_l} - p_k \frac{\partial h_l}{\partial x_l} + p_l \frac{\partial h_k}{\partial x_l}\right) n_k dS \qquad (13)$$

If here $h_k$ is put equal to $l_k$, $x_k$ and $-\varepsilon_{klm}x_m l_l$ and the equation of motion is used the integrals in eq. (9) are obtained. Equation (12) gives the general formula for the $I_{\underset{\sim}{h}}$ integral. It includes an infinite number of integrals.

In eq. (12) we may choose $\underset{\sim}{h} = \underset{\sim}{v}$, the actual displacement field in the body. Then the integral takes the form

$$I_{\underset{\sim}{v}} = \oint_S \underset{\sim}{n} \operatorname{curl}(\underset{\sim}{v}\times\underset{\sim}{p}) dS \qquad (14)$$

We may also carry through the above derivations with $\underset{\sim}{p}$ replaced by $dp = \sigma_{ij}v_i$ and obtain then e.g.:

$$I^o_{\underset{\sim}{h}} = \oint \underset{\sim}{n} \operatorname{curl}(h\times d\underset{\sim}{p})\, dS$$

The integrals $I$ in eqs. (12) and (14) may be evaluated for part of the surface. We use here eq. (12) for demonstration. The surface $S$ is divided into two parts $S_o$ and $S_f$ (Fig. 1).

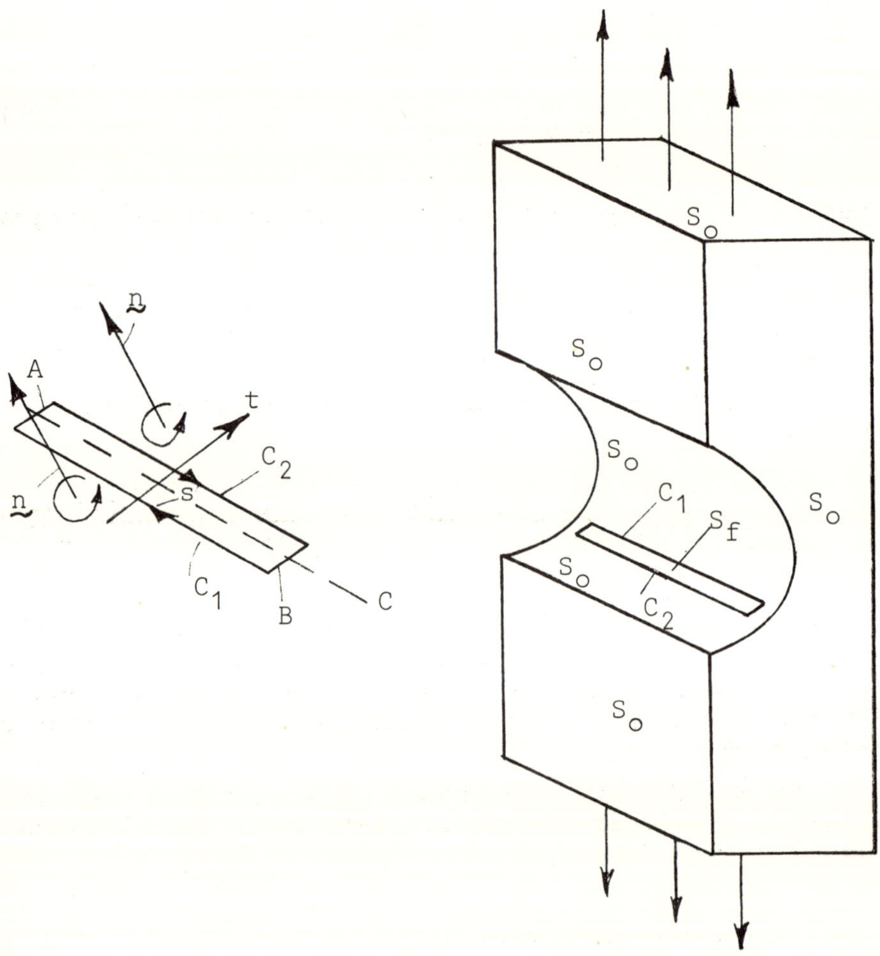


Fig. 1 Geometry for evaluation of $I_1^{S_f}$

Here subscript f in $S_f$ is chosen since in the succeeding analysis $S_f$ is that part of the considered surface where fracture or instability initiates.

Evidently

$$I_{\underset{\sim}{h}}^{S} = I_{\underset{\sim}{h}}^{S_o} + I_{\underset{\sim}{h}}^{S_f} \quad \text{with} \quad I_{\underset{\sim}{h}}^{S_o} = \int_{S_o} \underset{\sim}{n} \text{ curl } (\underset{\sim}{h} \times \underset{\sim}{p}) dS$$

Using Stokes theorem we find

$$I_{\underset{\sim}{h}}^{S_o} = \oint_C \underset{\sim}{\hat{s}} \cdot \underset{\sim}{h} \times \underset{\sim}{p} \, ds \qquad I_{\underset{\sim}{h}}^{S_o} = - I_{\underset{\sim}{h}}^{S_f} \tag{15}$$

Here C is the curve dividing S into $S_o$ and $S_f$. Now we choose C to consist of two lines $C_1$ and $C_2$ at constant distance from each other. The lines are connected with each other at their two ends to form a closed curve. We will mainly be interested in the case when $C_1$ and $C_2$ are infinitesimally close, so that the contributions from the cross connecting lines at the two ends are negligible. Eq. (15) then takes the form

$$I_{\underset{\sim}{h}}^{S_o} = \int_{C_1} \underset{\sim}{\hat{s}} \times \underset{\sim}{h} \cdot (\underset{\approx}{\sigma}^1 \cdot \underset{\sim}{u}^1 - \underset{\approx}{\sigma}^2 \cdot \underset{\sim}{u}^2) ds =$$

$$= \int_{C_1} \underset{\sim}{\hat{s}} \times \underset{\sim}{h} \cdot \Big( (\underset{\sim}{\hat{t}} \text{ grad}) (\underset{\approx}{\sigma} \cdot \underset{\sim}{u}) \Big) (\Delta t) ds \tag{16}$$

Since the lines are infinitesimally close the integration is made only along one line. Superscript index denotes to which line, $C_1$ or $C_2$, the respective value refers. $\underset{\sim}{\hat{t}}$ is a normal vector along the coordinate axis t falling in the same plane as $C_1$ and $C_2$ and perpendicular to $\underset{\sim}{\hat{s}}$. $\Delta t$ is the distance between the lines $C_1$ and $C_2$ along t . Parenthesis denotes that integration is not performed over t . In index notation the second integrand is

$$\varepsilon_{jkl} \hat{s}_k h_l \hat{t}_m (\sigma_{ij} u_i)_{,m} \, (\Delta t)$$

In the $I_{\underset{\sim}{l}}$ integral $\underset{\sim}{h}$ is equal to $\underset{\sim}{l}$ and a constant vector not related to the strain and stress field. The direction of $\underset{\sim}{l}$ may be chosen arbitrarily. To maximize

$$I_{\underset{\sim}{l}}^{S_o}$$

we choose $\underset{\sim}{l}$ perpendicular to the $C_1$-$C_2$-plane. If the equation is normalized, $\underset{\sim}{l}$ is a unit vector. Then $\underset{\sim}{s}\times\underset{\sim}{h} = \hat{t}$ and the integrand becomes

$\hat{t}_j\hat{t}_m(\sigma_{ij}u_i)_{,m}\ \Delta t$ .

Now consider the case when $C_1$ and $C_2$ fall on an unloaded part of the surface of a body and

$$I_{\underset{\sim}{l}}^{S_o}$$

is maximized by choosing $\underset{\sim}{l}$ perpendicular to the surface. If it further is required that surface tractions are balanced across $C_1$, in the $\underset{\sim}{t}$-direction, i.e.

$$T_i = t_j\sigma_{ij}^1 = t_j\sigma_{ij}^2$$

then eq. (16) takes the form

$$I_{\underset{\sim}{l}}^{S_o} = \int_{C_1} \underset{\sim}{T}(\hat{\underset{\sim}{t}}\ \mathrm{grad})\underset{\sim}{u}(\Delta t)ds = \int_{C_1} \hat{t}_j\sigma_{ij}\hat{t}_m u_{i,m}(\Delta t)ds \quad (17)$$

The integrand may be further reduced to $\sigma_{tt}\varepsilon_{tt}(\Delta t) = \sigma_{tt}\delta_t$ where $\delta_t$ is the relative displacement of the two lines $C_1$ and $C_2$ .

## RELATION OF CONSERVATION LAWS TO VARIATIONAL PRINCIPLES

The virtual work equation or the variational equation of motion reads

$$0 = \int_V \left(\delta\varepsilon_{ij}\sigma_{ij} + \rho\delta u_i(\ddot{u}_i - b_i\right)dV - \int_S \delta u_i\sigma_{ij}n_j dS \quad (18)$$

For the static case eq. (18) is also the variational equation for the strain energy potential of an hyperelastic material

$$\mathcal{V} = \int_V \left(W - b_i u_i\right) dV - \int_{S_T} T_i u_i dS \tag{19}$$

It is thus evident that the J-integrals, eqs.(3), (4) and (5), may be interpreted as variational equations with variational fields corresponding to the $f_i$ used in the respective derivation. Thus $\underset{\sim}{J_1}$ is the variational equation for $\mathcal{V}$ with the variational field

$$\delta u_i = - \frac{\partial u_i}{\partial x_k} l_k$$

This may be interpreted as a similarity variation of the field implying a test of equilibrium with respect to neighbouring configurations similar to the equilibrium state.

The variational procedure applied above requires that

$$\partial^2 W / \partial \varepsilon_{ij} \partial \varepsilon_{kl} \geq 0 \tag{20}$$

We define the fracture or instability zone in a body as the volume where this requirement is not fulfilled. Evidently equality sign in eq. (20) defines the boundary $S_f$ of the fracture zone.

We divide the surface $S$ in eq. (3) into two parts $S_f$ and $S_o$ where $S_f$ is the boundary between the fracture process zone and the sound material and $S_o$ is an outer boundary which may be chosen arbitrarily. We have (Fig. 2)

$$\underset{\sim}{J_1}^{S_o} = \int_{S_o} \left(W n_k - T_i \frac{\partial u_i}{\partial x_k}\right) l_k = \underset{\sim}{J_1}^{S_f} \tag{21}$$

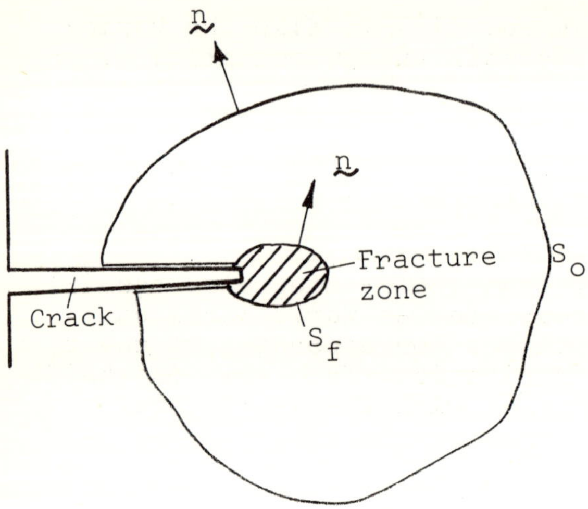


Fig. 2 . Division of $S$ into $S_f$ and $S_o$. The fracture zone is bounded by $S_f$.

On $S_f$ we have chosen $n_j$ pointing outwards from the fracture zone. Eq.(21) is an equilibrium condition as long as

$J_{\underset{\sim}{1}}^{S_f}$ increases with load and it is an instability condition for $J_{\underset{\sim}{1}}^{S_o} = J_c$ where $J_c$ is a critical value or a saturation value for

$J_{\underset{\sim}{1}}^{S_f}$

From eq. (19) we conclude that for this approach to be valid we require that at the boundary of the fracture zone stresses are prescribed and are independent of displacement or strain.

This is in agreement with the requirement that on the boundary $S_f$ we require that

$$\frac{\partial^2 W}{\partial \varepsilon_{ij} \partial \varepsilon_{kl}} = \frac{\partial \sigma_{ij}}{\partial \varepsilon_{kl}} = 0$$

This makes the perfectly soft hyperelastic material the prototype material for the $J_{\underset{\sim}{1}}$-integral.

For a body containing a plane crack with normal direction $\underset{\sim}{n} = (0,1,0)$ and with a straight crackfront along direction $(0,0,1)$ this variational field corresponds, for $\underset{\sim}{l} = (1,0,0)\cdot\delta a$, to a virtual extension of the crack. For the special case of a crack with a Dugdale-Barenblatt-model cohesive zone (Fig.3)the inner integral corresponds to the work done on the cohesive zone during the crack extension i.e.:

$$J_{\underset{\sim}{l}}^{S_f} = -\int_{S_f} n_j \sigma_{ij} \delta u_i \, dS = \int_0^{\delta a} \sigma_{22} \frac{\partial u_2}{\partial x_1} l_1 dx_1$$

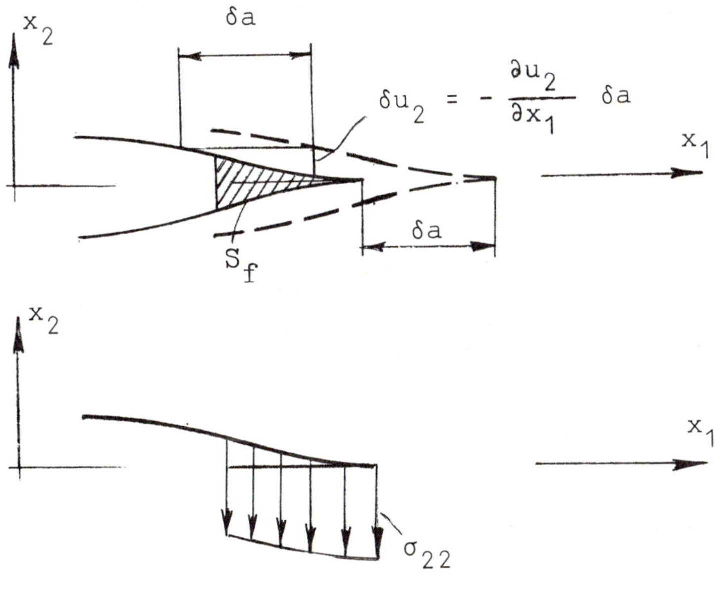


Fig. 3. Virtual displacement $\delta a$ of crack front.

If in eq. (2) we instead put $f_i = u_i$ and $g = \delta\sigma_{ij}$ we obtain the variational principle for the virtual complementary work. With also the inertia forces varied it takes the form

$$0 = \int_V \Big( \varepsilon_{ij}\delta\sigma_{ij} + \rho u_i \delta(\ddot{u}_i - b_i) \Big) dV - \int_S u_i \delta\sigma_{ij} n_j \, dS \qquad (22)$$

For a hyperelastic material with

$$\varepsilon_{ij} = \frac{\partial \overline{W}}{\partial \sigma_{ij}}$$

this is also the variational equation for the complementary potential

$$\overline{V} = \int_V \overline{W} dV - \int_{S_u} T_i u_i dS \tag{23}$$

We then conclude that the $\overline{J}$-integrals of eq.(6) to (8) may be interpreted as variational equations for the complementary potential, eq. (23), of an hyperelastic material. Thus e.g. the $\overline{J}_{\underset{\sim}{1}}$-integral relates to a variational field

$$\delta\sigma_{ij} = - \frac{\partial \sigma_{ij}}{\partial x_k} l_k$$

This variational approach leads to the requirement that $\delta\sigma_{ij} = 0$ on those parts of the boundary where $\sigma_{ij}$ is prescribed. This requirement applies also to the boundary between the fracture process zone and the sound material. As was the case for the J-integrals also here the fracture process zone has to be excluded from the volume considered. This is because the variation of eq. (23) requires that

$$\frac{\partial^2 \overline{W}}{\partial \sigma_{ij} \partial \sigma_{kl}} \geq 0$$

which by definition is not the case in the fracture zone. Making the same division of the bounding surface $S$ as for the $J_{\underset{\sim}{1}}$-integral we find

$$\overline{J}_{\underset{\sim}{1}}^{S_o} = \int_{S_o} \left( \overline{W} n_k - u_i \frac{\partial \sigma_{ij}}{\partial x_k} n_j \right) l_k dS = \overline{J}_{\underset{\sim}{1}}^{S_f} \tag{24}$$

The same type of restrictions, though complementary, apply to the $\overline{J}_1$ integral as to the $J_1$-integral. Thus we require that on the boundary of the fracture process zone the displacements are prescribed and are independent of the tractions. This makes the ideally locking hyperelastic material the prototype material for the $\tilde{J}_1$-integral. In order to give this integral a physical interpretation we would have to introduce another type of fracture zone than in the Dugdale-model.

We may combine the strain energy and the complementary strain energy potential to get the potential

$$\mathcal{V} + \bar{\mathcal{V}} = \int_V \sigma_{ij}\varepsilon_{ij} dV - \int_S T_i u_i dS \tag{25}$$

which is equal to zero.

The variation of this potential when both stress and displacement are varied leads to the conservation laws of eq. (9) if displacement variation is chosen equal to $f_i$ of the corresponding $\mathcal{V}$-variation and stress variation equal to $g_{ij}$ of the corresponding $\bar{\mathcal{V}}$-variation. This means for the $\underset{\sim}{I_1}$-integral, e.g.,

$$\delta u_i = -\frac{\partial u_i}{\partial x_k} l_k \quad \text{and} \quad \delta\sigma_{ij} = -\frac{\partial \sigma_{ij}}{\partial x_k} l_k$$

Here no restrictions of the form used for $J$ and $\bar{J}$ are put on the constitutive material properties.

Thus no strain or stress potential has to exist. However, the formulation is in total strain.

For a hyperelastic material we may exclude the fracture zone from the region considered in a crack problem. Then we find as above

$$\underset{\sim}{I_1}^{S_o} = I_1^{S_f} \tag{26}$$

It should be pointed out that in the above discussion of the relation of the conservation laws to variational principles it has been assumed that $S_f$ and $V_f$, i.e. the form and size of the fracture process zone, does not change due to the variational field. This is an approximation except for the instability moment.

If the variational equation of motion is used then also the inertia terms of eqs. (3) to (9) are obtained through the present approach. For the J- and $\bar{J}$-integrals the inertia terms are transformable to a surface integral only when steady state conditions are at hand as was the case in the derivation of eqs. (3) to (8). It is evident from the derivations above that the $I_h$-integral is a combination of the $J_h$- and $\bar{J}_h$-integrals. We have for hyperelastic materials

$$I_{\underset{\sim}{h}} = J_{\underset{\sim}{h}} + \bar{J}_{\underset{\sim}{h}} \qquad (27)$$

For an integration path surrounding a crack-tip in a linearly elastic material $J_{\underset{\sim}{1}} = 0$ and $J_{\underset{\sim}{1}} = -\bar{J}_{\underset{\sim}{1}}$

## APPLICATIONS TO LINEARLY ELASTIC MATERIALS

For materials which are mainly elastic two types of fracture criteria have come to extensive use. These are the stress criterion of Mohr and others for non-defect bodies and the energy criterion of Griffith for bodies with crack like defects. The Griffith criterion has been mathematically refined by Irwin in the $K_I$-criterion and by Rice in the J-criterion for crack extension.

There is no analytical connection between the Mohr-type stress criterion and the Griffith-type energy criterion.

We will now investigate the $I_{\underset{\sim}{1}}$- and $J_{\underset{\sim}{1}}$-integrals from these viewpoints. First we will study the case of a plane crack in a twodimensional body. The stress and displacement state at the crack tip is composed of three modes of loading:tension (I), in plane shear (II) and out of plane shear (III). The corresponding analytic expressions for the field may be found in most handbooks on elastic crack analysis, see e.g.[8].

As concluded earlier $I_{\underset{\sim}{1}}=0$ for an wholly linearly elastic body containing a crack and for a path enclosing the crack-tip.

For this case we thus have $J_{\underset{\sim}{1}} = -\bar{J}_{\underset{\sim}{1}}$ with:

$$J_{\underset{\sim}{1}} = \frac{(\varkappa+1)}{8\mu}\left((K_I^2 + K_{II}^2 + \frac{4}{\varkappa+1} - K_{III}^2)\cos\alpha - \right.$$

$$\left. - 2K_I K_{II} \sin\alpha\right) \tag{28}$$

Here $\underset{\sim}{1} = (\cos\alpha, \sin\alpha)$ is the direction of virtual displacement of the field around the crack tip. We may by hypotheses assume $J(\alpha)_{max}$ to be the direction of preferred crack growth, as is further elaborated in [9]. Comparison with experimental results from [10] and [11] shows that this is a good hypotheses for crack extension in directions close to the direction of the original crack, Fig. 4. This is an expected result.

Stress intensity factors do not apply for blunted notches. For such situations we use eq. (17) to predict crack growth and crack growth direction. We then find that the $I_{\underset{\sim}{1}}$ integral has its maximum on the line where $\sigma_{tt}\varepsilon_{tt}$ is maximal. (For definition of direction $t$ see Fig. 1.)For a linearly elastic material with $\varepsilon_{tt}$ proportional to $\sigma_{tt}$ this corresponds to a stress criterion for fracture.

The "strain energy density" criterion for fracture given by Sih may also be deduced from $I_{\underset{\sim}{1}}$. To this end we cut away material around the crack tip corresponding to a cylindrical volume of radius $r_o$. We assume that the state of stress and deformation is not changed outside $r = r_o$ due to this procedure. Now the same procedure is used as when dealing with blunted notches above, i.e. eq. (17) is applied to a line $(r_o, \Theta)$ on the surface of the cylindrical hole. Then the following expression for $I_{\underset{\sim}{1}}^{S_o}$ is obtained

$$I_{\underset{\sim}{1}}^{S_o} = \frac{1}{16\pi\mu}\left(f_{11}(\Theta)K_I^2 + 2f_{12}(\Theta)K_I K_{II} + f_{22}(\Theta)K_{II}^2 + \right.$$

$$\left. + 4f_{33}(\Theta)K_{III}^2\right)r_o\,d\Theta \tag{29}$$

where

$$\begin{cases} f_{11}(\Theta) = (1+\cos\Theta)(\varkappa-\cos\Theta) \\ f_{12}(\Theta) = \sin\Theta(2\cos\Theta - \varkappa+1) \\ f_{22}(\Theta) = \left[(\varkappa+1)(1-\cos\Theta)+(1+\cos\Theta)(3\cos\Theta-1)\right] \\ f_{33}(\Theta) = 1+\cos\Theta \end{cases} \tag{30}$$

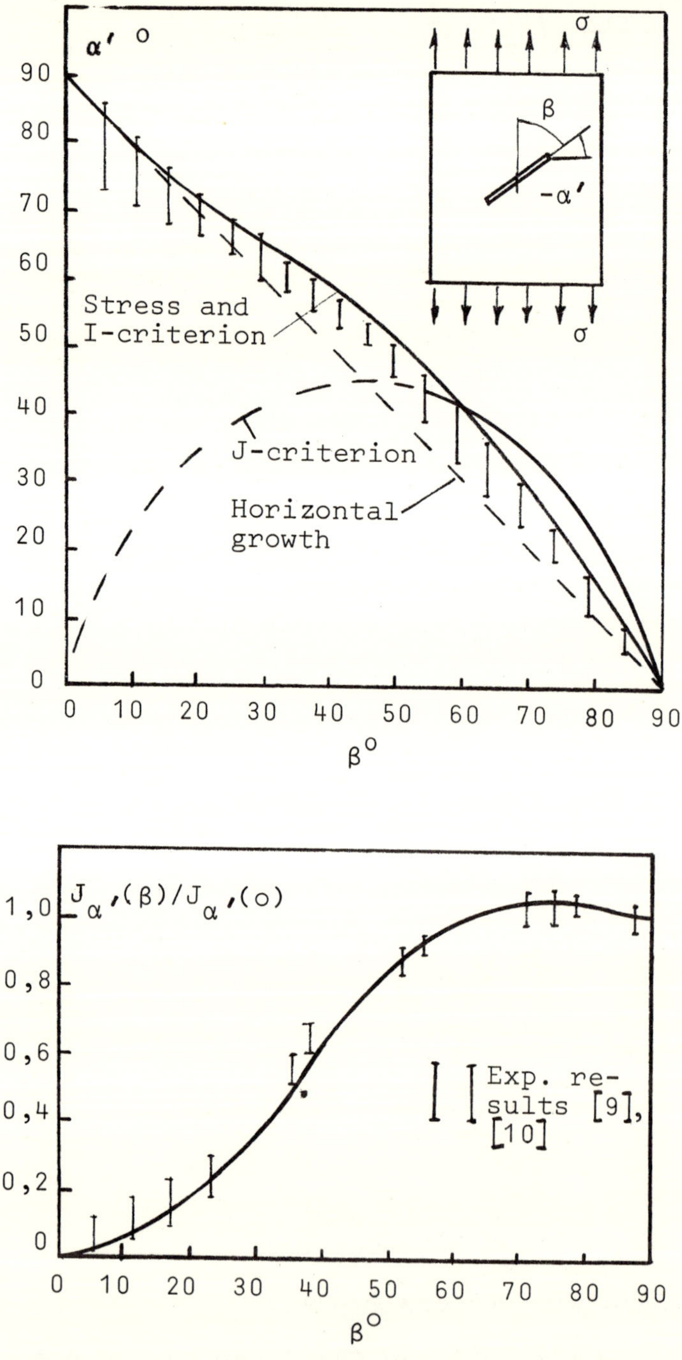


Fig. 4. Angle of crack growth and critical J value. J(α) has its maximum for α=α′.

These coefficients also enter the fracture criterion given by Sih, which has proven very accurate in predictions of crack growth in elastic bodies as regards direction of growth as well as critical load level (Fig. 4). This condition gives for many cases results coinciding with those of the stress criterion.

REFERENCES

[1] Eshelby, J.D., Solid State Physics, Vol.3, Academic Press, 1956.

[2] Rice, J.R., J.Appl.Mech., Vol.35, pp.379-386, 1968.

[3] Griffith, A.A., Phil.Trans.Roy.Soc.London, Ser. A 221, pp.163-168, 1920.

[4] Irwin, G.R., J.Appl.Mech., Vol.24, pp.361-364,1957.

[5] Signorini, A., Rend.Acad.Naz.Lincei, Ser.6, 18, pp.388-394, 1933.

[6] Knowles, J.K. and Sternberg, E., Archive for Rational Mechanics and Analysis, Vol.44, No.3, pp.187-211, 1972.

[7] Budiansky, B. and Rice, J.R., J.Appl.Mech., Vol.40, pp.201-203, 1973.

[8] Liebowitz, H., Fracture. Part II, Ch.2, Academic Press, 1968.

[9] Strifors, H., Dissertation. Dept.Strength of Materials and Solid Mechanics, Royal Inst.Techn., Stockholm, 1974.

[10] Williams, J.G. and Ewing, P.D., Int.J.Fracture Mech. 8, 441, 1972.

[11] Erdogan, F. and Sih, G.C., Trans. ASME, J.Basic Engng, Vol. 85D, pp.519-527, 1963.

DISCUSSION by D. Radenkovic, Laboratoire de Mécanique des Solides, Ecole Polytechnique, Paris.

I would like to make a comment on this very interesting paper. I think it is really comforting that some of the results, that the author obtained independently, coincide with the work of H.D. Bui[1] and of Bergez[2]; this shows that the results were to be expected and that they were needed. Indeed, the importance of complementary (or dual) path-independent integrals and of complete relations between these integrals and stress-intensity factors can be easily appreciated when we come to the effective calculation, using numerical methods, of the factors $K_I$ and $K_{II}$, for the given shape and loading of a cracked piece. On the other hand, the interpretation of path independent integrals as a kind of fracture criteria is in fact rather doubtful, as has often been stressed; a detailed analysis of this problem is also given in the reference [3]. I may add that the corresponding generalizations to the case of plates and shells were obtained in a recent thesis[3]; some more details about this subject will be presented in the near future[4].

REFERENCES

[1] H.D. Bui - Dualité entre les intégrals indépendantes du contour dans la théorie des solides fissurés. Ctes. Rend. Acad. Sc. Paris, t. 276 pp. 1425-1428 (1973).

[2] D. Bergez - Determination of stress-intensity factors by use of path-independent integrals. Mech. Res. Commun. t 1, vol. 2 (1974).

[3] D. Bergez - La rupture des plaques et des coques fissurées. Thesis, Paris VI, June 1974.

[4] D. Bergez, H.D. Bui, D. Radenkovic - Facteurs d'intensité des contraintes et intégrales indépendantes du contour. 2d Ital. Nat. Congr. Mechanics; Naples, October 1974.

# SECTION III

# FATIGUE CRACK PROPAGATION

# THE INFLUENCE OF FRACTURE TOUGHNESS ON SUB-CRITICAL FLAW GROWTH

P. K. Poulose, J. E. Morral, and A. J. McEvily

Department of Metallurgy, University of Connecticut, Storrs, Connecticut 06268 USA

## ABSTRACT

The influence of fracture toughness on sub-critical growth characteristics during stress-corrosion cracking and fatigue crack growth is considered. In bicrystal specimens of an Al-5.5 Zn-2.5 Mg alloy it is found that fracture toughness can markedly affect crack growth processes. Geometrical irregularities along grain-boundaries can also influence the process of crack growth. In polycrystalline material grain boundary complexity effects serve to increase fracture toughness and permit dissolution processes to become rate controlling. In the aluminum alloy studied, grain boundary precipitates of $MgZn_2$ act as sacrificial anodes; the greater the amount of $MgZn_2$ per unit of grain boundary area, the slower is the rate of crack growth at a given strength level.

In fatigue the operation of static modes of separation, particularly in materials of low toughness, can accelerate the rate of crack growth and alter the relationship between the stress intensity factor and growth rate. For materials of low toughness there is also a greater sensitivity to mean stress effects. Improvement in fracture toughness can have a beneficial effect on fatigue crack growth; however, in the case of stress-corrosion cracking, this may not always be the case.

## INTRODUCTION

Fracture mechanics provide a useful framework for the systematic study of sub-critical flaw growth during stress-corrosion and fatigue. Within this framework it is clear that the fracture toughness represents a terminal upper bound to these processes.

However, this may not be the only role of fracture toughness, for recent research has shown that static modes of crack growth which depend upon fracture toughness can couple with the basic modes of growth in the sub-critical range to affect the processes of crack propagation. The present paper presents examples of this interaction in stress corrosion cracking and also in fatigue.

## STRESS-CORROSION CRACKING

The process of intergranular stress-corrosion cracking in materials such as the high strength Al-Zn-Mg alloys is often visualized in terms of bicrystalline rather than polycrystalline behavior. Most of the data on crack growth appropriately enough, however, have been obtained on polycrystalline materials. It seemed, therefore, that it would be of some interest to study actual bicrystals in the somewhat naive expectation that the relative simplicity of such specimens would lead to additional insight into the mechanism of crack growth. Such an approach is not unique; for example, while this study was underway, a description of an investigation into bicrystal behavior was published by Kovacs and Low [1]. Their specimens were prepared by a strain-anneal technique and they found that most of the boundaries were of the large-angle type, but some low angle boundaries were present. In their stress corrosion tests crack growth rates were reduced whenever the boundary was of a low angle or coincidence boundary type. With respect to the present study, it is pointed out at the outset that despite the conceptual simplicity of the ideal bicrystal, experimental difficulties also encountered by Kovacs and Low in obtaining truly planar boundaries detracted from the usefulness of such specimens.

### Material

The Al-Zn-Mg alloy used in this investigation was kindly supplied by the Olin Corporation. The material was received in sheet form, 0.04 inches thick. The chemical composition was Al-5.32 w/o Zn, 2.53 w/o Mg, 0.03 w/o Si, 0.04 w/o Cu, 0.02 w/o Mn, 0.01 w/o Cr, 0.02 w/o Fe, and 0.003 w/o Ti. In order to develop the large grains needed to machine bicrystal specimens from the sheet, a strain-anneal procedure as developed by Carpenter and Elam [2] was used. This procedure involves a pre-anneal at 510°C for 40 minutes followed by room temperature extension of 1.6-1.8%. The alloy is then annealed in a staircase procedure starting at 380°C and terminating at 530°C. In traversing this range temperature steps of 15°C were used with a hold time of eight hours at each step, with a final hold of 24 hours at 530°C. This procedure resulted in grains of several centimeters in diameter. An x-ray diffraction check indicated that most of the grain boundaries were probably of the large angle type observed by Kovacs and Low. The boundaries themselves were found to be quite irregular for reasons which are

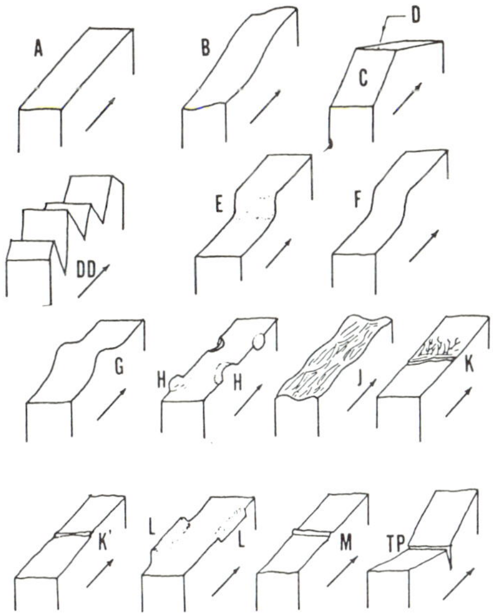


Fig. 1. Types of geometrical intergranular irregularities observed in bicrystals. Direction of crack growth indicated by arrow.

not clear. A qualitative indication of nature of these irregularities arranged by type is given in Fig. (1). From point to point along any given boundary the local type of irregularity might vary from any one of these types to another.

Final heat treatment of specimens machined from the sheet consisted of solution treatment at 510°C, water quenching, and then aging at 120°C for 24 hours. Some sheet material was not strain-annealed, but was given this final heat-treatment to provide polycrystalline specimens of a 0.25 mm grain size for comparison purposes. In a sense the material may not be truly representative of a polycrystal because it is only four grains in thickness. The final heat treatment resulted in a matrix containing G.P. zones and η' precipitates, a precipitate free zone along the grain bound- of 0.04 μ in total width, and a precipitate of η($MgZn_2$) in the grain boundary. The mean diameter of this grain boundary precipitate was 0.04 μ, and the area fraction of the grain boundary covered by precipitate was 0.136, with 70 precipitates per square micron of grain boundary area.

Mechanical Properties

Tensile tests of bicrystal specimens were carried out to determine yield and ultimate strengths. The average value of the

0.2% yield strength of five specimens was 51.4 ksi. The average value of the tensile strength of five specimens was 51.5 ksi. However, these values do not relate directly to the strength of the boundaries, since failure occurred by necking at a location remote from the boundary. It is probable that the strength values of the boundary were more like the 60 ksi level achieved in polycrystalline material. The increment of approximately 10 ksi is due to the complexity effect at grain boundaries. This effect results from the constraints imposed at a large angle boundary with respect to the accommodation of slip in each of the adjacent grains with their different slip systems. Compatibility of displacements requires the operation of additional slip systems with a corresponding increase in strain hardening due to their interaction as compared to regions remote from the boundary.

Bicrystal specimens, therefore, failed in a ductile, transgranular mode in tension tests. However, if a notch were introduced at the grain boundary, then a crack could be made to run in the boundary and lead to completely intercrystalline failure. Polycrystalline material, even in an unnotched condition, failed predominantly in an intercrystalline mode, with only occasional patches of transgranular (dimpled) fracture being observed. Intercrystalline fracture surfaces in both bicrystal as well as polycrystalline specimens were quite smooth in appearance except for slip lines, even as viewed in the scanning electron microscope at 8000X. Microstructural features such as grain boundary precipitates of $MgZn_2$ were not resolved on the fracture surfaces.

## Test Specimen and Tests

The test specimen was of a compact tension type as shown in Fig. (2), with typical values of the dimensions H and W being 0.35 inch and 1.1 inch, respectively. The K calibration as a function of crack length for this specimen type was determined by means of an experimental compliance method and the relation:

$$K = \left(\frac{P^2E}{2B}\,\frac{\delta C}{\delta a}\right)^{1/2} \qquad (1)$$

where K is the stress intensity factor, P is the load which can be related to the displacement, $\Delta$, of the loading points through $\Delta = CP$, C is the specimen compliance, E is Young's modulus, and a is the crack length measured from the line of action of the applied load.

The load-deflection relationship was linear up to K levels of 30 ksi$\sqrt{\text{in}}$ and became non-linear at higher values. Only values below 30 ksi$\sqrt{\text{in}}$ are therefore considered to be significant.

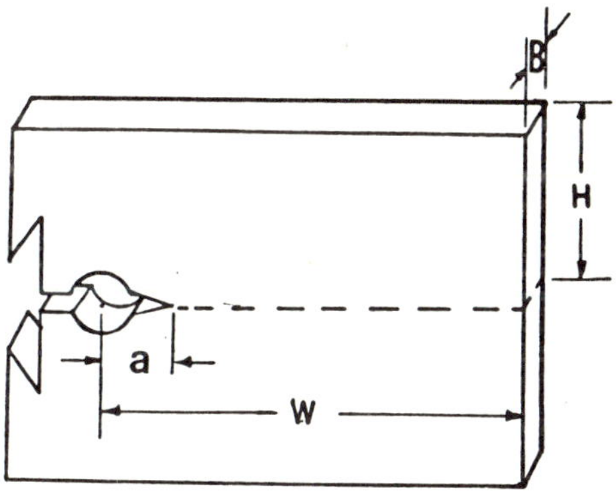


Fig. 2. Test specimen. Typical dimensions, B = .04 inch; W = 1.1 inch; H = 0.35 inch.

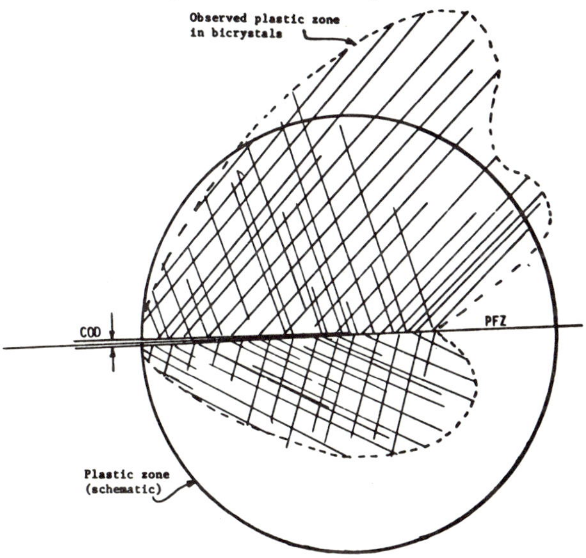


Fig. 3. Comparison of the sizes of actual and computed plastic zones in a bicrystal, the crack opening displacement and the precipitate-free zone (PFZ). K-level 15 ksi$\sqrt{\text{in}}$.

Since the specimen was 0.04 inch thick, the ASTM suggested plane strain condition ($t \geq 2.5\ K_1^2/\sigma_y^2$) was exceeded for stress intensity values in excess of about 8 ksi$\sqrt{\text{in}}$. Figure (3) provides an indication of relative dimensions of interest in the crack growth process at a stress-intensity of 15 ksi$\sqrt{in}$. These include the width of the precipitate free zone (0.04 μ) present in these specimens, the crack opening displacement and the actual and computed plastic zone sizes.

The stress-corrosion tests were carried out in a fixed-displacement mode by inserting a tapered steel pin at the loading points to maintain the displacement as determined by an auxiliary clip gauge. Fine scribe marks transverse to the grain boundary and of predetermined spacing were used as reference points in the determination of the length of the crack as a function of time. A 30X telescope was used in making the observations. Most of the stress-corrosion tests were carried out in 3% NaCl solution. Additional tests were carried out in ambient air or in a vacuum ($10^{-5}$ torr). In the NaCl tests only the region of the specimen into which the crack would grow was immersed in the solution. Acoustic emission readings were obtained to supplement the information obtained by visual examination.

Results

The first finding of interest was that crack growth was not a smooth function of time in bicrystal specimens. In fact growth was often quite erratic as indicated in Fig. (4), and both crack jumps as well as crack arrests were observed. This type of behavior was largely due to the geometrical discontinuities present in grain boundaries. The crack jumping process could be detected by means of the acoustic emission monitoring equipment. During periods of slow growth, no acoustic signals were detected.

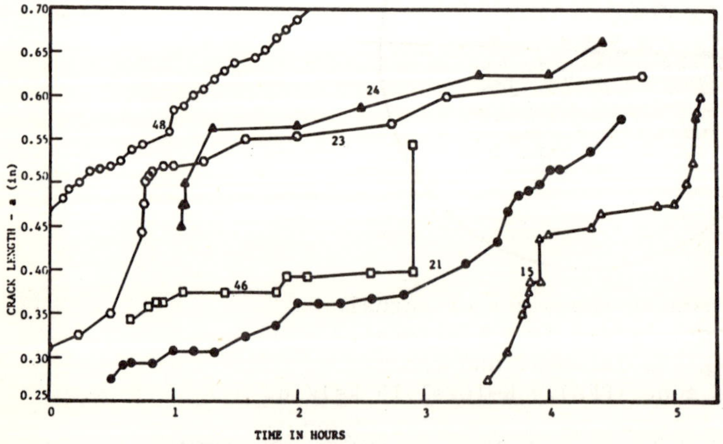


Fig. 4. Examples of crack length as a function of time in bicrystals.

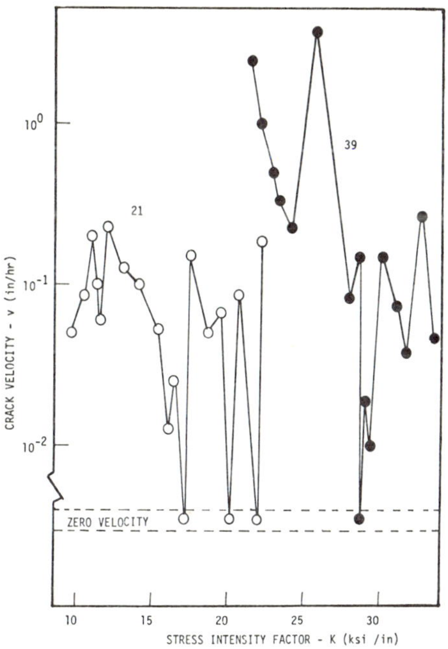


Fig. 5. Crack velocity as a function of stress intensity factor for two bicrystals.

The variation in rate of growth as a function of stress intensity factor is shown in Fig. (5) for two bicrystals. The designation "zero velocity" indicates that the crack was stopped at least as observed macroscopically for a period of time before it again began to advance. It should be recalled that as the crack lengthens, the stress-intensity factor decreases in these constant-deflection tests.

Figure (6) shows the growth characteristics in ambient air (72°F, 28% relative humidity). Again an erratic pattern is noted. Overall, the average rate of crack growth in these tests was about $4 \times 10^{-3}$ in/hr. Cracks did not grow at all in the vacuum tests ($10^{-5}$ torr), a result which supports the view that moisture is a requisite for stress-corrosion crack growth [3]. A main effect of the NaCl solution is to increase the rage of growth over the air tests by at least an order of magnitude.

In an attempt to reduce the data to an interpretable form, the rates of crack growth were correlated with the corresponding type of boundary given in Fig. (1). As a further simplification, herein we will consider only the simplest types of boundary, i.e., A, B, C types. On this basis the rates of crack growth are plotted

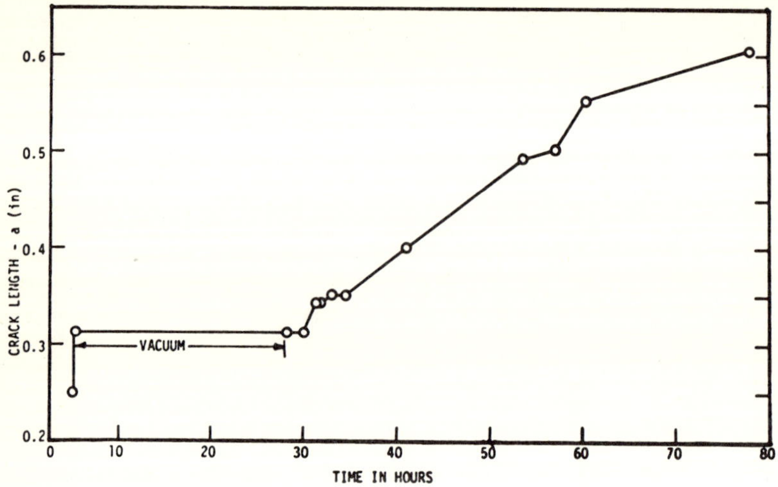


Fig. 6. Influence of vacuum and air environments on crack growth in a bicrystal specimen.

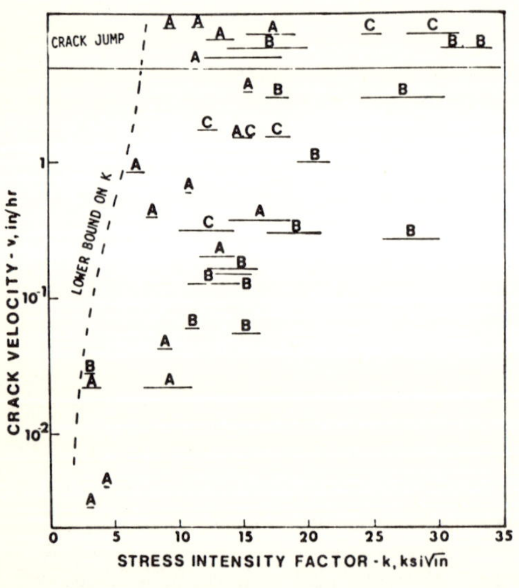


Fig. 7. Crack velocity as a function of stress intensity factor for bicrystals of Types A, B, and C.

as a function of K in Fig. (7). There is still a wide variation in growth rate at a given K value, but trends are more apparent than if all data for all boundary types are considered. Some of the remaining variation for a grain boundary type may be due to the presence of minor geometrical effects or to the presence of low angle or coincidence type boundaries. In addition, the amount of grain boundary precipitate, an important feature as subsequently discussed, can vary with the misfit across a boundary [4] and lead to variations in growth rate. Nevertheless, it is seen that there is a tendency for results obtained for A type boundaries to lie at the lower K values. These boundaries most nearly approach the concept of the ideal bicrystal boundary, and an estimated lower bound curve for this type of boundary is indicated in Fig. (7). Observable crack growth starts at a K value of about 2.5 ksi$\sqrt{\text{in}}$, and crack jumps are first observed at a K value of 8.5 ksi$\sqrt{\text{in}}$ for the A type boundary.

Crack growth in polycrystalline specimens was less erratic than in the bicrystals, Fig. (8). Figure (9) shows that the surfaces are quite smooth and that some micro-branching of the cracks takes place during crack growth. An average rate of crack growth (K independent) of about 5 x $10^{-3}$ inches per hour was determined for the polycrystalline specimens tested in the 3% NaCl solution.

Examination of fracture surfaces by scanning electron microscopy did not reveal detail of significance. Fracture surfaces were generally quite smooth, and no metallurgical structural detail such as the precipitates was observed. At locations where the crack had been arrested for a period of time, evidence of localized pitting attack of the fracture surface was present. Slip lines were also observed at some of the geometrical discontinuities.

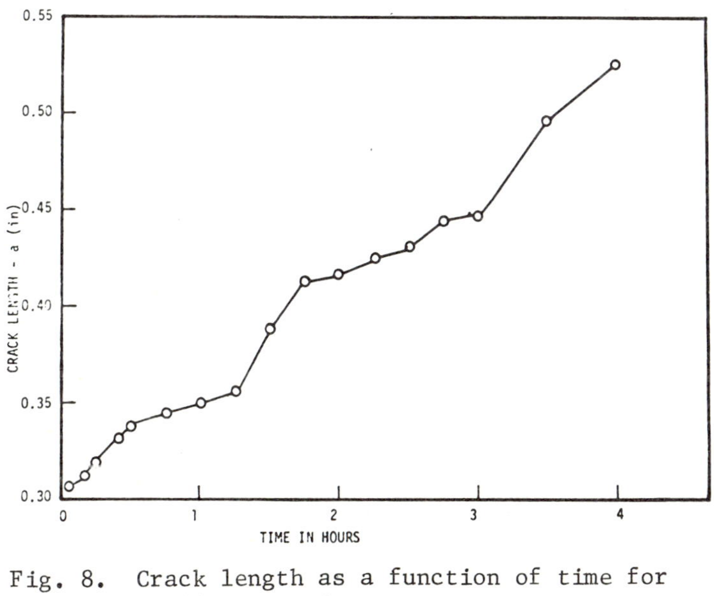


Fig. 8. Crack length as a function of time for a polycrystalline specimen.

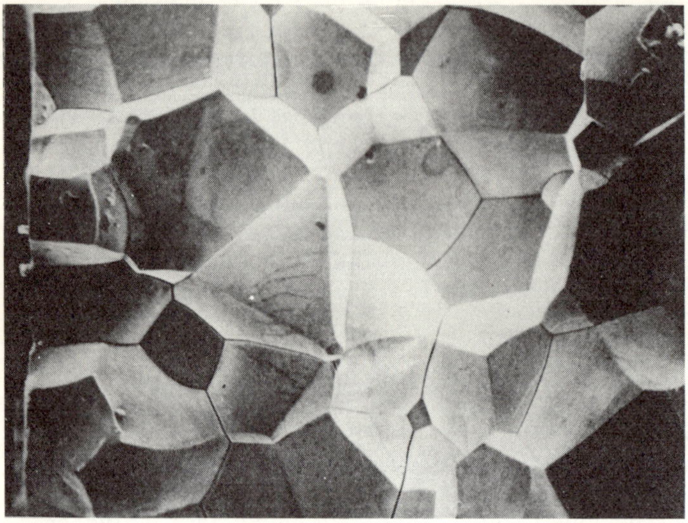

Fig. 9. Fracture surface of polycrystalline specimen after stress corrosion cracking. X120.

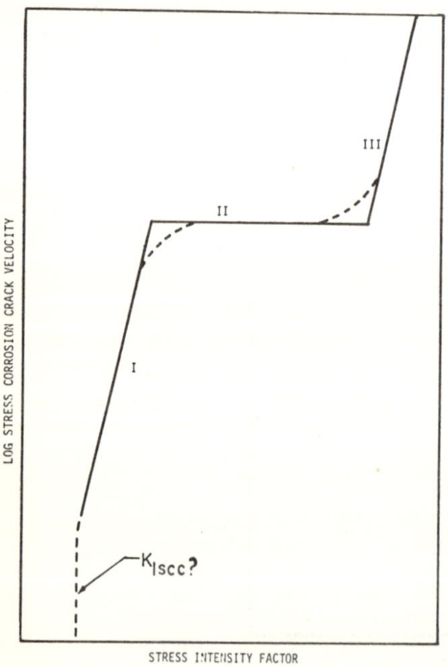


Fig. 10. Schematic representation of the three stages of the stress corrosion fracture process [3].

## Discussion

The process of stress-corrosion cracking can be considered to consist of three stages [3] as shown in Fig. (10). Both Stage I and Stage III are K-dependent, whereas Stage II is not.

The present lower-bound results on A type bicrystals indicate that Stage I begins at a stress intensity of about 2.5 ksi$\sqrt{in}$, and that Stage III terminates at a stress-intensity of 8.5 ksi$\sqrt{in}$, and perhaps at an even lower value in the case of the ideal boundary. There does not appear to be any marked extent of the Stage II process in such specimens, rather there is a smooth transition from Stage I into Stage III over the relatively narrow K-range involved. On the other hand, the stress-corrosion crack growth in the polycrystalline specimens appeared to be largely of the Stage II type, that is, the growth rate was independent of the stress intensity. An increase in the number of grains in the cross section is therefore attended by an increase in the fracture toughness which permits the development of the Stage II growth process. This increase in toughness may be due to another grain boundary complexity effect, analogous to that which is present during plastic deformation. In the case of crack advance, this new complexity effect inhibits the tendency for crack jumping along the crack front as a whole, at least during Stage II of the crack growth process. In this stage, micro-crack branching and crack tip blunting may also serve to stabilize the crack mechanically and allow dissolution-controlled kinetics to determine the Stage II rate of cracking over a wider range of K values.

Our view of the processes involved in the various stages of stress-corrosion cracking is as follows. Stage II involves the breakdown of protective films by slip to allow dissolution processes to operate. The K-dependency stems from the variations in resistance to slip and crack advance encountered at the crack tip. These variations arise because of orientation differences as well as grain boundary morphological effects. A reported activation energy associated within this stage of 27 K cal/mole for aluminum alloys [3] may relate to the activation energy for dissolution, which is rate controlling in this stage. As the K level is increased, the rate of cracking becomes controlled by diffusion of reactants in the liquid at the crack tip, and the K-independent Stage II is observed provided the fracture toughness is sufficiently high. The activation energy for this stage is about 4 K cal/mole [5]. At higher K levels static modes of failure may occur during sub-critical growth, and Stage III is observed. Stage III is therefore considered to be a combination of localized static rupture combined with dissolution processes. The activation energy for this stage should be similar to that of Stage II. It also appears that the slope of the rate plot in Stage I and Stage III

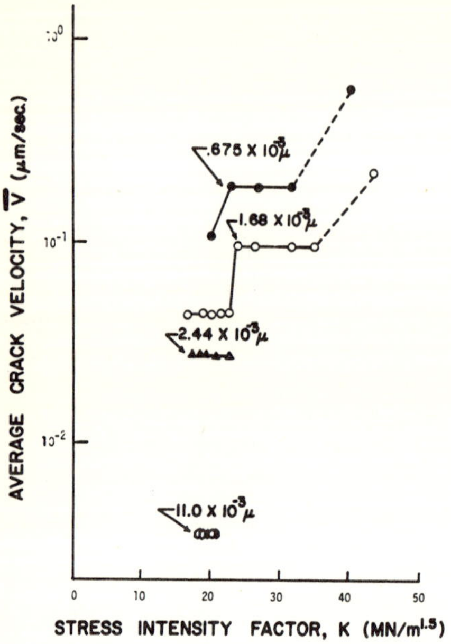


Fig. 11. Stress corrosion-crack velocities as function of stress intensity factor in Al-5.5 Zn-2.5 Mg polycrystalline specimens of 40 ksi yield strength. Numbers in figure refer to volume of $MgZn_2$ per unit of grain boundary area.

may be dependent on the degree of micro-structural inhomogeneity. The more homogeneous the structure, the steeper may be the slope of the rate vs K relationship in these ranges. Fracture toughness plays an active role in Stage III, a passive role in Stage II, and may be related to the $K_{ISCC}$ level of Stage I [5].

A companion study to this work [6] has indicated that in Stage II dissolution of the $MgZn_2$ grain boundary precipitate coupled with concentration polarization at the cathode is rate controlling. The $MgZn_2$ is anodic to the surrounding aluminum and therefore acts as a sacrificial anode. Figure (11) is a plot of the crack growth rate as a function of K for the same polycrystalline Al-5.5 Zn-2.5 Zn alloy described herein. The alloy was heat treated by different paths to a common yield strength level of 40 ksi, but with different amounts of $MgZn_2$ in the grain boundaries. The experimental results show that as the amount of $MgZn_2$ increases, the rate on cracking decreases in accord with the sacrificial anode concept. Once the sacrificial anodes dissolve, crack advance on a micro-level can take place either by tearing or dissolution of the surrounding material.

Although an increase in the amount of $MgZn_2$ in the grain boundaries has a beneficial effect on Stage II kinetics, such an increase may have an adverse effect on $K_{IScc}$ and Stage III if the intercrystalline fracture toughness is reduced with increase of $MgZn_2$ content. If so, a compromise between resistance to stress-corrosion cracking and fracture toughness may be desirable in a particular case. Alternatively, thermal-mechanical processing treatments which disrupt easy paths for cracking along grain boundaries, combined with overaging heat treatment, may be appropriate.

## FATIGUE

In this section some of the recent research which bears upon the influence of static modes of fracture on the fatigue crack growth process will be discussed. In order to provide a quantitative basis for interpreting effects due to static modes, the crack-opening displacement (COD) approach to the analysis of fatigue crack-growth will be utilized [7]. In this approach the rate of crack growth per cycle, $\Delta a/\Delta N$, is expressed as:

$$\frac{\Delta a}{\Delta N} = A(COD-COD_{TH}) = \frac{A}{\sigma_y E}(\Delta K^2-\Delta^2_{TH}) \tag{2}$$

where A is a material constant, $\sigma_y$ is the yield strength, E is Young's modulus, $\Delta K$ is the range of the stress intensity factor, and $\Delta K_{TH}$ is the range of the stress intensity factor which must be exceeded for crack growth to occur. This latter quantity is also a function of R which is defined as the ratio of the minimum to the maximum stress in a loading cycle. In the absence of environmental effects the constant A has been found empirically to be equal to $\frac{8}{\pi}\frac{\sigma_y}{E}$, which leads to

$$\frac{\Delta a}{\Delta} = \frac{8}{\pi}\left[\left(\frac{\Delta K}{E}\right)^2 - \left(\frac{\Delta K_{TH}}{E}\right)^2\right] \tag{3}$$

This relationship indicates that except for the threshold term the rate of crack growth in the linear elastic range is structure insensitive. (It is noted that a similar expression can be obtained if the crack is considered to be an elastic ellipse. In this case the COD, taken as twice the tip radius $\rho$, is equal to

$$\frac{4\sigma^2 a}{E^2} \simeq \frac{K^2}{E^2}$$

This relationship suggests that the plastic strains at the tip of a crack where blunting is taking place may be of the same order as the elastic strains).

Before considering the influence of static modes on the above expressions for crack growth, it is of interest to examine the degree of correspondence between the COD and the crack advance per cycle. For example, in the case of an aluminum alloy of 50 ksi yield strength the constant A in Eq. (2) would be approximately equal to $10^{-2}$, which might suggest that the rate of growth per cycle is but one-hundredth of the COD value. However, a number of considerations contribute to a reduction of this apparent discrepancy. Among these is that the plastic zone size in plane strain is about 1/4 that of the plane stress zone size used in obtaining Eq. (2); the yield strength under cyclic conditions is twice the static value [8]; crack branching along two 45° shear zones at the crack tip could introduce a factor of one-half; cyclic strain hardening is neglected, and any reversibility of the flow process during unloading is not included in the analysis. In view of these considerations, it appears that the degree of correspondence between the COD and the increment of crack advance per cycle may be better than suggested by Eq. (2).

In comparing experiment and Eq. (2) it has been found that for R = 0 loading reasonable agreement is obtained at growth rates below $10^{-4}$ inches per cycle. At higher growth rates experimental results and predictions diverge because of the introduction of static separation modes not considered in the model. Recent experimental studies by Richards and Lindley [9], Ritchie and Knott [10], and Pearson [11] provide evidence that in materials of low fracture toughness in particular, the rate of crack growth can be accelerated by the operation of both cleavage, intergranular and transgranular static separation processes. The sensitivity to these static modes depends upon the difference between the fracture toughness and the peak stress intensity of a loading cycle, and hence is a function of mean stress. To account for these effects, the following empirical modification to Eq. (2) has been proposed [12]:

$$\frac{\Delta a}{\Delta N} = \frac{A}{\sigma_y E} (\Delta K^2 - \Delta K_{TH}^2) \left(1 + \frac{B\Delta K}{K_c - K_{max}}\right) \qquad (4)$$

In comparing this expression with experimental results, it was found that the constant B could be set equal to unity. Mean stress effects are incorporated, since $K_{max} = \Delta K/1-R$. Figure (12) presents a comparison of Eq. (4) with the experimental results of Pearson for aluminum alloys. It is seen that for the L64 material, a high toughness alloy, the contribution of static modes is minimal over the range investigated. On the other hand in the low toughness alloy, RR58, the crack growth process is markedly affected by these modes. It is also seen that mean stress effects become more important as the fracture toughness decreases. As in the case of stress-corrosion cracking, the fracture toughness can therefore exert a considerable influence on the sub-critical growth process in fatigue.

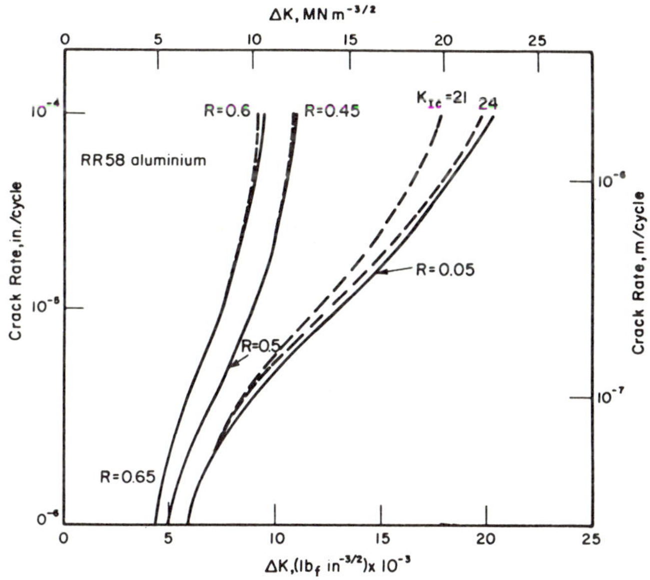


Fig. 12a

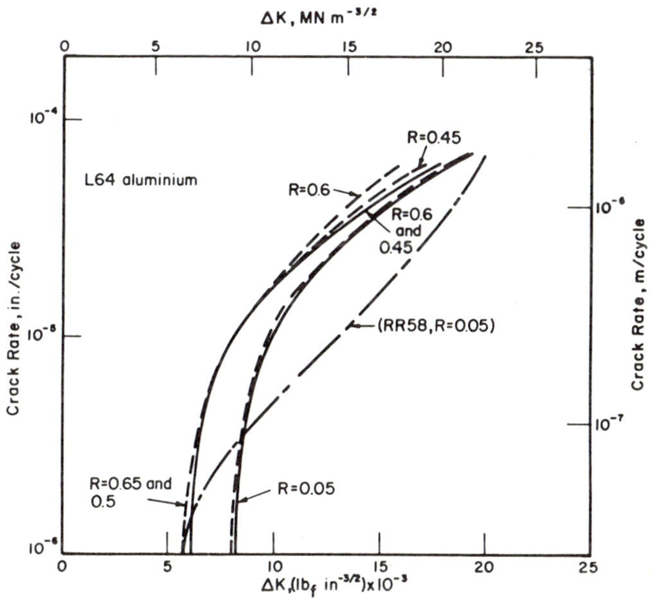


Fig. 12b

Figs. 12a and 12b. Comparison of experimental (11) and calculated (dashed curves) growth rates for two aluminum alloys: (a) a low toughness alloy, RR58; (b) a high toughness alloy, L64, with RR58 test results shown for comparison.

## CONCLUSIONS

The process of sub-critical crack growth has been found to be influenced by fracture toughness both in stress-corrosion and in fatigue. In the case of fatigue an improvement in toughness will have a beneficial effect on fatigue crack growth characteristics. However, for the Al-5.5 Zn-2.5 Mg alloy considered, the same may not be true of stress corrosion, at least over the entire growth range.

An important aspect of both of these processes in need of better understanding is the threshold level. Does it exist, and, if so, why? In fatigue a growth rate of an interatomic spacing per cycle may impose a lower limit to the growth rate in the absence of environmental effects, but in stress corrosion the situation is less clear. Since the threshold level is history dependent, much of the fatigue crack retardation effect due to overloads may be associated with an increase in the level of $\Delta K_{TH}$. In addition, the role of environment in fatigue, i.e., corrosion fatigue, is in need of further clarification.

## ACKNOWLEDGMENT

The authors express their thanks to Dr. J. Ford of the Olin Corporation for providing the material for this investigation and to the Air Force Office of Scientific Research for providing financial support (Contract No. F44620-69-C-0011).

## REFERENCES

[1] Kovacs, W. J. and Low, J. R., Jr., Intergranular fracture in an Al-15 wt pct Zn alloy, Met. Trans., 2, 3385, 1971.

[2] Carpenter, H. H. and Elam, C. F., The production of single crystals of aluminum and their tensile properties, Proc. Roy. Soc., London, A, 1921, 100, 329.

[3] Speidel, M. O., Current understanding of stress corrosion crack growth in aluminum alloys, Theory of Stress Corrosion Cracking in Alloys, Scully, J. C., ed, NATO, Brussels, 1971, 289.

[4] Unwin, P.N.T. and Nicholson, R. B., The nucleation and initial stages of growth of grain boundary precipitates in Al-Zn-Mg and Al-Mg alloys, Acta Met., 13, 1379, 1969.

[5] Krafft, J. M. and Mulherin, J. H., Tensile-ligament instability and the growth of stress-corrosion cracks in high-strength alloys, Trans. ASM, 62, 64, 1969.

[6] Poulose, P. K., Morral, J. E. and McEvily, A. J., Stress corrosion crack velocity and grain boundary precipitates in an Al-Zn-Mg alloy, <u>Met. Trans.</u>, to be published.

[7] Donahue, R. J., Clark, H. McI., Atanmo, P., Kumble, R. and McEvily, A. J., Crack opening displacement and the rate of fatigue crack growth, <u>Int. J. of Frac. Mech.</u>, 8, 209, 1972.

[8] Rice, J. R., The mechanics of crack tip deformation and extension by fatigue, <u>ASTM STP 415</u>, Philadelphia, 1967.

[9] Richards, C. E. and Lindley, T. C., The influence of stress intensity and microstructure on fatigue crack propagation in ferritic materials, <u>Eng. Fracture Mech.</u>, 4, 951, 1972.

[10] Ritchie, R. O. and Knott, J. F. Brittle cracking processes during fatigue crack propagation, <u>Third Int. Conf. on Fracture</u>, Munich, 1973, Paper V-434A.

[11] Pearson, S., The effect of mean stress on fatigue crack propagation in half-inch (12.7 mm) thick specimens of aluminum alloys of high and low fracture toughness, <u>Eng. Fracture Mech.</u>, 4, 9, 1972.

[12] McEvily, A. J., Phenomenological and microstructural aspects of fatigue, <u>Third Int. Conf. on the Strength of Metals and Alloys</u>, Cambridge, 1973.

# EXPERIMENTALLY OBSERVED PLASTICITY EFFECTS ON CRACK CLOSURE AND SUBCRITICAL CRACK GROWTH

H. L. Marcus, W. L. Morris, O. Buck and J. D. Frandsen

Science Center, Rockwell International, Thousand Oaks, California, 91360, USA

## ABSTRACT

This paper summarizes a wide range of ongoing investigations relating various aspects of fatigue crack growth to the local deformation in the vicinity of a propagating crack tip. Such deformation dominates the fatigue crack growth process in all but the most severe environments. The subject of the threshold stress intensity is discussed and evidence presented to challenge its validity as an intrinsic material parameter. The $\Delta K_{eff}$ and crack closure concepts are presented along with surface and bulk acoustic wave techniques to measure crack closure. Results of experiments to assess effects of environment on crack propagation rate are described. For mildly aggressive environments, changes in crack closure account for the differences in crack growth rates. In an aggressive hydrogen environment, a change in fracture mode is observed and can account for the change in crack growth rates. Finally, fatigue crack growth at high $\Delta K$ values is related to the R-curve characterization of crack growth.

## INTRODUCTION

Fatigue crack growth in ductile alloys is dominated by the cyclic plastic deformation that occurs ahead of the propagating crack. The residual strain resulting from the deformation cycle in turn affects the mechanics of the crack opening so that a crack may not be completely opened to the tip even under a tensile load unless a critical crack closure stress is exceeded. This "crack closure" phenomenon has the effect of reducing crack propagation rate, since the net plastic work per cycle that is

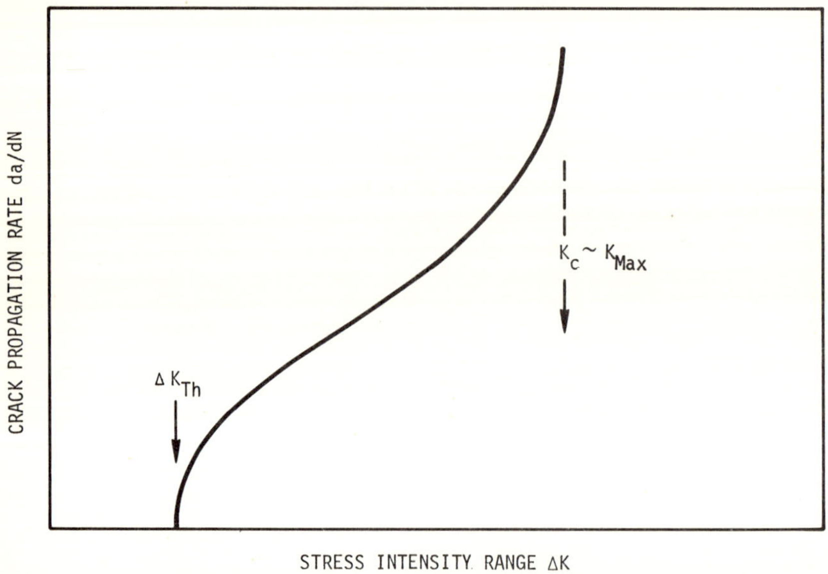


Fig. 1 Typical dependence of crack propagation rate upon stress intensity range.

done on material at the crack tip is lowered. Accordingly, definition of the cyclic crack closure and the effective stress range can become one of the important requirements in the assessment of crack growth processes, and can serve as a unifying basis for a more informative evaluation of the effects of other factors such as environment, load ratio, maximum load and load spectra.

Crack closure effects are particularly important for the evaluation of competing and synergistic factors. For example, during fatigue of certain aluminum alloys, environmental acceleration of crack growth rate with increasing humidity can be entirely attributed to an environmental alteration of the crack closure stress.

In what follows, we report on the results of a wide range of experiments aimed at further understanding the mechanisms of fatigue crack growth, and, where appropriate, the results are related to the crack closure concept. For purposes of discussion, the regimes investigated will be related to the normal fatigue crack growth representation where crack growth rate per cycle, da/dN, has a sigmoidal shape, Fig. 1, when log-log plotted against stress intensity range, $\Delta K$.

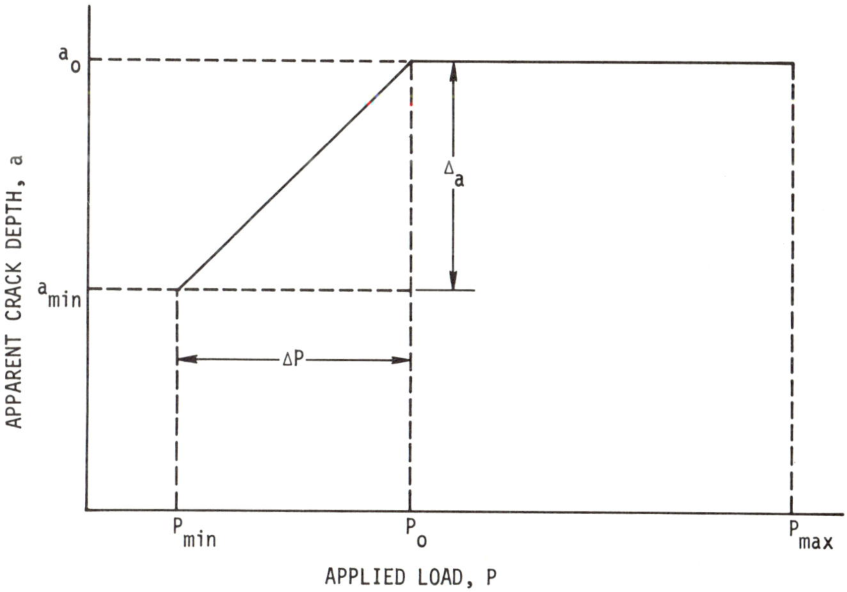


Fig. 2 Apparent crack depth as a function of applied load from $P_{min}$ to $P_{max}$. Crack is closed for loads less than $P_o$.

## CRACK CLOSURE CONCEPT AND MEASUREMENT

Elber[1,2] first introduced the concept of crack closure and the resulting relationship between fatigue crack growth rate and effective stress intensity range, $\Delta K_{eff}$. Crack closure under tensile loads is expected when cyclic deformation and final tensile separation produce sufficient strain, in which case the fracture surfaces touch and the stress singularity is eliminated on the low load portion of the fatigue cycle. The definition of $\Delta K_{eff}$ is illustrated schematically in Fig. 2, where apparent crack length is plotted as a function of instantaneous cyclic load. In this schematic, $P_o$ is the "crack closure load." For an applied load less than $P_o$ the crack tip is closed, and no crack growth associated with the crack tip singularity can occur for this portion of the fatigue cycle. The effective stress intensity range $\Delta K_{eff}$ is then defined to be $K(P_{max}) - K(P_o)$.

This $\Delta K_{eff}$ is an approximation since the plastic zone may be in compression at the crack tip up to some load $P_c > P_o$ on the loading part of the cycle. For an ideal crack, $P_c$ might be theoretically estimated by an analysis of the actual plastic strain at the crack tip, but this is presently impractical.

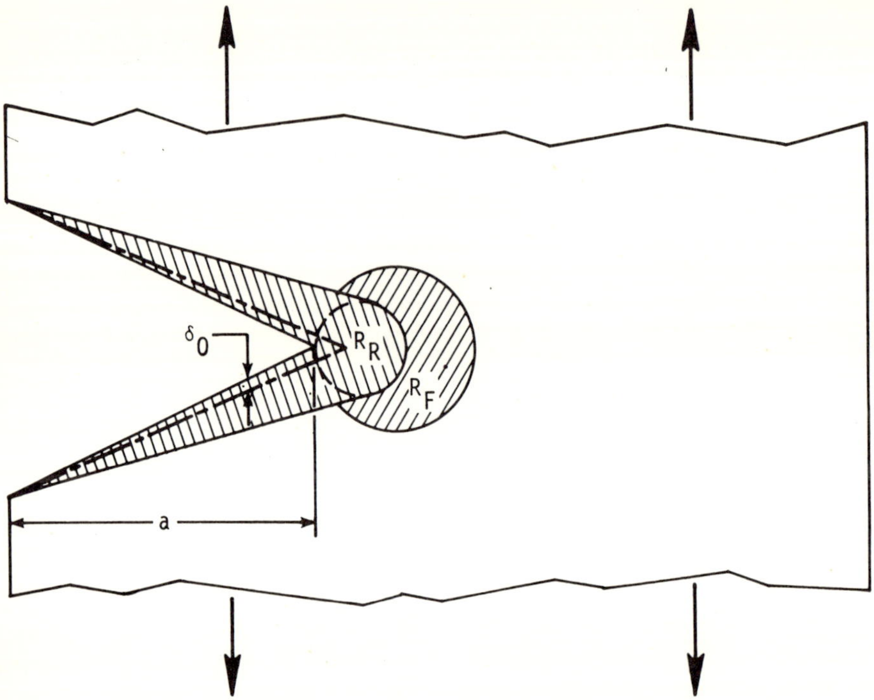


Fig. 3 Representation of near crack tip deformation zones and residual strain. $R_F$ and $R_R$ are the forward and reverse yielding spaces.

Conceptually, closure results from residual strains developed near a crack tip and is schematically illustrated in Fig. 3. The regions of $R_F$ and $R_R$ are the forward and reverse plastic zones, respectively, and the region marked $\delta_0$ is the residual displacement (strain) due to the massive deformation which occurs during final separation of the material. Indeed, it is this residual strain which produces closure at positive applied loads.

The authors have found that ultrasonic techniques are particularly useful for determining onset of crack closure. Ultrasonics can potentially be applied to a variety of specimen configurations and is not restricted to just sampling the specimen surface, as is the case for an Elber gauge.

A surface wave technique has been used to sample the crack front for a part through crack (PTC) specimen, [3,4] as shown in Fig. 4. The signal intensity is calibrated to crack depth, and $P_0$ may be determined from the apparent change in crack depth at low load segment of the cyclic loading. More recently, crack closure loads have been measured for the compact tension (CT)

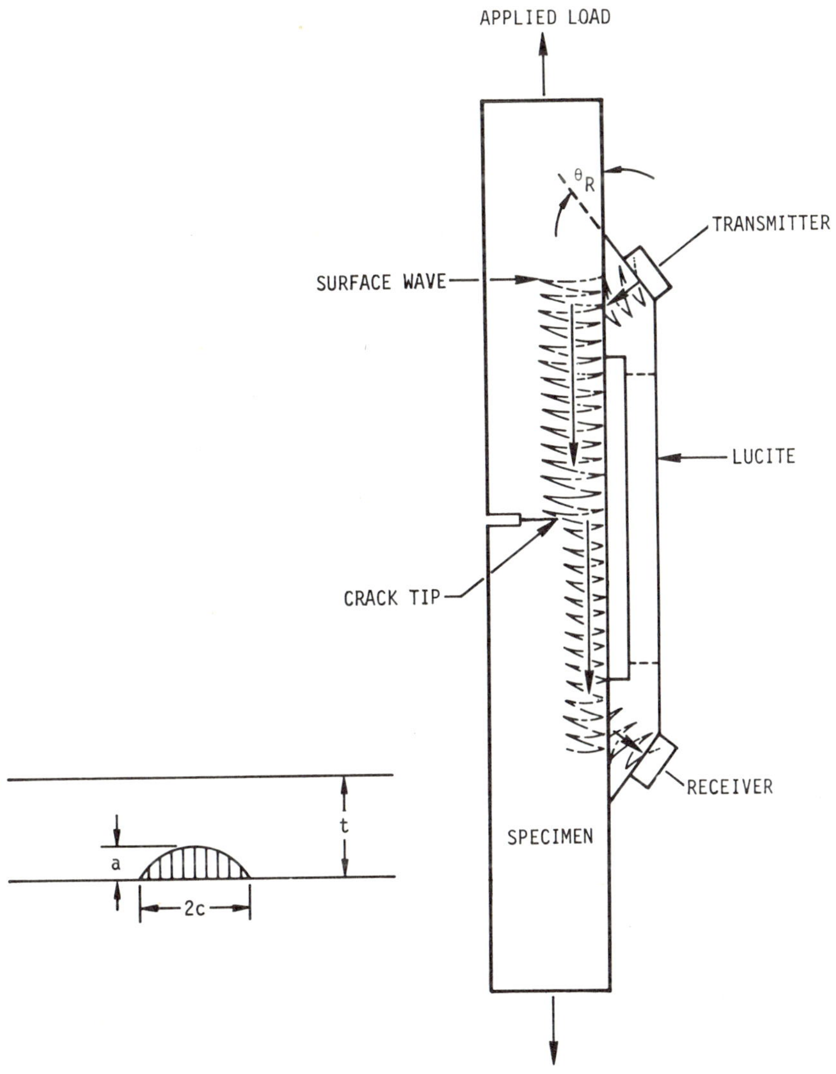


Fig. 4 Surface wave acoustic configuration utilized to measure crack closure for the part through crack specimen geometry.

specimen geometry using bulk waves as illustrated schematically in Fig. 5. This allows a direct comparison of the variation in closure stress with crack depth for the stiff, high dK/da, PTC specimen geometry to that for the more compliant, low dK/da, CT geometry.

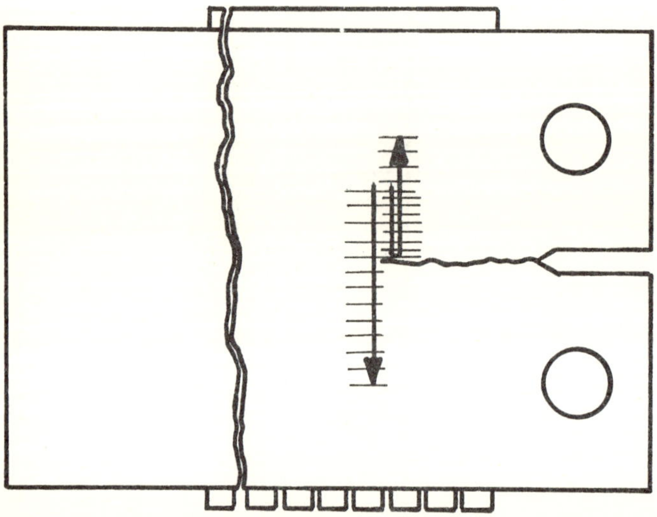


Fig. 5 Sketch of bulk wave generator-detector on a compact tension specimen used to measure crack closure.

A remaining problem confronting all such closure measurements is the definition of closure load in cases where the variation of apparent crack depth with applied load is significantly nonlinear, i.e. $P_o$ is not well defined. At least one factor in such nonlinear behavior may be the consequence of the two-dimensionality of the crack closure front. Future experiments on this aspect of closure are planned. At present, extrapolation of the two segments of the experimental curves is used to define $P_o$.

## FATIGUE CRACK INITIATION AND ITS RELATION TO THRESHOLD STRAIN INTENSITY

Fatigue crack initiation studies clearly show that no prior precrack is necessary for a crack to initiate, propagate and ultimately fail a component. The artificial separation of fatigue initiated cracks from fracture mechanics controlled cracks after the former reach a given stress intensity is arbitrary and unsatisfactory. "Stage I" versus "Stage II" crack propagation is also only definitive in terms of the mechanism of crack growth but does not define a loading limit below which crack propagation cannot take place. It is only logical that there should not exist such an intrinsic material property as $\Delta K_{Th}$ for a precracked specimen since the root of the crack undergoes plastic flow. In fact, crack closure and residual compressive stresses appear to be the concepts that can draw together what at first glance appears to be different phenomena.

Initiation and Stage II propagation of fatigue cracks at a surface is known to occur for a $\Delta K$ range significantly below threshold values measured with fracture mechanics specimens. In the absence of environmental and grain size effects this disparity can be attributed to crack closure. A threshold value obtained for a particular geometry can be considered a determination of $P_c$ for that geometry's crack length and applied load history. In terms of a closure model, threshold occurs when the maximum applied load $P_M \leqslant P_c$ and when continued cycling to $P_M$ does not reduce the value of $P_c$. Thus, threshold measurement is geometry dependent and is dependent upon cyclic load history prior to obtaining the "threshold" limiting value.

The authors are currently engaged in a detailed study of the fatigue behavior of Al 2219, where the emphasis is in the connection between the initiation, the early stage of fatigue crack growth, and the fatigue life for the material as measured with an S-N curve. Fatigue behavior and lifetime is found to be dependent upon the material's rolling direction, i.e. they are anisotropic. The fatigue of smooth bar specimens in the rolling direction produces Stage II propagation immediately after initiation at (Cu, Al) intermetallic inclusions at the bar surface. For peak stresses above 0.65 $\sigma_y$, these cracks propagate essentially in a plane normal to the loading direction, and each crack typically stops when it encounters a grain boundary, with failure eventually occuring by the processes of crack coalescence. For the lower stress regime (< .65 $\sigma_y$), propagation of a single crack is responsible for specimen failure. Current research objectives are to define the statistics for these processes and the manner in which they are affected by specimen surface machining, humidity, and such material parameters as grain size and amount of intermetallic inclusions. This picture is entirely altered, however, when Al 2219 is fatigued normal to the rolling direction. Cracks are not initiated at the surface intermetallics, but by grain boundary cracking. An important implication of these results is that regardless of crack initiation and growth mechanism, be it "Stage I" or "Stage II," the crack will always propagate below $\Delta K_{Th}$.

## EQUILLIBRIUM CRACK PROPAGATION REGIME

For the intermediate portion of fatigue life, crack propagation rate is typically relatable to stress intensity range by a power law, $da/dN = C(\Delta K)^n$ [5]. The physics involved in this relationship are not well understood. Even less well understood are the mechanisms by which C and n are altered by environment as a function of material condition. In addition, very little is known of the effects of spectrum loading, i.e. non constant $\Delta P$.

Our limited work on the effects of changes in load ratio and overloads on crack closure have been previously reported. [4] One main result is that after a given change in loading condition a distinct number of cycles is required to re-establish the steady state crack closure condition. When the minimum load, $P_{min}$, is increased, a higher $P_o$ is established after the crack propagates into the new residual strain field. When the upper load $P_{max}$ is reduced, $P_o$ slowly decreases to a value consistent with the new equilibrium strain field before the crack resumes the steady state propagation rate. If the upper load is reduced to a value below the original $P_o$, no growth is observed until the prior residual strain is modified by the cyclic loading such that the crack is able to open again. Such studies, of course, included only a small range of possible variables for this type of investigation. It is obvious that definition of a predictive model for spectrum loading requires a great deal more understanding of local deformation processes, particularly for nonuniform loading conditions.

A second phase of the authors' effort involves investigating the mechanisms by which environment can influence crack growth. [6,7] Fatigue crack propagation in dry and humid environments has been investigated for aluminum 2024 and 7075. It is found that humidity reduces the crack closure stress, which results in increases of $\Delta K_{eff}$. When plotted against $\Delta K_{eff}$, and not $\Delta K$, the fatigue crack growth data for both alloys fall in a single scatter band, Fig. 6, irrespective of environment. One can infer that this convergence in behavior is a result of a reduction in residual strain of the material in the plastic zone as humidity increases. There are at least two means by which this could occur: a) the exposed material at the crack tip, as a consequence of surface chemical reaction, physically separates at a much lower strain producing less residual strain, and thus, a lower closure stress; and b) the environment is introduced directly into the specimen during forward and reverse yield cycling by a "dislocation sweep-in" mechanism resulting in reduced local ductility and limited residual strain.

Some direct evidence for the activity of the latter mechanism has been obtained through investigations of the influence of hydrogen, oxygen and nitrogen on fatigue crack growth in nickel based alloys. [8,9] In addition, transmission electron microscopy and scanning electron microscopy examinations indicate that

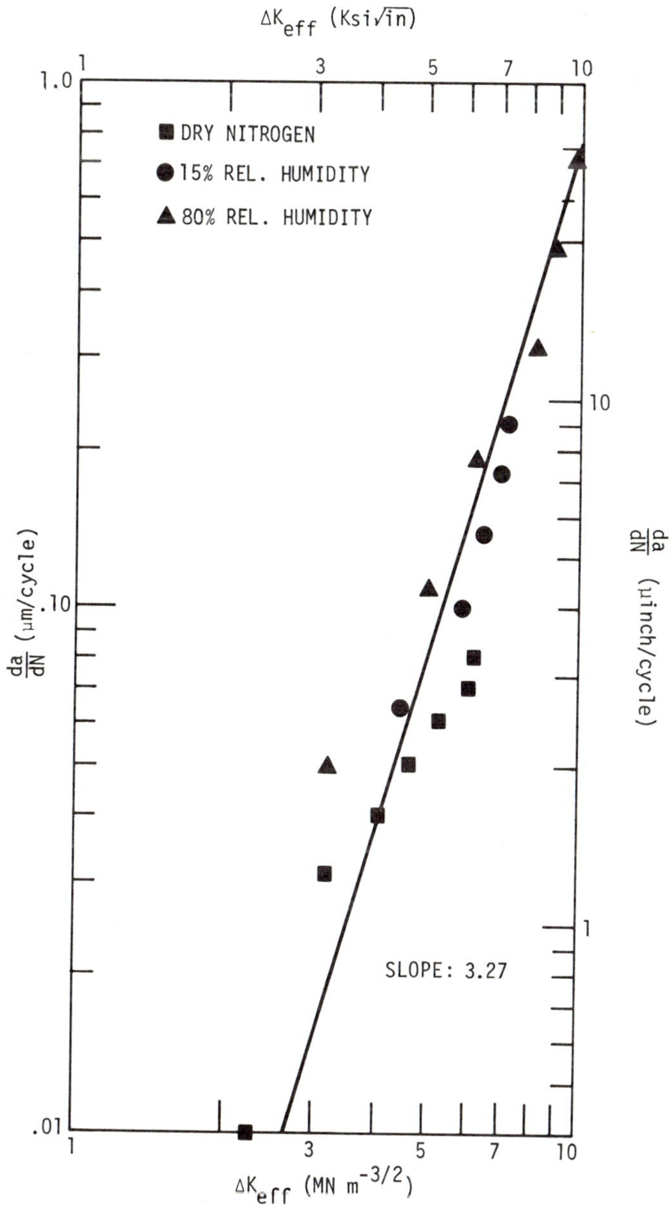


Fig. 6 Crack propagation rate vs. $\Delta K_{eff}$ in Al 7075-T651.

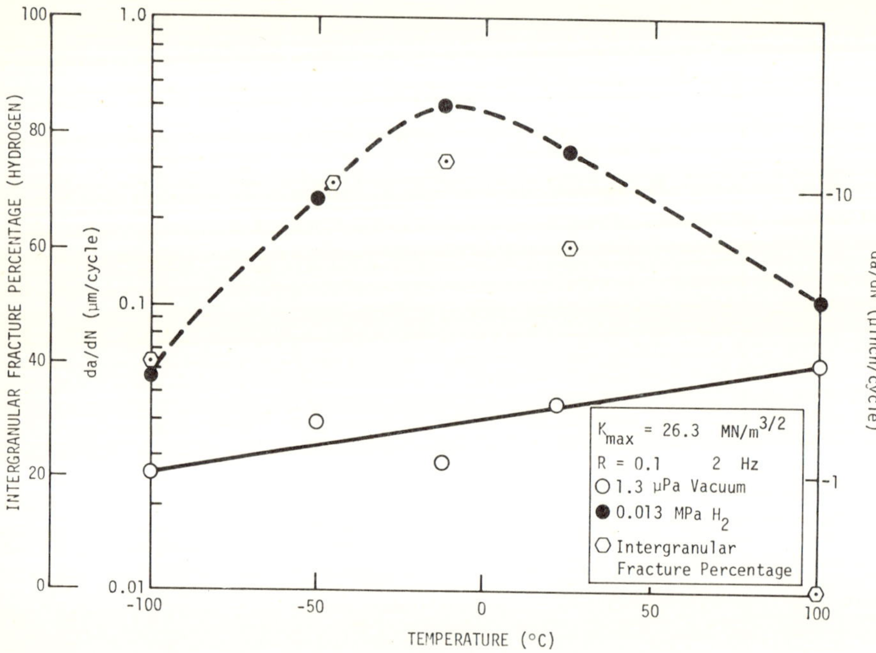


Fig. 7 Fatigue crack propagation rate in hydrogen and vacuum percentage of intergranular fracture in hydrogen as a function of specimen temperature for Monel-400.

oxygen, hydrogen and nitrogen can alter the slip mode near the fracture surface.

Much of the information resulted from thin film transmission examination of foils cut normal to and including the fracture surface, using a controlled coating and etching technique developed by Morris to preserve the fracture surface. [10] It is observed that cross slip is inhibited during fatigue in oxygen and hydrogen, when compared to vacuum, resulting in slip bands, with larger band spacing and deeper penetration of the effected zone in hydrogen than in oxygen. Further, in hydrogen the fracture mode is also seen to change from transgranular to intergranular. The increase in crack growth rate due to hydrogen is temperature dependent and is attributed to competition between $H_2$ association and absorption kinetics. As seen in Fig. 7, the effect peaks at about -12°C, and is consistent with a concomitant peak in the fraction of intergranular fractures and not with changes in crack closure. Figure 8 compares observed surface morphology at -12°C for vacuum, hydrogen, oxygen and nitrogen.

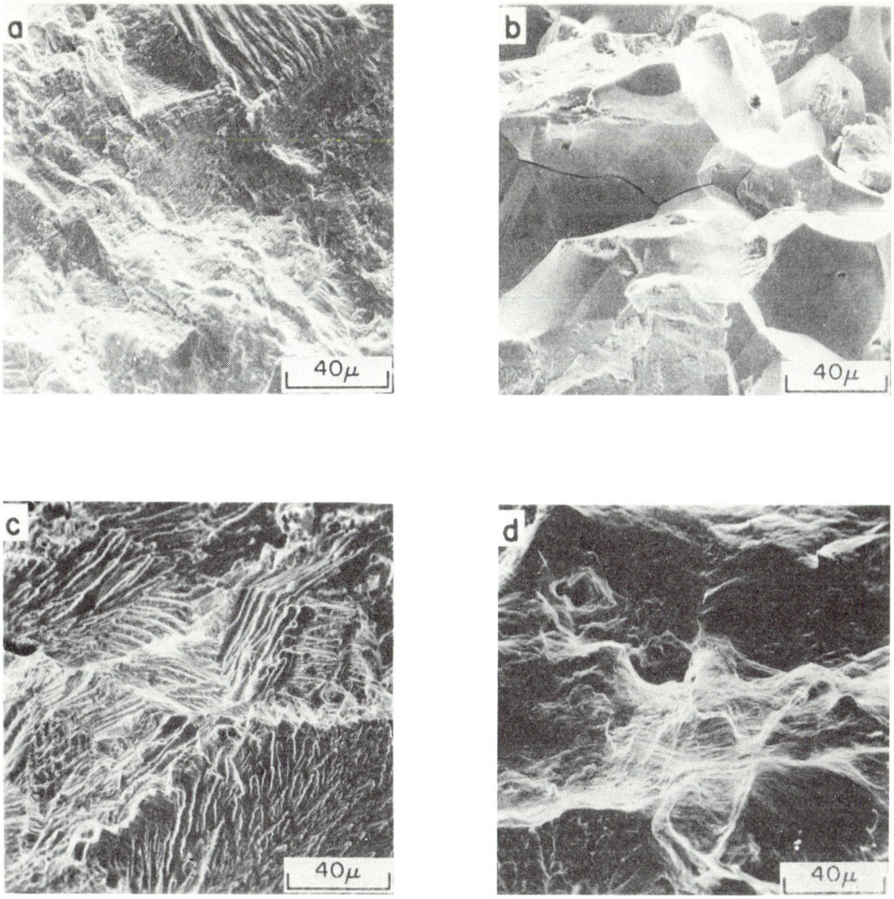


Fig. 8 Scanning electron micrographs of fatigue crack surfaces for Monel-400, -12°C, $K_{max}$ = 26.3 MN/m$^{3/2}$, R = .1 at 2Hz: in vacuum; in 100 torr hydrogen; in 100 torr oxygen; in 100 torr nitrogen.

HIGH $\Delta K$ REGIME

The third distinct region of Fig. 1 is the high $\Delta K$ regime where onset of instability dominates. Here the results of Vroman [11] are of interest, particularly as it relates to crack behavior during cyclic proof test. It was observed that the crack growth during each load cycle can be directly related to the monotonically loaded R-curve. Thus, for high loads, crack growth per cycle is determined by stress intensity related tearing. This growth per cycle is then directly obtainable from an R-curve shown in Fig. 9. Each cycle yields the growth associated with a given $K_{max}$, and accumulation of residual damage is found not to be

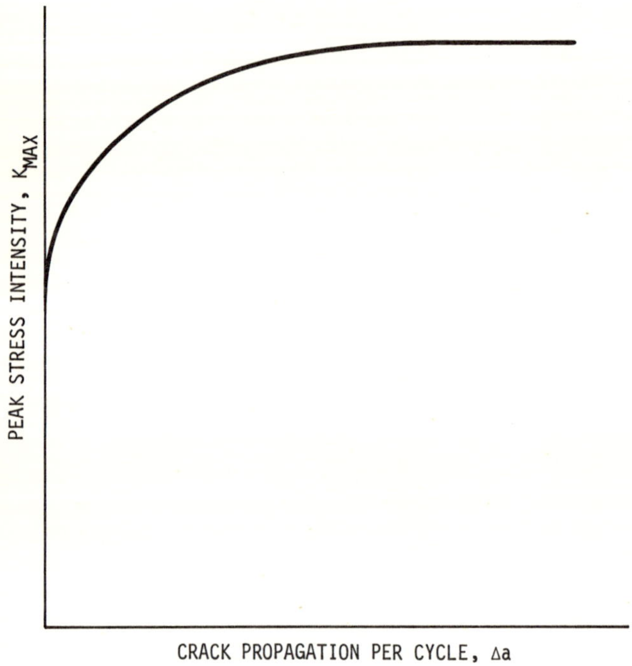


Fig. 9 Schematic resistance curve for structural material.

important. However, the change in ΔK due to the incremental crack growth must be accounted for after each cycle. The mechanism of crack arrest associated with the R-curve, however, still remains to be defined.

CONCLUSIONS

This paper has discussed a broad range of experimental results relating fatigue crack growth. We conclude that a great deal of additional effort is necessary to establish the controlling factors that define growth rate under variable amplitude loading and environmental conditions. Further, effects of metallurgical variable on fatigue crack growth have yet to be ascertained. The effects of relaxation effects due to long time holds at fixed load must also be investigated.

1. Elber, W., Engineering Frac. Mech., 2, 37 (1970).

2. Elber, W., ASTM STP 486, ASTM, 230 (1971).

3. Buck, O., Ho, C. L., Marcus, H. L. and Thompson, R. B., ASTM STP 513, ASTM, 280-291 (1972).

4. Buck, O., Ho, C. L., and Marcus, H. L., J. Eng. Fract. Mechs. 5, 23 (1973).

5. Paris, P. C. and Erdogan, F., J. Basic Engrg., Trans. ASME D 85, 528 (1963).

6. Buck, O., Frandsen, J. D., Ho, C. L. and Marcus, H. L., Proc. 3rd International Conf. on the Strength of Metals and Alloys, 1, The Institute of Metals, and Iron and Steel, 462 (1973).

7. Buck, O., Frandsen, J. D. and Marcus, H. L., to be published in J. of Engrg. Frac. Mech.

8. Frandsen, J. D., Morris, W. L. and Marcus, H. L., to be published in Proc. of "Hydrogen Embrittlement: Effect on Metal Properties, Selection and Design," Seven Springs, Penn. September 1973.

9. Frandsen, J. D., Paton, N. E. and Marcus, H. L., to be published in Met. Trans.

10. Morris, W. L., to be published.

11. Vroman, G. A., Rocketdyne Publication 572-K-16, Canoga Park, California, September 1973.

ACKNOWLEDGEMENTS

The authors would like to acknowledge the experimental assistance of R. V. Inman in developing the acoustic measuring techniques and the critical review of the manuscript by Dr. John Tien.

# FATIGUE CRACK GROWTH UNDER VARIABLE-AMPLITUDE LOADING

J. Schijve

Delft University of Technology, Aeronautical Department

## ABSTRACT

An attempt is made to give a full description of fatigue damage, including the amount of cracking, crack front orientation, crack tip blunting, crack closure and cyclic-strain hardening in the crack tip zone. Examples of interaction effects during variable-amplitude loading are given. The usefulness and limitations of the stress intensity factor for stationary variable-amplitude loading are considered. Recent models for crack growth are surveyed. Finally the paper presents requirements and a framework for a new model, accounting for observed interaction mechanisms.

## NOTATION LIST

| | |
|---|---|
| CA | constant amplitude loading |
| COD | crack opening displacement |
| FST | flight-simulation test |
| $K_{mF}$ | stress intensity for flight-simulation test, Eq. (6) |
| $K_{rms}$ | root mean square of K during random load test |
| $S_{mF}$ | mean stress in flight during FST |
| VA | variable-amplitude loading |

## INTRODUCTION

The propagation of fatigue cracks in an aircraft structure is an essential aspect of the fail-safe quality of the structure.

Consequently it is a design problem to ensure slow crack propagation. The main variables to achieve this goal are: (i) material selection (type of alloy, heat treatment), (ii) production techniques (e.g. integral structure or built up structure) and (iii) type of structure, dimensious (crack stopping features). After having made a choice for these variables the following problem is to predict the crack propagation rate assuming that crack nucleation has occurred. Alternative approaches are:

(a) calculations based on available data
(b) testing full-scale components
(c) compromises, such as combinations of calculations and simple tests on the material selected.

With respect to calculations considerable progress has been made for constant-amplitude loading [1, 2]. Actually the problem in that case has been reduced to calculating the stress intensity factor for a crack in a structure. The correlation between crack growth under constant-amplitude loading in a simple sheet specimen and in a complex structure by employing the stress intensity factor seems to be well established. However, in many practical cases service load histories are very much dissimilar from constant-amplitude loading, see for an example Fig. 1 [3].

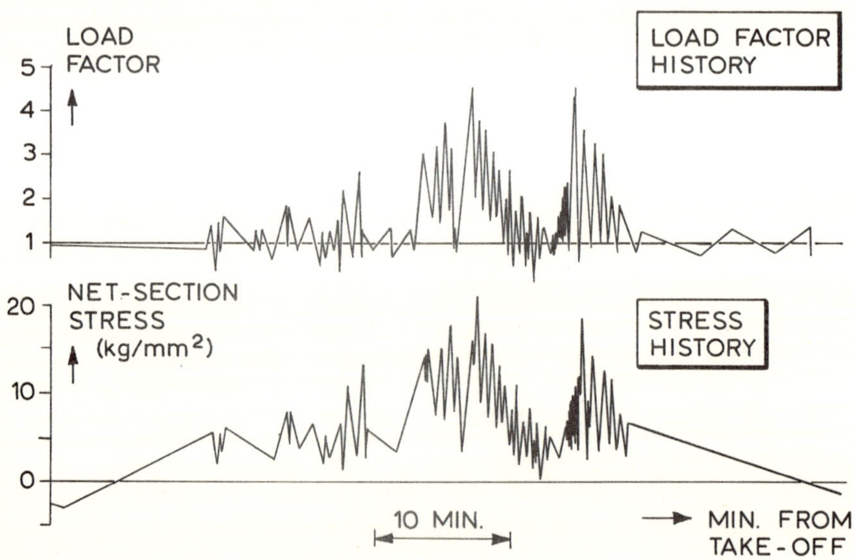


Figure 1. Service load history during a typical F-104 G flight. Maneuver loads, stress in wing. Record of minima and maxima connected by straight lines [3].

Estimation of the crack growth rate under such a loading, having available only constant-amplitude data as basic information, requires the solution of two intricate problems:
(A) The random load history has to be decomposed into a collection of discrete load cycles or ranges for which constant-amplitude data are available.
(B) A cumulative damage theory is required for predicting the crack extension in each cycle.
For both topics proposals were made in the literature [3 - 6]. Nevertheless the problem is still far from being solved.

The present paper is primarily dealing with incremental crack growth under veriable-amplitude loading, which is the background of problem B. By surveying the various aspects involved it will be shown that a theoretical crack propagation model for prediction purposes has to cope with a rather complex phenomenon. Basic requirements for a new model are indicated.

## FATIGUE DAMAGE

Fatigue damage is most generally defined as the changes of the material caused by cyclic loading. Apparently the amount of cracking is a prominent aspect of these changes, but there are more changes than cracking alone. A survey is given in Fig. 2, which shows that cracking should be replaced by crack geometry, characterized by a number of complex aspects. Moreover, strain-hardening and residual stress in the crack tip zone have to be added. As a consequence a complete description of fatigue damage implies that all aspects listed in Fig. 2 should be known in quantitative terms. This must be an extremely complex description since several aspects cannot simply be indicated by a single number, consider as an example the plastic strain distribution in the crack tip zone.

The picture of fatigue damage accumulation would considerably simplify if there was a unique relation between the aspects in Fig. 2. That means if for a certain amount of cracking all other aspects b - f would always be the same, irrespective of the magnitude of the cyclic load that produced the amount of cracking. Unfortunately this situation does not apply. For example, the crack front orientation at a certain length will be predominantly in the tensile mode for low-amplitude loading, whereas it is in the shear mode for high-amplitude loading. Accordingly the other aspects will differ as well. It will be shown that this is important for crack growth under variable-amplitude loading.

| | CRACK GEOMETRY | Additional aspects |
|---|---|---|
| 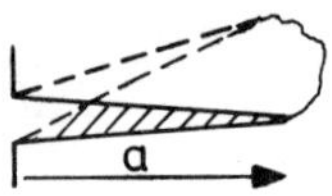 | a. Amount of cracking | - crack length<br>- shape of crack front, straight or curved |
| 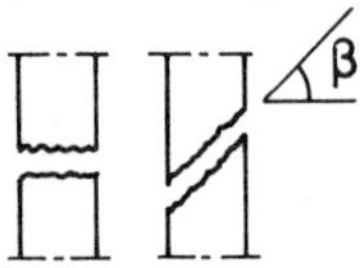 | b. Crack front orientation | - tensile mode<br>- shear mode, single shear or double shear<br>- mixed mode |
|  | c. Crack blunting (and sharpening) | - shape of crack tip |
| 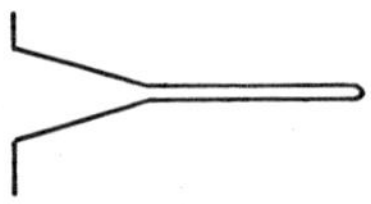 | d. Crack closure | - plastic deformation in the wake of the crack |
| | MATERIAL CONDITION AT TIP OF CRACK | |
| 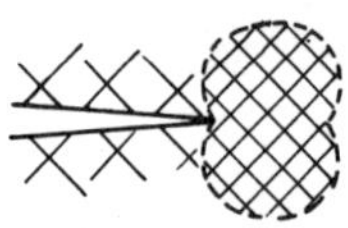 | e. (cyclic) strain-hardening | - distribution in crack tip zone |
| 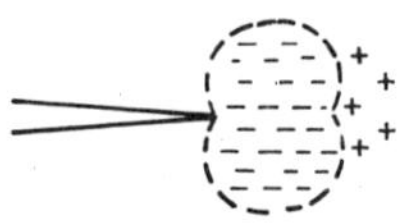 | f. Residual stress | - plastic deformation in crack tip zone |

Figure 2. Aspects of fatigue damage.

INTERACTION EFFECTS

In the previous section it was pointed out that fatigue damage accumulation and crack growth are not synonymous. The crack extension in a load cycle, $\Delta a$, is only one part of the damage increment in that cycle. The damage increment will be a function of: (i) the crack geometry being present before the cycle started, (ii) the condition of the crack tip material before the cycle started and (iii) the external load or stress variation to be applied in that cycle. Arguments (i) and (ii) imply that the damage increment, including $\Delta a$, will be dependent on the preceding cyclic-load history. Similarly the stress cycle applied will effect damage increments in subsequent load cycles. In the literature the dependence on the preceding history and the effect on future damage increments are referred to as "interaction effects".

Before considering the significance of interaction effects another question will be discussed first. If interaction effects would <u>not</u> occur, which theoretical consequences can be indicated? The absence of interaction effects implies that the damage increment cannot be dependent on all aspects listed in Fig. 2. A load cycle will simply add more damage to the damage already present. For convenience the concept "fatigue damage" may then be replaced by crack length. The size of this crack length (a) should give a sufficient indication for further damage accumulation. The damage increment in a cycle can be indicated by $\Delta a$ and since it is independent of the prehistory:

$$\Delta a = da/dn \tag{1}$$

where the crack rate da/dn is a function only of the cyclic stress applied in that particular cycle and the crack length being present. For constant-amplitude loading (CA tests, constant $S_a$ and constant $S_m$), it was amply shown that the dependence can be accounted for by the stress intensity factor $\Delta K = C\ \Delta S \sqrt{\pi a} = C\ 2S_a \sqrt{\pi a}$

$$da/dn = f_R\ (\Delta K) \tag{2}$$

The subscript R indicates that the relation is depending on the stress ratio $R = S_{min}/S_{max}$. Crack growth under variable-amplitude loading then follows from:

$$a = a_o + \Sigma\ \Delta a_i = a_o + \Sigma\ (da/dn)_i \tag{3}$$

$a_o$ is the initial crack length and i refers to the successive cycles of different magnitude. Equation (3) is a simple summation process of crack length increments, which are obtained as results from CA tests Eq. (2).

Some illustrations of interaction effects will now be given.
1. The crack growth delaying effect of an occasional peak load inserted into a CA test is well documented in the literature. Original explanations were based on residual stresses in the crack tip zone only. Later it became clear that crack closure [7] can also explain the delay and its contribution was clearly shown in [8].
2. In a recent test series [9] crack-opening-displacement (COD) measurements were made during flight-simulation tests. Load sequences as shown in Fig. 3 were applied. COD measurements during the larger load amplitudes are shown in Fig. 4. When the load

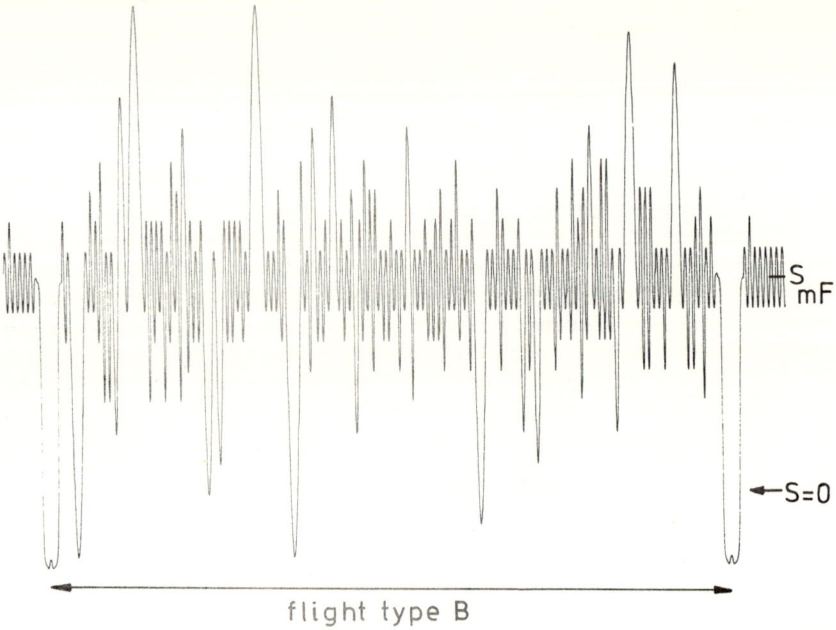


Figure 3. Load history during flight-simulation test. Flight with severe gust loads [9].

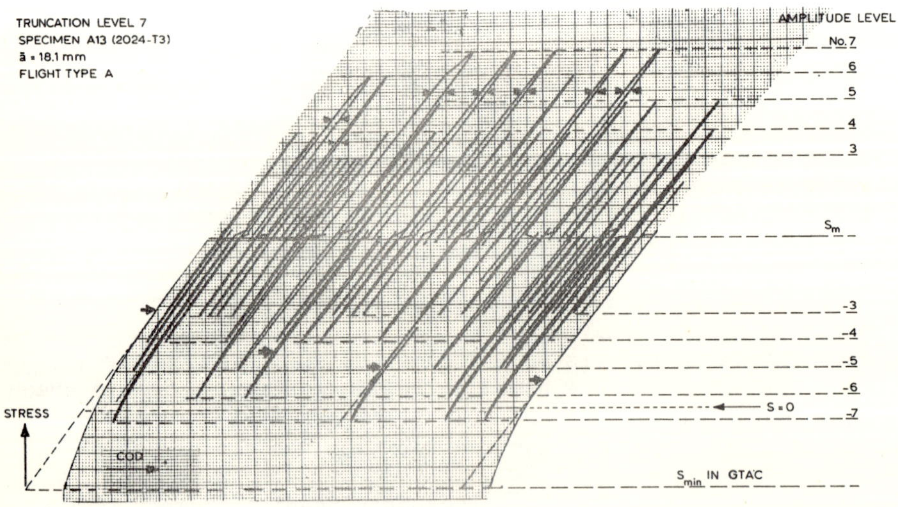


Figure 4. COD measurements during a severe flight of a flight-simulation test [9].

passed the mean stress level in flight small horizontal shifts were given to the recorder for separation of the loops of the individual loads. Two interesting observations can be made. First, the five positive peak loads occurring in the same flight do not have the same effect. The non-linearity at the top of the former one is mainly due to crack extension as shown by crack growth observations. Apparently the other peak loads give a much smaller contribution. This is in agreement with results from a more elementary study [10], where it was shown that $\Delta a$ during the first peak load was much larger than da/dn corresponding to the size of load cycles (accelerated crack growth). It was argued that incompatible crack front orientation and crack sharpening and strain-hardening during the preceding cycles (aspects b, c and e in Fig. 2) may have contributed to this phenomenon.

The second observation is related to crack closure which is clearly evident from the non-linearity in the lower part of the recording. Arrows indicate the stress levels below which the crack is partly closed. The first high load, causing crack extension and plastic elongation of the crack tip zone, has decreased the crack closure stress. After further crack growth during subsequent flights the crack closure stress was again restored to a higher level. This once again confirms the occurrence of crack closure.

3. In Fig. 5 an example of incompatible crack front orientation is shown.

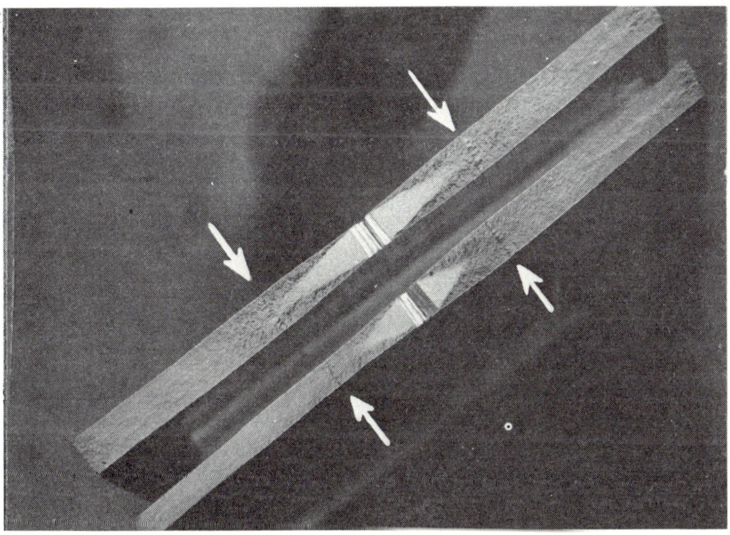

Figure 5. Fracture surfaces (specimen thickness 4 mm) showing incompatible crack front orientation (arrows) during an intermittent block of low-amplitude cycles [10].

The high-amplitude loading caused the transition from the tensile mode to the single shear mode. When the latter one was present a block of low-amplitude cycles was applied. In a CA test with the latter cycles only the crack front at the crack lengths indicated by arrows was still in the tensile mode. In other words there was a mismatch between the low amplitude loading and crack front orientation. The crack rate during the low-amplitude cycles was far below the corresponding value of the CA tests (delayed crack growth). In addition to the incompatibility of the crack front orientation other aspects may have contributed also (aspects c - f in Fig. 2).
4. Sequence effects have been observed in variable-amplitude test series. In a recent investigation [11] the same variable-amplitude cycles were applied in a random sequences and in a programmed sequence. Crack growth was about three times faster for the former case.

## PREDICTION PURPOSES

Some remarks should be made on the goals to be achieved and the accuracies required. Two classes of problems can be defined:
1. Prediction of trends. In these types of problem the effect of a certain variable is looked for. Some examples will be mentioned.
1a. The load spectrum in service may differ from the load spectrum applied in a full-scale fatigue test, during which crack propagation data were obtained. For instance, it may contain relatively less high-amplitude cycles, or relatively more low-amplitude cycles. The question is, how should the full-scale test data be corrected for these effects.
1b. Comparing two alternative designs for the same structure, for instance an old design and a redesign. The question is how much slower the crack propagation will be in the redesigned structure.
1c. Considering two alloys for material selection a relevant question is how the propagation rates under the service load history will compare.
In these practival problems a comparison has to be made between two or more different cases. It would be very helpful already if the order of magnitude of the effect of the variable can be indicated. This implies that high accuracies appears to be unnecessary.

2. Estimation of the crack propagation rate. For a specific structure the crack propagation rate for the anticipated service load spectrum has to be estimated. This has to be known in view of fail-safe consideration. In this case a better accuracy should be achieved than for predicting the effect of trends.

## INTERACTION MODELS FROM THE LITERATURE

Non-interaction models are incorrect from a physical point of view, but one might hope that predictions with Eq. (3) could still give useful indications for problems as listed before. Unfortunately the non-interaction concept can highly misjudge the effect of certain variables and even the sign of the predicted trend may be incorrect [9, 12]. The simple conclusion is that there is an apparent need for a more reliable prediction method.

Interaction models for crack propagation were proposed by Willenborg, Engle and Wood [4], by Habibie [5] and by Wheeler [6]. Although these models are different they have some aspects in common:

1. Crack growth increments ($\Delta a$) are calculated cycle by cycle.
2. The crack growth increment in each cycle is correlated to the increment that would have been obtained in a constant-amplitude test (CA data) at the same crack length.
3. Mathematically Habibie and Wheeler employ a delay factor $\varphi$ defined by

$$\left(\frac{da}{dn}\right)_{VA,\ a=a_i} = \varphi \cdot \left(\frac{da}{dn}\right)_{CA,\ a=a_i} \tag{4}$$

(VA = variable amplitude test.) Willenborg et al employ a reduced $K_{max.}$ value. All three models basically consider delay effects due to high peak loads only.

The essential elements of the models were recently compared by Broek [13]. He indicated the similarity of the models of Habibie and Wheeler, expressing some preference for the Wheeler model in view of its more clear physical conception. He also mentioned the doubtful assumption about the reduced K value in the Willenborg model.

Habibie developped his method especially for flight load conditions. He introduces several empirical constants ($\alpha$, $\gamma$, $\xi_i$, $\eta_i$) for calculating the delay factor $\varphi$. These constants were derived from comparisons between flight-simulation test data and non-interaction predictions. In view of the many empirical constants the method should be expected to be more flexible in predicting results for similar loading conditions. However, for other conditions its usefulness have to be proven again, because the method is largely an empirical method.

In the Wheeler model and the Willenborg model the dimensions of the plastic zones are calculated with K-values, assuming that the same yield stress applies in any case. The Willenborg model does not employ any additional constants, whereas the Wheeler model includes one more constant m. The latter constant gives some flexibility for matching data, but unfortunately the m-value required for that purpose may be somewhat unsystematic [14].

The three models consider delay effects only, which are assumed to be due to crack tip plasticity. This involves severe limitations, examples being listed below.

1. Crack closure does not occur in the above models. This implies that the observations of Van Euw et al [8] on the effect of overloads cannot be explained. The significance of this observation for the comparison between programmed load sequences and random sequences was recently stressed by Katcher [14].

2. Incompatible crack front orientation as discussed before does not occur in the above models. However, it was shown to be significant [10].

3 Crack tip blunting and sharpening should have some effect on the distribution of stress and strain in the crack tip zone, but they do not occur in any model.

4. Cyclic strain-hardening is another aspect ignored by the above models. According to these models multiple overloads, applied in one block, should have a harmfull effect as compared to a single overload. The delay would be the same, but the crack growth increment of the overloads would be larger. However, test results of Hudson and Raju [15] indicated that a batch of overloads gives a longer crack growth delay than a single overload. This may well be attributed to an increased cyclic hardening, although the full explanation may be more complex.

A physically sound model should be able to deal with all above observations and not with just one of them. This requirement leads to a complex formalization of a model as will be shown later on.

## THE STRESS INTENSITY FACTOR APPLIED TO VARIABLE-AMPLITUDE LOADING

The application of K for correlating data from constant-amplitude tests has been very succesful so far. Unfortunately crack growth delays cannot be reconciled with Eq. (2), unless further refinements are introduced [16]. However, it is difficult to see how simple adaptions to the K-concept could explain the observations discussed before. It is then a promising result that crack propagation under stationary random loading can be correlated if $S_{rms}$ is substituted into K [17-19] , as proposed by Paris [20].

$$K_{rms} = C\, S_{rms} \sqrt{\pi a} \tag{5}$$

The basic idea is that the random variations of the stresses in the crack tip zone are fully described by $K_{rms}$.

Recently [21] an attempt has been made to apply the same idea to flight-simulation tests (Fig. 3). Actually the loading is a mixture of random gust loads and deterministic ground-to-air cycles. Nevertheless the loading has a statistically stationary character. New load histories were obtained by amplifying all stress levels with the same factor, without changing the load sequence. In engineering terms this means that the design stress level was changed. For that reason one stress value fully describes the intensity of the load history in these test series. Taking the mean stress in flight $S_{mF}$ (Fig. 3), as such a value it was hoped that $K_{mF}$ would correlate the crack propagation data.

$$K_{mF} = C\ S_{mF} \sqrt{\pi a} \qquad (6)$$

Results for the 7075-T6 alloy are shown in Fig. 6. A similar graph was obtained for the 2024-T3 alloy. Unfortunately the results indicate that a unique correlation

$$(da/dn)_{FST} = f\ (K_{mF}) \qquad (7)$$

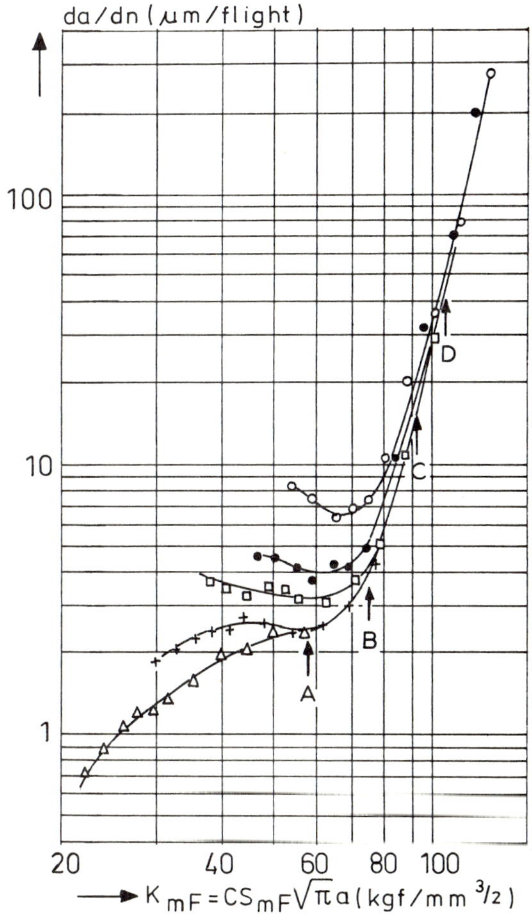


Figure 6. Crack propagation rates (microns per flight) as obtained in flight-simulation tests [21].

apparently does no longer hold true. This is indeed most unfortunate because the validity of Eq. (7) would have allowed a simple extrapolation from one stress level to another level. Several practical applications could have been foreseen.

Since $\Delta K$ and $K_{rms}$ are applicable to crack growth under CA and random loading, it remains to be explained why $K_{mF}$ fails to correlate the data in Fig. 6. The basic idea of the application of K to fatigue crack growth is: two specimens with different crack length and different cyclic loads, but the same instanteneous K-values, should have the same crack growth rate [20]. This similarity, however, will only be valid if the material of the two specimens is the same. In more detail, the material in the plastic zone at the tip of the crack should be the same. This requires that cyclic-strain hardening and residual stresses at the tip of the crack should be the same. However, they are depending on the previous cyclic-strain history. The significance of this aspect will be illustrated by discussing Fig. 7.

Figure 7 shows the relation between K and a for two specimens, fatigue tested at a high and a low stress level respectively. At a crack length $a_1$ in specimen 1 and a crack length $a_2$ in specimen 2 the same K-value applies and one might expect similar crack rates. At $a_1 + \Delta a_1$ and $a_2 + \Delta a_2$ similar crack rates should again be expected. However, since $\Delta a_2 > \Delta a_1$ the cyclic-strain histories at the tips of the two cracks cannot have been the same. A similar cyclic-strain history would require the two curves to be parallel.

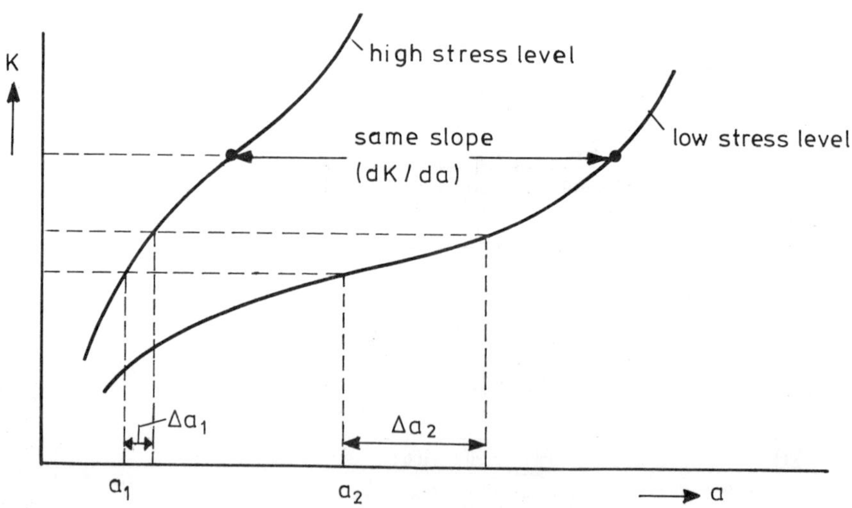


Figure 7. K as a function of crack length for two different stress levels [21].

More specifically, similar crack rates in two specimens may be expected if both K and dK/da are similar. Unfortunately these two requirements are generally incompatible, and hence Eq. (2) should be rewritten as:

$$da/dn = f_R\ (\Delta K,\ dK/da) \tag{8}$$

As illustrated by Fig. 7 equal dK/da values will occasionally occur for the same K-value and two different stress levels. Such K-values, calculated in [21], are indicated in Fig. 6 by arrows. It is noteworthy that equal da/dn values are apparently obtained then. This gives some support to Eq. (8).

It follows from the previous reasoning that the validity of $da/dn = f_R\ (\Delta K)$ for CA tests, independent of dK/da, should not be appreciated as an obvious law, but rather as a kindness of nature. In [20] Paris pointed out that the load history at a certain point in the material through which the crack will pass may be important. He emphasized that a relation between da/dn and K for any type of loading should be confirmed by tests. If da/dn is uniquely correlated to K this implies that the effect of dK/da in Eq. (8) is negligible. Apparently this applies to CA loading, although dK/da values may be highly different. An encouraging result [22] was obtained in a comparison between specimens with end-loading and specimens with wedge-force loading. The value of dK/da is positive for the former loading and negative for the latter one. Nevertheless da/dn could well be correlated by K alone. The cyclic strain-history in the plastic zone at the tip of the crack is changing only slowly from cycle to cycle in these tests. That may explain why differences in the cyclic-strain history (due to different dK/da values) do not affect the crack rate.

Another encouraging result was the applicability of $K_{rms}$ in Eq. (5). In [21] it was pointed out that random loading with a Rayleigh distribution for the higher peak loads could be relatively free from interactions effects. In flight-simulation tests, however, such effects are known to be present [12]. Under such conditions the load history in the plastic zone becomes important and an effect of dK/da should be expected.

## REQUIREMENTS FOR A NEW MODEL

The previous discussion indicates that significant interaction effects do occur during variable-amplitude loading. These effects are caused by a number of different mechanisms. Basically a reliable model should incorporate all these mechanisms. An attempt will now be made to indicate the analytical consequences of this requirement.

Crack extension occurs during a load increase only. The amount of crack extension ($\Delta a$) is supposed to be a function of the plastic deformation in the crack tip zone ($\varepsilon^+$).

$$\Delta a = f_1 (\varepsilon^+) \tag{9}$$

The plastic deformation in the crack tip zone will be a function of the local stress variation ($\Delta S_{tip}$) and the local yield stress ($S_{yield,\, tip}$), which is depending on the local strain hardening.

$$\varepsilon^+ = f_2 (\Delta S_{tip}, S_{yield,\, tip}) \tag{10}$$

As a first approximation crack closure will be accounted for by assuming that the crack tip zone remains unaffected by the external loading until the stress has exceeded the stress level to fully open the crack ($S_{c.o.}$). It will also be affected by the crack tip geometry ($r_{tip}$) and the crack front orientation (angle $\beta$).

$$\Delta S_{tip} = f_3 (S_{max} - S_{c.o.}, r_{tip}, \beta) \tag{11}$$

In order to calculate $\Delta a$ from Eq. (9) by substituting Eqs. (10) - (11) the instantaneous values of $S_{yield,\, tip}$, $S_{c.o.}$, $r_{tip}$ and $\beta$ have to be known. They all depend on the preceding history. $S_{yield,\, tip}$ and $r_{tip}$ will change from cycle to cycle, while both are dependent on the local strain history, indicating that the two quantities are interdependent.

$$S_{yield,\, tip} = f_4 \text{ (local strain history)} \tag{12}$$

$$r_{tip} = f_5 \text{ (local strain history)} \tag{13}$$

On the other hand the crack opening stress is a function of the plastic strain history in the wake of the crack. The same will apply to $\beta$.

$$S_{c.o.} = f_6 \text{ (strain history along the crack)} \tag{14}$$

$$\beta = f_7 \text{ (strain history along the crack)} \tag{15}$$

The local strain history is an accumulation of positive and negative strain increments. Indicating the local position by x=a and the cycle number by i the local strain history is characterized by the sum of the strain increments during load increases ($\varepsilon^+$) and decreases ($\varepsilon^-$).

$$\text{local strain history} \mathrel{\hat{=}} \sum_{i=1}^{n} \varepsilon^+_{x,i}, \varepsilon^-_{x,i} \quad (x=a) \tag{16}$$

Making the summation for all positions along the crack one obtains the strain history along the crack.

$$\text{strain history along crack} \mathrel{\hat{=}} \sum_{i=1}^{n} \varepsilon^+_{x,i}, \varepsilon^-_{x,i} \quad (x) \tag{17}$$

In order to calculate strain histories an equation for $\varepsilon^-$ similar to Eq. (9) is still required. Assuming that all above equations were available the crack extension in a load cycle ($\Delta a$) can be calculated. First the strain histories of the preceding load cycles are calculated, Eqs. (16) and (17). Substitution in Eqs. (12) - (15) gives the input data for Eqs. (10) and (11) and $\Delta a$ then follows from Eq. (9).

Obviously the model outlined above requires highly complex calculations depending on the functional relations. In the above equations $\varepsilon^+$, $\varepsilon^-$ and $S_{yield, tip}$ were tacitly supposed to be single-valued variables. However, in reality they should be characterized by a spatial distribution. Consequently, to arrive at a feasible model some simplifications will be necessary. It then should be kept in mind whether the simplification appears to be acceptable in view of interaction mechanisms, which can be active. A first simplification may be the omission of $\beta$, because the model could still be relevant to crack propagation occuring predominantly in plane strain ($\beta$=0). More simplifications may be considered, however, the models presented in the literature were simplified too rigorously. As a consequence crack growth accelerations became impossible.

## CONCLUDING REMARKS

1. During variable-amplitude fatigue several interaction mechanisms can exert a significant effect on crack growth. As a result non-interaction techniques are incorrect for the prediction of crack propagation rates. These techniques are even unreliable for indicating trends when comparing different conditions. Recently published crack growth models have been based on crack growth delays caused by high peak loads. Unfortunately, the models do not pay much attention to the physical mechanism of this interaction. Actually the starting point is too restrictive for having confidence in a broader application beyond the conditions of the tests, which served to determine the empirical constant(s) of the model.
2. A reliable model for crack growth should account for all interaction mechanisms that are known to exert significant effects. The aim of the model outlined in this paper is a preliminary framework for this purpose. An attempt was made to incorporate the present knowledge about fatigue damage accumulation, but it cannot be denied that the problem is still open to discussion. The derivation of the various function is still another tremendous problem. Nevertheless, it is thought that a more rational model is a challenging possibility by now.
3. The stress intensity factor $\Delta K$ produced excellent correlations for crack growth under constant-amplitude loading. For stationary random loading the same applies to $K_{rms}$, although the empirical evidence is still somewhat limited. Unfortunately for stationary flight-simulation loading $K_{mF}$ was unsuccessful in correlating the

crack growth data. This has led to a re-evaluation of the K-concept indicating that dK/da may also effect crack growth for stationary load histories, especially if significant interaction effects occur.

## REFERENCES


[1] Poe, C.C., Fatigue crack propagation in stiffened panels. ASTM STP 486,79, 1971.

[2] Salvetti, A. and Del Puglia, A., Fatigue crack propagation in stiffened panels. ICAS Paper 72-40, 1972.

[3] Van Dijk, G.M., Statistical load data processing. NASA SP-309, 1972, 565.

[4] Willenborg, J.D., Engle, R.M. and Wood, H.A., A crack growth retardation model using an effective stress concept. AFFDL-TM-FBR-71-1, 1971.

[5] Habibie, B.J., Eine Berechnungsmethode zur Voraussage des Fortschritts von Rissen unter beliebigen Belastungen und Vergleiche mit Entsprechenden Versuchsergebnissen. Paper DGLR meeting, Baden-Baden, 1971.

[6] Wheeler, O.E., Spectrum loading and crack growth. J. Basic Engng, Trans. ASME, 181, 1972.

[7] Elber, W., The significance of fatigue crack closure. ASTM STP 486, 230, 1971.

[8] Von Euw, E.F.J., Hertzberg, R.W. and Roberts, R., Delay effects in fatigue crack propagation. ASTM STP, 513, 230, 1972.

[9] Schijve, J., Jacobs, F.A. and Tromp, P.J., Flight-simulation tests on notched elements. NLR TR 74033, 1974.

[10] Schijve, J., Fatigue damage accumulation and incompatible crack front orientation. Engng. Fracture Mech. (to be published shortly).

[11] Schijve, J., Effect of load sequences on crack propagation under random and program loading. Engng. Fracture Mech., 5, 269, 1973.

[12] Schijve, J., The accumulation of fatigue damage in aircraft materials and structures. AGARDograph No. 157, 1972.

[13] Broek, D., Faile-safe design procedures, NLR MP 72021, 1972.

[14] Katcher, M., Crack growth retardation under aircraft spectrum loads. Engng. Fracture Mech., 5, 793, 1973.

[15] Hudson, C.M. and Raju, K.N., Investigation of fatigue-crack growth under simple variable-amplitude loading. NASA TN D-5702, 1970.

[16] Crews, J.H. and White, N.H., Fatigue crack growth from a circular hole with and without high prior loading. NASA TN D-6899, 1972.

[17] Paris, P.C., The fracture mechanics approach to fatigue, in Fatigue, an Interdiciplinary Approach, Syracuse Un. Press, 1964, 107.

[18] Smith, S.H., Fatigue crack growth under axial narrow and broad band random loading, in Acoustical Fatigue in Aerospace Structures, Syracuse Un. Press, 1965, 331.

[19] Swanson, S.R., Cicci, F. and Hoppe W., Crack propagation in Clad 7079-T6 aluminium alloy sheet under constant and random amplitude fatigue loading. ASTM STP 415, 312, 1967.
[20] Paris, P.C., The growth of cracks due to variations in load. Doctor thesis, Lehigh Un., 1962.
[21] Schijve, J., The prediction of fatigue crack propagation under service load-time histories. Paper ICAF Symp. London, July '73.
[22] Figge, I.E. and Newman, J.C., Fatigue crack propagation in structures with simulated rivet forces. ASTM STP 415, 71, '67.

# MODELING OF THE PHENOMENA OF FRACTURE IN SOLIDS

L. I. Sedov, G. P. Cherepanov, V. Z. Parton

Moscow, USSR

## INTRODUCTION

The distinctive feature of the present-day solid mechanics consists in actual development of the macroscopic theory of motion and fracture of deformable solids in close connection with physicochemical processes inside a particle considered as well as with surrounding particles of the body and external objects.

These defects being looked upon as indivisible objects, plastic deformation can be considered as the result of motion and interaction of dislocations in a crystal. It should be emphasized that, physically, plastic deformation acts as a process of accumulation of defects, i.e. as a process of microfracture which is gradually paving the way for a macroscopic fracture.

The analysis of fracture of solids is based on the methods of classical elasticity theory with allowance made for some peculiar physical effects involving formation of discontinuities inside the deformable solids considered. Of critical importance here is the energy required for the formation of cracks. It is a fundamentally new characteristic of the body behaviour not found in the classical theory of elasticity.

In a theoretical study of strength and propagation of a strong discontinuity of displacements in deformable solids it is convenient to proceed from the universal thermodynamic equation which expresses the energy conservation law [1]

$$dE + dU = dA^{(e)} + dQ^{(e)} + dQ^{**} \qquad (1)$$

Here E is the kinetic energy of the body, U is the total internal energy, $dA^{(e)}$ is the total inflow of energy from the outside due to the work of macroscopic body and surface forces, $dQ^{(e)}$ is the total external supply of heat, $dQ^{**}$ is the external macroscopic energy inflow due to special microscopic mechanisms ( such as chemical processes at the body surface, external electromagnetic effects, etc.).

## PHENOMENOLOGICAL CRITERIA AND THE THEORY OF A LIMITING STATE

The presence of structural formations like grains, microcracks, dislocations, packets of molecules, etc. causes the strength of real materials to decrease by two or three orders of magnitude as compared to the theoretical strength corresponding to an ideal molecular order.

Various kinds of materials are characterized by appropriate types of structural formations which are responsible for distinguishing features of their deformation and fracture. Parallel with a physical analysis of a microstructure and microfracture of materials, it is expedient to carry out phenomenological studies of fracture using some models taking account of the most essential features of this phenomenon. We confined ourselves with "close-range" systems*) with no latent parameters (related, say, to chemical reactions, phase transitions, electromagnetic effects, etc.). The behaviour of such systems is described by one of the main types of rheological bodies (elastic-plastic body, viscous and hereditary bodies) or by their combination. In practice, the reasonable choice of the simplest model represents the major difficulty in the study of a concrete rheological

*) For these systems the parameters entering into the governing equation for an arbitrary elementary volume are independent of the state of any other elementary volume.

phenomenon and plays a vital part when solving a problem of strength and fracture.

There exist two different approaches to the modeling of fracture phenomena. In the first place we shall speak about the one which dates back to Galileo. According to this approach, the fracture of a body is assumed to occur at the moment when a certain combination of physical parameters ( such as stress $\delta_{ij}$ , strain $\varepsilon_{ij}$ , temperature (T) and time (t)) attains a critical value at some point of the body. In this case the fracture process itself is ignored. Within the framework of this approach, the problem of strength is, obviously, solved by the appropriate choice of a rheological model and a fracture criterion (in engineering sciences the latter is usually referred to as a strength theory).
Such an approach immediately follows from the assumed phenomenological analysis based upon the above parameters.

As a rule, the role of a criterion is usually played by the maximum main stress, maximum main strain, maximum main tangential or octahedral stress, specific energy of form changes, total specific energy of deformation. Each criterion is suitable for a certain class of materials under quite definite conditions. The proper use of these criteria depends largely on the practical skill of a scientist or an engineer.

The theory of a limiting state and the theory of ideal elastic-plastic continuum offer an idealized description of principal features of the process of deformation and fracture of most solids within the viscous fracture region over a wide range of time, rate of deformation and so forth. The practical significance of these theories is far beyond the scope of evaluation of strength and bearing capacity of constructions (suffice it to mention applications to the processing of metals, soil and rock mechanics as well as recent applications to the problem of fluidization in a chemical technology).

## THE THEORY OF A LIMITING STATE OF CRACKS

Study of fracture criteria on the basis of the approach specified above retains nowadays its main practical importance in many particular cases of the strength design. Nevertheless, investigation of the strength and fracture only in such a manner is insufficient. Numerous examples of fracture of metal structures under action of stresses smaller than the yielding stress as well as some other fracture phenomena are in principle unaccountable within the scope of these criteria. This caused many investigators to give up the Galileo strength conception where the strength is thought about as a certain material constant (it goes without saying that external conditions are assumed to be fixed). This new trend stems from known works by Griffith, Taylor, Orowan, Irwin and by others and is based on analysis of the fracture process in itself [1-20].

The following considerations are commonly used in order to start with. The fracture of a solid body occurs almost always as a consequence of the developing within it of some surfaces where displacements are discontinuous. If discontinuity of the displacement component normal to a surface takes place, the surface is identified as an opening mode crack or a crack simply; if the tangential displacement component undergoes a discontinuity, then the corresponding surface is defined as a shear mode crack or a dislocation. Relative voles of discontinuities of the two types mentioned are different under various specific conditions. The importance of dislocations and slip bends increases, as a rule, when the strength of a material decreases or temperature grows, and also under compression. On the contrary, increase in the strength, temperature reduction as well as occurrence of cyclic loads, aggressive media and radiation lead usually to increase of the role of opening mode cracks.

The development of discontinuity surfaces begins from structural defects within a material which are bound to be considered in an initial moment as certain given finite perturbations. Such perturbations are always present in any system and are usually represented as certain initial cracks or cavities.
This is in a good agreement with direct experimental observations. The subsequent evolution of initial

perturbations may be rather different. The growth of dislocations is characterized by almost simultaneous and stable extension of many dislocations at once so that slip bands and even plastic regions are formed. Therefore the theory of dislocations represents a physical background for the phenomenological theory of plasticity. The growth of cracks is characterized for the most part by developing of a single crack which is more 'dangerous' in a sense*) and by its capability to grow quite rapidly and in an unstable way, which results usually in rupture and clearage of a body into pieces. While working out a strength criterium on the basis of the crack theory, routine strength theories happen to be obtained in the majority of cases, but constants involved should be regarded as quantities depending upon size of initial cracks as well as upon their shape and location. However, for a rather broad number of fracture phenomena the strength of microinhomogeneous bodies does not depend on the size of an original perturbation, that is, of an initial crack, and is determined by specific structural parameters of a body, e.g., by the grain size [2]. Thus, from the formal point of view one is free to look upon these theories as upon the simplest generalization of the usual strength theories due to introduction of one additional inner structural parameter which does not appear in formulation of the rheological model. Such an approach is related to a general idea of including of additional structural parameters into the state equation [1]. Investigation of the fracture process is rather often of independent interest, irrespective of the problem of bearing capacity.

The theory of dislocations and the crack theory turned into existence independently and distinction between their formal mathematical formulations may be explained in the following way. Within the scope of the dislocation theory displacement discontinuities are considered directly so that the corresponding linear theory deals with logarithmic singularities whereas in the crack theory certain force balance conditions are assumed to be valid at a discontinuity surface and exponential singularities arise. However, there exists a fundamental affinity

---

*) However, there are some exceptions, e.g., the crack growth under the conditions of nearly isotropic compression.

of these theories which lies in attributing to singularity factors of both theories the meaning of key parameters governing the fracture process. If there is a singularity of elastic stresses in an arbitrary point at the crack boundary

$$\delta_{ij}= \frac{1}{\sqrt{2\pi r}}\left[K_I \varphi_{ij}(\theta)+K_{II} \psi_{ij}(\theta)+K_{III} \chi_{ij}(\theta)\right] \tag{2}$$

$r$ and $\theta$ being the polar co-ordinates, then coefficients $K_I$, $K_{II}$ and $K_{III}$ named stress intensity factors, corresponding to opening and shear modes of a crack.

The most principal problem in the crack theory consists in formulation of a condition of local fracture at a given point of the crack contour. This problem is solved in the most simple manner in the theory of so-called "quasi-brittle" cracks when the largest dimension of the region of plastic deformation in the vicinity of the contour point under consideration is small as compared with the crack length or with the distance between this point and the closest boundary of the body. The simplest version of this condition was put forward by Irwin on the basis of ideas by Griffith and Neuber. According to this version, the singularity coefficient at a given point at the moment of local fracture (and of beginning of crack propagation at this point) is presumed to be equal to a certain material constant, stresses being calculated under the assumption of ideal elasticity.

The relation (1) at $dQ^{**} = 0$ together with the expression (2) and the Hooke's law gives the Griffith-Irwin condition. In the most important case of opening mode crack under the condition of plane deformation the latter is of the following form

$$K_I \leqslant K_{IC} \left(K_{IC}^2= \frac{2E\gamma}{1-\nu^2}\right) \tag{3}$$

where $K_{IC}$ is a constant of a brittle material, $E$ and $\nu$ are the Young's modulus and Poisson coefficient, respectively, and $\gamma$ is irreversible work per unit surface of a growing crack. A generalization of condition (3) to the case of dynamic crack propagation is to be written as follows:

$$K_I = \left[ 2\pi R(m,\nu)\ F(m,\nu)\ E\gamma \right]^{1/2}$$
$$m = \frac{v}{c_2}, \quad c_2^2 = \frac{E}{2(1+\nu)\delta} \qquad (4)$$

$R(m,\nu)$, $F(m,\nu)$ being known functions.

So far as the coefficient $K_I$, representing a function of external loads, the crack length and geometry of a body, can be found from solution of elastic problem on the whole, the condition of local fracture at the crack boundary gives in principle an opportunity to determine the evolution of the crack and, particularly, to find that combination of external loads which separates the regions of stability and instability.

Various models describing in detail the fracture mechanism at the tip of a quasibrittle crack were previously brought into action; their total number amounts to ten. Nonetheless, the known models which differ by a specific (right or wrong) manner of description of the local fracture in the close vicinity of a brittle crack tip, are equivalent in the sense that all of them lead to the Griffith-Irwin condition while dealing with the limiting state of cracks [2-4].

The scientific domain concerning the strength has been developed fairly intensively during the last half-century. First of all this is due to the progress in producing new materials and alloys with higher strength. If the tensile strength of casting crude iron, utilized usually in structures, has been increased from 30 up to 40 kg/mm$^2$ approximately, during the 19th century, steel alloys with the tensile strength of the order of 200-300 kg/mm$^2$ exist at present. Values of the tensile strength of about 400-600 kg/mm$^2$ seem to be quite within the reach in the following years. Clearly, physical theories concerning the nature of strength and fracture have affected the choice of a possible way of producing more strong alloys to a considerable extent (in the first place, this assertion bears on the theories of dislocations and cracks).

Let us distinguish between two conceptions: the metallurgical strength and the structural strength.

The former one is understood as a tensile strength value obtained while experimenting in a laboratory with smooth samples of a standard size cut off from a supplied material. The real strength of a structure made of this material (i.e. the structural strength) is far less sometimes. The influence of the structure configuration upon the structural strength is defined here as the "scale effect".

The scale effect is absent if the tough fracture occurs. In this case the strength dependence upon the body configuration is to be determined by a calculation, based on a chosen model of the body and the local condition of fracture, in accordance with some theory of strength. For ideal elastic-plastic bodies, the necessity to make use of a special strength theory does not arise so that the strength is calculated within the scope of the very model. If the brittle fracture is realized, the scale effect is always present. In this case the strength dependence on the configuration and size of a body (form and size of cracks and other crack-like defects being included) is to be evaluated according to the theory by Griffith and Irwin on the basis of the model of an elastic body.

The physical nature of the scale effect.

The evolution of ideas concerning the physical character of the scale effect was rather lengthy and tedious. This was caused by the fact that phenomenological notions about brittle fracture and plasticity were purely descriptive and had a relevance to visual observations of the fracture process and of the form of breakage surfaces. The brittle fracture is a rather rapid process characterized by absence of necking and by the orientation of cleavage surfaces along planes with maximal tensile stress. Under the condition of the tough fracture, when considerable plastic deformations occur, a neck is formed on a sample and the cleavage surface is oriented along a plane with maximal tangential stress. However, for real materials some combination of the brittle and tough types of fracture is always realized.

When the scale of a sample increases and there are slits and concentrators a tendency toward rise of the probability of brittle fracture is observed.

Therefore the first problem is how to compare the brittle type of fracture of materials with the tough one, i.e. in what possible manner to determine the resistance with respect to breakage and to define the degree of inclination of a material to be fractured according to the brittle type. In spite of the fact that this problem had been solved theoretically as early as 1920 in the classical paper by Griffith, this paper was not influencing as is well known; practical solving of the problem for structural materials during nearly thirty years.

A difficulty of comparison is caused by the following reason. The scale effect is not inherent to the ideal plasticity[x]) whereas for the brittle fracture the strength depends also on latent structural parameters whose dimension coincides with that of length.

In compliance with the Griffith's theory, the latter are understood as lengths of certain original cracks which are always present within a real material. Actually local concentrators of stresses, i.e. various inclusions, cavities voids, pores etc. can play a role of original cracks.

Defects providing a reason for fracture of a sample or a structure can be conventionally subdivided into two classes. The first one includes defects appearing during a metallurgical or technological process. Defects of the second class are being originated or are developing in the course of the exploitation and working with a structure (for example, corrosion and fatigue cracks). An alloy resulting from a metallurgical process has usually a rather complex structure. It is heterogeneous, anisotropic and is characterized by a complicated distribution of inner stresses. The strength is equal approximately to $K_{IC}/\sqrt{d}$ , where d is the diameter of the most dangerous crack-like defect and $K_{IC}$ represents a material complex function of coordinates (at the moment of local fracture $K_I = K_{IC}$). The main objective

---

[x]) From the purely theoretical view point, the scale effect is insusceptible to be explained within the scope either the models for an ideal elastic-plastic body or the model of the limiting state. The same assertion is true for any other model if the fracture is being described with the help of a theory not dealt with constants of dimension of length.

of the metallurgy in this respect is not only to assure certain conditions of chemical and temperature stability of an alloy but also to provide for occurrence of structural microcells of minimal possible size uniformly distributed throughout the space inside the alloy, their boundaries playing a role of energetic strength barriers. These cells represent, as a rule, grains of a basic metal and of chemically active impurities which are formed at crystallization centres while hardening of a melt. The role of the barriers seems to be played by intercrystalline films built up with atoms of chemically inactive admixtures which have been brought away to the grain boundary during grain growth. An initial crack-like defect is developing during a loading process so that at the fracture moment the quantity $d$ equals the diameter of the largest grain. This explains that the strength of even very brittle alloys changes within a comparatively narrow interval as compared with the strength of amorphous material of the type of glass. Thus the main method of increasing of the metallurgical strength consists as viewed by the mechanics of fracture in increase of $K_{IC}$ and in diminishing of the largest crack length and grain size. The former can be achieved with the help of alloyed additives and by way of heat treatment influencing phase transitions occurring especially at grain boundaries. The latter can be provided for by means of homogenezation of the crystallization process.

## The statistical nature of the scale effect.

The strength of a material represents always a certain random quantity because, first, the exact location of all defects is unknown a priori and, second, if even this location were known, solving of the corresponding mathematical problem would be impossible due to its complexity. A simple consideration that the probability to come across the most large-scale and dangerous defect is larger for a larger volume of a sample forms a background for explanation of the scale effect in terms of the statistical theory.

While constructing the statistical theory one may choose between the following alternatives:

i) By using experimental evidence or intuition to concentrate one's attention on one or several of the most dangerous defects and to look upon other defects as though they were evenly distributed over the volume. Properties of the continuum obtained in such a way are presumed to be known on the basis of macroscopic experimenting, the length and, perhaps, some other parameters, determining the location of the most dangerous defect, being regarded as random quantities with given distribution functions.

ii) To distribute all defects without exception by means of the volume averaging and to consider the resulting medium as a continuum with no defects. Then the local strength of this continuum as well as local stresses are considered as random functions of coordinates whose distribution functions are given for any point within a body whereas mean values of the strength and stresses are to be found from the macroscopic theory or experiment. This approach needs many further assumptions to be made for derivation of final expressions.

The first method is more close to the crack theory because emphasis is laid on the physical nature of the strength and of the scale effect. The second approach is more formal and closer to ideas of strength common in the theory of material resistance.

## FATIGUE FRACTURE IN CHEMICALLY INERT MEDIA

Practically all constructions designed for long service fail due to slow development of cracks resulting in a rapid fracture after the critical state has been attained.

The larger portion of the period of crack development is related to the relatively low load level when the main concept of fracture mechanics concerning the stress intensity factors is valid. Theory of subcritical crack growth constitutes the physical basis of strict engineering design of the lifetime of structures.

### Fatigue crack growth in inert environment (metals).

Consider a structure under a nonstationary load. Let the crack growth under constant as well as diminishing subcritical loadings be neglegibly small (assumption 1). This supposition is quite satisfactory for the most of metallic structures in inert environment. Let us also suppose that in the case of increasing subcritical load the crack length grows (assumption 2). The assumption 3 holds that the body is elastic-plastic while all constants of elastic and plastic state being of dimension of stress (or are dimensionless).

Assumptions 2 and 3 result in the following energy equation [2]

$$\gamma_* = \alpha_2 \frac{K_I^2}{E} + \alpha_3 \frac{K_I^3}{\delta_s^3} \frac{dK_I}{dl} \quad (dl > 0,\ dK_I > 0) \tag{5}$$

where $\gamma_*$ is energy dissipation per unit area of the surface of newly formed nonstationary crack, $E$ and $\delta_s$ are characteristic Young's modulus and yielding constant, $dl$ is the crack length increment with growing coefficient $K_I$ at $dK_I$. Dimensionless values $\alpha_2$ and $\alpha_3$ are some constants of a material which depend on the ratio of elastic and plastic constants only.

Let us formulate the following physical statement [2]: the value $\gamma_*$ is a certain constant of a given material ( $\gamma_*$ - concept). This concept generalized the well known energetic conception by Irwin-Orowan for the nonstationary case.

It allows one to integrate Eq. (5) and to find the crack increment at any interval of monotonous growth of the stress intensity factor. Besides, the use of assumption 1 allows to find the crack increment for any programmed loading (say, when external loads are periodic functions of time, stochastic stationary function etc.).

Eq. (5) may be rewritten as

$$K_C^2 - K_I^2 = 2\beta\ K_I^3 \frac{dK_I}{dl} \tag{6}$$

$$\left(K_C^2 = \frac{E\gamma_*}{\alpha_2}; \quad \beta = \frac{\alpha_3 E}{2\alpha_2 \delta_s^3}\right)$$

Thus the theory contains only two material constants.

E.g., in the case of cyclic load one has the following relation [2]

$$\frac{dl}{dn} = -\beta \left(\frac{K_{Imax}^2 - K_{Imin}^2}{K_C^2} + \ln\frac{K_C^2 - K_{Imax}^2}{K_C^2 - K_{Imin}^2}\right) \qquad (7)$$

Here $n$ is the number of loading cycle, $K_{Imax}$ and $K_{Imin}$ are the maximum and minimum values of the function during a cycle. A large number of experimental investigations were done for the case of cyclic loads which have given various empirical formulae. The most widely known are those by Paris, Forman, Broek and Schijve, Erdogan, Pearson etc. Analysis of published experimental data has shown that Eq. (7) describes experimental results well enough. Further investigations must be done in order to find out limits of validity of the theory and to construct more precise theories. A modified theory must be, to our opinion, based upon energy Eq. (5) or (6) and experimental data giving $\gamma_*$ (or $K_C$) and $\beta$ as functions of frequency and cycle characteristics, body and crack geometry, the type of material etc.

A quite analogous approach can be used in the case when plastic domains are so large that the concept of stress intensity factor is senseless.

## Fatigue crack growth in inert media (polymers)

The mechanisms of subcritical crack growth in structurally unstable materials to which belongs the majority of polymers is based upon a sudden diminishing of activation energy in the region near the end of a crack where high stresses are the case. Theoretical modeling of this phenomenon for stationary loading results in the following relationship [2]

$$\frac{dl}{dt} = V_0 e^{\lambda K_I} \qquad (8)$$

( $V_0$ and $\lambda$ are temperature-dependent material constants and $t$ is time). This relation served as a basis for study of various cases of varying loads and analysis of experimental data. The analysis showed that the model can be used for many materials in a wide temperature range under various loads.

## THE EFFECT OF AN AGGRESSIVE ENVIRONMENT UPON STRENGTH PROPERTIES OF SOLIDS

Tests show that some environments have a selective influence upon the strength of solids. This points to the chemical nature of the phenomenon. However in free bodies an aggressive environment leads only to general corrosion which can be easily observed and designed. In the case of extension the region of the most intensive physical and chemical effect of an aggressive environment is localized at the end of a crack. Thus the environment composition influences first of all the subcritical crack growth. The theory of subcritical cracks growth due to environmental effects serves as a basis of strict engineering design of life time and resistance of solids. Notice once again that the notion of aggressivity is commonly selective. E.g., in some cases an ordinary atmosphere may be very aggressive what is not accounted by experimentators in some cases.

Theoretical analysis of various physical and chemical mechanisms of subcritical crack growth due to environmental effects and their comparative analysis allowed to select the following main mechanisms which are mentioned in the order of their practical importance [2]:

a) local hydrogen embrittlement,
b) electrochemical corrosion,
c) chemical corrosion,
d) adsorption mechanism (in "metal-metal" systems mostly).

In nature there are known many other mechanisms of crack growth due to environmental effects [2].

The theoretical description of these mechanisms was done [2] by means of various mathematical models (the local diffusion model, the microtunnel model, the film model, etc.) which allowed to find fracture diagrams for the mentioned mechanisms. Analysis of experimental data based upon these diagrams allows one to verify and propose hypotheses concerning the reasons of fracture in the difficult problem of diagnostics. The statement of the leading mechanism of the local fracture makes it possible to predict the life time by the help of these theories.

## FATIGUE CORROSION FRACTURE

Simultaneous action of an aggressive environment and nonstationary load furthers the subcritical crack growth even at low load level, when each separate factor is incapable to produce fracture. Theoretical modeling of fatigue corrosion [2] is based upon the use of various combinations of above mentioned mechanisms of subcritical crack growth. The growth of fatigue corrosion microcracks determines the life time of different structures. When operating stresses are much lower than the strength.

## FRACTURE DUE TO EXPLOSION

One of the main problems of fracture mechanics is the study of a fracture process of solids acted upon by short but very intensive pressures developed during an explosion. The most part of a destructed (that is divided into parts) volume of a body is formed due to development of cracks produced by the shock wave. Here we have both the process of propagation and bifurcation of the main cracks and development of initial microcracks present practically in every piece of material.

The theory of fracture of solids by an explosion is based upon introduction of an additional wave of fracture separating the destroyed volume from the intact one. Mechanical properties of material before and after fracture are quite different and are described by different models of continuum. Besides the ordinary conservation laws, an additional condition is put upon the fracture wave: the specific energy dissipation at the front is a certain constant of the given material. This condition is used to determine motion of the front of fracture in space. Under high pressures the fracture front is transformed into the wave of crumbling which in the limiting case corresponds to the selfsustained destruction.

SOME NEW PROBLEMS OF FRACTURE MECHANICS

The study of the problem of strength of ideal lattices is of fundamental significance for mechanics of fracture and its applications. By an ideal lattice a strictly periodic arrangement of atoms in space is meant. The first estimates of theoretical strength had a revolutionary influence upon development of physics of strength and mechanics of fracture. The further progress may be evidently attained by means of the use of quantummechanical ideas from the very beginning. Here great successes may be expected not only in a more profound and precise study of theoretical strength of the crystalline structures which are already known, but in finding of new structures possessing, say, much higher strength.

Nowadays mechanics of fracture finds ever more unexpected and interesting applications in new spheres of rapidly developing technology. Here are some of them.

1. Rocketry and space exploration give a practical importance to the questions of fracture of solid propellents which take place under complex conditions of combustion and sometimes are the reason of unstable regime of functioning. It often is due to presence of too large crack-like defects inside the propellent which may result in unstable combustion with the following mechanism [2]. When the front of combustion approaches to the edge of a crack, its surface starts to burn because the chamber pressure is much higher than that inside the crack. The local pressure and temperature may sharply rise. Besides, due to specific structure of propellent near crack tip, a volume combustion may start at the end of a crack which may result in burning through, i.e. in undersigned regime. The use of fracture mechanics allows to find the conditions of stability of such systems.

2. Alongside with processes of local fracture and volume destruction of great practical and theoretical interest are the processes of slow destruction of the body's surface or its part. Of the greatest importance here are the erosion fractures of the surface layer under force action of flow of gas or liquid (e.g., wearing down of solids due to friction, wearing down of a body in flow with solid particles and other phenomena).

3. Intense study of electrophysical methods of fracture of various materials requires a detailed study of electrophysical and other properties of materials acted upon with intensive highfrequency electromagnetic fields and electric discharges.

4. A pronounced change of physical, chemical and mechanical properties of materials, especially polymers, acted upon with neutron radiation required not only development of technological measures of protection but also stimulated development of new methods of strength design of structural elements (in atomic piles, artificial satellites, space ships and stations).

5. New aspects of study of strength and fracture of solids are seen in connection with the use of lazers capable of generating extremely intensive light beams.

6. At large enough rates of deformation a liquid behaves as a solid body capable of brittle fracture. Here we have new prospects in investigation of such phenomena as cavitation, destruction of liquid polymers, etc.

REFERENCES

1. Sedov, L. I., Mechanics of Continuum, 1, 2, Nauka, Moscow, 1973 (in Russian).

2. Cherepanov, G. P., Mechanics of Brittle Fracture, Nauka, Moscow, 1974 (in Russian).

3. Cherepanov, G. P. and Parton, V. Z., Fracture mechanics, in Mechanics in the USSR for 50 years, Nauka, Moscow, 1972 (in Russian).

4. Morozov, E. M. and Parton, V. Z., Mechanics of Elastic-Plastic Fracture, Nauka, Moscow, 1974 (in Russian).

5. Fracture. An advanced Treatise, 1-7, Liebowitz, H., Ed., Academic Press, New York - London, 1968.

6. Panasyuk, V. V., Limiting Equilibrium of Brittle Bodies with Cracks, Naukova Dumka, Kiev, 1968 (in Russian).

7. Sih, G. C., Methods of analysis and solutions of crack problems, in Mechanics of Fracture, 1, Sih, G. C., Ed., Noordhoff International Publishing, Leyden, 1973.

8. Sneddon, I. N. and Lowengrub, M., Crack Problems in the Classical Theory of Elasticity, J. Wiley, New York, 1969.

9. Tetelman, A. S. and MacEvily, A.J., Fracture of Structural Materials, Wiley, New York, 1967.

10. McClintock, F. A. and Argon, A. S., Mechanical Behavior of Materials, Addison-Wesley Publishing Company, Inc., Reading. Massachusetts, USA, Addison-Wesley (Canada) Limited, Don Milles, Ontario, 1966.

11. Brown, W. F., Jr. and Srawley, J. E., Plane strain crack toughness testing of high strength metallic materials, ASTM Special Technical Publication, 410, 1969.

12. Fracture toughness testing and its applications, A Symposium Presented at the Sixty-Seventh Annual Meeting American Society for Testing and Materials, Chicago, June 21-26, 1964, ASTM Special Technical Publication, 381.

13. Proceedings of the 3rd International Congress on Fracture, Munich, 1973.

14. Irwin, G. R., Fracture, in Handbuch der Physik, 6, Springer-Verlag, Berlin, 1958.

15. Impact Testing of Metals, ASTM Special Technical Publications, 466, 1970.

16. Johnson, H. H. and Paris, P. C., Engineering Fracture Mechanics, 1, 1, 1968.

17. Vitvitsky, P. M., Panasyuk, V.V. and Yarema, S. Ya., Plastic deformations near crack tips and fracture criteria, Strength Problems, 2, 1973 (in Russian).

18. Rice, J. R., The elastic-plastic mechanics of crack extension, International Journal of Fracture Mechanics, 4, 1, 1968.

19. Willis, J. R., Self-similar problems in elastodynamics, Phil. Trans. Roy. Soc. London, 1973.

20. Yokobori, T., An Interdisciplinary Approach to Fracture and Strength of Solids, Wolters-Noordhoff Scientific Publications Ltd., Groningen, 1970.

# LOAD AND ENVIRONMENT INTERACTIONS IN FATIGUE CRACK GROWTH*

T. T. Shih and R. P. Wei

Lehigh University, Bethlehem, Pennsylvania 18015 USA

## ABSTRACT

The influences of environment on delay and on fatigue crack growth under programmed loading were examined for a mill annealed Ti-6Al-4V alloy. Test environments included dehumidified argon, air (with 30-60% relative humidity), distilled water, and 3.5% NaCl solution. The effects of load sequence and block size on fatigue under programmed loads were investigated also. Experimental results show that the mildly aggressive environments (atmospheric moisture and distilled water) have little effect on delay. Salt solution, on the other hand, has a significant effect on delay and on crack growth under programmed loading; the effect on delay being dependent on frequency or hold-time. Both load sequence and block size can significantly alter fatigue life under programmed loading. Their effects are accountable phenomenologically in terms of delay. The need for further fundamental understanding of load and environment interactions in fatigue is discussed.

## I. INTRODUCTION

The importance of delay (or, retardation in the rate of fatigue crack growth) produced by load interaction in variable amplitude loading, on the accurate prediction of fatigue lives of engineering structures has been well recognized for some time [1-10]. Delay properly refers to the period of abnormally low

---

*Based on research performed in partial fulfillment of requirements for the Ph.D. degree in Applied Mechanics.

rate, or approximately zero rate, of fatigue crack growth between a decrease in load level and the establishment of a rate of growth commensurate with that for constant amplitude loading at the prevailing (lower) load, that is, between points a and c in a simplified schematic diagram shown in Fig. (1). It is usually measured in terms of the number of elapsed load cycles. For experimental accuracy and potential engineering utility, however, it is more convenient to define delay ($N_D$) artificially as a period of zero crack growth, represented by a-b in Fig. (1), which is obtained by extrapolating the constant amplitude growth curve cd to b. Following Jonáš and Wei [5], this definition for delay will be used here. The actual crack growth response of materials following a high load excursion is more complex and is incorporated indirectly in this definition for delay.

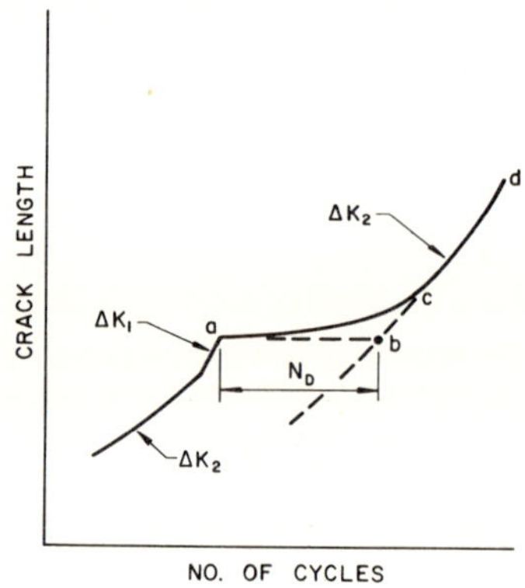


Fig. (1): Schematic Illustration of Delay in Fatigue Crack Growth and Definition of $N_D$ [5].

Systematic studies on a Ti-6Al-4V alloy [5-8] showed that the phenomenon of delay is very complex and is affected by a broad range of loading variables. The possible influence of environment, both chemical and thermal, has received little attention thus far. Although the influence of environment on fatigue crack growth under constant amplitude loading is relatively well established [11-13], it is not clear that this information can be applied directly to the variable amplitude loading situation. Since engineering structures can be expected to encounter a wide range of operating conditions, a systematic study of load and environment interactions on delay is needed. In this paper, only the effects of chemical environment on delay at room temperature is examined. The environments used in this investigation include air, distilled water, and 3.5% NaCl solution. Dehumidified argon is used as the reference environment. Experiments were performed within the framework of linear fracture mechanics. The crack tip stress intensity factor K or $\Delta K$ was used to characterize the mechanical crack driving force. Results from the delay studies were used for estimating fatigue crack growth behavior under programmed loads for comparison with experimental observations on the effects of loading sequence and block size. The need for further fundamental understanding of the load/environment interaction problem is discussed.

## II. MATERIAL AND EXPERIMENTAL WORK

A 0.2-inch-thick (5.08 mm) mill annealed Ti-6Al-4V alloy plate was used in this investigation. The chemical composition, heat treatment, and longitudinal and transverse tensile properties of this alloy are given in [6]. Constant load-amplitude fatigue crack growth data in several test environments at room temperature are shown in Fig. (2) for reference.

Two-inch-wide (50.8 mm) wedge-opening-load (WOL) specimens, oriented in the long transverse (TL) direction [14], were used in this investigation. The specimens were precracked in air at a stress ratio R of 0.05 through a sequence of loads that reduced $K_{max}$ to a level that is equal to or less than the selected starting $K_{max}$ level for the actual experiments [15]. This precracking procedure provided fatigue cracks of at least 0.08 inch (2 mm) in length from the end of the starter notch, such that subsequent fatigue crack growth will be through material that has not been altered by the notch preparation procedure and will be unaffected by the starter notch geometry. The stress intensity factor, K, for the WOL specimens was computed from Eq. (1) [16, 17].

$$K = \frac{P\sqrt{a}}{B\ W} \left[30.96 - 195.8 \left(\frac{a}{W}\right) + 730.6 \left(\frac{a}{W}\right)^2 - 1186.3 \left(\frac{a}{W}\right)^3 + 754.6 \left(\frac{a}{W}\right)^4\right] \qquad (1)$$

where P = applied load; a = crack length; B = specimen thickness; and W = specimen width. A continuous recording electrical potential system was used for monitoring crack growth [8, 18]. With a working current of 2.1 amperes, resolution in the measurement of crack length was estimated to be about 0.001 in. (0.025 mm) for this material and specimen configuration. The accuracy in crack length measurement was better than 1%.

The following environments were used in this investigation: dehumidified argon, air, distilled water, and 3.5% salt solution. Dehumidified argon was used as the reference environment, and was maintained around the crack by flowing argon through chambers clamped to the faces of the specimen. Dehumidification was accomplished by passing ultra-high-purity grade argon (99.999% purity) through a molecular-sieve drier, cold traps at less than

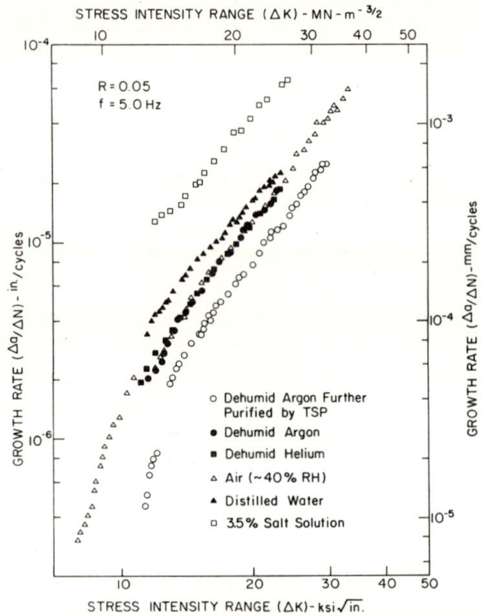


Fig. (2): Kinetics of Fatigue Crack Growth in Various Environments at Room Temperature. (Residual moisture in the dehumidified environments were well below 30 ppm [19].

-140°C, and a titanium sublimation pump (TSP), before admitting the gas into the environment chamber through a high-conductance coupling. (The TSP served a dual role - as a getter, and as a pump for active residual impurities in the chamber.) The effluent from the chamber was passed through cold traps and a silicone fluid back-diffusion trap before being discharged. The effectiveness of this purification system has been demonstrated by Wei and Ritter [19]. No environmental control was used for tests in air. Relative humidity was monitored, however, and ranged between 30 to 60%. Triple-distilled water, and 3.5% NaCl solution, made from reagent grade NaCl and triple-distilled water, were used as the aqueous environments in this study. The appropriate environment was deaerated continuously in a reservoir, and was circulated through Plexiglas environment chambers clamped onto the faces of the specimen.

Delay experiments were carried out under tension-tension (sinusoidal) loading in a MTS Systems Corporation closed-loop electrohydraulic testing machine operated in load control at 5 Hz. Deviations from this loading frequency were made only when a single (manual loading) or a small number of high load cycles were to be applied. Load control was estimated to be better than ± 1%. Because previous experiments indicated the existence of a significant effect on delay when a high load excursion occurs during delay produced by a previous high load excursion [5, 6], and that this influence can be minimized or eliminated by allowing at least 0.06 in. (1.52 mm) of crack extension between the successive high load excursions, such a procedure was used in this investigation and permitted an average of eight individual delay experiments to be performed on each test specimen. To minimize error, delay ($N_D$) data were determined directly from the electrical potential (crack length) versus time (cycles) records.

Several tests were carried out under programmed loading in air and in 3.5% NaCl solution to assess the utility of delay data, obtained under simple (elemental) loading conditions, for estimating fatigue crack growth behavior under more complex load spectra. A four-layer (or four load levels) block spectrum was used. Maximum loads of 420, 558, 687, and 829 lb. (1.87, 2.48, 3.06, and 3.69 kN), with a minimum load of 20 lb. (0.089 kN), were selected. These loads correspond to starting ΔK levels of approximately 9, 12, 15, and 18 ksi √in. (9.9, 13.2, 16.5, and 19.8 MN-$m^{-3/2}$), respectively. Two different block sizes were examined. The number of load cycles in the respective layers in each block (in ascending order of loads) was selected to be 22,000, 1,400, 600, and 300 cycles for the long block, and 1,245, 80, 34, and 16 cycles for the short block. High-to-low, low-to-high, and low-high-low sequences were used in the study. The total number of cycles per block in the low-high-low sequence was equal to that for the other load sequences. The load profiles were established with the aid of a DataTrak programmer on the MTS machine. A cycling frequency of 5 Hz was chosen for the long block tests, and 5.63 Hz was used for the short block tests. (The choice of frequencies was dictated by the fixed gear ratios that were available with the DataTrak programmer used. The difference in frequency, however, is not expected to alter the comparative value of the results significantly.)

## III. RESULTS AND DISCUSSIONS

Fatigue crack growth test results under elemental load spectra (delay) and under programmed loads will be reported and discussed separately.

### A. Fatigue Crack Growth under Elemental Load Spectra

To determine the effect of loading variables on delay, it is most convenient and meaningful to isolate the applied loads to their elemental form. Such elemental load spectra consist of a block of high loads (preload) of specified character followed by cyclic loads at a lower level until the equilibrium growth rate is established. The resulting delay can be attributed unequivocally to each particular load interaction. The effect of a broad range of loading variables on delay in fatigue crack growth in air (relative humidity - 30 to 60%) has been examined for this mill annealed Ti-6Al-4V alloy and has been reported [5, 6]. The results showed that the delay phenomenon is quite complex, and that delay ($N_D$) is a function of the magnitude and range of the high loads, as well as of the subsequent lower loads, the number of high-load cycles, hold-time at high load, dwell-time (or rest period) at an intermediate load level, ratio between the high and low loads, etc. Denoting the preloading and the subsequent (lower) loading conditions by subscripts 1 and 2 respectively, delay ($N_D$) may be expressed implicitly as a function of the pertinent loading variables.

$$N_D = F[K_{1max}/K_{2max},\ K_{2max} \text{ or } K_{1max},\ R_1,\ R_2,\ N_1,\ t_1] \tag{2a}$$

or,

$$N_D = F[\Delta K_1/\Delta K_2,\ \Delta K_2 \text{ or } \Delta K_1,\ R_1,\ R_2,\ N_1,\ t_1] \tag{2b}$$

$K_{max}$ and $K_{min}$ are the maximum and minimum values of stress intensity in a given load cycle; $\Delta K = K_{max} - K_{min}$ is the stress intensity range; $R = K_{min}/K_{max}$ is the stress ratio; N is the number of load cycles; and t is the time at load. In addition, delay ($N_D$) depends on the state of stress, temperature and/or thermal history, and environment [7, 8].

In this investigation, only the influence of chemical environment on delay at room temperature was examined. Experimental work was limited to an examination of delay as a function of $K_{2max}$ for single and multiple high-load cycles (that is, for various $N_1$) at prescribed ratios of $K_{1max}/K_{2max}$, $R_1$, and $R_2$. Dehumidified argon, air (relative humidity - 30 to 60%), distilled water, and 3.5% NaCl solution were used as the test environments. Experimental results from this investigation are shown in Figs. (3) through (6) and may be summarized as follows:

1. Data obtained in dehumidified argon, air, and distilled water indicate that the mildly aggressive environments (atmospheric moisture and distilled water) have little effect on delay (Fig. 3), although the rate of fatigue crack growth under constant-amplitude loading differs by about a factor of two between dehumidified argon (further purified by a titanium sublimation pump, TSP) and the other two environments (Fig. 2).

2. For a single high-load cycle, delay ($N_D$) in dehumidified argon and in the mildly aggressive environments decreases with increasing $K_{2max}$ for a given value of $K_{1max}/K_{2max}$, and increases

with $K_{1max}/K_{2max}$ for fixed values of $K_{2max}$ (Fig. 3). These results are consistent with the previous observations [5, 6].

3. Delay produced by a single high-load cycles in the 3.5% NaCl solution, however, is quite different (Fig. 4). For $K_{1max}/K_{2max} = 2$, delay ($N_D$) is generally longer than that observed in air at comparable values of $K_{2max}$ even though the constant-amplitude growth rate is some three times faster. Instead of a smooth variation with $K_{2max}$, $N_D$ exhibits a trough in the range of $K_{2max}$ from 10 to 15 ksi $\sqrt{in.}$ (11 to 16.5 $MN\text{-}m^{-3/2}$) with a minimum at $K_{2max}$ of about 12 ksi $\sqrt{in.}$ (13.2 $MN\text{-}m^{-3/2}$). The observed behavior is quite reproducible and has been verified by tests on six specimens. Data scatter is quite large. Possible reasons for this trough and the large data scatter will be discussed later.

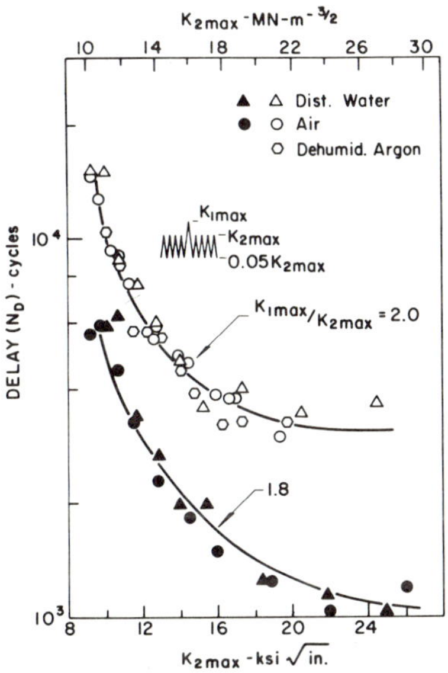


Fig. (3): Effects of $K_{2max}$ and $K_{1max}/K_{2max}$ on Delay in Various Environments.

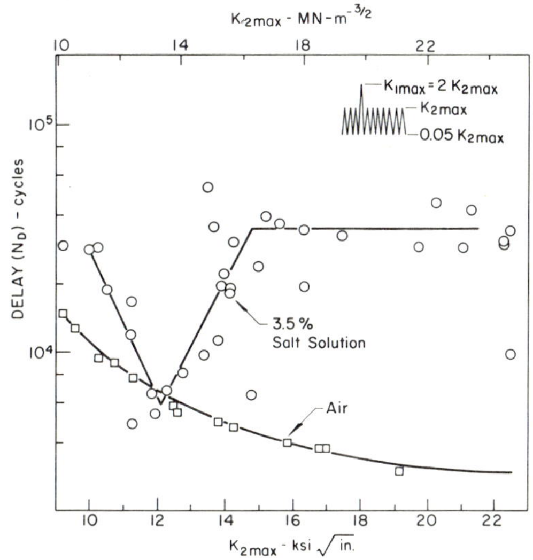


Fig. (4): Effect of $K_{2max}$ on Delay in Air and in 3.5% NaCl Solution for $K_{1max}/K_{2max} = 2.0$.

4. For multiple cycles of high loads, delay ($N_D$) in air increases with increasing number of high-load cycles ($N_1$) up to an apparent limiting value, Fig. (5). For $N_1$ = 12 cycles, delay is about twice that for a single cycle of high load. For $N_1$ greater than 100 cycles, a limiting value equal to about four times the single cycle delay is obtained. These ratios hold over the entire range of $K_{2max}$ values investigated. The variation of

delay with $K_{2max}$ follows the same trend as that for a single cycle of high load excursion.

5. Delay ($N_D$) for multiple cycles (500 cycles) of high loads in 3.5% NaCl solution decreases monotomically with increasing $K_{2max}$, and did not exhibit a "trough" at $K_{2max}$ of about 12 ksi √in. (13.2 MN-m$^{-3/2}$) as was observed in the single high-load excursion tests (Figs. 5 and 6). A strong effect of cyclic-load frequency for the high loads ($f_1$), however, was observed. Limited data indicate that by decreasing $f_1$ from 5.0 Hz. to 1.0 Hz. delay ($N_D$) was increased by about a factor of 5 (Fig. 6). Reducing $f_1$ from 1.0 Hz. to 0.1 Hz. produced a further increase in delay by a factor of about 2. These results suggest that there may be a significant hold-time ($t_1$) effect on delay for this alloy in 3.5% NaCl solution. This aspect of delay will be discussed further, in particular in relation to the single high-load excursion results. Delay is expected to increase with the number of high load cycles ($N_1$) in a similar manner as that observed in air, provided that the loading frequency and waveform are maintained constant.

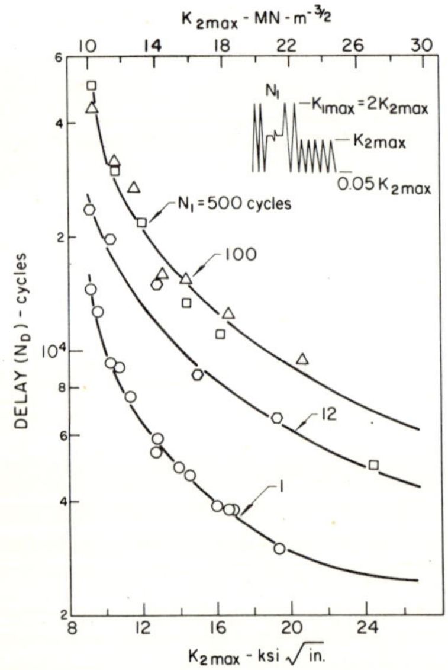


Fig. (5): Effects on $K_{2max}$ and $N_1$ on Delay in Air for $K_{1max}/K_{2max}$ = 2.0.

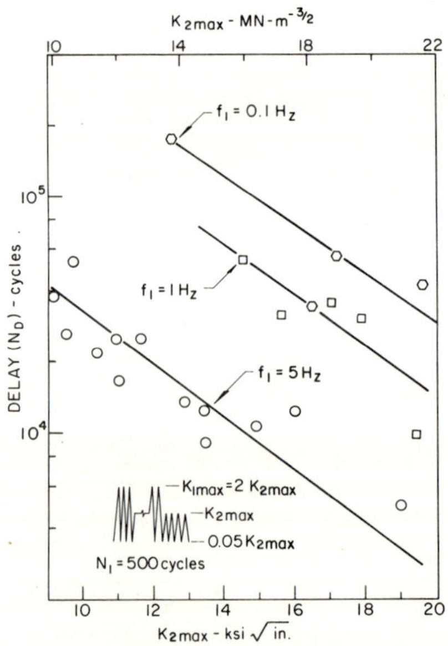


Fig. (6): Effects on $K_{2max}$ and $f_1$ on Delay in 3.5% NaCl Solution for $K_{1max}/K_{2max}$ = 2.0 and $N_1$ = 500 cycles.

The results of single and multiple high-load excursions on delay in dehumidified argon and in the mildly aggressive environments are generally in agreement with the results of previous studies [5-6]. Comparison of the results for single high-load excursions in these environments with delay data on 2024-T3 and 7075-T6 aluminum alloys indicate that the variation of delay ($N_D$) with $K_{2max}$ in this titanium alloy is directly opposite to the behavior exhibited by the aluminum alloys (Fig. 3 and [4,7,9]). For multiple high-load tests, delay ($N_D$) increased with the number of high-load cycles ($N_1$) in all three alloys and tended toward a maximum (saturation) in each case (Fig. 5 and [4,9]). The amount of increase and the number of high load cycles required to produce maximum delay, however, varied among the three alloys. For example, saturation was reached after 100 cycles of high load in the Ti-6Al-4V alloy, with a four-fold increase in delay over that obtained for a single high load excursion (Fig. 4); a three-fold increase in delay was observed in the 7075-T6 aluminum alloy after only 30 cycles of high load excursions [4]; and a maximum value had not been obtained in the 2024-T3 aluminum alloy after 9,000 high-load cycles with an increase in delay by about ten-fold [9]. The precise values are expected to be dependent on a host of loading variables (such as, $K_{1max}/K_{2max}$, $K_{2max}$, $R_1$, $R_2$, etc.), and on specimen thickness, temperature, and environment [5-8]. No satisfactory model has been advanced to account for the various aspects of delay. The differences in behavior among the various alloys, in particular between the aluminum and titanium alloys, suggests that simplistic modeling would not be totally adequate, and that suitable models must incorporate the effects of material property variations. Extensive research is needed to provide a better understanding of the mechanism for delay. In the mean time, extensive extrapolation of available experimental data to cover various design situations (involving different alloys, material thickness, etc.) should be carefully avoided.

Delay results in 3.5% NaCl solution are much more complex, and are not clearly understood at this time. Delay produced by multiple cycles ($N_1$ = 500 cycles) of high load at 5.0 Hz. decreased with increasing $K_{2max}$ in a well behaved manner, and is consistent with results obtained in the other environments (Figs. 5 and 6). The small reduction in delay (at $f_1$ = 5.0 Hz.) in comparison with the other environments is consistent with the higher rates of fatigue crack growth in 3.5% NaCl solution. The extensive increase in delay for multiple cycles of high load at the lower frequencies (Fig. 6) and for single high-load cycles (Fig. 4), and the presence of a "trough" in the single high-load cycle data are unique to the 3.5% NaCl environment and are more difficult to explain. The multiple high-load cycle data suggested that there may be a significant hold-time ($t_1$) effect on delay for this alloy in 3.5% NaCl solution. Preliminary data on the effect of hold-time, from 0 to 100 seconds at $K_{1max}/K_{2max}$ = 2.0 and $K_{min}$ = 0.05 $K_{2max}$, indicate that delay ($N_D$) can vary from about $10^3$ to $4 \times 10^4$ cycles at $K_{2max}$ = 21 ksi $\sqrt{in.}$ (23.1 MN-m$^{-3/2}$) and from about

$7 \times 10^3$ to $4 \times 10^4$ cycles at $K_{2max}$ = 12 ksi $\sqrt{in.}$ (13.2 $MN\text{-}m^{-3/2}$). Since the high-load cycles in the single high-load excursion experiments were applied manually, some variations in hold-time would be expected. Typical hold-time for manual loading was estimated to be in the range of about 0.5 seconds to 30 seconds. This variation in hold-time could well account for the presence of the "trough" (Fig. 4) and for the extensive amount of scatter in delay data for the tests in 3.5% NaCl solution. The "trough" indicated in Fig. (4) may be, in fact, associated with some systematic variation introduced by manual loading (for example, greater caution, hence longer effective hold-time, was required for loading at the higher K levels because of the possibilities of rapid stress corrosion cracking and of fracture). A more extensive investigation of hold-time effects in 3.5% NaCl solution is being made and will be reported separately. The precise cause for the observed frequency (or hold-time) effect has not been established. It may be associated with changes in fracture morphology produced by stress corrosion cracking or corrosion fatigue, the build-up of corrosion products, crack-tip blunting by corrosion, and/or penetration of the "damaged zone" by stress corrosion or corrosion fatigue cracking. Extensive studies will be needed to elucidate the mechanism of load interactions in the presence of an aggressive environment. In the mean time, the significant influences of environment induced frequency and hold-time effects on delay must be recognized and be taken into account in the development of life prediction procedures. It is likely that each material-environment system will need to be treated separately. Thus, generalization of existing data should again be avoided.

### B. Fatigue Crack Growth under Programmed Loads

Fatigue crack growth experiments under block-programmed loads were carried out in air and in 3.5% NaCl solution to assess the effects of loading sequence, block size, and environment. The experimental results are used in evaluating estimates of fatigue crack growth behavior obtained from considerations of crack growth kinetics (Fig. 2) and of delay under elemental load spectra (Figs. 3 to 6, for example). Typical experimental results are shown in terms of crack length versus number of load blocks in Fig. (7). For ease of comparison, fatigue lives (arbitrarily defined here as the number of load cycles to extend the crack from an initial length of 0.7 inch (17.8 mm) to 1.4 inches (35.6 mm)) were determined from these tests and are shown in Fig. (8). Estimated fatigue crack growth curves and fatigue lives, made with and without considering delay (load interaction), are also shown in these figures for comparison (see Appendix). (Note that these estimates serve primarily as indications of trend; particularly for the tests conducted in 3.5% NaCl solution where the basic

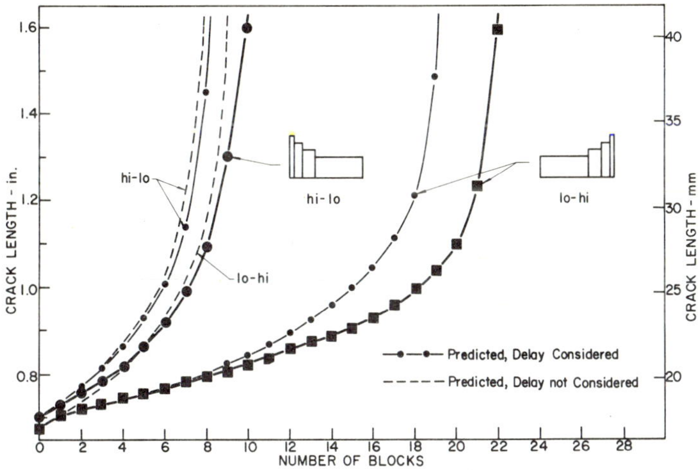


Fig. (7): Fatigue Crack Growth under Block-Programmed Loads in Air (24,300 cycles per block).

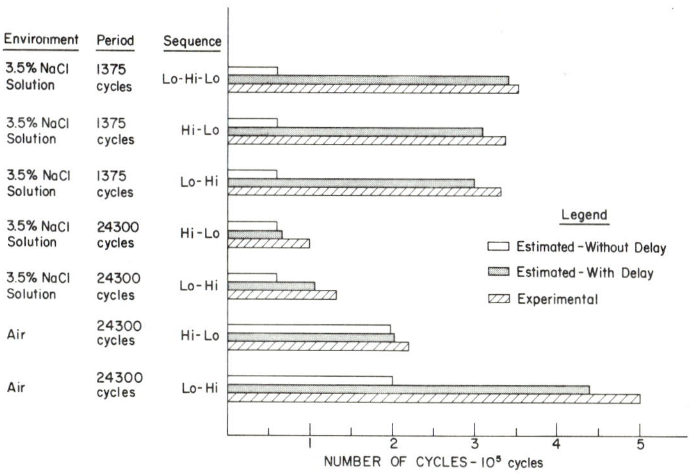


Fig. (8): Effects of Environment, Load Sequence, and Block Size on Fatigue Life.

crack growth and delay data are limited. Estimates for the tests conducted in air are supported by a more extensive data base [5, 6] and are expected to be more accurate).

The essential results of this series of experiments may be summarized below, and will be discussed separately.

1. Load sequence has a significant effect on fatigue crack growth under block-programmed loading (Fig. 8). High-to-low load sequence tends to be more severe for the longer loading block, and results in higher rates of fatigue crack growth and shorter fatigue lives (Fig. 8). The effect of load sequence is diminished or reversed when the block size (number of load cycles per block) is reduced (Fig. 8).

2. There is a significant effect of block size on fatigue crack growth under block-programmed loads. Reducing the block size from 24,300 cycles to 1,375 cycles increased fatigue lives by about a factor of three for the tests in 3.5% NaCl solution (Fig. 8). (A similar trend is expected for tests in the other environments.) As indicated in item 1, reductions in block size tended to decrease the influence of loading sequence on fatigue crack growth.

3. For a given loading spectrum, exposure to 3.5% NaCl solution increases the rate of fatigue crack growth and reduces fatigue life (Fig. 8). This observation is based on data obtained at a test frequency of 5.0 Hz. In view of the observed frequency (or hold-time) effect on delay, this observation will require further consideration.

4. Estimates of fatigue crack growth behavior and of fatigue lives on the basis of data on fatigue crack growth kinetics and on delay are in good agreement with the experimental observations (Figs. 7 and 8). Thus, examinations of delay under elemental loading spectra can provide insight into the problem of crack growth under programmed loads, and serve as a phenomenological basis for the prediction of lives and crack growth behavior under more complex loading spectra.

The effects of load sequence and block size on fatigue crack growth and fatigue lives under block-programmed loading may be explained in terms of delay (load interaction) on a phenomenological basis. The effect of load sequence is more easily explained for the longer of two loading blocks utilized (24,300 cycles), since the interaction effects were limited to adjoining layers (load levels) in a given load block and to adjoining layers in adjacent blocks. In the low-to-high load sequence, no delay would be encountered during the first block since the loads appeared in ascending order (crack acceleration will not be considered in this discussion and is not expected to alter the results significantly). In the first layer of the second block, and of subsequent blocks, a large number of delay cycles (depending on $K_{2max}$, Figs. 3 to 6), would be encountered because of the abrupt drop in load from the highest to the lowest level in this sequence. Crack growth, therefore, would occur only over the remaining portion of this layer. No delay would be expected in the subsequent load layers in each load block. Since the largest number of load cycles were applied to the first (lowest load) layer, a significant increase in fatigue life over that obtained in the absence of delay would be expected, and is consistent with the estimated behavior (Fig. 8). For the high-to-low sequence, no delay would be encountered in the first (highest load) layer of the first block and of the subsequent blocks. Delay in the succeeding layers in each block will represent a relatively small portion of the load cycles in each layer because of the relative small reduction in load between each layer (approximately 20 to 25%). As such, each level would contribute almost fully to crack growth, thus producing a higher rate of crack growth and shorter fatigue life as compared with the low-to-high sequence (Fig. 8).

As the block size is reduced, changes in load level can occur well before the delay period is completed. Thus, it becomes necessary to consider not only load interactions with the adjoining layers, but with the subsequent layers, and from block to block. For the low-to-high sequence considered (4 layers with a total of 1,375 cycles per block), there was (almost) no crack growth in the first layer of the second load block and of the succeeding blocks, since delay ($N_D$) produced by load interaction was longer than the number of load cycles in this layer. Some delay would, therefore, be encountered in the second layer and/or third layer of each block, and would be dependent on the interaction between the fourth and the second and third layers [6]. Because of the foregoing considerations, the fatigue life for the shorter block would be longer than that for the longer load block (Fig. 8). For the high-to-low load sequence, the delay in each of the low load layers will become a larger fraction of the total number of load cycles in each layer. For a sufficiently short load block, it is conceivable that crack growth would take place only during the highest load layer in each block, and that the amount of crack growth per block can become comparable to, or less than, the corresponding low-to-high sequence. Such a reduction in the severity of the high-to-low load sequence, or reversal in severity between the high-to-low and the low-to-high sequences, is indicated by the estimates and is confirmed by the experimental data (Fig. 8). The low-high-low sequence may be treated in a similar manner, and the estimated behavior is again in conformance with the experimental observation (Fig. 8).

Thus, it appears that the load sequence and block size effects can be explained satisfactorily, in a phenomenological way, in terms of the fatigue crack growth kinetics and delay data obtained from elemental load interactions. It should be noted, however, that the influences of load sequence and block size for a 2024-T3 aluminum alloy are completely opposite to these results on mill annealed Ti-6Al-4V alloy [20].* A reversal in trend has been observed between the delay behaviors for these two alloys. It is not clear that the difference in behavior under programmed loading can be rationalized in terms of the differences in delay behavior. Additional work on the aluminum alloy will be needed to provide a quantitative accounting of these differences.

Although the experimental data on fatigue crack growth under block-programmed loads indicated that 3.5% NaCl solution produced an increase in the rate of crack growth and a corresponding reduction in fatigue life, this observation cannot be readily generalized in view of the observed effect of frequency (or hold-time) on delay (Figs. 4, 6, and 8). A decrease in cyclic loading frequency (or increase in hold-time), for example, increases delay (Fig. 6), and would be expected to increase fatigue lives under block-programmed loads. Any improvement in fatigue life may be negated, however, by the increased propensity for corrosion fatigue crack growth at the lower frequencies (or stress corrosion cracking at long hold-times). The net result, therefore, can be determined only by a careful consideration of these competing effects. A systematic examination of this problem is needed, and should be considered in design.

## SUMMARY

The effect of environment on delay and on fatigue crack growth under programmed loading was examined on a mill annealed Ti-6Al-4V alloy. Delay tests were carried out in dehumidified argon, air (with 30-60% relative humidity), distilled water and 3.5% NaCl solution. Fatigue crack growth tests under programmed loading were carried out only in air and in 3.5% NaCl solution, and included examinations of the effects of load sequence and block size.

Comparison of delay data obtained in dehumidified argon, air, and distilled water indicate that the mildly aggressive environments (moist air and distilled water) have little effect on delay. These results are generally in agreement with the results of previous studies, and showed that delay ($N_D$) increased with increasing $K_{1max}/K_{2max}$ and $N_1$, and decreased with increasing $K_{2max}$. It was observed that the variation of delay ($N_D$) with $K_{2max}$ in this

*This comparison must be tempered by the fact that the influence of test environment and of distribution of high loads could not be ascertained for the 2024-T3 aluminum alloy.

titanium alloy is directly opposite to the behavior exhibited by the 2024-T3 and 7075-T6 aluminum alloys. This observation suggests that suitable models for delay must incorporate not only the effects of loading variables, but also account for material property variations. Experimental results of delay tests in 3.5% NaCl solution showed that this environment had a striking effect on delay which was dependent on the frequency of overload cycles. Extensive studies will be needed to elucidate the mechanism of this load/environment interaction effect.

Experimental results from programmed load tests show a significant effect of load sequence and block size on fatigue crack growth. For a fatigue test with long blocks, the high-to-low load sequence tended to be more severe than the low-to-high sequence. The effect of load sequence on fatigue crack growth under programmed loading diminished as the block size was reduced. Fatigue tests (in 3.5% NaCl solution) exhibited longer lives for those tests with a shorter loading block. These observed trends were explained satisfactorily, in a phenomenological way on the basis of data on crack growth kinetics and delay. The influences of load sequence and block size on fatigue crack growth in a 2024-T3 aluminum alloy are again completely opposite to the results on mill annealed Ti-6Al-4V alloy. Additional work on the 2024-T3 aluminum alloy will be needed to provide a quantitative accounting of these differences. For a given load spectrum, fatigue crack growth rate is increased by the exposure to 3.5% NaCl solution. In view of the observed frequency effect on delay in this material/environment system, this observation will require further consideration.

It was shown that fatigue under programmed loading is very complex and is affected by load sequence, block size, maximum load, etc. It is further complicated by material property variations and load/environment interactions, which is frequency dependent. Nevertheless, fatigue crack growth under programmed loading can be accounted for phenomenologically based on data of crack growth kinetics and delay. Because of the differences in response between the aluminum and titanium alloys, extensive extrapolation of existing data should not be made in developing life prediction procedures.

## APPENDIX

### Procedure for Estimating Crack Growth under Programmed Loading

A numerical integration procedure was used for estimating fatigue crack growth under programmed loading. This procedure utilized piece-wise power law representations of the constant load-amplitude fatigue crack growth data, and delay data from the experimental program. A power law of the following form was used:

$$\frac{\Delta a}{\Delta N} = A\ (\Delta K)^n \qquad \text{(A-1)}$$

where A and n are empirical constants. For $\Delta a/\Delta N$ and $\Delta K$ to be expressed in units of in./cycle and ksi $\sqrt{in}$. respectively, the empirical constant A assumes the dimension of (in./cycle) $(\text{ksi}\ \sqrt{in})^{-n}$. The following estimated values of A and n were used data from Fig. (2), with $R \approx 0$):

| Environment | ΔK range | Empirical Constant | |
|---|---|---|---|
| | | N | A |
| air | $\Delta K<14\text{ksi}\sqrt{in}$ | 3.6 | $3.38\text{x}10^{-10}$ (in/cycle)$(\text{ksi}\sqrt{in})^{-3.6}$ |
| air | $\Delta K\geq14\text{ksi}\sqrt{in}$ | 2.6 | $5.18\text{x}10^{-9}$ (in/cycle)$(\text{ksi}\sqrt{in})^{-2.6}$ |
| 3.5% NaCl | $\Delta K<14\text{ksi}\sqrt{in}$ | 3.14 | $3.78\text{x}10^{-9}$ (in/cycle)$(\text{ksi}\sqrt{in})^{-3.14}$ |
| Solution | $\Delta K\geq14\text{ksi}\sqrt{in}$ | 2.23 | $4.17\text{x}10^{-8}$ (in/cycle)$(\text{ksi}\sqrt{in})^{-2.23}$ |

Procedurally, delay encountered in the $i^{th}$ load layer, $N_{Di}$, was estimated on the basis of $\Delta K$ at the start of the $i^{th}$ layer, $\Delta K_i$, and $\Delta K$ at the end of the previous layer, $\Delta K_{(i-1)}$. The increment of crack growth during the $i^{th}$ layer was determined numerically by integrating Eq. (A-1) between $N_{Di}$ and $N_i$, where $N_i$ = total number of cycles in the $i^{th}$ layer. Integration was performed only when $N_i$ exceeded $N_{Di}$. If no crack growth took place during the $(i - 1)^{th}$ layer, $N_{Di}$ was estimated on the basis of the next previous layer, that is, the (i - 2) layer.

## ACKNOWLEDGMENT

Partial support of this work by the Office of Naval Research under Contract Nonr N00014-67-A-0370-0008, NR 036-097 is gratefully acknowledged.

## REFERENCES

1. Christensen, R. H., Cracking and Fracture in Metals and Structures, Proceedings - Crack Propagation Symposium, Cranfield, The College of Aeronautics, 2, 326, 1961.

2. Schijve, J., Fatigue Crack Propagation in Light Alloy Sheet Material and Structures, Advances in Aeronautical Sciences, 3-4, 387, 1962.

3. Hudson, C. M., and Hardrath, H. F., Effects of Changing Stress Amplitude on the Rate of Fatigue-Crack Propagation in Two Aluminum Alloys, NASA TN D-960, 1961.

4. Hudson, C. M., and Raju, K. N., Investigation of Fatigue-Crack Growth under Simple Variable-Amplitude Loading, NASA TN D-5702, 1970

5. Jonas, O. and Wei, R. P., An Exploratory Study of Delay in Fatigue-Crack Growth, Int. Journ. of Fracture Mech., 7, 116, 1971.

6. Wei, R. P. and Shih, T. T., Delay in Fatigue Crack Growth, Int. Journ. of Fracture, 10, 77, 1974.

7. Shih, T. T. and Wei, R. P., The Effect Specimen Thickness on Delay in Fatigue Crack Growth, Journ. of Testing and Evaluation, to be published.

8. Shih, T. T., Fatigue Crack Growth under Variable Amplitude Loading, Ph.D. Dissertation, Lehigh University, Bethlehem, Pa., 1974, (in preparation).

9. von Euw, E. F. J., Hertzberg, R. H., and Roberts R., Delay Effect in Fatigue Crack Propagation, Stress Analysis and Growth of Cracks, ASTM STP 513, 230, 1972.

10. Trebules, V. W., Roberts, R., and Hertzberg, R. W., Effect of Multiple Overloads on Fatigue Crack Propagation in 2024-T3 Aluminum Alloy, Progress in Flaw Growth and Fracture Toughness Testing, ASTM STP 536, 115, 1973.

11. Wei, R. P., Fatigue-Crack Propagation in High-Strength Aluminum Alloy, Int. Journ. of Fracture Mech., 4, 159, 1968.

12. Wei, R. P., Some Aspects of Environment-Enhanced Fatigue-Crack Growth, Engineering Fracture Mechanics, 1, 633, 1970.

13. Hartman, A., On the Effect of Oxygen and Water Vapor on the Propagation of Fatigue Cracks in 2024-T3 Alclad Sheet, Int. J. of Fracture Mech. 1, 167, 1965.

14. Goode, J., The Slow Growth and Rapid Propagation of Cracks, Materials Research and Standards, 1, 389, 1961.

15. FitzGerald, J. and Wei, R. P., A Test Procedure for Determining the Influence of Stress Ratio on Fatigue Crack Growth, Journ. of Testing and Evaluation, 2, 67, 1974.

16. Wilson, W. K., Analytical Determination of Stress Intensity Factors for the Manjonie Brittle Fracture Test Specimen, Report No. WERL-0029-3, Westinghouse Research Laboratories, 1965.

17. Wilson, W. K., Optimization of WOL Brittle Fracture Test Specimen, Report No. 66-1B4-BTLFR-R1, Westinghouse Research Laboratories, 1966.

18. Li, C. Y. and Wei, R. P., On Calibrating the Electrical Potential Method for Studying Slow Crack Growth, Materials Research and Standards, 6, 392, 1966.

19. Ritter, D. L. and Wei, R. P., The Influence of Temperature on Fatigue Crack Growth in a Mill Annealed Ti-6Al-4V Alloy, Journ. of Materials, 7, 240, 1972.

20. Schijve, J., Jacobs, F. A., and Tromp, P. J., The Effect of Load Sequence on Fatigue Crack Propagation under Random Loading and Program Loading, NLR TR 71014U, 1971.

# Section IV

# DYNAMICS AND ARREST

# AN ANALYSIS OF DYNAMIC CRACK PROPAGATION AND ARREST FOR A MATERIAL HAVING A CRACK SPEED DEPENDENT FRACTURE TOUGHNESS

M. F. Kanninen

Battelle
Columbus Laboratories
505 King Avenue
Columbus, Ohio 43201

ABSTRACT

The beam-on-elastic foundation model for the DCB test specimen is extended to treat dynamic crack propagation with a fracture energy that is some given function of crack speed. Calculations are performed that simulate a wedge-loaded DCB test using a speed-toughness relation suggested by data for low-carbon steel at low temperatures. Results are obtained for $K_{Ia}$ , the static stress intensity factor corresponding to the crack arrest point. It is found that $K_{Ia}$ is not a constant but depends upon the applied stress intensity factor at the initiation of crack growth and upon the geometry of the test piece. It is concluded that crack arrest can be sensibly treated only as part of an energy-based dynamic fracture mechanics formalism.

## INTRODUCTION

Evidence in support of the "energy balance" approach to dynamic fracture mechanics has been obtained in a combined program of experiment and analysis of dynamic crack propagation and crack arrest pursued at Battelle-Columbus.[1,2] This work has used the double cantilever beam (DCB) test specimen. Because of the complications that arise from the necessity to include inertia forces, a simple beam-on-elastic foundation model has been developed to treat dynamic crack propagation in the DCB specimen. Comparisons in both the static [3] and dynamic [4] situations have shown that the predictions of the model are completely consistent with experimental observations. In this paper, the previous restriction of the model to treat only a constant dynamic fracture energy is removed. Calculations can now be performed for any specified single-valued function of the crack speed. To demonstrate, some heuristic calculations are performed in this paper by taking a parabolic fracture toughness-crack speed dependence suggested by data appearing in the literature for low-carbon steel near the ductile-brittle transition temperature. [1,2,5-7] For comparison, calculations are also performed for speed-independent behavior.

The arrest of a rapidly propagating crack is currently a controversial topic in fracture mechanics. One point of view is that crack arrest can be completely characterized in terms of the static condition following arrest. If this approach is to be valid, there must be an "arrest toughness" $K_{Ia}$ that is a material property to the same extent that $K_{Ic}$ is a material property. While some experimental evidence has been offered in support of this idea, [8,9] it has not previously been possible to scrutinize this approach analytically. The reason is that dynamic analyses capable of treating a speed-dependent fracture energy did not exist. Hence, with the analytical approach developed here, it is of some interest to subject this approach to a critical examination. Accordingly, the crack-arrest calculations given in this paper are extended to include a determination of $K_{Ia}$.

## BASIS OF A DYNAMIC FRACTURE MECHANICS

There is a dichotomy evidenced in the literature concerning the fundamental way to characterize crack growth. Some believe that a critical stress state must be achieved ahead of the crack tip, others that a critical energy must be made available. In the static problems of ordinary fracture mechanics, of course, either point of view leads to a correct formulation. This is true because the relation

$$K^2 = E\mathcal{G} \qquad \text{(plane stress)} \qquad (1)$$

makes the stress intensity factor K equivalent to $\mathcal{G}$, the rate of energy supplied to the crack tip. The situation is significantly different in the case of dynamic crack propagation. A relation such as Eq. (1) may exist, but the proportionality factor will be a geometry-dependent function of the crack speed and, possibly, derivatives of crack speed as well. Consequently, a critical value of $\mathcal{G}$ cannot correspond to a unique critical value of K that is a property only of the material.* A distinction between K and $\mathcal{G}$, academic in the static situation, therefore becomes of considerable importance for dynamic crack propagation.

It is probably not possible to devise an experiment that could directly determine whether a stress criterion or an energy criterion is the more fundamental. The best alternative is to accept each criterion axiomatically to test its logical consequences. This procedure would seem to favor the energy point of view. In particular, while $\mathcal{G}$ is associated with the actual amount of energy flowing to the crack tip region, K is not associated with any physical quantity that actually exists in the material. That is, the singularity (that K is the coefficient of) is merely a mathematical abstraction that cannot be tolerated in a real material.

The abstract nature of the crack tip singularity is well known. The rationalization usually given is that K nevertheless "represents" the actual stress field closely enough for practical purposes. But, above the small scale-yielding regime, where the critical K parameter no longer suffices for practical purposes, parameters derivable from energy considerations remain applicable. These are the critical crack opening displacement (COD) and the critical J integral. It can be shown that the J integral is a generalization of $\mathcal{G}$ valid for the case where plastic yielding can be described by nonlinear elasticity or deformation-type plasticity. [10] A connection between the COD approach and an energy approach to fracture mechanics can also be demonstrated. This suggests that the K parameter representation of fracture mechanics works as well as it does--not because it represents the crack tip stresses in any physical sense--but only because it is connected to the energy supplied to the crack tip.

The energy balance approach to dynamic fracture mechanics, taking account of a possible kinetic energy contribution, was suggested originally by Mott. [11] It can be conveniently expressed in terms of a dynamic energy release rate $\mathcal{G}$ defined as

$$\mathcal{G} = \frac{dW}{dA} - \frac{dU}{dA} - \frac{dT}{dA} \tag{2}$$

* This does not mean that a dynamic parameter $K_{Id}$ cannot be defined via Eq. (1), as will be done in the following, as long as it is recognized that this value may have no particular relation to the stresses ahead of a rapidly propagating crack.

where W is the work done on the structure, U and T are the strain energy and kinetic energy of the structure, respectively, and A is the crack area. The energy balance idea is fulfilled by setting $\mathcal{G}$ equal to an intrinsic energy absorption rate at the crack tip. If the rate of energy dissipation is a function of crack speed, this is

$$\mathcal{G} = \mathcal{R}(V) \tag{3}$$

where $\mathcal{R}$ is the energy absorbed at the crack tip per unit area of crack advance and V is the crack speed. Notice that the dynamic energy release rate defined in Eq. (2) generalizes its static counterpart in two ways: first, by supplementing the strain energy release rate with a kinetic energy release rate, and, second, by using dynamic values of the quantities involved.

## THE BEAM-ON-ELASTIC FOUNDATION MODEL

The beam-on-elastic foundation model of the DCB test specimen has evolved through a series of successive improvements. Currently, Timoshenko beam theory and a generalized or Pasternak foundation is employed. A general formulation of the governing equations is given in Reference [4]. For present purposes, it suffices to take the special form for the situation where the specimen has a uniform rectangular cross section. These are

$$\frac{\partial^2 \Psi}{\partial x^2} + \frac{4}{h^2}\left(\frac{\partial w}{\partial x} - \Psi\right) - \frac{2}{h^2} H^*(\theta_c - \theta)\Psi = \frac{1}{C_o^2}\frac{\partial^2 \Psi}{\partial t^2} \tag{4}$$

and

$$\frac{\partial^2 w}{\partial x^2} - \frac{\partial \Psi}{\partial x} - \frac{6}{h^2} H^*(\theta_c - \theta) w = \frac{3}{C_o^2}\frac{\partial^2 w}{\partial t^2} \tag{5}$$

where
- $x$ = coordinate along the beam axis,
- $t$ = time from onset of unstable propagation,
- $w = w(x,t)$ = mean deflection of a beam cross section,
- $\Psi = \Psi(x,t)$ = mean rotation of a beam cross section,
- $\theta = \theta(x,t)$ = crack extension parameter (see below),
- $\theta_c$ = critical value of the crack extension parameter,
- $2h$ = beam height,
- $b$ = specimen thickness,
- $E$ = elastic modulus,
- $C_o = \sqrt{E/\rho}$ = bar wave speed (5000 M/sec in steel).

In Eqs. (4) and (5) the function $H^*(\theta_c-\theta)$ denotes the ordinary Heaviside step function modified such that once $\theta$ exceeds $\theta_c$, its value remains fixed at zero irregardless of subsequent values of its argument.

Expressions for the strain energy and kinetic energy for the Timoshenko beam-generalized foundation model, given in general form in Reference [4], similarly specialize to

$$U = \frac{Ebh^3}{12} \int_0^L \left\{ \left(\frac{\partial \Psi}{\partial x}\right)^2 + \frac{4}{h^2}\left(\frac{\partial w}{\partial x} - \Psi\right)^2 + \frac{24}{h^2} H^*(\theta-\theta_c) \left[\frac{w^2}{h^2} + \frac{\Psi^2}{12}\right]\right\} dx \qquad (6)$$

$$T = \frac{\rho bh^3}{12} \int_0^L \left\{ \frac{12}{h^2}\left(\frac{\partial w}{\partial t}\right)^2 + \left(\frac{\partial \Psi}{\partial t}\right)^2 \right\} dx \qquad (7)$$

These expressions serve a double purpose in this work. They are used to monitor the relative contributions to the crack extension process during the computation itself and, in addition, to determine the crack extension parameter $\theta$ in accord with an energy balance. The latter is accomplished by assuming that there is a unique crack speed V at each point in time. Then

$$\frac{dU}{dA} + \frac{dT}{dA} = \frac{1}{bV}\left\{\frac{dU}{dt} + \frac{dT}{dt}\right\} \qquad (8)$$

By, (i) substituting for U and T in the right-hand side of Eq. (8) from Eqs. (6) and (7), (ii) performing the indicated differentiations and integrations, (iii) using Eqs. (4) and (5) to simplify the resulting expressions, then (iv) comparing with Equation (2), noting that W = 0 for the wedge-loading being considered, it is found that

$$\mathcal{G} = 2Eh\left\{\frac{1}{h^2} w^2(a,t) + \frac{1}{12}\Psi^2(a,t)\right\} \qquad (9)$$

In Eq. (9), $a = a(t)$ denotes the position at which an increment of crack growth is presumed to occur.

For crack propagation to proceed in accord with the dictates of energy conservation, Eq. (3) must be satisfied. To make the model embody this principle, guided by Eq. (9), the definition of the crack extension parameter $\theta$ can only be

$$\theta = 2Eh \left\{ \frac{1}{h^2} w^2 + \frac{1}{12} \Psi^2 \right\} \tag{10}$$

Clearly, $\theta_c$ must then be taken as equal to $\mathcal{R}$.

In all previous work, computations were performed for the special case where $\theta_c = \mathcal{R}$ is a prescribed constant. The computation was performed such that at every time step in the finite difference solution of Eqs. (4) and (5), values of $\theta$ were computed at every mesh point and compared with $\theta_c$. The amount of crack growth as a function of time was therefore determined by monitoring the "switches" of the step function at points along the crack path as a function of time. Crack arrest occurred when the process could no longer continue.

Because the comparison between $\theta$ and $\theta_c$ was routinely made at every position along the length of the specimen, crack growth was free to take place discontinuously; i.e., with bridges of unbroken material being left behind. But, in previous work discontinuous crack growth occurred only under exceptional conditions. Hence, it is believed that there will be little loss of fidelity if this feature is removed from the model. This simplifies the generalization of the model to consider speed-dependent values of $\mathcal{R}$. While it would be awkward to associate a crack speed with every possible pattern of crack advance, it is simple to determine a crack speed when the extension takes place in a well-defined manner from an existing crack at its tip.

The computational procedure using a given dependence of the fracture energy upon the crack speed $\mathcal{R} = \mathcal{R}(V)$ is as follows. At each time step the crack speed $V^*$ that would exist if crack extension were to occur at that time is determined. This hypothetical crack speed is then used to calculate a critical value of the crack extension parameter via $\theta_c^* = \mathcal{R}(V^*)$. This is compared with the current value of $\theta$. An increment of crack growth occurs when the hypothetical value $\theta_c^*$ first becomes less than the instantaneous value of $\theta$ at the crack tip. Note that at the earliest time t when $\theta(a,t) \geq \theta_c^*(t)$, $V = V^*(t)$ and $\mathcal{R} = \theta_c^*(t)$ where V and $\mathcal{R}$ denote the actual values for that point of the crack propagation process.

The computational procedure is designed to simulate the manner in which experiments with the wedge-loaded DCB are carried out. [1] [2] In these experiments, the wedge is inserted slowly, continually forcing apart the load pins, until unstable crack propagation commences. Subsequent motion of the wedge is neglected. A blunted initial crack tip is used to allow the stress intensity factor at the onset of propagation, $K_q$, to be arbitrarily greater than $K_{Ic}$. (The radius of the blunted region is varied to control $K_q$ systematically.) The boundary and initial conditions used in the computation reflect these conditions, as described fully in Reference [4].

A complete crack propagation computation can be performed provided values of the specimen dimensions, the initial crack length, and the initial displacement of the load pins (fixed via a specified value of $K_q$) together with a function relating $\mathcal{R}$ to V are specified. Notice that no arbitrary specification of the arrest condition is made.

## COMPUTATIONAL RESULTS FOR AN ASSUMED FRACTURE TOUGHNESS-CRACK SPEED DEPENDENCE

The data of Eftis and Krafft, [5] Shoemaker and Rolfe, [6] Burns and Bilek, [7] together with that of Hahn, et. al., [2] for low-carbon steel at temperatures in the vicinity of the static ductile-brittle transition temperature suggest that the fracture toughness behavior $K_{Id} = K_{Id}(V)$ can be approximated by a parabolic relation such as

$$K_{Id} = K_{Ic} \left\{1 + \left[1 - \left(\frac{K_{Im}}{K_{Ic}}\right)\right] \left[\left(\frac{V}{V_m}\right)^2 - 2\left(\frac{V}{V_m}\right)\right] \right\} \tag{11}$$

where $K_{Ic}$ has its usual meaning, $K_{Im}$ is the minimum value of the fracture toughness and $V_m$ is the crack speed at which the minimum toughness occurs.* The best fit of these data is obtained by taking $K_{Im} = 0.7\ K_{Ic}$ and $V_m = .05\ C_o$. Only these values will be

---

* As discussed above, the K parameter is inappropriate for dynamic crack propagation from a rigorous point of view. It is nevertheless useful because, (1) experimental results are invariably given in terms of it, and (2) it facilitates comparison with crack growth initiation data. In this work, it will be taken as a definition that $\mathcal{R}/\mathcal{G}_{Ic} = (K_{Id}/K_{Ic})^2$; cf, Eq. (1).

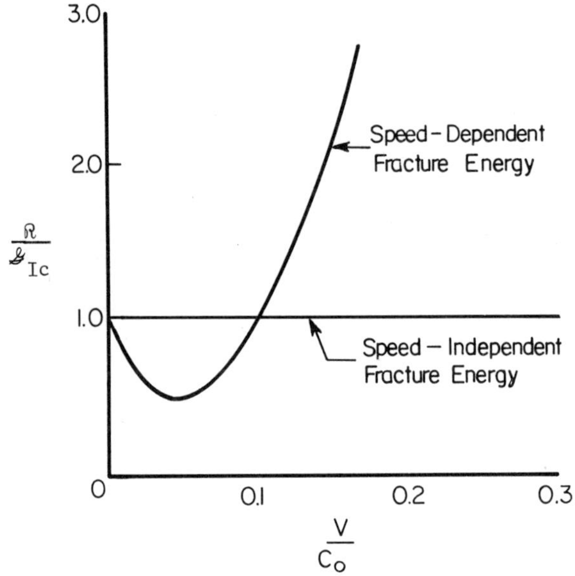


FIGURE 1. ASSUMED CRACK SPEED--FRACTURE ENERGY RELATIONS USED IN THE DYNAMIC CRACK PROPAGATION COMPUTATIONS

considered for the calculations given in this paper. The effect of different toughness-crack speed relationships (obtained by changing the values of the parameters in Eq. (11) and/or changing the form of the relation) will be examined in a subsequent paper. The fracture energy-crack speed relation given by Eq. (11) together with a speed-independent relation obtained by setting $K_{Im} = K_{Ic}$, are shown in Fig. (1).

In both experiments and computations, it is easy to impose an initial stress intensity that is high enough to completely sever the specimen. While these cases have their uses, the primary purpose of this paper is to examine the characterization of arrest. Hence, the specimen length in the calculations reported here was made quite long in order to extend the range of parameters for which crack arrest was obtained within the specimen.* To set the initial conditions and to characterize the

---

* In the notation of Reference [4], the computations reported in this paper were carried out for a specimen having the following dimensions: h = 63.5 mm, L = 600 mm, e = 20.3 mm, b = 12.7 mm, $\ell$ = 88.0 mm, and D = 25.4 mm.

arrest point, the relation for the stress intensity factor derived from a finite-length beam-on-elastic foundation model can be used. [3] This relation can be written

$$K_I = \frac{\sqrt{3}}{2} \frac{Eh^{3/2}\delta}{a^2} \Phi(a,c) \tag{12}$$

where $\delta$ is the displacement of the load pins, $a$ is the distance from load pins to the crack tip, c is the distance from crack tip to free end of specimen, and $\Phi$ denotes a function of the specimen geometry given by

$$\Phi = \left\{\left(\sinh^2\lambda c + \sin^2\lambda c\right) + \frac{1}{\lambda a}\left(\sinh\lambda c\cosh\lambda c - \sin\lambda c\cos\lambda c\right)\right\} \cdot$$

$$\cdot \left\{\left(\sinh^2\lambda c - \sin^2\lambda c\right) + \frac{3}{\lambda a}\left(\sinh\lambda c\cosh\lambda c + \sin\lambda c\cos\lambda c\right) + \right.$$

$$\left. + \frac{3}{\lambda^2 a^2}\left(\sinh^2\lambda c + \sin^2\lambda c\right) + \frac{3}{2\lambda^3 a^3}\left(\sinh\lambda c\cosh\lambda c - \sin\lambda c\cos\lambda c\right)\right\}^{-1}$$

where $\lambda = (6)^{1/4}/h$.

The parameter $K_{Ia}$, by definition, is just the static stress intensity factor following arrest. It can therefore be related to the material constant $K_{Im}$ via Eq. (12). This is

$$\frac{K_{Ia}}{K_{Im}} = \frac{\Phi(a_r)}{\Phi(a_o)}\left(\frac{a_o}{a_r}\right)^2\left(\frac{K_q}{K_{Ic}}\right)\left(\frac{K_{Ic}}{K_{Im}}\right) \tag{13}$$

where $a_r$ is the crack length at arrest, $a_o$ is the initial crack length, and the ratios $K_q/K_{Ic}$ and $K_{Ic}/K_{Im}$ are given. Hence, it is only necessary to determine $a_r/h$ in the dynamic calculation to evaluate the ratio $K_{Ia}/K_{Im}$ from Eq. (13). It might be remarked that although Eq. (12) differs slightly from conventional formulas, it is used here because, being derived from the beam-on-elastic foundation model, it is consistent with dynamic results obtained with the model. In addition, it is the only closed-form relation known for a finite-length specimen.

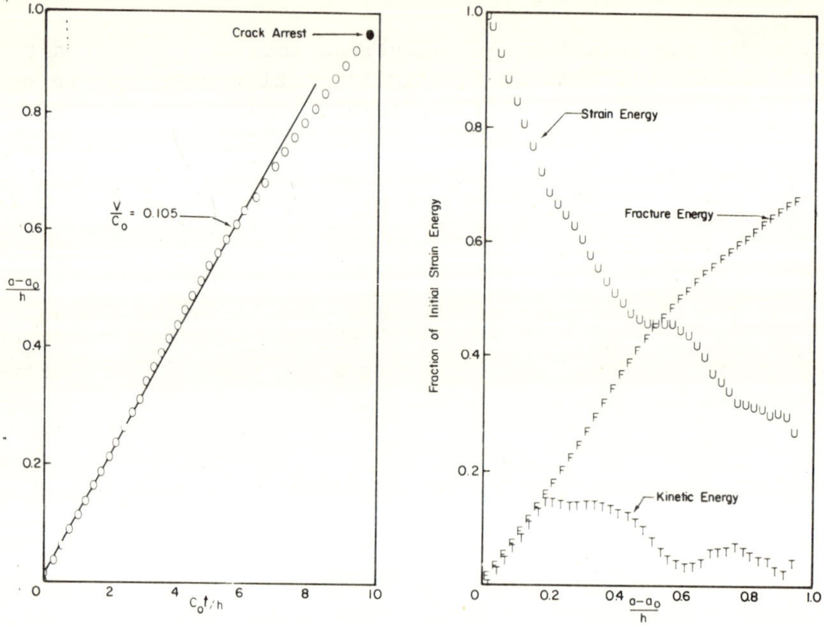


(a) Crack Growth as a Function of Time

(b) Partition of Energy as a Function of Crack Growth

FIGURE 2. COMPUTATIONAL RESULTS FOR $K_q/K_{Ic}$ = 1.5 AND $a_0/h$ = 1.0 USING CRACK SPEED DEPENDENT FRACTURE ENERGY GIVEN BY EQ. (11) WITH $K_{Im}/K_{Ic}$ = 0.7 AND $V_m/C_o$ = .05

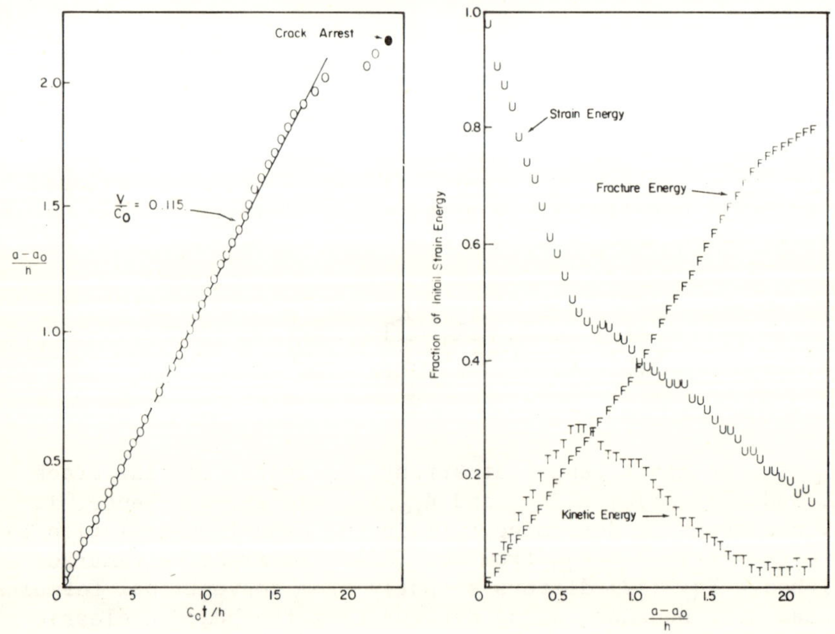


(a) Crack Growth as a Function of Time

(b) Partition of Energy as Function of Crack Growth

FIGURE 3. COMPUTATIONAL RESULTS FOR $K_q/K_{Ic}$ = 2.0 AND $a_0/h$ = 2.0 USING A CRACK SPEED DEPENDENT FRACTURE ENERGY GIVEN BY EQ. (11) WITH $K_{Im}/K_{Ic}$ = 0.7 AND $V_m/C_o$ = .05

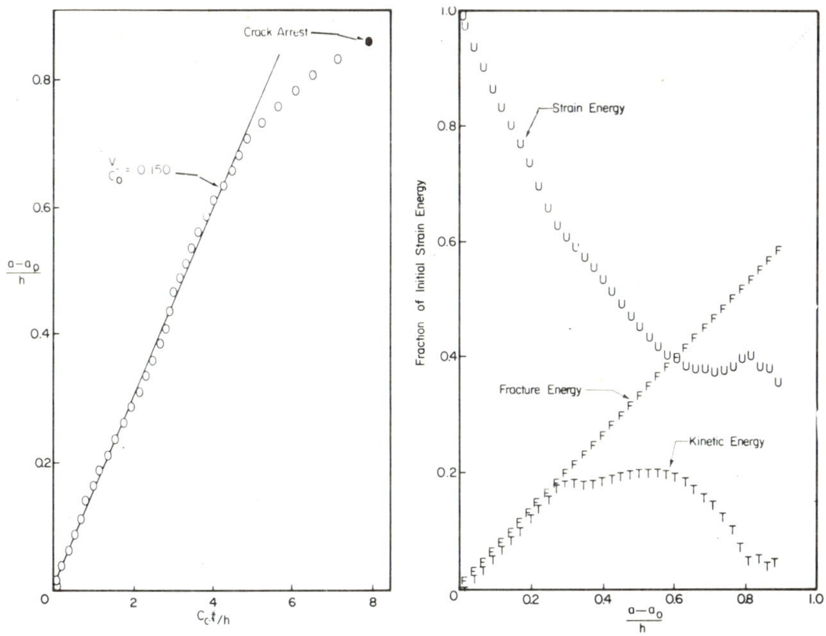


(a) Crack Growth as a Function of Time

(b) Partition of Energy as a Function of Crack Growth

FIGURE 4. COMPUTATIONAL RESULTS FOR $K_q/K_{Ic}$ = 1.5 AND $a_0/h$ = 1.0 USING A CONSTANT FRACTURE ENERGY

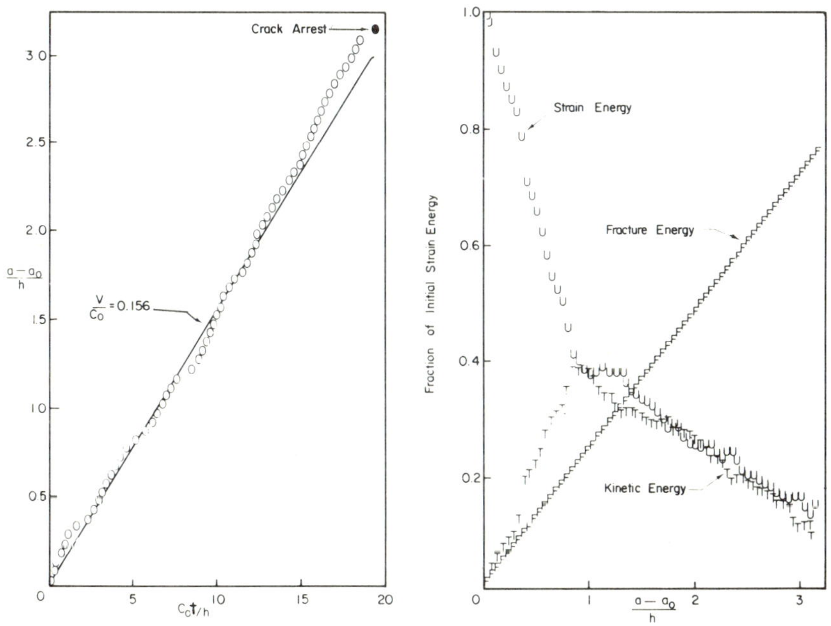


(a) Crack Growth as a Function of Time

(b) Partition of Energy as a Function of Crack Growth

FIGURE 5. COMPUTATIONAL RESULTS FOR $K_q/K_{Ic}$ = 2.0 AND $a_0/h$ = 2.0 USING A CONSTANT FRACTURE ENERGY

Computational results for a material with a crack speed-dependent fracture toughness for two different initiation conditions are given in Figs. (2) and (3). Figs. (4) and (5) show the corresponding results for a speed-independent material. It can be seen that the most striking feature observed in the experiments--the virtual constancy of the crack speed from the onset of motion until shortly before arrest--is well reproduced in Figs. (2) and (3). By comparing these with Figs. (4) and (5), it can be seen that the speed-dependence acts to smooth out the oscillations in crack speed that occur in the calculations where a constant fracture energy is used. The "start-stop" propagation behavior occurring in the latter case at high values of $K_q$ (see Reference [4]) tends to be suppressed in the speed-dependent case. (This can be seen to some extent by comparing Figs. (3) and (5).) Also shown in Figs. (2)-(5) are the individual values of the dynamic strain energy, the dynamic kinetic energy, and fracture energy monitored during the process. These indicate the significant role that is played by the kinetic energy in supporting crack propagation.

Table 1 shows the results of a series of computations in which the mesh size and time step were varied in the finite-difference solution. It can be seen that the parameters of most interest--$\bar{V}$, the crack speed over the essentially constant speed portion of the event, and $K_{Ia}$--are evidently computed very consistently. The crack speed just prior to arrest, $V_r$ (which might be anticipated to approach $V_m$), is evidently obtained less precisely. On the basis of these results, all subsequent computations were performed using the parameters corresponding to the third entry in Table 1.

The determination of the arrest point in these computations always involves some uncertainty. This arises from the necessity of completing the computation in a reasonable amount of time. In this work the computation is arbitrarily terminated when the hypothetical crack speed (which decreases monotonically with time since the last previous increment of crack advance) falls below a certain pre-set number. In the results reported here, this threshold speed value was .015 $C_0$. As a check some special computations in which the termination point was drastically postponed were performed. It was found that a spurt of additional growth can in fact occur long after arrest has ostensibly taken place. But, because detecting the small amount of delayed growth increases the computer time required by about an order of magnitude, we have decided to ignore this possibly spurious effect. Hence, the calculations reported here likely underestimate the amount of growth at arrest. Because $K_I$ decreases monotonically with crack length, the procedure therefore slightly overestimates $K_{Ia}/K_{Im}$.

TABLE 1. AFFECT OF FINITE-DIFFERENCE PARAMETERS ON THE COMPUTATIONAL RESULTS USING A CRACK-SPEED DEPENDENT-FRACTURE TOUGHNESS GIVEN BY EQ. (11) FOR $K_q/K_{Ic} = 1.5$ AND $a_o/h = 1.0$

| Input Parameters | | Computational Results | | | |
|---|---|---|---|---|---|
| $\Delta x/h$ | $C_o\Delta t/\sqrt{12}h$ | $a_r/h$ | $\bar{V}/C_o$ | $K_{Ia}/K_{Im}$ | $V_r/V_m$ |
| 0.100 | .0200 | 1.900 | .106 | 0.91 | 1.70 |
| 0.100 | .0100 | 1.900 | .107 | 0.91 | 1.65 |
| 0.050 | .0100 | 1.950 | .104 | 0.88 | 1.31 |
| 0.050 | .0050 | 1.950 | .105 | 0.88 | 1.52 |
| 0.025 | .0050 | 1.925 | .104 | 0.89 | 1.44 |
| 0.025 | .0025 | 1.950 | .105 | 0.88 | 1.34 |

The results from a series of computations in which $K_q$ and $a_o$ were varied for both the crack speed-dependent and crack speed-independent fracture energy cases are shown in Table 2. Of most interest in these results are the calculated values of the ratio $K_{Ia}/K_{Im}$. On the basis of a quasi-static argument, these would be expected to be unity. Contradicting the expectation of Crosley and Ripling, [8-9] the calculated values of $K_{Ia}$ are not equal to $K_{Im}$ nor, in fact, do they correspond to $K_{Im}$ in a one-to-one fashion.

## DISCUSSION AND CONCLUSIONS

It is important to emphasize that the conclusions that can be drawn from the analysis described in this paper are directly applicable only to the DCB test specimen. To what extent they are true for large-scale structures is not known at the present time. It seems certain, however, that analyses of large-scale structures cannot be based solely upon quasi-static considerations, but must be based upon the concepts of dynamic fracture mechanics as described and applied in this paper.* In other words, in a crack arrest design analysis, attention must be given to the details of

* Crack propagation in pressurized pipelines has been attacked in this way by Kanninen and Sampath. [12]

TABLE 2. COMPUTATIONAL RESULTS FOR CRACK ARREST IN THE DCB SPECIMEN FOR VARIOUS DIFFERENT GEOMETRIES AND INITIAL STRESS INTENSITY FACTORS

| Initiation Conditions | | Computational Results: Speed-Dependent Fracture Energy | | | Computational Results: Speed-Independent Fracture Energy | | |
|---|---|---|---|---|---|---|---|
| $a_o/h$ | $K_q/K_{Ic}$ | $a_r/h$ | $\overline{V}/C_o$ | $K_{Ia}/K_{Im}$ | $a_r/h$ | $\overline{V}/C_o$ | $K_{Ia}/K_{Im}$ |
| 1.0 | 1.00 | 1.45 | .074 | 0.89 | 1.00 | 0 | 1.00 |
| 1.0 | 1.25 | 1.75 | .094 | 0.86 | 1.40 | .086 | 0.82 |
| 1.0 | 1.50 | 1.95 | .104 | 0.88 | 1.90 | .149 | 0.64 |
| 1.0 | 1.75 | 2.05 | .116 | 0.95 | 2.55 | .192 | 0.47 |
| 1.0 | 2.00 | 2.15 | .122 | 1.01 | 3.35 | .203 | 0.35 |
| 1.0 | 3.00 | 2.60 | .146 | 1.13 | 6.70 | .262 | 0.15 |
| 1.0 | 4.00 | 2.90 | .163 | 1.26 | * | .308 | * |
| 2.0 | 1.00 | 2.90 | .063 | 0.80 | 2.00 | 0 | 1.00 |
| 2.0 | 1.50 | 3.60 | .097 | 0.83 | 3.50 | .097 | 0.61 |
| 2.0 | 2.00 | 4.20 | .115 | 0.85 | 5.15 | .156 | 0.42 |
| 2.0 | 3.00 | 5.35 | .138 | 0.84 | * | .214 | * |
| 2.0 | 4.00 | 6.55 | .149 | 0.78 | | | |
| 3.0 | 1.00 | 3.55 | .067 | 1.08 | 3.00 | 0 | 1.00 |
| 3.0 | 1.50 | 5.60 | .089 | 0.73 | 4.80 | .072 | 0.67 |
| 3.0 | 2.00 | 6.95 | .106 | 0.66 | 6.90 | .124 | 0.47 |

* Crack did not arrest.

various conceivable unstable crack propagation processes that could occur in order to delineate the severity of the conditions in which the crack must be arrested. These conditions must be characterized with a fully dynamic analysis and properly determined dynamic material properties.

The results that indicate the inapplicability of the quasi-static approach are the calculated values of $K_{Ia}/K_{Im}$ in Table 2. These results show that the "arrest toughness" $K_{Ia}$ varies when the crack arrest experiment in the wedge-loaded DCB test specimen is performed in different ways. Specifically, by varying the initial load level (through the parameter $K_q/K_{Ic}$) and/or the geometry of the test piece (through the parameter $a_o/h$), the value of $K_{Ia}/K_{Im}$ that is obtained is seen to change. The fact that the change is not always marked is consistent with the experimental findings of Crosley and Ripling [8-9] and is the likely reason that they have concluded that $K_{Ia}$ is a material constant. The variation in the speed-dependent results, at least for the smaller values of $a_0$, would likely be attributed to normal experimental scatter.

It might be argued that the variation in $K_{Ia}$ displayed in Table 2 is small enough that, for "practical engineering work", the variations can be neglected and $K_{Ia}$ taken as if it were a constant. This would be a suitable approximation for structures having an applied loading and geometrical configuration comparable to a laboratory test specimen. However, for application of small-scale test results to structures that involve orders of magnitude extrapolation, it is very possible that even slight discrepancies in laboratory test data could be magnified catastrophically. The business of extracting fundamental data from laboratory-scale experiments and applying it to service structures absolutely requires that such data have a firm logical basis. Thus, while it is possible that $K_{Ia}$ can be a useful structural design parameter under certain restricted conditions,* as the results of this paper show, it is not a rigorous concept and must be used with great caution. The only viable alternative is, therefore, the energy-based dynamic fracture mechanics approach in which crack arrest is treated as the termination of a dynamic process.

ACKNOWLEDGEMENTS

The author is greatly indebted to his colleagues at Battelle-Columbus, G. T. Hahn, R. G. Hoagland, and A. R. Rosenfield, for making their experimental results available and for their contributions to the ideas embodied in this paper. The author would also like to thank Mr. M. Tikson of the Battelle Computer Center for providing the necessary computer time.

---

* A distinction can be made between the use of $K_{Ia}$ as a material property input to a dynamic calculation of crack arrest, which may be acceptable in certain circumstances, and the use of $K_{Ia}$ in a static calculation as the sole criterion of crack arrest.

REFERENCES

[1] Hahn, G. T., Hoagland, R. G., Rosenfield, A. R., and Sejnoha, R., "Rapid Crack Propagation in a High-Strength Steel", Met. Trans., Vol. 5, p 475, 1974.

[2] Hahn, G. T., Hoagland, R. G., Kanninen, M. F., and Rosenfield, A. R., "The Characterization of Fracture Arrest in Structural Steel", in Pressure Vessel Technology, Part II, ASME, New York, p 971, 1973.

[3] Kanninen, M. F., "An Augmented Double-Cantilever Beam Model for Studying Crack Propagation and Arrest", Int. J. Fracture, Vol. 9, p 83, 1973.

[4] Kanninen, M. F., "Dynamic Analysis of Unstable Crack Propagation and Arrest in the DCB Test Specimen", Int. J. Fracture, in press, 1974.

[5] Eftis, J., and Krafft, J. M., "A Comparison of the Initiation With the Rapid Propagation of a Crack in a Mild Steel Plate", J. Basic Eng., Vol. 87, p 257, 1965.

[6] Shoemaker, A. K., and Rolfe, S. T., "Static and Dynamic Low-Temperature $K_{Ic}$ Behavior of Steels", J. Basic Eng., Vol. 91, p 512, 1969.

[7] Burns, S. J., and Bilek, Z. J., "The Dependence of the Fracture Toughness of Mild Steel on Temperature and Crack Velocity", Met. Trans., Vol. 4, p 975, 1973.

[8] Crosley, P. B., and Ripling, E. J., "Crack Arrest Toughness of Pressure Vessel Steels", Nuc. Eng. Design, Vol. 17, p 32, 1971.

[9] Crosley, P. B., and Ripling, E. J., "Crack Arrest in an Increasing K-Field", in Pressure Vessel Technology, Part II, ASME, New York, p 995, 1973.

[10] Rice, J. R., "A Path Independent Integral and the Approximate Analysis of Strain Concentration by Notches and Cracks", J. Applied Mech., Vol. 35, p 379, 1968.

[11] Mott, N. F., "Fracture of Metals: Theoretical Considerations", Engineering, Vol. 165, p 16, 1948.

[12] Kanninen, M. F., and Sampath, S. G., "Crack Propagation in Pressurized Pipelines", in Pressure Vessel Technology, Part II, ASME, New York, p 981, 1973.

# TEMPERATURE DEPENDENCE OF RAPID CRACK PROPAGATION AND ARREST IN A517F STEEL

G. T. Hahn, R. G. Hoagland, and A. R. Rosenfield

Battelle
Columbus Laboratories
505 King Avenue
Columbus, Ohio 43201

## ABSTRACT

The velocity dependence of the dynamic fracture toughness associated with a rapidly propagating crack in A517F steel has been measured over the temperature range -35 to -196°C. The data obtained using wedge-loaded double-cantilever beam specimens, are analyzed using the beam-on-elastic-foundation model. According to both theory and experiment, the stress intensity at crack arrest is found to vary with the initial displacement of the loading points and is not a material constant.

## INTRODUCTION

There is renewed concern today over the conditions required for the arrest of a fast moving crack. This concern arises from the possibility of large overloads in flawed structures, such as when two ships collide. Under these circumstances unstable fractures can be initiated by the very large crack-tip stress-intensities produced at local flaws in the area of impact. As another example, particular concern has been voiced over the possibility of unstable crack extension in the wall of a nuclear pressure vessel as a result of high thermal stresses associated with emergency core cooling. In both cases, there is a need for a reliable analysis of the likelihood of crack arrest.

In view of the importance of crack arrest, it is essential that proper criteria be developed. Currently, three material parameters are receiving most attention:

(a) The dynamic toughness approach utilizes the toughness value measured for a rapidly-loaded stationary crack, called $K_{ID}$. It is based on the idea that these conditions simulate a propagating crack and on Irwin and Wells' suggestion [1] that crack arrest is the reverse of crack extension. One difficulty with this approach is that the $K_{ID}$ values measured so far seem to decrease continuously with increasing loading rate, and that it is not possible to determine a minimum value which could be used in design. The relation between loading rate and crack velocity is also not clear.

(b) The arrest toughness approach involves the assumption that the stress intensity at the point of crack arrest is a material property, called $K_{Ia}$, [2] which can be calculated from a static analysis. Such an assumption is valid on an experiment where little kinetic energy is generated or where the kinetic energy that is generated is not converted into fracture energy. In general, these conditions are not likely to be achieved. For example, Kanazawa [3] has shown a significant effect of specimen geometry on crack arrest in wide-plate tests (Fig. 1). To explain these results he suggests that the static K value at arrest must be modified to account for the previous crack velocity and the characteristics of the loading system. This conclusion is reinforced by the results obtained in our laboratory, [4-7] such as the calculations presented by Kanninen at this Conference [7] showing the effects of geometry, grip displacement, and material properties on the stress intensity at arrest.

(c) The energy balance approach, utilized here is based on the actual energy dissipation rate for crack propagation, $K_{Id}$. This is derived from a dynamic analysis of the test procedure in which both kinetic energy and specimen inertia are accounted for. [4-7] In particular, the beam-on-elastic-foundation model [7] makes several predictions which have been verified using the wedge-loaded double-cantilever-beam specimen:

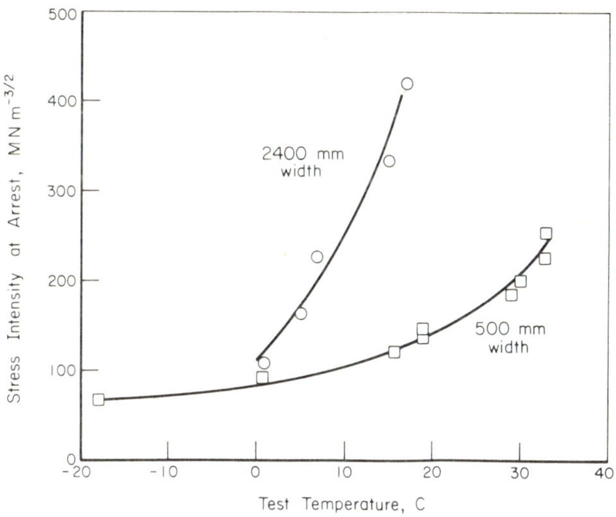


FIGURE 1. CRACK ARREST RESULTS FROM THE WIDE-PLATE TESTS OF KANAZAWA [37]

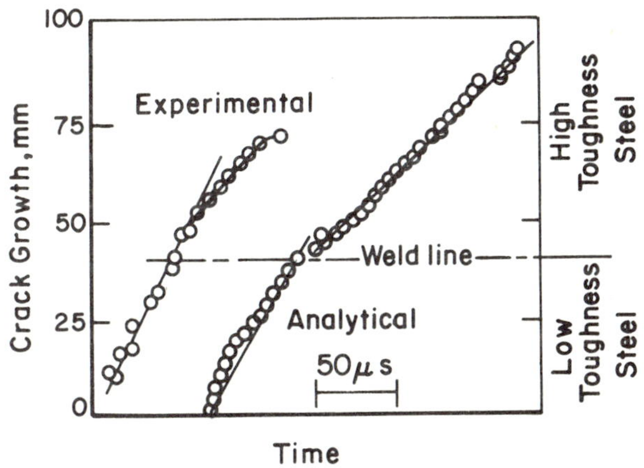


FIGURE 2. COMPARISON BETWEEN EXPERIMENTAL AND THEORETICAL CRACK PROPAGATION CURVES FOR THE WEDGE-LOADED DCB TEST SPECIMEN. Experimental data are for a duplex DCB specimen consisting of 4340 and A517F steels (specimen 3VY10--Table 1). Calculation is for $K_q/K_{d(\text{starter section})} = 2$, $K_{d(\text{test section})}/K_{d(\text{starter section})} = 1.4$. After Reference [5]

(i) The crack travels at a steady-state velocity which is determined by the material, test temperature, specimen geometry, and stress intensity at the onset of propagation, $K_q$.

(ii) For any given geometry there is a unique relation between steady-state velocity and crack length at arrest.

(iii) The stress intensity at arrest, $K_a$, for fixed grips is given by:

$$K_a = \frac{K_{Id}^2}{K_q} \qquad (1)$$

Since $K_q$ can be varied by changing the initial conditions (e.g., root radius of the starter notch), $K_a$ clearly depends on the testing system as well as the material. This conclusion is identical to that reached by Kanazawa [3] and, more recently by Bergkvist [8] and by Poynton and Fearnehough [9].

(iv) In duplex specimens, which consist of two steels of differing toughness welded together, a crack initiated in the starter section will attain the steady-state velocity characteristic of that material. Upon passing into the test section, the velocity will quickly assume a new steady-state value characteristic of the test section steel as shown in Fig. 2.

This paper reviews earlier data on A517F steel [4,5] reports new data on the same steel, and reanalyzes earlier results with the aid of recent developments in dynamic beam-on-elastic-foundation model. [7] Particular attention is paid to the velocity dependence of dynamic fracture toughness over a wide range of low temperatures and to the conditions for crack arrest.

## PROCEDURE

The wedge-loaded double-cantilever-beam (DCB) specimen was used in these experiments and the results were analyzed with the beam-on-elastic-foundation model. Both the experimental and analytical procedures have been described extensively elsewhere*. [4-7] In tests at -196°C ordinary DCB specimens were

* The only modification was that for temperatures below -78°C, the velocity measuring grid was insulated from the specimen by A12 Epoxy (Techkits, Demarest, N.J.).

used while at higher temperatures duplex and arrestor specimens were used. For specimens tested at -78°C, the duplex specimens consisted of SAE4340 starter sections, in which rapid cracks are initiated, and which are electron beam welded to a A517F steel test section. The specimens tested at temperatures between -35 and -54°C were arrestor specimens in which a length of A517F was welded between two sections of 4340. The crack path was 12 to 40 mm in the A517F for the different specimens. The specimen tested at -152°C consisted of an A517 starter section electron-beam welded to a 9Ni steel test specimen. With the exception of two 25.4 mm thick specimens tested at -78°C, all specimens were 12.7 mm thick.

There were several early experiments on the A517 steel at -196°C for which the crack velocities were not measured. In this case they were estimated from the relation:

$$K_{Id} = K_{Ic} \left\{ 1 - 9.3 \left( \frac{v}{c_o} \right) + 78 \left( \frac{v}{c_o} \right)^2 \right\} \qquad (2)$$

where $c_o$ is the bar wave speed. Eq. (2) is of the form generally accepted for cleavage failure: with increasing velocity $K_d$ first decreases, passes through a minimum and then begins to increase rapidly. [10] The coefficients of Eq. (2) were determined from the more recent data where crack velocities were measured. Note also that this equation is identical in form to the rate-dependent material of Kanninen's [7] calculation with the minimum value of $K_d = 0.72\ K_{Ic}$ occurring at a velocity of $0.06\ C_o$.

In some of the specimens tested at -196°C crack arrest occurred by a perpendicular deviation of the crack from its original path so that one arm broke off the specimen (Fig. 3). However, this occurred after the crack had begun to decelerate. As shown in Fig. (4) the arrest point calculated from the model is close to the deceleration region. As a result the reported arrest values represent deviations of the crack an amount 10-20% of the beam height from its original path. The veering is due to lateral movement of the loading pins which relieves the compressive load due to the wedge, and can be eliminated by the use of side grooves.

## RESULTS

Table 1 lists the experimental results, while the data are plotted in Figs. (5) - (7). $K_{ID}$ values are taken from the data of Barsom and Rolfe [11]. Although there are insufficient points to define the full $K_{Id}$ versus velocity curve at any one temperature, several trends are clear. Chief among these is the increasingly sharp rise in the curves of Fig. (5) at the higher temperatures. This effect may be due partially to the loss of

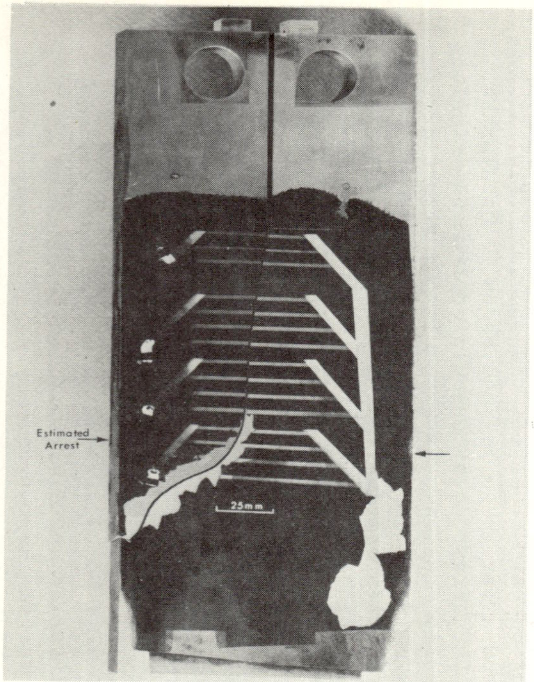


FIGURE 3. A517F TEST SPECIMEN BROKEN AT -196°C. Specimen number 3Y3. $K_q/K_{Ic}$ = 2.44

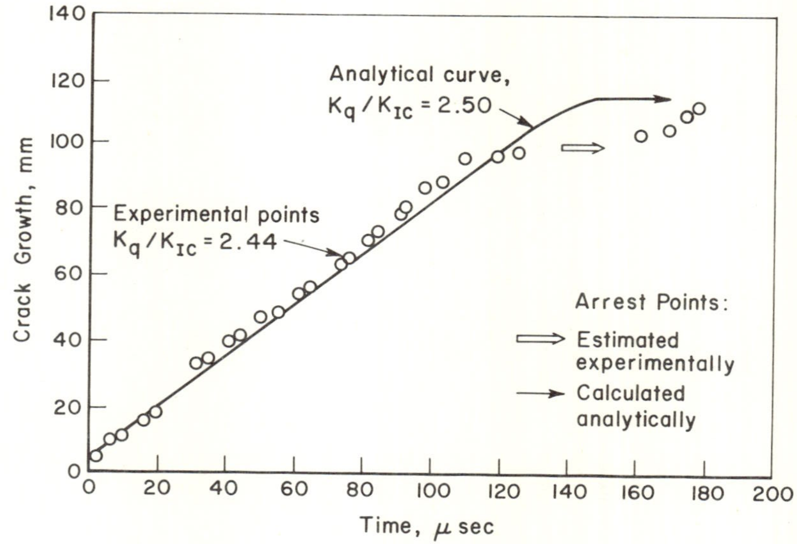


FIGURE 4. COMPARISON OF EXPERIMENTAL RESULTS ON SPECIMEN 3Y3 WITH ANALYTICAL PREDICTIONS OF THE DYNAMIC BEAM-ON-ELASTIC FOUNDATION MODEL [7]

TABLE 1. CRACK PROPAGATION AND ARREST DATA FOR A517F STEEL

| Specimen No. | Temperature (°C) | Crack Velocity (m/s) | Dynamic Toughness, $K_d$, $MNm^{-3/2}$ | Stress Intensity at Arrest, $K_a$, $MNm^{-3/2}$ |
|---|---|---|---|---|
| | | Duplex Specimens | | |
| 3VYV-1 | -48/-50* | 700 | 161 | - |
| -3 | -52/-54* | 155 | 225 | - |
| -5 | -35/-44* | 370 | 276 | 207 |
| -6 | -44/-48* | 300 | 242 | 217 |
| 3VY-10 | -78 | 475 | 180 | 114 |
| -12 | -78 | 560 | 159 | 102 |
| -13 | -78 | 625 | 170 | 68 |
| 3Y4a-2 | -152 | 830 | 73 | - |
| | | Ordinary Specimens | | |
| 517-1 | -196 | 0** | 34 | 34 |
| -2 | -196 | (500)** | 31 | 20 |
| -3 | -196 | (500)** | 26 | 13 |
| -4 | -196 | (660)** | 40 | 18 |
| 3Y-2 | -196 | 640 | 49 | 33*** |
| -5 | -196 | 870 | 61 | - |
| -3 | -196 | 830 | 48 | 28*** |
| -4 | -196 | 890 | 56 | 27*** |

* Temperature variation over the crack path.

** Estimated from dynamic beam-on-elastic-foundation model.

*** Crack ran to one side.

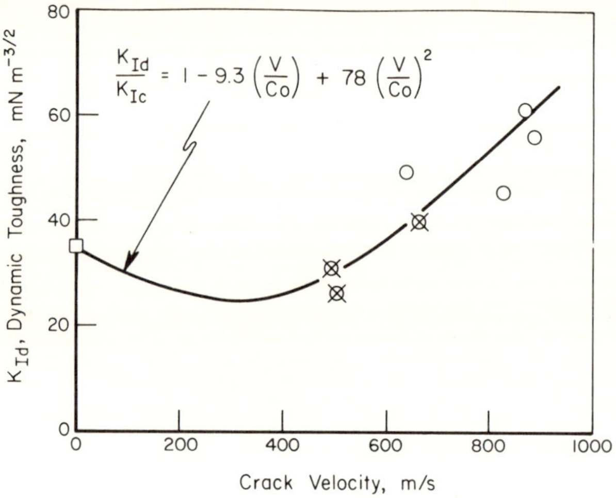


FIGURE 6. EFFECT OF CRACK VELOCITY ON DYNAMIC TOUGHNESS OF A517F STEEL AT -196°C

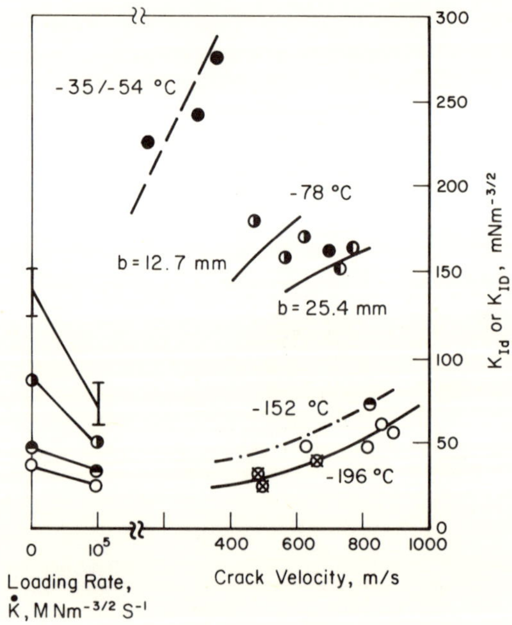


FIGURE 5. IMPACT RATE AND CRACK VELOCITY DEPENDENCE OF DYNAMIC TOUGHNESS OF A517F STEEL OVER A RANGE OF TEMPERATURES. Points marked ⊗ represent calculated velocities. All others were measured. $K_{ID}$ data from Barsom and Rolfe [11]

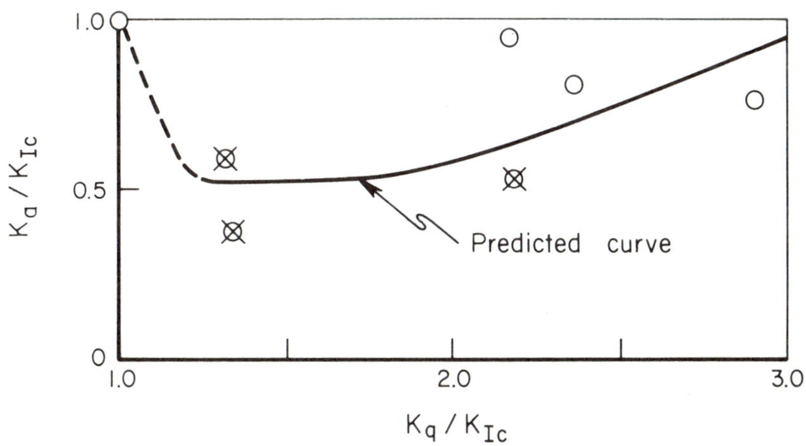


FIGURE 7. RELATION BETWEEN STRESS INTENSITY AT CRACK ARREST AND STRESS INTENSITY AT THE ONSET OF RAPID CRACK PROPAGATION IN THE WEDGE-LOADED DCB SPECIMEN. Solid curve is calculated from the Dynamic Beam-On-Elastic Foundation-Model [7] using Eq. (2) to specify the velocity dependence of dynamic toughness $K_{Id}$. Points are from experimental results on A517F Steel

constraint as the temperature is raised and, consequently, as the yield stress is lowered. Some loss of constraint is evident from the thickness effect at -78°C and constraint is clearly lost at the higher temperatures. There is some added uncertainty to the results at the highest temperatures (-35 to -54°C). In three of these specimens the crack arrested within the arrestor strip. The velocities plotted appear to be steady-state, but might actually represent deceleration. Therefore, the plotted velocities might be underestimated, possibly by as much as 200 m/s. At the same time, $K_d$ is calculated from the velocity via Fig.(5c) of Reference [5]. If the actual velocity is 200 m/s higher than the reported value, the dynamic fracture toughness would be 10-20% lower. Taking both of these effects into account, it is possible that the curve for this temperature range is closer to being parallel to that for -78°C. The fourth point of this series exhibited an unusually high velocity and low toughness for reasons that are not clear.

The data at -196°C are shown on an expanded scale in Fig.(6). The zero velocity value is $K_{Ic}$. Our value for reinitiation from a sharp crack agrees with the values reported by Barsom and Rolfe [11] and by Tetelman, et al [12]. The parabolic curve is a plot of Eq. (2). The points here represent two different size samples with the velocity measurements having been made on larger samples [5] and the velocity estimates on smaller samples. [4] In particular the ratio of initial crack length to beam height ($a_o/h$) was $>1$ and $<1$, respectively. While it was not possible to locate the toughness minimum precisely using the present data, there is no reason to modify the view [8] that it occurs at a few hundred m/s.

The crack arrest data at -196°C, shown in Fig.(7), are of interest for several reasons. In the first place, $K_a$ is clearly not constant but a function of the initial conditions as expressed by $K_q$ (the stress intensity at crack extension) and of the material as expressed by $K_{Ic}$. Secondly, the $K_a$ values associated with rapidly moving cracks are larger than for more slowly moving cracks. This is in accord with Kanazawa's [3] analysis of his data in Fig.(1) since the velocities in his large plates are larger than in the small plates. This result arises partially from the higher energies needed to sustain more rapid cracks. Indeed, if the material were rate independent $K_a$ would decrease with increasing velocity. [7]

## DISCUSSION

Taken together the experimental results define a consistent picture of the fracture process and provide added support for the energy balance approach. The data are also similar to the collected results on low-carbon ship steel in Fig. 8.

Because of the varying viewpoints on crack arrest, the results in Fig.(7) are of particular importance. The raw data suggest a variation of approximately a factor-of-two in stress intensity at arrest for specimens tested at -196°C. This is far beyond experimental error and is inconsistent with the concept of arrest toughness being a material property. Rather, the stress intensity at arrest contains an integrated record of the entire crack propagation event.

The question of application of these results to practical geometries has needs to be resolved. What is required is a combined analytical and experimental program which should include:

(a) A fully dynamic analysis of structures capable of yielding the strain energy release rate and kinetic energy as a function of geometry and loading, and

(b) A full description of the dynamic fracture energy as a function of velocity and temperature along the lines of Figs. (3) and (8).

These two ingredients must be combined in order to provide safe and reliable predictions of crack arrest. Until this has been done successfully, it is not possible to declare any particular criterion as conservative.

There are several additional problems which also arise. Among these is the significance of $K_{ID}$, the toughness associated with the initial extension of a rapidly loaded stationary crack. The relation between loading rate and crack velocity is not clear, nor have the assumptions underlying the application of this parameter been verified. Secondly, methods are needed for defining the toughness minimum ($K_{Id,min}$) more precisely. It is hoped that an improved analysis of the deceleration region of our experimental curves will aid in this respect. It is clear both from our experiments and from Kanninen's calculations [7] that $K_{Ia}$ will equal the toughness minimum only under special circumstances (see Table 2 of Reference (7)). In general for the DCB specimen, $K_{Ia} \neq K_{Id,min}$, and $K_{Ia}$ can be either greater than or less than or less than $K_{Id,min}$, depending on specimen geometry, loading conditions, and the $K_{Id}$ versus velocity relation.

## CONCLUSIONS

(1) Rapid crack propagation and crack arrest in A517 steel are governed by an energy balance.

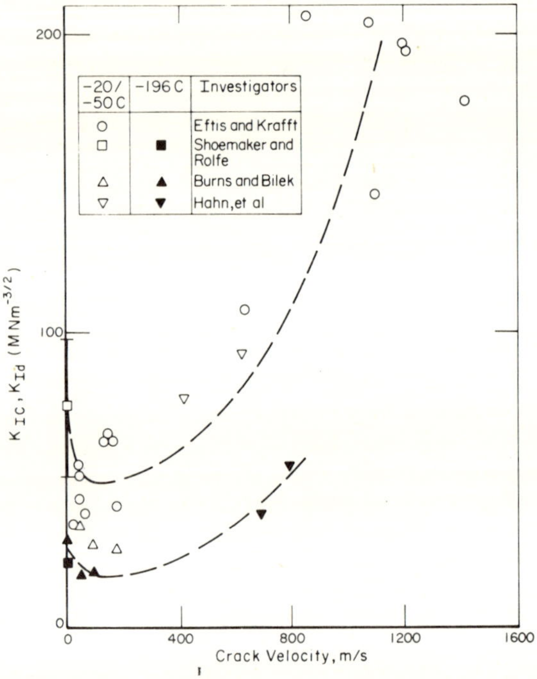


FIGURE 8. A SUMMARY OF DIRECT MEASUREMENTS OF THE DYNAMIC FRACTURE TOUGHNESS OF PLAIN CARBON STEELS. Data from References [10, 13-14]

(2) The stress intensity at arrest after cleavage failure is not constant but varies in accord with conservation of energy.

## ACKNOWLEDGMENTS

The work described in this paper was sponsored by the Ship Structure Committee by means of the Naval Ship Systems Command, Contract N00024-72-C-5142. The authors are grateful for the Committee's support. The authors also wish to thank P. Mincer for important contributions to the experimental work, and C. Pepper for valuable assistance with the manuscript. The opinions and conclusions presented in this paper are those of the authors and not necessarily those of the Ship Structure Committee nor of the Department of the Navy.

## REFERENCES

[1] Irwin, G. R. and Wells, A. A., Met Rev., 10, 223, 1965.

[2] Crosley, P. B. and Ripling, E. J., Pressure Vessel Tech., Part II, ASME, New York, 1973, 995.

[3] Kanazawa, T., Proceedings Int. Conf. on Dynamic Propagation, Lehigh U., July, 1972 (to be published).

[4] Hoagland, R. G., Rosenfield, A. R., and Hahn, G. T., Met. Trans., 3, 123, 1972.

[5] Hahn, G. T., Hoagland, R. G., Kanninen, M. F., and Rosenfield, A. R., Pressure Vessel Technology, Part II, ASME, New York, 1973, 971.

[6] Hahn, G. T., Hoagland, R. G., Rosenfield, A. R., and Sejnoha, R., Met. Trans., 5, 475, 1974.

[7] Kanninen, M. F., Conference on the Prospects of Fracture Mechanics, Delft, Netherlands, June, 1974.

[8] Bergkvist, H., Engg. Fract. Mech. (in press).

[9] Poynton, W. A., Fearnchough, C. D., Proceedings Int. Conf. on Dynamic Propagation, Lehigh U., July, 1972 (to be published).

[10] Eftis, J. and Krafft, J. M., Trans ASME, 87D, 257, 1964.

[11] Barsom, J. M. and Rolfe, S. T., Engg. Fract. Mech., 2, 341, 1971.

[12] Tetelman, A. S., Wullaert, R. A., and Ireland, D., Conf. on Practical Applications of Fracture Mechanics to Pressure Vessel Technology.

[13] Burns, S. J. and Bilek, Z. J., Met. Trans., 4, 975, 1973.

[14] Shoemaker, A. K. and Rolfe, S. T., Trans ASME, 91, Series D, 512, 1969.

# APPLICATION OF THE STRAIN ENERGY DENSITY FRACTURE CRITERION TO DYNAMIC CRACK PROBLEMS

R. J. Hartranft and G. C. Sih

Institute of Fracture and Solid Mechanics, Lehigh University, Bethlehem, Pennsylvania 18015 USA

## ABSTRACT

A plane wave impinging nonorthogonally on a flat crack in a plane strain analysis results in mode II as well as in mode I stress intensity factors. The mixed mode stress state cannot be used in the conventional energy release rate fracture criterion, for the actual direction of crack growth is not along the plane of the initial crack. Instead, the strain energy density fracture criterion is used to determine the direction of crack growth as well as the amplitude of the wave necessary to produce the crack growth. Results are presented for various values of the angle of incidence of a standing compressional wave and of its frequency.

## INTRODUCTION

Various types of dynamic effects on crack tip stress fields have been investigated. For example, several papers by Sih and Loeber have dealt with steady plane wave diffraction by a finite crack. Noteworthy for this application is reference [1]. A similar problem in which a semi-infinite crack extends in a specified way through the wave field has been treated by Chen and Sih [2]. Impulsive loading of cracks in an initially undisturbed material has also been considered [3].

Many workers have provided excellent dynamic stress analyses. Until recently, however, only those cases in which the symmetry of loading and geometry is sufficient to insure that the crack will run straight ahead were of interest in applications. Situations in which lack of symmetry leads to a mixture of the basic modes of crack deformation are not so easily applied. The mixed mode crack tip stress intensity factors can be computed with only slightly more effort than the single mode symmetric case. However, utilizing these stress intensity factors to evaluate the physical response of the material has not been straightforward due to lack of a theory of crack propagation which can be used with relative ease. Of primary interest are the direction of crack growth and the conditions under which it initiates.

An appealing extension of the energy release rate analysis of cracks which grow straight ahead requires the solution of a problem in which the main crack has a small branch at one end. For each position of the small branch, form an energy release rate by calculating the work required to close the branch and arrive at the state of the main crack with no branch. Through a limiting process, the energy release rates for infinitesimal branch crack lengths are, in principle, available for arbitrary angles between the main crack and the branch. It may be postulated that the unbranched crack will propagate in the direction of the largest energy release rate when the loads are sufficiently large that the energy release rate has reached the maximum value tolerable to the material. In practice, such a suggestion requires the solution of problems so extremely difficult that it may be regarded as impractical, especially in dynamic cases.

A less difficult, and more widely used, approach requires only the maximum normal stress in the neighborhood of the crack tip. It is then postulated that the crack will grow in the direction of the plane on which the maximum normal stress acts. Kerkhof [4] has used this theory to correlate his experimental work on wave fractography. Conceptually, the postulated crack growth law implies a particular model of crack growth involving cleavage of the material along a plane. There is evidence that, instead of cleavage, the crack does not run, but rather creates voids which enlarge to join with the initial crack [5].

A new, easily useable fracture criterion has been applied quite successfully to several problems of crack growth initiation under static loads. The basic theory and several applications are described in [6]. Further details of the theory are given in [7]. The basic idea is that it is the strain energy density in the neighborhood of the crack tip which governs crack growth.

A recent photoelastic analysis [8] of a problem of dynamic fracture investigated the path followed by a crack as its tip approaches a circular hole. A type of strain energy release rate was obtained from the photoelastic data. The strain energy density was similarly obtained. It was found that the path followed by the crack was in the direction of maximum strain energy release rate and minimum strain energy density.

This paper will extend the application of the strain energy density fracture criterion to the case of standing plane waves interacting with a finite crack. The analogous static case, discussed in [6], is one in which a finite crack is subjected to a uniaxial stress field at some angle to the plane of the crack. The dynamic case considered here differs in that the stress field varies in time. Because of the much greater complexity for a uniaxial stress field, a biaxial one is used. The biaxial stresses reduce the displacement field to one component only.

The conditions which, according to the strain energy density criterion, lead to crack growth are computed and presented graphically. The effect of frequency and crack position are examined in detail.

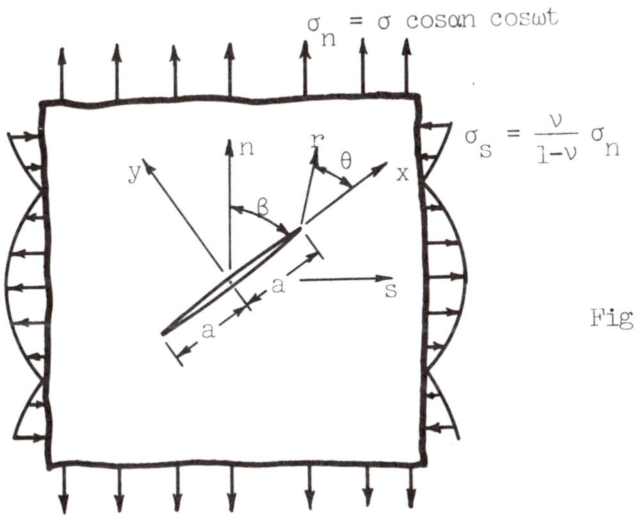


Fig. (1) - Cracked medium subjected to varying biaxial stresses at infinity

DYNAMIC STRESS ANALYSIS

An infinite body contains a crack of finite width, 2a, oriented as shown in Fig. (1). At large distanes from the crack, the vertical displacement field imposed is given by

$$u_n = \frac{\alpha}{\rho\omega^2}\sigma \sin\alpha n \cos\omega t \tag{1}$$

and the horizontal displacement, $u_s$, is zero. The wave number, $\alpha$, and the circular frequency, $\omega$, are related by

$$\omega = \alpha c_1 \tag{2}$$

where $c_1$ is the speed of compressional waves in the medium. It is convenient also to introduce the speed of shear waves,

$$c_2 = \sqrt{\frac{\mu}{\rho}} \tag{3}$$

and the ratio of the two wave speeds,

$$\kappa = \frac{c_2}{c_1} = \sqrt{\frac{1-2\nu}{2(1-\nu)}} \tag{4}$$

In these relationships $\nu$ is Poisson's ratio, $\mu$ is the shear modulus, and $\rho$ is the mass density of the material. The other parameters of the wave motion may be expressed in terms of $\omega$: period = $\tau = 2\pi/\omega$ and wavelength = $\lambda = 2\pi/\alpha = 2\pi c_1/\omega = c_1\tau$, where the wavelength, $\lambda$, is twice the distance between nodes of the standing wave.

The parameter $\sigma$ in the displacement field is the amplitude of the stress oscillation very far from the crack. The stress field associated with this uniaxial displacement field is given by

$$\begin{aligned} \sigma_n &= \sigma \cos\alpha n \cos\omega t \\ \sigma_s &= \frac{\nu}{1-\nu}\sigma_n, \quad \tau_{ns} = 0 \end{aligned} \tag{5}$$

The nonuniform stresses acting on the vertical sides of the infinite body of Fig. (1) are necessary in order to prevent horizontal displacements.

The solution of this problem may be extracted from [1], where it is shown that the crack tip stress fields have the form of the static solution except that the stress intensity factors are functions of time,

$$\begin{aligned} k_1 &= \sigma\sqrt{a}\,(1-2\kappa^2\cos^2\beta)\,\mathrm{Re}\{K_1(\omega,\beta)e^{-i\omega t}\} \\ k_2 &= \sigma\sqrt{a}\,\kappa^2\sin 2\beta\,\mathrm{Re}\{K_2(\omega,\beta)e^{-i\omega t}\} \end{aligned} \tag{6}$$

The complex functions $K_1$ and $K_2$ are determined from the solutions of the Fredholm integral equations ($j = 1,2$)

$$\Phi_j(\xi) - \int_0^1 \Phi_j(\eta)\, F_j(\xi,\eta)d\eta = \sqrt{\xi}\, J_0(\alpha\xi\cos\beta) \tag{7}$$

by

$$K_j(\omega,\beta) = \Phi_j(1) \tag{8}$$

In the integral equations,

$$F_j(\xi,\eta) = \frac{\sqrt{\xi\eta}}{1-\kappa^2} \int_0^\infty s f_j(s)\, J_0(s\xi)\, J_0(s\eta) ds \tag{9}$$

where

$$f_j(s) = \frac{2\kappa^2}{\alpha^2 a^2 p_j s} \left[\left(s^2 - \frac{\alpha^2 a^2}{2\kappa^2}\right)^2 - s^2 p_1 p_2\right] + 1-\kappa^2 \tag{10}$$

where

$$p_1 = \begin{cases} \sqrt{s^2-\alpha^2 a^2}, & |s|>\alpha a \\ -i\sqrt{\alpha^2 a^2 - s^2}, & |s|<\alpha a \end{cases} \tag{11}$$

and

$$p_2 = \begin{cases} \sqrt{s^2 - \frac{\alpha^2 a^2}{\kappa^2}}, & |s| > \frac{\alpha a}{\kappa} \\ -i\sqrt{\frac{\alpha^2 a^2}{\kappa^2} - s^2}, & |s| < \frac{\alpha a}{\kappa} \end{cases} \tag{12}$$

and $J_0$ is the zeroth order Bessel function of the first kind.

The functions $K_1$ and $K_2$ have the symmetry properties

$$K_j(\omega,\beta) = K_j(\omega,-\beta) = K_j(\omega,\pi-\beta) = K_j(\omega,\beta-\pi) \tag{13}$$

and the limiting behavior

$$K_j(\omega,\beta) \to 1 \text{ as } \omega\to 0,\ K_j(\omega,\beta) \to 0 \text{ as } \omega\to\infty \tag{14}$$

Curves in [1] of the magnitudes of $K_1$ and $K_2$ show them initially increasing from unity and subsequently decreasing toward zero as the frequency, $\omega$, is increased.

## DYNAMIC FRACTURE CRITERION

The strain energy density field in the neighborhood of the crack tip is a function of time for the case analyzed here. The singular part of the stress field is

$$\sigma_x = \frac{k_1}{\sqrt{2r}} \cos\frac{\theta}{2}[1 - \sin\frac{\theta}{2}\sin\frac{3\theta}{2}] - \frac{k_2}{\sqrt{2r}}\sin\frac{\theta}{2} \times$$

$$\times [2 + \cos\frac{\theta}{2}\cos\frac{3\theta}{2}]$$

$$\sigma_y = \frac{k_1}{\sqrt{2r}} \cos\frac{\theta}{2}[1 + \sin\frac{\theta}{2}\sin\frac{3\theta}{2}] + \frac{k_2}{\sqrt{2r}}\sin\frac{\theta}{2}\cos\frac{\theta}{2}\cos\frac{3\theta}{2}$$

$$\tau_{xy} = \frac{k_1}{\sqrt{2r}} \cos\frac{\theta}{2}\sin\frac{\theta}{2}\cos\frac{3\theta}{2} + \frac{k_2}{\sqrt{2r}}\cos\frac{\theta}{2} \times \qquad (15)$$

$$\times [1 - \sin\frac{\theta}{2}\sin\frac{3\theta}{2}]$$

$$\sigma_z = \nu(\sigma_x + \sigma_y)$$

$$\tau_{xz} = \tau_{yz} = 0$$

as in [6] for the static problem. Except for the fact that the stress intensity factors vary with time, the strain energy density factor

$$S = a_{11}k_1^2 + 2a_{12}k_1k_2 + a_{22}k_2^2 + a_{33}k_3^2 \qquad (16)$$

has the same form as in [6]. Furthermore, the coefficients

$$\begin{aligned} 16\mu a_{11} &= (3-4\nu-\cos\theta)(1+\cos\theta) \\ 16\mu a_{12} &= 2\sin\theta(\cos\theta-1+2\nu) \\ 16\mu a_{22} &= 4(1-\nu)(1-\cos\theta) + (1+\cos\theta)(3\cos\theta-1) \\ 16\mu a_{33} &= 4 \end{aligned} \qquad (17)$$

are also the same. Note that the polar coordinates r and θ are measured relative to the end of the crack as in Fig. (1).

The fracture criterion based on the strain energy density factor, S, requires first an investigation of its behavior as a function of θ. It is found that, in general, there are two distinct local maxima and two distinct local minima. The angles corresponding to the two minima give the directions of crack growth, provided that the loads are large enough to cause the crack to grow. One direction is associated with tensile loading conditions at the crack tip, while the other is for compressive loading. In the dynamic case, one can observe how one set of angles merges into the other as the loading changes from tensile to compressive. Thus, for any set of parameters and any value of the time, the direction of crack growth can be calculated.

The second consideration in the strain energy density fracture criterion is of the load required to produce fracture. The value of the strain energy density factor in the direction of crack growth, $S_0$, obtained in the first step depends on load and geometry. For fixed geometry, the value of $S_0$ increases with load until it reaches a certain value, characteristic of the material used, and then the crack grows. In the dynamic case, as the load fluctuates, the value of $S_0$ may vary from below to above the critical value. Thus, at the time when $S_0$ first reaches this level, the crack will run.

## RESULTS OF THE CALCULATIONS

The integral equation (7) was solved for values of the frequency $\omega = \alpha c_1$ such that $\omega a/c_1$ is between 0.1 and 1.0. The value of Poisson's ratio used in all of the calculations was $\nu = 0.25$. Once the integral equation (7) has been solved, the complex functions given by Eq. (8) can be inserted into Eqs. (6) for the stress intensity factors. Subsequently, the strain energy density factor given by Eqs. (16) and (17) can be examined for minima and used to discuss possible crack growth.

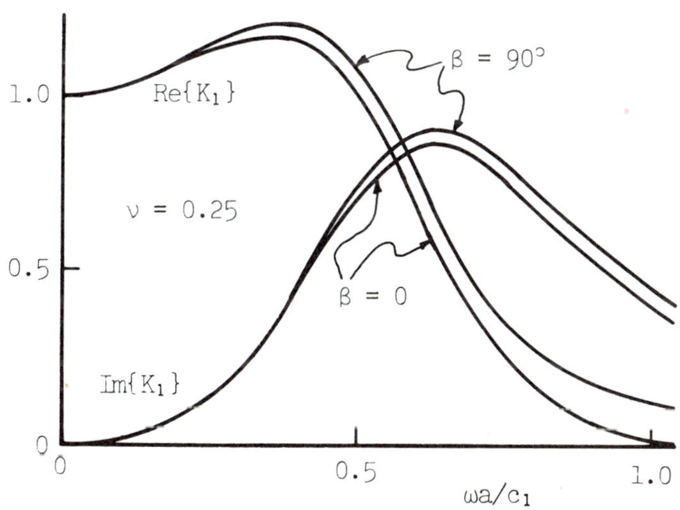


Fig. (2) - Correction factor for mode I stress intensity factor

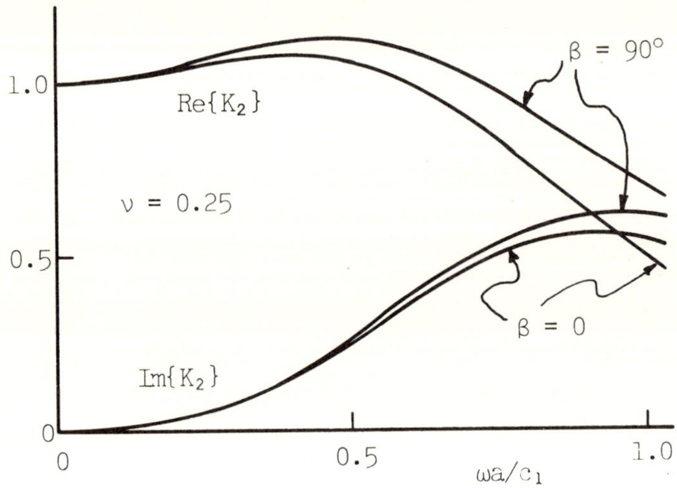


Fig. (3) - Correction factor for mode II stress intensity factor

Stress Intensity Factors. The correction factors of Eq. (8) which are to be used in Eqs. (6) are shown in Figs. (2) and (3). At low frequencies (long wavelengths), the angle of the crack has very little effect on the correction factors. They reach peaks at different frequencies, as pointed out in [1] before decreasing to zero.

The stress intensity factors of Eqs. (6) vary sinusoidally with time as shown in Fig. (4). The two values of frequency, $\omega = 0.1\ a/c_1$ and $\omega = 0.7\ a/c_1$, result in stress intensity factors of different amplitudes and phases. For the lower frequency, $k_1$ and $k_2$ are nearly in phase and reach their maximum at about the same time that the remote stress reaches its maximum. However, for the higher frequency wave, there is a delay. At a time of $0.075\tau$ after the remote stress peaks, the mode II stress intensity factor, $k_2$, reaches its maximum. The delay until the mode I stress intensity factor reaches its first maximum is $0.177\tau$. This phase difference between the stress intensity factors has as much effect on the fluctuating strain energy density field as their amplitudes.

The stress intensity factors of Eqs. (6) can be written in terms of their amplitudes and phases as

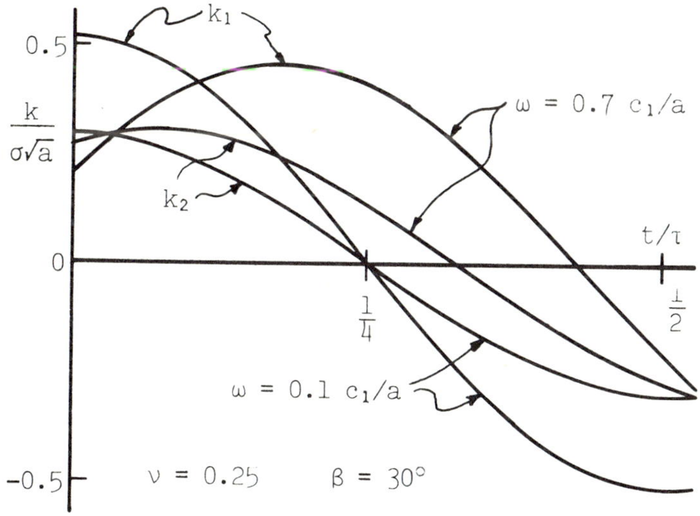


Fig. (4) - Dynamic stress intensity factors

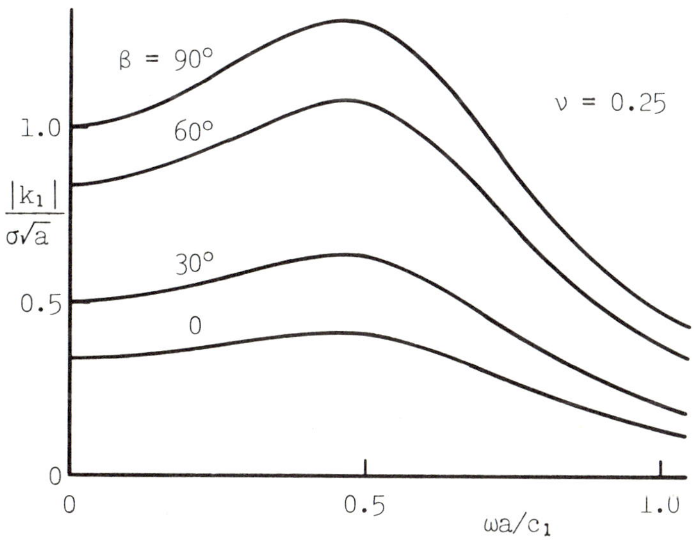


Fig. (5) - Amplitude of mode I stress intensity factor

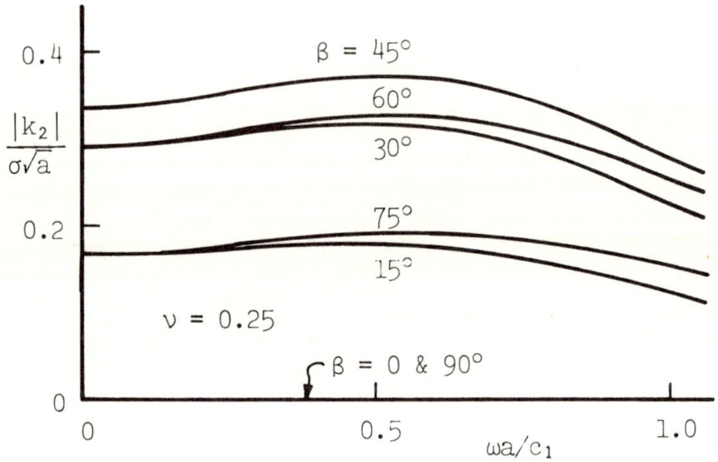


Fig. (6) - Amplitude of mode II stress intensity factor

$$k_1 = |k_1| \cos\omega(t-\delta_1)$$
$$k_2 = |k_2| \cos\omega(t-\delta_2) \tag{18}$$

where the phase is written in terms of the delay, $\delta$, between the peak applied stress and the peak stress intensity factor. The dependence of the amplitude and delay on input frequency and crack angle is depicted in Figs. (5) through (7). The amplitudes of the oscillating stress intensity factors initially increase as the frequency increases from zero. Subsequently, they reach maxima and then decrease steadily to zero. The largest mode I stress intensity factors occur when the crack is parallel to the nodal planes of the standing wave. The 45° position results in the largest mode II stress intensity factor.

The times, $\delta_1$ and $\delta_2$, at which the stress intensity factors reach their peak values are shown in Fig. (7). For low frequencies the stress intensity factors are nearly in phase with each other and with the oscillating stress field far from the crack. However, as the frequency increases the delays increase so that after the load peaks, the mode II stress intensity factor peaks, followed by the peak of the mode I stress intensity factor. The behavior for high frequencies is relatively less important because of the decrease in the amplitudes. But as the curves show, the delay decreases, at least for one curve, after reach-

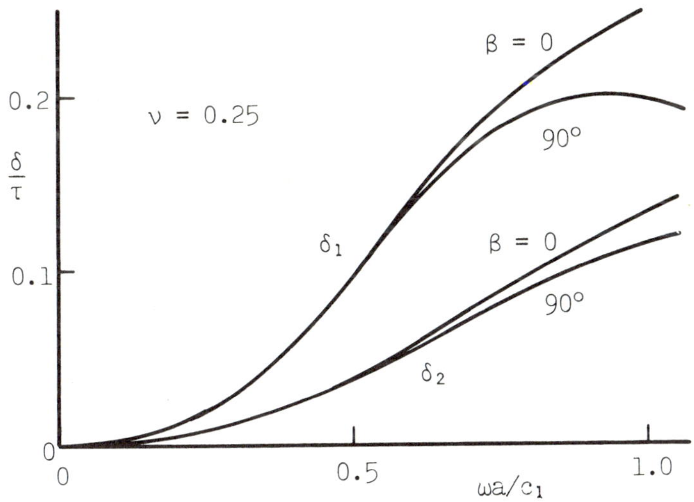


Fig. (7) - Time at which stress intensity factors reach maximum

ing a maximum. Thus, for high frequencies the stress intensity factors may again oscillate in phase.

Strain Energy Density. The numerical results for the stress intensity factors may be used in Eqs. (16) and (17) to obtain the strain energy density factor, S. The direction, $\theta_o$, in which the minimum value, $S_o$, of the strain energy density factor occurs varies with time because the stress intensity factors are out of phase. At low frequencies, for which the phase difference is small, the direction switches quite abruptly from a value appropriate for tensile loading to one for compressive loading when the load varies from tension to compression. For higher frequencies, the change of angle with time is more gradual, as shown in Fig. (8).

While the load varies and the direction, $\theta_o$, changes, the magnitude, $S_o$, of the minimum value of the strain energy density factor also fluctuates as shown in Fig. (9). The value of $S_o$ drops as the load drops after a delay which depends on frequency. As the load passes into compression, $S_o$ becomes and remains small until the load becomes tensile again. According to the strain energy density fracture criterion, the crack extends in the di-

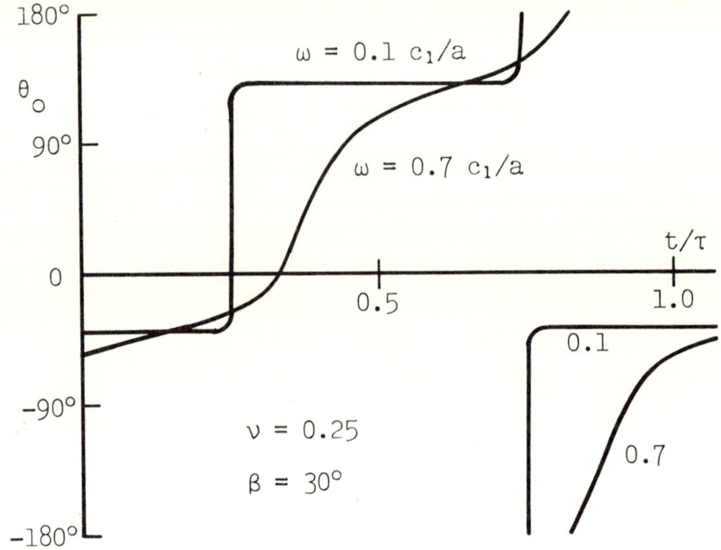


Fig. (8) - Direction of crack growth

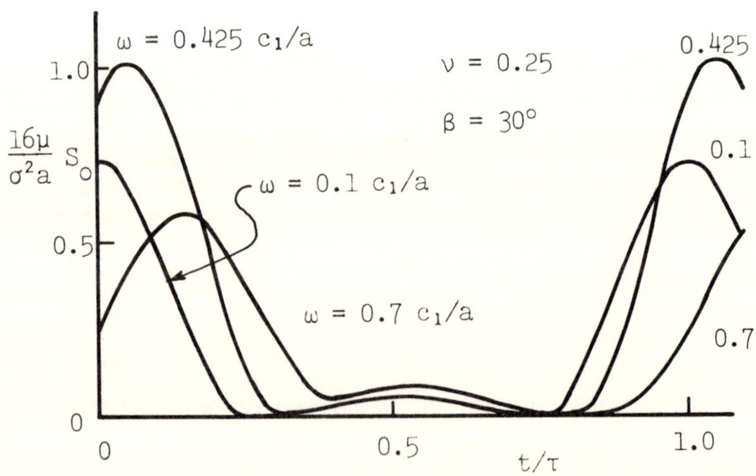


Fig. (9) - Variation of the minimum strain energy density factor rection $\theta_o$ when $S_o$ reaches the critical value for the material, $S_{cr}$. As Fig. (9) shows, the value of $S_o$ may, depending on the amplitude of the load, be larger than $S_{cr}$ for a long time, short time, or never. In the following, it will be assumed that the crack runs at the time when $S_o$ first increases above $S_{cr}$ and that the direction of crack growth is given by the corresponding angle.

The amplitude of the far field stress variation may be assigned arbitrary values. Suppose that the reference value is

$$\sigma_s = \sqrt{\frac{4\mu}{1-2\nu}\frac{S_{cr}}{a}} \tag{19}$$

which is the failure load in the static case of a uniform stress at infinity perpendicular to a crack of length 2a. The failure criterion may be expressed as

$$\frac{16\mu S_o}{\sigma^2 a} = 4(1-2\nu)\left(\frac{\sigma_s}{\sigma}\right)^2 \tag{20}$$

That is, when the left hand side, which is plotted in Fig. (9), reaches the value on the right, which depends on the load amplitude, the crack will run. When Poisson's ratio is held fixed, there are three independent parameters to consider: frequency, load amplitude, and orientation of crack. For various combinations of these parameters, one can determine the time and direction of crack growth.

From curves similar to those of Fig. (9), the peak values of the left hand side of Eq. (20) may be taken. Stress levels just high enough to cause crack growth at the peak values of $S_o$ are shown in Fig. (10). These stress amplitudes, denoted by $\sigma_m$, are the minimum which will cause fracture. Higher amplitudes will cause the curves of Fig. (9) to reach the critical point sooner, and a different angle of fracture will be obtained. At the 90° position of the crack, perpendicular to the remote displacement field, the load required to cause crack growth is smallest. At very low frequencies, $\sigma_m = \sigma_s$, and so the static failure stress of Eq. (19) is the minimum required for crack growth. As the frequency increases the minimum stress for failure, $\sigma_m$, decreases and then increases again. For example, in the 90° crack position, the frequency at which the crack would grow at the smallest stress amplitude is $\omega \approx 0.5\ c_1/a$. As the crack position is rotated toward the direction of the applied displacement field,

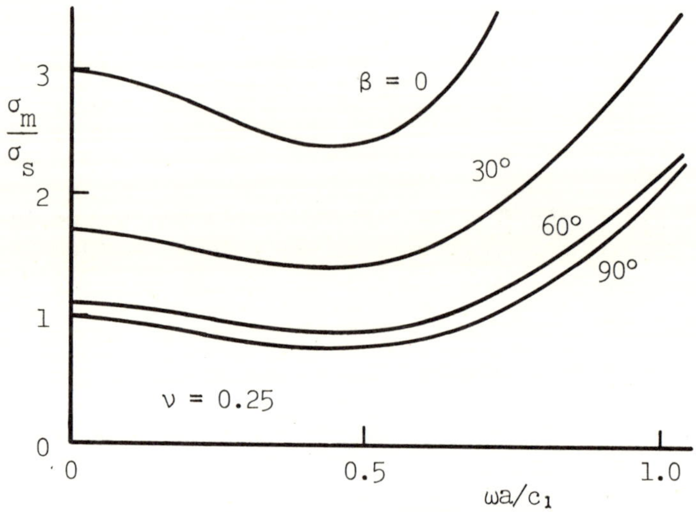


Fig. (10) - Minimum stress amplitude to cause crack growth

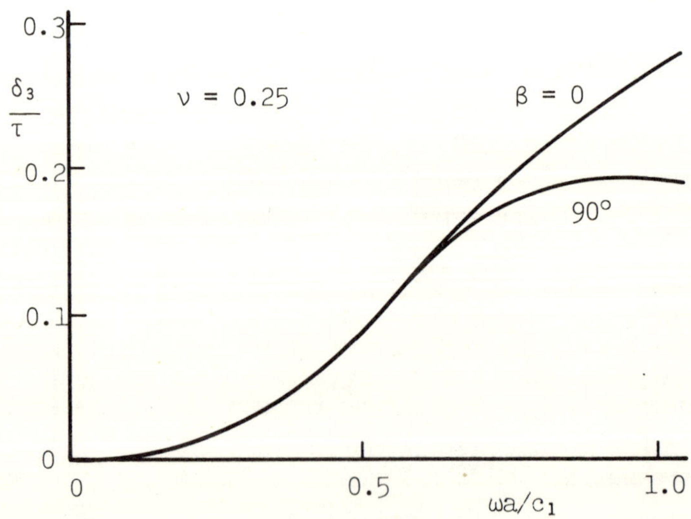


Fig. (11) - Time between maximum stress and maximum $S_o$

the load required for failure increases. It should be noted that the limit in these cases for low frequencies is not the static failure stress obtained as in [6].

The delay, $\delta_3$, the interval of time from the peak of the oscillating load to the peak value of $S_o$, is shown in Fig. (11). The behavior is similar to the delay curves in Fig. (7). The angle of crack growth, $\theta_o^*$, when the minimum amplitude for failure is applied is shown in Fig. (12). Larger amplitudes would lead to values of $\theta_o^*$ having larger absolute values. As Fig. (8) shows, the angle becomes smaller in magnitude for $\omega = 0.7\ c_1/a$. The angle shown in Fig. (12) for $\beta = 30°$, $\omega = 0.7\ c_1/a$ is taken from Fig. (8) at $t \approx 0.125\tau$, which is the time at which the corresponding curve in Fig. (9) reaches its peak.

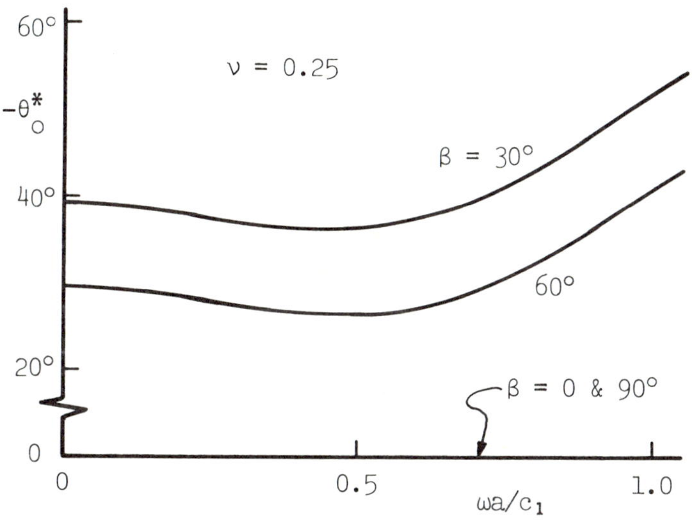


Fig. (12) - Angle of crack growth associated with maximum $S_o$

CONCLUSIONS

The detailed stress analysis of a crack in a standing, plane, compressional wave field shows that the stress intensity factors are out of phase with the remote stress field. The phase differences, as well as the amplitudes, depend on the frequency of the input wave. The peak values of Figs. (5) and (6) occur at a frequency of approximately 0.5 $c_1/a$. For steel and aluminum

$c_1 \approx 6350$ m/s (250,000 in/sec). This frequency is, therefore, very large. If a small crack of length $2a_o = 5.0$ cm (2.0 in) is used as a reference, the frequency can be expressed as 20,000 $(a_o/a)$ Hz.

This same frequency results in the smallest stress field amplitude required to cause failure. In Fig. (10), for the case of the crack normal to the remote displacement field, the minimum amplitude is about 75% of the low frequency value. At higher frequencies, the curves in Fig. (10) show that higher stress amplitudes are required to initiate crack growth. The direction of crack growth depends on the time at which it begins in Fig. (8). If the fracture initiates at the stress amplitude of Fig. (10), the angles shown in Fig. (12) result.

REFERENCES

[1] Sih, G. C. and Loeber, J. F., "Wave propagation in an elastic solid with a line of discontinuity or finite crack", Quarterly of Applied Mathematics, Vol. 27, 1969, pp. 193-213.

[2] Chen, E. P. and Sih, G. C., "Running crack in an incident wave field", International Journal of Solids and Structures, Vol. 9, 1973, pp. 897-919.

[3] Sih, G. C., Ravera, R. J. and Embley, G. T., "Impact response of a finite crack in plane extension", International Journal of Solids and Structures, Vol. 8, 1972, pp. 977-993.

[4] Kerkhof, F., "Wave fractographic investigations of brittle fracture dynamics", Dynamic Crack Propagation, G. C. Sih, ed., Noordhoff International Publishing, Leyden, 1973, pp. 3-29.

[5] Bauer, R. W. and Wilsdorf, H. G. F., "The mechanical behavior in tension and fracture characteristics of stainless steel thin rolled foils", Dynamic Crack Propagation, G. C. Sih, ed., Noordhoff International Publishing, Leyden, 1973, pp. 197-213.

[6] Sih, G. C., "A special theory of crack propagation", Methods of Analysis and Solutions of Crack Problems, G. C. Sih, ed., Noordhoff International Publishing, Leyden, 1973, pp. XXI-XLV.

[7] Sih, G. C., "Handbook of stress-intensity factors", Vol. I, Institute of Fracture and Solid Mechanics Publication, Lehigh University, 1973.

[8] Kobayashi, A. S., Johnson, B. H. and Wade, B. G., "Crack approaching a hole", ONR TR No. 16, University of Washington, March 1973.

# CRITERIA FOR STEADY STATE CRACK EXTENSION IN GAS PIPELINES

H. C. van Elst
Physical Department, Metal Research Institute TNO,
Apeldoorn, Netherlands

## ABSTRACT

When work A is done by external forces on a cracked specimen or constructional member and crack extension results, concurrent with the energy dissipation $W_R$ as mostly required by the plastic flow connected with the fracture surface realization, a complimentary absorption of energy can take place. This can be partly stored as recoverable energy $E_I$ and (or) is partly dissipated as well ($W_P$). This complimentary absorption has often a more global character than the local energy consumption required for crack extension. The recoverable energy $E_I$ stored in parallel with the dissipated crack energy can be present as potential elastic energy U or kinetic energy V; the dissipated energy $W_P$ is the result of plastic flow, as developed by the concomitant deformation. As a consequence of this latter deformation a geometrical change of the cracked part can occur, which in turn can promote the additional energy dissipation $W_P$ or storage $E_I$ during the crack extension, while it can also provide a mechanism for its (partial) feed back to the crack, i.e., a (partial) transfer of $E_I$ into $W_P$. This phenomenon complicates the interpretation of the energy balance; in particular, the indication of fracture toughness tends to be confused and its concept will sometimes need an adjusted definition.

This fracture aspect enters when analyzing the long range steady state ductile fracture propagation in a pipe under gas pressure. Here the constant velocity appears to require a crack driving force which decreases with increasing velicity - at least more progressively than the fracture toughness.

It is suggested that work done by the gas pressure which was converted into kinetic energy and strain energy can be fed back to the crack via a lever action provided by the connected deformation of the split pipe parts ("flaps").

NOTATION

| | |
|---|---|
| $A$ | = work done by pressurized gas [Nm] |
| $A_B$ | = work done by pressurized gas on bulged part of fracturing pipe [Nm] |
| $A_F$ | = work done by pressurized gas on "flaps" part of fracturing pipe [Nm] |
| $W_R$ | = dissipated energy by crack extension [Nm] |
| $E_I$ | = stored energy [Nm] |
| $E_B$ | = energy absorbed in bulge [Nm] |
| $W_P$ | = dissipated energy not required for crack extension [Nm] |
| $U$ | = strain energy [Nm] |
| $U_B$ | = strain energy of bulged part of fracturing pipe [Nm] |
| $U_F$ | = strain energy of "flaps" part of fracturing pipe [Nm] |
| $V$ | = kinetic energy [Nm] |
| $V_B$ | = kinetic energy of bulged part of fracturing pipe [Nm] |
| $V_F$ | = kinetic energy of "flaps" part of fracturing pipe [Nm] |
| $T$ | = work done by inertia forces [Nm] |
| $T_B$ | = work done by inertia forces in bulge development [Nm] |
| $T_F$ | = work done by inertia forces in "flaps" development [Nm] |
| $G$ | = crack extension force [$Nm^{-1}$] |
| $G_R$ | = fracture toughness [$Nm^{-1}$] |
| $v$ | = fracture velocity [m $sec^{-1}$] |
| $v_e$ | = steady state fracture velocity [m $sec^{-1}$] |

| | |
|---|---|
| $l^{*}$ | = effective crack length [m] |
| $\sigma_H$ | = hoop stress [$Nm^{-2}$] |
| $\sigma_{H,o}$ | = initial hoop stress [$Nm^{-2}$] |
| P | = gas pressure [$Nm^{-2}$ = Pa] |
| $P_o$ | = initial gas pressure [$Nm^{-2}$ = Pa] |
| $c_g$ | = sound velocity of pressurized gas [m $sec^{-1}$] |
| $\gamma$ | $= \frac{c_p}{c_v}$, with $c_p$ = specific heat at constant pressure of the gas; $c_v$ = specific heat at constant volume of the gas |
| $\sigma_f$ | = stress at failure of the pipe [$Nm^{-2}$] |
| Y | = yield strength of pipe material [$Nm^{-2}$] |
| $Y_{eff.}$ | = effective Y $\approx$ flow stress of Hahn et al. [$Nm^{-2}$] |
| $R_o$ | = pipe radius [m] |
| $B_o$ | = pipe wall thickness [m] |
| K | = stress intensity factor [$Nm^{-3/2}$] |
| M | = stress concentration factor of Folias |
| $\phi$ | = polar angle in planes // XOY |
| r | = polar radius in plane // XOY |
| $h_\phi$ | = bulge height in plane through Y-axis (= pipe axis) making an angle $\phi$ with XOY-plane [m] |
| | N.B. XOZ contains crack tip; $-\frac{\pi}{2} \leq \phi \leq \frac{\pi}{2}$ |
| s | = arc length in meridional plane [m] |
| b | = length of pipe part showing bulging [m] |
| f | = length of split pipe part ("flaps") showing variation in momentaneous shape behind crack tip [m] |
| $\phi_{ty}$ | = cf. p. |
| $\Theta$ | = crack opening angle of "flaps" |
| $\rho$ | = density (mass units pro volume unit) [kg mass $m^{-3}$] |
| $\alpha$ | $= \frac{Y}{\sigma_{H,o}}$ |
| $k^2$ | = cf. p. 311 |

$\chi$ $= \frac{c_g}{Y}$ [$m^3$ $N^{-1}$ $sec^{-2}$]

$\eta$ = cf. p. 314

$\xi$ = cf. p. 315

$\zeta$ = cf. p. 315

$D_o$ $= \frac{2}{3}$ Charpy specimen thickness [m]

$A_c$ = area of fracture surface of Charpy specimen [$m^2$]

$C_v$ = Charpy-V value [Nm]

$C_{v,2/3}$ = Charpy-V value for $\frac{2}{3}$ Charpy specimen [Nm]

c = constant cf. p. 315

## INTRODUCTION

If external work A is done on a cracked body the energy balance of the terms conditioning crack extension can generally involve absorbed energies, both dissipated and stored, concurrent with the dissipated part $W_R$ of this work A required for crack extension.

This tends to complicate or even confuse the determination of the "effective fracture surface energy" or fracture toughness $G_R$. The stored energy can be present as elastic potential energy U or kinetic energy V. It will be possible that both these stored energies at least partly will become recoverable for the benefit of the crack extension.

In the classical case of a large cracked plate under constant load it is well known that the work done at quasi-static crack extension is shared in equal amounts between both the increase of elastic potential energy and the crack extension - if the dissipated energy is only by crack extension. (For the constant grip situation and in the quasi-static elastic case the decrease of elastic potential energy equals the dissipated crack extension energy).

If the crack extension becomes unstable the released elastic potential energy is partly used for the generation of kinetic energy of the vibrational type (i.e., with zero algebraic momentum), presumably initially present as stress waves emitted by the moving crack. For the fixed ends situation this kinetic energy cannot realize displacement and in the end it will totally be dissipated in a global way. For the free ends situation it will lose to the surroundings; at least partially, it can however be converted into strain energy.

The external work done on a cracked constructional member or specimen can also generate translational and (or) vibrational kinetic energy from which the crack extension can profit after the external forces disappeared and this energy has been converted into potential energy.

Note: When opening a soldered tin by winding the cover around a pin, the tearing process is accompanied by plastic energy dissipation in the winding process, offering a suitable lever function, which has not to be confused with the crack extension energy. In general, when an energy source is available, but the derivable force for using it to achieve an appreciable displacement and (or) to drive a crack is insufficient, a lever for gaining up to the force amplitude required for this, might be constructed. The enlarged force can evoke irreversible and reversible displacements, which the original force could not, leading to an absorption of energy, which without the lever action was not possible. Part of the work done will be dissipated in the construction and operation of the lever.

If this lever construction is produced by the fracture itself, e.g., through the concomitant deformation under load of separated parts resulting from crack extension, this might entail a large additional energy dissipation.

If the fracture energy is defined as the difference between external work done and kinetic energy generated, this fracture energy will in general not be equivalent to the crack extension energy.

The ductile fracture propagation on a large scale as can occur in a pipeline under gas pressure implies the analysis of the above mentioned aspects. In particular:

a) the energy dissipated for construction of a mechanism that contributes to the crack driving force by a deformation accompanying the fracture and which has not to be confused with crack extension energy;

b) the tapping off by the same construction of "stored" kinetic energy for driving the crack.

## FORMULATION OF CONDITIONS PROVIDING FOR A STATIONARY SHEAR FRACTURE PROPAGATION IN A GAS PIPELINE

It is assumed that one is dealing with a state of dynamic equilibrium for which no kinetic energy is generated by the crack extension occurring with constant velocity; present kinetic energy is caused by the external forces. The rate in elastic potential energy depends on the crack extension rate and external work rate;

a priori its sign is not even clear.

This "quasi-static" extension must then be a self-stabilizing phenomenon in the sense that an increase or decrease of fracture velocity is countered.

A crack extension force G decreasing more progressively with increasing fracture velocity v than a possible decreasing crack extension resistance $G_R$ with increasing fracture velocity, offers a possibility, as was remarked by Poynton and Fearnehough [1], Poynton [2], Shannon and Wells [3]. The velocity $v_e$ at which $G = G_R$ will determine the stationary case of an indefinite growing stable crack with constant velocity.

For if v becomes smaller (larger) than $v_e$, $G_R$ becomes smaller (larger) than G, which as a consequence increases (decreases) v. If the situation can be described by an effective cracklength $l^*$ and if $\frac{dl^*}{dv} < 0$ this also means that $|\frac{dl^*}{dv}| > \frac{2l^*}{\sigma_H}\frac{d\sigma_H}{dv}$ provides for the stationary case of interest.

An alternative way for formulating this stationary propagation behaviour can be given by considering the dependence of crack velocity on (hoop) stress $\sigma_H$. Intuitively - and from experimental evidence - the relevant fracture velocity v will increase with applied stress $\sigma_f$.

The pressurized gas will show a decompression by leakage through the crack. If the crack proceeds this will imply that the actual hoop stress will increase with crack velocity.

The value of the pressure P at the cracktip can be taken as described by:

$$P = P_o \left[\frac{\frac{v}{c_g}(\gamma-1) + 2}{\gamma+1}\right]^{2\gamma/(\gamma-1)} \qquad \text{(cf. e.g., Wright [4])} \qquad (1)$$

with v = fracture velocity

$P_o$ = undisturbed pressure in the infinite gas reservoir.

For $\gamma = 1.4$ and $c_g = 450$ m/sec this implies for the decompressed hoop stress $\sigma_H$:

$$\sigma_H = \frac{\sigma_{H,o}}{279936} \left(\frac{v}{450} + 5\right) \quad [7]$$

For $\frac{d\sigma_H}{dv} < \frac{d\sigma_f}{dv}$ steady state crack propagation can occur for that v value at which $\sigma_f = \sigma_H$.

For $\frac{d\sigma_H}{dv} > \frac{d\sigma_f}{dv}$ the crack propagation will become unstable if $\sigma_f$ reaches the $\sigma_H$ value, cf., Mimura and Ogasawara [5].

Reviews of presented theories with respect to pipeline fracture, both brittle and ductile, providing also a bibliography of relevant references were recently given by Fearnehough [7] and Dick et al [8]. In particular, attention was paid to the relation between full scale and small scale (Charpy-V) tests.

Most work until now is due to the Battelle Memorial Institute (e.g., Hahn et al [9], Maxey et al [10]) and British Gas [1,2,3]. The former use a quasi-static approach with an effective crack-length $l^*$ and a for the bulging adjusted hoop stress. Plotting

$$\frac{\sigma_H}{Y_{eff.}} \text{ versus } \frac{K^2}{Y^2_{eff.}\sqrt{R_oB_o}} = \frac{8}{\pi}\frac{l^*}{\sqrt{R_oB_o}} \ln \sec \frac{\pi}{2}\frac{M\sigma_H}{Y_{eff.}}$$

appears to provide a propagation-arrest boundary, meeting full scale test results, for the stress concentration factor M of Folias equal to 3.3, corresponding with

$$\frac{l^*}{\sqrt{R_oB_o}} = 3.$$

Poynton and Fearnehough [1], Poynton [2] and Shannon and Wells [3] use an energy balance analysis. A similar interpretation is used here, but certain dissipative energy terms are incorporated, which should not be ignored in our opinion. These are aimed at separating the dissipative energy in that corresponding to the G-concept and otherwise. Accordingly, the relation between G and the pipe parameters changes.

## THE DEFORMATION ACCOMPANYING THE STEADY STATE CRACK IN A GAS PIPE

The following assumptions based on or suggested from experimental evidence as reported in [2,3,6] are made:

a) the fracture in the pipe moves with a constant velocity below the sound velocity of the pressurized gas and parallel to the axis;

b) in an (Eulerian) co-ordinate system for the movement of the crack a constant crack profile exists, being a bulging ahead of the cracktip and an involute of the "flaps" behind the cracktip, which finally can become flat. With increasing "back fill" resistance the involute will develop less e.g., take a U-shape.

The pipe axis is chosen as Y-direction, the meridional plane through the pipe axis and crack is the YOZ plane, with OZ along the diameter of the cross section, in which the cracktip is present. XOY is the equator plane, through the axis and perpendicular to the crack meridional plane, see Fig. (1).

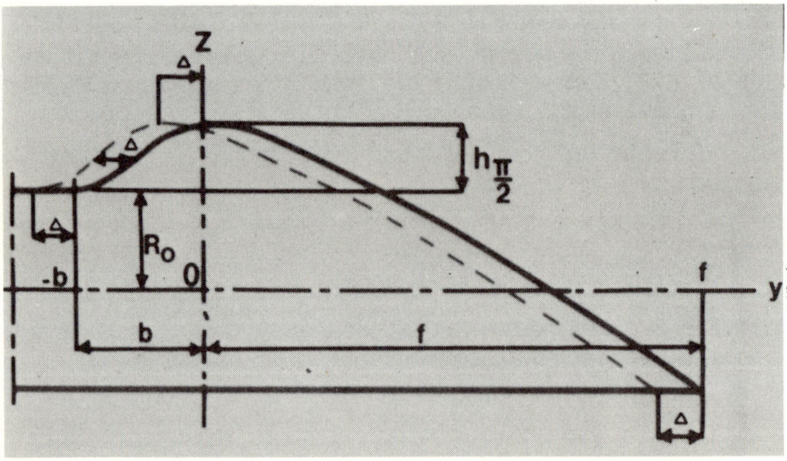


Fig. (1) - Assumed geometry of ductile shear fracture in pipelines

Cylindrical co-ordinates r, y, $\phi$ or $r_{y\phi}$, with the Y-axis as reference, are used; YOY is the $\phi$=0 plane. The bulge occurs for negative y and towards the cracktip the diameter parallel to OX increases monotonically; the diameter parallel to OZ first increases and then decreases. The distortion of the bulged pipe by its interactions with the "flaps" induces an ovalized cross section, which near the tip shows a larger diameter in the x- than in the z-direction. (This was observed by Shoemaker et al [6]).

In y-direction the bulged part measures from -b to 0. For the bulged part a sine like description was chosen. The height decrease just ahead of the cracktip in and near the crack meridional plane was ignored to simplify calculations. It was further assumed that the bulge height increased linearly with the angle

$\frac{\pi}{2} - |\phi|$ between the crack meridional plane and the equatorial plane, somewhat different, but also rather in analogy with [1,2]. The bulge description then reads:

$$r_{y\phi} = R_o + \frac{1}{2} h_\phi \left(1 + \cos \frac{\pi y}{b}\right) \tag{2a}$$

$$h_\phi = h_o - (h_o - h_{\pi/2}) \frac{2|\phi|}{\pi} \tag{2b}$$

$$h_{y\phi} = r_{y\phi} - R_o = \frac{1}{2} h_\phi \left(1 + \cos \frac{\pi y}{b}\right) = \frac{1}{2} \{h_o - (h_o - h_{\pi/2}) \times$$

$$\times \frac{2|\phi|}{\pi}\} \left(1 + \cos \frac{\pi y}{b}\right) \tag{2c}$$

$$r_{y\phi} = R_o + \frac{1}{2} \{h_o - (h_o - h_{\pi/2}) \frac{2|\phi|}{\pi}\} \left(1 + \cos \frac{\pi y}{b}\right) \tag{2d}$$

$$r_{yo} = R_o + \frac{1}{2} h_o \left(1 + \cos \frac{\pi y}{b}\right) \tag{2e}$$

N.B.

$$\dot{r}_{y\phi} = - \frac{1}{2} \{h_o - (h_o - h_{\pi/2}) \frac{2|\phi|}{\pi}\} \sin \frac{\pi y}{b} \frac{\pi}{b} v \tag{3a}$$

$$\ddot{r}_{y\phi} = - \frac{1}{2} \{h_o - (h_o - h_{\pi/2}) \frac{2|\phi|}{\pi}\} \cos \frac{\pi y}{b} \frac{\pi^2}{b^2} v^2 \tag{3b}$$

For positive y the "flaps" of the split pipe take an involute shape. Neglecting the deformation of the cross section containing the cracktip this can be pictured between the following geometries:

1°. The cross section behind the cracktip is developed linearly with distance behind the cracktip (Fig. (2a)). Thus, if the involute finally becomes flat at distance f behind the cracktip, for $0<y<f$ the length of the assumed straight involute cross section is $\frac{y}{f} \cdot \pi R = (\frac{\pi}{2} - \phi_{ty})R$, with $\phi_{ty}$ the angle it makes with the crack meridional plane, being equal to the polar angle of its tangency point. For the description of the involute one has then:

$$\begin{aligned} \phi_y &= \phi_{ty} + \text{arc tg}(\phi_o - \phi_{ty}); \quad \phi_{ty} = \frac{\pi}{2} - \frac{y}{f}\pi \\ r_{y\phi} &= R_o\sqrt{1+(\phi_o-\phi_{ty})^2} \end{aligned} \tag{4}$$

This apparently implies an elongation of the original describing lines of the pipe.

2°. The "flaps" of the split pipe describe a line surface with as generators a) the cross section containing the cracktip and b) the final involute of this, being complete, i.e., a straight line on a distance f in x-direction behind the cracktip, if there is a suitable sufficiently weak back fill resistance. The description of this line surface will be: (cf. Fig. (2b)).

$$r^2_{y\phi} = R_o^2 \left\{\frac{f-y}{f}\cos\phi_o + \frac{y}{f}\left(\frac{\pi}{2}+\phi_o\right)\right\}^2 + R_o^2\left\{\frac{f-y}{f}\sin\phi_o - \frac{y}{f}\right\}^2$$

$$\mathrm{tg}\phi_y = \frac{\frac{f-y}{f}\sin\phi_o - \frac{y}{f}}{\cos\phi_o + \frac{y}{f}\left(\frac{\pi}{2}+\phi_o - \cos\phi_o\right)} = \frac{(f-y)\sin\phi_o - y}{(f-y)\cos\phi_o + y\left(\frac{\pi}{2}+\phi_o\right)} \tag{5}$$

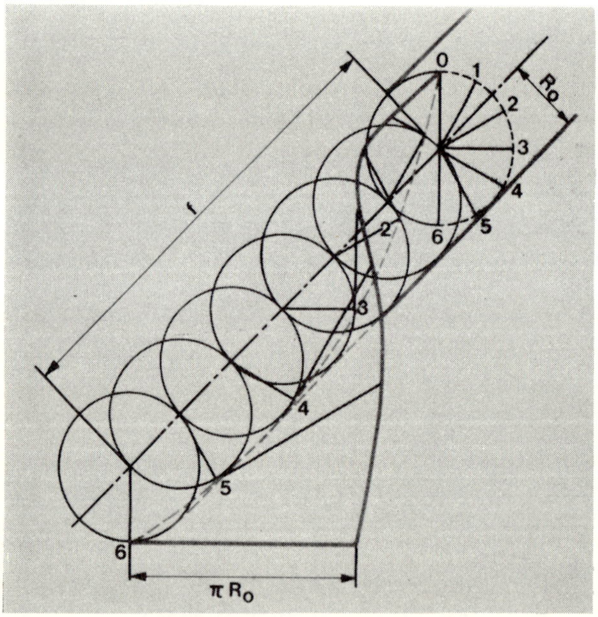


Fig. (2a) - Geometry approximation (1°) of split pipe (linear involute)

For the crack opening angle Θ between the flaps one will have:

$$\mathrm{tg}\,\frac{\Theta}{2} = \int_{-\pi/2}^{\pi/2} \frac{d\phi r_{o\phi}}{\sqrt{f^2+(r_{o,\pi/2}+r_{o,-\pi/2})^2}} \simeq \frac{\pi R_o}{\sqrt{f^2+4R_o^2}} \tag{6}$$

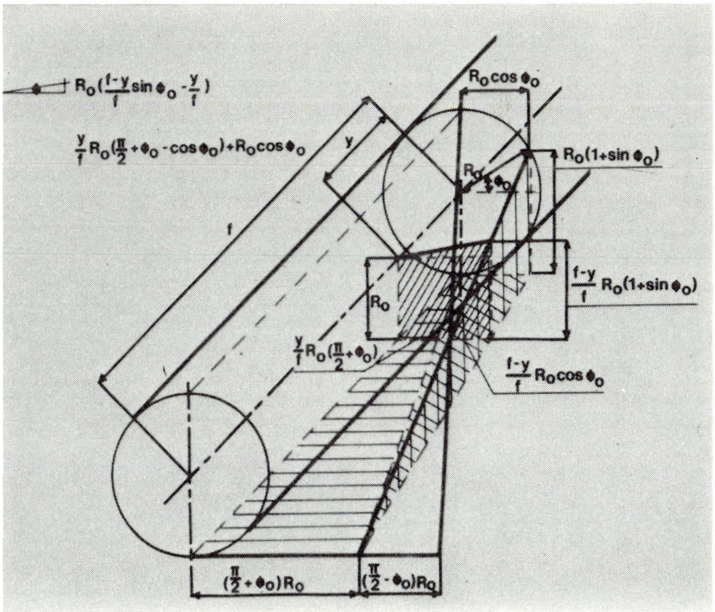


Fig. (2b) - Geometry approximation (2°) of split pipe (line surface)

The describing lines show an elongation with respect to the original cylinder (only the line $z = -r_{o,-\pi/2}$; x=0 is not elongated). The line originally determined by $z=R_o$; x=0 will have the largest elongation; its distorted length is

$$\simeq \sqrt{f^2 + \{\int_{-\pi/2}^{\pi/2} d\phi r_{o\phi}\}^2 + (r_{o,\pi/2} + r_{o,-\pi/2})^2} \times$$

$$\times \simeq \sqrt{f^2 + (\pi^2+4)R_o^2} \tag{7a}$$

$$\text{Thus } \varepsilon_m \simeq \frac{1}{2}\ln\{1 + (\pi^2+4)\frac{R_o^2}{f^2}\} \simeq (\frac{\pi^2}{2} + 2)\frac{R_o^2}{f^2} \tag{7b}$$

The resistance of the material against the elongation of these describing lines implies stress components parallel to the axis and in circumferential sense. These account for the described ovalization of the cross section ahead of the cracktip showing a reduction of diameter in the crack meridional plane, when moving towards the cracktip, [6].

The value of f will anyway have to comply with the condition that $\varepsilon_m$ does not exceed a certain value, e.g., the uniform elongation. If this is 20-30%, this implies:

$$f \leq 6R_o \text{ and } tg\,\frac{\Theta}{2} \leq 0.5 \text{ or } \Theta \leq 50°.$$

In reality, the "flaps" geometry is anticipated between 1°. and 2°.; there appears no sufficient experimental evidence up to now to justify more exact conclusions in this respect.

THE ENERGY BALANCE ANALYSIS OF THE PIPE WITH INTERNAL GAS PRESSURE, FRACTURING UNDER STEADY STATE CONDITIONS

The movement of the "flaps" behind the cracktip towards their involuted shape, caused by the gas doing or having done work, will imply an additional loading action on the crack. This eccentrically "crack line loaded" situation, as well as the aforementioned stress distribution by the elongation in y- direction of the flaps has to be superposed on that resulting from the gas pressure in the uncracked pipe. The resulting enlarged K-value does not appear directly disintangible.

For a fully yielded net ligament situation, however, Erdogan and Ratwani [11] gave a crack-opening displacement treatment from which in combination with the depressurizing aspect it could be concluded that the crack arrest possibility increases with $\delta_c$ and decreasing v (or crack acceleration). See also Kanninen and Sampath [12] who discussed the stress intensity factor of a long axial crack in a pipe.

The here proposed separation of the dissipated energy into a part required for crack extension and into a part accompanying the fracture in an alternative way, remains a priori somewhat debatable. For the latter is taken the plastic work for bulging ahead of the cracktip and the plastic work required for unwinding the split pipe into its flat involute.

Stored energies of the external work done are kinetic energy and elastic potential energy. To what degree these are recoverable for crack extension or dissipated in another way is not directly clear either. Apparently the kinetic energy of the flaps of the described line surface will be used for making the involute. Its difference with the plastic work required for that can be used for crack driving, presumably via an increase of potential energy.

The Energy Balance Analysis of the Bulge

1) The work $A_B$ done by the pressurized gas in forming the bulge will be (assuming for the moment a constant pressure not affected by the fracture):

$$A_B = 2 \int_{-\pi/2}^{\pi/2} d\phi \int_{-b}^{o} dy \int_{R_o}^{r_{y\phi}} P\underline{r_{y\phi}}\, d\underline{r_{y\phi}} = \int_{-\pi/2}^{\pi/2} d\phi \int_{-b}^{o} dy\ P(r_{y\phi}^2 - R_o^2)$$

$$= \pi P b \,\{R_o(h_o - \frac{\Delta h}{2}) + \frac{3}{8}\, k^2\} \qquad (8)$$

with $\Delta h \equiv h_o - h_{\underline{\pi}}$ and $k^2 \equiv h_o^2 - h_o \Delta h + \frac{(\Delta h)^2}{3}$ .

2) The energy $E_B$ absorbed and presumably mostly dissipated in the bulge will be:

$$E_B = 2 \int_{-\pi/2}^{\pi/2} d\phi \int_{-b}^{o} dy \int_{R_o}^{r_{y\phi}} (P\underline{r_{y\phi}} - \rho B_o R_o \ddot{\underline{r}}_{\underline{y\phi}})\, d\underline{r_{y\phi}}$$

$$= A_B - T_B$$

with $T_B$ = work done by the inertia forces and $T_B = V_B$ = kinetic energy stored in the bulge.

$$T_B = V_B = \rho B_o R_o \int_{-\pi/2}^{\pi/2} d\phi \int_{-b}^{o} dy\ \dot{r}_{y\phi}^{\ 2}$$

$$= \frac{\rho}{8}\, B_o R_o v^2\, \frac{\pi^2}{b^2}\, \pi b k^2 \qquad (9)$$

[It is assumed that $B_o R_o\, d\phi dy = B_{y\phi} r_{y\phi} d\phi ds_{y\phi}$ with $ds_{y\phi}$ $= dy \sqrt{1 + (\frac{\partial r_{y\phi}}{\partial y})^2}$ ]

3) The part $W_B$ of the dissipated work in the bulge forming due to plastic flow not connected with the crack is approximated by:

$$W_B = 2 \int_{-\pi/2}^{\pi/2} d\phi \int_{-b}^{o} dy\ B_o R_o \frac{2}{3} \sqrt{3}\ Y \sqrt{\varepsilon_1^2 + \varepsilon_2^2 + \varepsilon_1 \varepsilon_2}$$

$$\text{with } \varepsilon_1 = \varepsilon_{y\phi} = \frac{1}{2} \ln \frac{r_{y\phi}^{\ 2} + \left(\frac{\partial r_{y\phi}}{\partial \phi}\right)^2}{R_o^2}$$

$$\varepsilon_2 = \varepsilon_{yy} = \frac{1}{2} \ln \left\{1 + \left(\frac{\partial r_{y\phi}}{\partial y}\right)^2 \right\}.$$

Neglecting terms of the order $\left(\frac{h_o}{R_o}\right)^n$ with $n \geq 1$, this reduces to:

$$W_B = 2\,\frac{2}{3}\,\sqrt{3}\;Y\,B_oR_o \int_{-\pi/2}^{\pi/2} d\phi\,\frac{1}{8}\left(h_o - \Delta h\,\frac{2|\phi|}{\pi}\right)^2 \times$$

$$\times \int_{-b}^{o} dy\,\sin^2\frac{\pi y}{b}\,\frac{\pi^2}{b^2} = \frac{1}{12}\,\sqrt{3}\;Y\,B_oR_o\,\frac{\pi^2}{b^2}\,\pi b k^2 \tag{10}$$

4) The increase of elastic potential energy $U_B$ in the bulge is approximated by:

$$U_B \approx \frac{Y^2(1-\alpha^2)}{2E}\,2\pi R_oB_ob \approx \frac{P_o^2R_o^2}{B_o^2}\,\frac{1-\alpha^2}{\alpha^2}\,\frac{\pi R_oB_ob}{E} \tag{11}$$

with $\frac{P_oR_o}{B_o} = \alpha Y = \sigma_{H,o}$ and $U_B << W_B$.

The Energy Balance Analysis of the "Flaps"

1) the work $A_F$ done by the pressurized gas in forming the "flaps" is estimated by using the description (4) for their geometry.

$$A_F = -\,2\int_o^f dy \int_{\pi/2}^{\phi_{ty}} P\,\frac{1}{2}\left(\frac{\pi}{2} - \phi\right)^2 R_o^2 d\phi$$

with $\phi_{ty}$ = the polar angle of the point of tangency of the assumed straight involute to the as circular approximated cross section of the pipe at distance y behind the cracktip; apparently $\frac{\pi}{2} - \phi_{ty} = \frac{y}{f}\,\pi$.

$$A_F = \int_o^f dy\,\frac{Py^3}{3f^3}\,\pi^3R_o^2 = \frac{\pi^3PR_o^2f}{12} \tag{12}$$

2) The kinetic energy $V_F$ stored in the flaps using their geometry description (4) is estimated as:

$$V_F = 2 \times \frac{1}{2} \rho \int_0^f B_o dy \int_{\pi/2}^{\phi_{ty}} - d\phi_o \, R_o R_o^2 \{1 + (\phi_o - \phi_{ty})^2\} \frac{v^2 \pi^2}{f^2}$$

$$= \pi^2 \rho B_o R_o^3 \frac{v^2}{f^2} \int_0^f dy \, \{(-\phi_{ty} - \frac{\pi}{2}) + \frac{(\frac{\pi}{2} - \phi_{ty})^3}{3}\}$$

$$= \pi^2 \rho B_o R_o^3 \frac{v^2}{f^2} \int_0^f dy \, (\frac{\pi y}{f} + \frac{\pi^3 y^3}{f^3})$$

$$= \frac{1}{2} \pi^3 \, (1 + \frac{\pi^2}{2}) \, \rho B_o R_o^3 \frac{v^2}{f^2} f \qquad (13)$$

3) The plastic energy $W_F$ dissipated in the "flaps" are connected with the crack, in fact, consumed after the crack passed, is estimated as only due to their unwinding, ignoring elongation or bending of the describing lines of the pipe, and thus:

$$W_F = 2 \, \frac{2}{3} \, V3 \, Y \int_0^f dy \, R_o \, (\frac{\pi}{2} - \phi_{ty}) \, B_o \frac{B_o}{2R_o}$$

$$= 2 \, \frac{2}{3} \, V3 \, Y \frac{B_o^2}{2} \int_0^f dy \, \frac{y}{f} \, \pi = \frac{1}{3} \, V3 \, \pi Y \, B_o^2 f \qquad (14)$$

4) The decrease in elastic potential energy in the "flaps" part is approximated by:

$$U_F = \frac{P^2 R_o^2}{B_o^2} \, \frac{2\pi R_o B_o}{2E} \, f = \frac{\pi P^2 R_o^3}{E B_o} \, f = \frac{\pi \alpha^2 Y^2 B_o R_o f}{E} \qquad (15)$$

and $U_F \ll W_F$.

## INTERPRETATION OF THE ENERGY BALANCE INTO FRACTURE TOUGHNESS REQUIREMENTS WITH RESPECT TO PIPE PARAMETERS, PREVENTING STEADY STATE FRACTURE IN GAS PIPELINE

The total energy balance, using energy rate considerations will read:

$$\dot{A} = \{\dot{A} - \dot{T}\} + \dot{V}$$

$$\dot{A}_B + \dot{A}_F = \{(\dot{W}_B + \dot{U}_B) + (\dot{W}_F + \dot{U}_F) + GB_o v\} + \dot{V}_B + \dot{V}_F$$

This offers for the crack extension force, using Eqs. (8) through (15):

$$G = \pi \frac{PR_o}{B_o} R_o \left\{\frac{h_o - \frac{1}{2}\Delta h}{R_o} + \frac{3}{8}\frac{k^2}{R_o^2} + \frac{\pi^2}{12}\right\}$$

$$- \pi Y R_o \left\{ \frac{\pi^2}{12} \sqrt{3} \frac{R_o^2}{b^2}\frac{k^2}{R_o^2} + \frac{1}{3}\sqrt{3}\frac{B_o}{R_o} + \frac{Y}{E}(1-2\alpha^2)\right\}$$

$$- \frac{\pi}{2}\rho v^2 R_o \left\{\frac{\pi^2}{4}\frac{R_o^2}{b^2}\frac{k^2}{R_o^2} + \pi^2\left(1+\frac{\pi^2}{2}\right)\frac{R_o^2}{f^2}\right\} \tag{16}$$

Taking $P = P_o\,(0.285 + 0.325\,\frac{v}{c_g} + 0.384\,\frac{v^2}{c_g^2})$ as best fitting quadratic approximation of (1), cf. [13], remembering $\frac{P_o R_o}{B_o} = \sigma_{H,o} = \alpha Y$ (with $\alpha \simeq 2/3$), noting: $\rho = \chi \frac{Y}{c_g^2}$ (with $\chi$ = 2 to 4 for Y = 800 to 400 MPa and $\rho$ = 8000 kg mass units pro $m^3$ and $c_g$ = 450 m/sec) then:

$$G = \pi Y R_o \left\{0.285\alpha\eta - \frac{\pi^2}{12}\sqrt{3}\frac{k^2}{b^2} - \frac{1}{3}\sqrt{3}\frac{B_o}{R_o} - \frac{Y}{E}(1-2\alpha^2)\right\}$$

$$+ \pi Y R_o \frac{v}{c_g} 0.325\alpha\eta + \pi Y R_o \frac{v^2}{c_g^2}\left\{0.384\alpha\eta - \frac{\pi^2}{8}\chi\frac{k^2}{b^2} - \pi^2 \times\right.$$

$$\left.\times \left(1+\frac{\pi^2}{2}\right)\chi\frac{R_o^2}{f^2}\right\} \tag{17}$$

with $\eta = \frac{h_o - \frac{1}{2}\Delta h}{R_o} + \frac{3}{8}\frac{k^2}{R_o^2} + \frac{\pi^2}{12}$.

Ignoring the dependence of $h_o$, b and f on fracture velocity v, the maximum crack extension force occurs for:

$$\frac{v}{c_g} = \frac{v_m}{c_g} = \frac{0.325\alpha\eta}{2\left\{\frac{\pi^2}{8}\chi\frac{k^2}{b^2} + \pi^2\left(1+\frac{\pi^2}{2}\right)\chi\frac{R_o^2}{f^2} - 0.384\alpha\eta\right\}} \tag{18}$$

(For $\alpha = \frac{2}{3}$; $\eta \simeq \frac{h_o}{R_o} + \frac{\pi^2}{12} \simeq 1$; $\chi \simeq 2$; $\frac{k^2}{b^2} \simeq \frac{h_o^2}{b^2} < 1$ and $\frac{R_o}{f} \simeq 0.15$ then $\frac{v_m}{c_g} \simeq \frac{1}{25}$).

According to this, one will have for the maximum G-value as a function of $\alpha$:

$$\frac{G}{\pi Y R_o} = \frac{3.072\frac{Y}{E}\alpha^3+(0.332\eta^2-8\frac{Y}{E}\xi)\alpha^2-(1.14\xi\eta+1.536\zeta\eta)\alpha+4\zeta\xi}{1.536\alpha\eta-4\xi} \tag{19}$$

with $\xi = \frac{\pi^2}{8}\chi\frac{k^2}{b^2} + \pi^2(1+\frac{\pi^2}{2})\chi\frac{R_o^2}{f^2}$

$$\zeta = \frac{\pi^2}{12}V3\frac{k^2}{b^2} + \frac{1}{3}V3\frac{B_o}{R_o} + \frac{Y}{E}$$

For $\frac{Y}{E} \leq \frac{1}{300}$; $\eta \simeq 1$; $\chi \simeq 2$; $\frac{k^2}{b^2} \simeq \frac{1}{25}$; $\frac{R_o}{f} \simeq 0.15$; $\frac{B_o}{R_o} \simeq \frac{1}{25}$; $\xi \simeq 3.5$ and $\zeta \simeq 0.09$ then:

$$\frac{G}{\pi Y R_o} = -\frac{0.01\alpha^3-0.24\alpha^2+4.25\alpha-1.3}{14.4-1.54\alpha} \simeq \frac{4.24\alpha-1.3}{14} \tag{19a}$$

Accepting one of the proposed correlations between $G_R$ and the uppershelf Charpy level, e.g.,

$$G_R = cC_v\left(\frac{B_o}{D_o}\right)^{0.5} \quad \text{Poynton and Fearnehough [1]} \tag{20a}$$

$$\text{or} \quad G_R = \frac{C_{v,2/3}}{A_c} \quad \text{Maxey et al [10]} \tag{20b}$$

$$\text{or} \quad G_R = 5\frac{Y}{E}(C_v - \frac{Y}{20}) \text{ or } G_R = 0.048\,C_v^{3/2} \quad \text{Barsom and Rolfe [15]} \tag{20c}$$

allows to express $G_R$-requirements into Charpy requirements as a function to pipe parameters preventing steady state fracture.

The obtained experimental data by Battelle as compiled in [14] by Kiefner et al all amply show fracture for $\frac{G_R}{\pi Y R_o}$ smaller than described by Eq. (19a) for relevant $\alpha$. Arrest data were not readily available. Presumably the fracture velocity exceeded the one for maximum crack driving force. Even if the introduced dimensionless pipe parameters at fracture remain largely independent on fracture velocity their quantitative estimate is vague and as a crack propagation arrest boundary, Eq. (19) suggests:

$$\frac{G}{\pi Y_{eff.} R_o} = \frac{a_3\alpha^3+a_2\alpha^2+a_1\alpha_1+a_0}{b_1\alpha+1} \qquad (21)$$

with $Y_{eff.}$ somewhat larger than Y and the a's and b's adjustable. $\alpha$ has to be such large that plastic flow in bulge forming and development of flaps occur.

This suggests a required $G_R$ linearly increasing with increasing $R_o$, increasing hoop stress level and increasing yield strength, though the change in fracture geometry of the pipe with yield strength might interfere. As remarked by Wells [15] the steady state fracture velocity has to have a value compatible with that of the flexural waves in the flaps responsible for their development by plastic flow.

It remains of course rather trivial that an adjustment of 5 parameters, viz. $a_3$, $a_2$, $a_1$, $a_0$ and $b_1$ - with $a_1$, $a_0$ possibly proportional to $B_o/R_o$ - for pipes with comparable flow stresses and otherwise not too widely different geometry, also at fracture will allow a prediction of failure stress from known experimental data of $\frac{G_R}{\pi YR_o}$ and $\alpha$ of such pipes (as e.g. reported in [13]).

Computer programs were run, using the relations (20) to determine the best fitting curve of 88 of the 92 in [14] reported pipe failure tests (excluded were tests nrs 6, 7, 48 and 54) according to the least square procedure of Banachiewitz. The smallest standard deviation was obtained with Eq. (20a). Taking $a_1$ and $a_0$ proportional to $\frac{B_o}{R_o}$ decreased the width of the describing band somewhat further. Presumably an estimate of the arrest-propagation boundary will be found from this description by shifting it downwards over a suitable distance proportional to the standard deviation.

An improved energy balance analysis, in particular of the $W_p$-terms, should match the adequate Battelle description [10]; plastic flow characteristics of the material as revealed by the stress-strain diagram are expected to play a role.

It has also to be remarked that the parametric grouping of terms from the energy balance interpretation as demonstrated by British work [1,2,3] proved to offer a satisfactory description of the arrest-propagation boundary by adjustment of the constants entering their analysis.

REFERENCES

[1] Poynton, W. A. and Fearnehough, G. D., "An analysis of shear fracture propagation in gas pipelines", Proc. of an International Conference on Dynamic Crack Propagation, Part III, p. 183, Lehigh University, Bethlehem, Pa., USA, July 1972.

[2] Poynton, W. A., "A theoretical analysis of shear fracture propagation in backfilled gas pipelines", Proc. of the 1st Int. Conference on Crack Propagation in Pipelines, Paper 14, England, March 1974.

[3] Shannon, R. W. E. and Wells, A. A., "A study of ductile crack propagation in gas pressurized pipelines", Proc. of the 1st Int. Conference on Crack Propagation in Pipelines, Paper 17, England, March 1974.

[4] Wright, J. K., "Shock tubes", London: Methusen & Co., Ltd., New York: John Wiley & Sons, Inc., (ch. 2 e.g. (2,13) p. 11), 1961.

[5] Mimura, H. and Ogasawara, M., "Theoretical considerations on crack propagation in pipelines", Proc. of the 1st Int. Conference on Crack Propagation in Pipelines, Paper 13, England, March 1974.

[6] Shoemaker et al., Preliminary oral communication at the 1st Int. Conference on Crack Propagation in Pipelines, to be published.

[7] Fearnehough, G. D., "Fracture propagation control in gas pipelines", Proc. of the 1st Int. Conference on Crack Propagation in Pipelines, Paper 1, England, March 1974.

[8] Dick, J. A., Jamieson, P. McK., and Walker, E. F., "The prediction of toughness requirements for the arrest of unstable fracture in pipelines", Proc. of the 1st Int. Con-

ference on Crack Propagation in Pipelines, Paper 15, England, March 1974.

[9] Hahn, G. T., Sarrate, M., Kanninen, M. F., and Rosenfield, A. R., "A model for unstable shear crack propagation in pipes containing gas pressure", Int. Journal of Fracture, Vol. 9, No. 2, 209, 1973.

[10] Maxey, W. A., Kiefner, J. F., Eiber, R. J. and Duffy, A. R., "Ductile fracture initiation, propagation and arrest in cylindrical vessels, fracture toughness", Proc. of the 1971 Nat. Symposium on Fracture Mechanics, Part II, ASTM STP 514, 70, 1972.

[11] Erdogan, F. and Ratwani, M., "Fracture initiation and propagation in a cylindrical shell containing an initial surface flaw", Proc. of the 2nd Int. Conference on Structural Mechanics in Reactor Technology, Germany, September 1973.

[12] Kanninen, M. F. and Sampath, S. G., "Crack propagation in pressurized pipelines", 2nd Int. Conference on Pressure Vessel Technology, San Antonio, Texas, USA, II-69, 1973, pp. 971-979.

[13] Trense, R. W. and Wiekema, B. J., "Mathematical analysis of British Gas model for ductile crack extension in gas pipelines", Report RVO-TNO, S3408, 1974.

[14] Kiefner, J. F., Maxey, W. A., Eiber, R. J. and Duffy, A. R., "Failure stress levels of flaws in pressurized cylinders, progress in flaw growth and fracture toughness testing", Proc. of 6th Nat. Symposium on Fracture Mechanics, Philadelphia, 1972, ASTM STP 536, 1973, pp. 461-481.

[15] Barson, J. M. and Rolfe, S. T., "Correlations between $K_{Ic}$ and Charpy-V notch results in the transition temperature range, Symposium Impact Testing of Metals, Atlantic City, USA, 1969, ASTM STP 466, pp. 281-302.

[16] Wells, A. A., "Fracture control; past, present and future", Experimental mechanics, October 1973, p. 401.

# ELASTODYNAMIC STRESS INTENSITY FACTORS FOR A BIFURCATING CRACK

J. D. Achenbach

Northwestern University, Evanston, Illinois 60201, USA

## ABSTRACT

It has been suggested by several authors that elastodynamic effects play an important role in the phenomenon of crack branching. In recent years, problems of elastodynamic crack propagation have been considered in considerable detail, but only for cases where either symmetry or antisymmetry exists with respect to the plane of crack propagation. In this paper a transient elastodynamic analysis is presented of the stress singularities which are generated in antiplane strain, at the propagating tips of the branches of a bifurcating crack. Each branch propagates with a different but constant velocity, and under a different angle. Numerical results are presented, showing the stress intensity factors versus the branching angles.

## NOTATION

| | |
|---|---|
| $c$ | $= (\mu/\rho)^{\frac{1}{2}}$, velocity of transverse waves |
| $f(\zeta)$ | $= dF(\zeta)/d\zeta$ |
| $i$ | $= \sqrt{-1}$ |
| $k_T$ | = dynamic stress intensity factor |
| $p,\psi$ | = polar coordinates centered at the crack tip in the $\theta$-$\beta$ plane |
| $r,\theta,z$ | = stationary cylindrical coordinates in the physical plane |
| $s$ | $= r/t$ |
| $t$ | = time |
| $v$ | = crack propagation velocity |
| $w$ | = antiplane displacement |

| | |
|---|---|
| $x, y, z$ | = rectangular cartesian coordinate system |
| $F(\zeta)$ | = analytical function in the $\zeta$-plane |
| $G(\gamma)$ | = analytical function in the $\gamma$-plane |
| $H(\ )$ | = Heaviside function |
| $K$ | = intensity factor |
| $R, \varphi$ | = moving polar coordinates centered at the crack tip in physical plane |
| $\alpha$ | = $v/c$ |
| $\beta$ | = arcosh $(c/s)$, Chaplygin's transformation |
| $\gamma$ | = $\omega(\zeta) = \beta + i\theta$ = mapping function which maps the $\theta$-$\beta$ plane onto the $\zeta$-plane |
| $\gamma_D$ | = position of the crack tip in the $\theta$-$\beta$ plane |
| $\zeta$ | = $\xi + i\eta$ |
| $\varkappa\pi$ | = angle of crack propagation |
| $\mu$ | = shear modulus |
| $\rho$ | = mass density |
| $\tau_{\theta z}, \tau_{rz}$ | = shear stresses |
| $\tau_{\varphi z}$ | = shear stress in the vicinity of the moving crack tip |
| $\Phi(\alpha, \varphi)$ | = function governing variation with angle $\varphi$ near crack tip |
| Re | = real part of complex quantity |
| Im | = imaginary part of complex quantity |

## INTRODUCTION

It has been known for some time that for a crack propagating in it's own plane, the maximum values of elastodynamic stresses move out of the plane of crack propagation when the crack propagation velocity exceeds a certain critical value. The first results of this kind were derived by Yoffe [1], who considered a steady-state elastodynamic problem in which a crack of fixed length moves at constant speed through a stretched elastic sheet. A transient elastodynamic problem in which a crack suddenly appears and extends at a constant velocity in a stretched elastic body was solved by Baker [2]. Baker obtained the same angular variation of the circumferential stress as in Ref [1], although the complete expressions for the stresses near the propagating crack tip were quite different from those of Ref [1], because of the different character of Baker's problem. The result that the maximum values of pertinent stresses move out of the plane of crack propagation suggest an elastodynamic explanation of the phenomenon of crack branching.

In this paper stress intensity factors are computed for the case that branching actually takes place. The particular example is concerned with the case that the propagating crack bifurcates into two cracks propagating in different directions.

The paper consists of two parts. In the first part the transient elastodynamic stress field in the immediate vicinity of the crack tip is analyzed for a crack propagating along an arbitrary trajectory. The method of analysis used in the first part of the paper is analogous to the one introduced in static elasticity by Knein [3], and later applied by Williams [4] and Karp and Karal [5]. In the present treatment the considerations are restricted to the case of antiplane shear. A full account of the in-plane case by the author and Z. P. Bazant will be published elsewhere [6]. The results for the near-tip fields obtained by this method do give the angular variation of the fields, but they do not provide any information on the dependence of the stress-intensity factor on the time, and on the geometrical and material parameters. The latter information must be obtained by considering the complete formulation of the problem. The second part of the paper is, therefore, devoted to a complete transient elastodynamic analysis for a branching crack.

In recent years problems of elastodynamic crack propagation have been investigated in considerable detail, but only for cases where either symmetry or antisymmetry exists with respect to the plane of crack propagation. For a survey of recent work we refer to a paper by Achenbach [7]. The present paper appears to be the first elastodynamic analysis for a bifurcating crack. The two branches emanate from the tip of a semi-infinite crack. Each branch propagates with a different but constant velocity, and at a different angle. The loading which gives rise to crack propagation consists of uniform antiplane shear tractions, which are suddenly applied to the surfaces of the semi-infinite crack.

The method of solution is based on the observation that for the wave motion generated by step-stresses, the particle velocity shows the property of self-similarity. The crucial step in the analysis is the application of Chaplygin's transformation, which reduces the wave equation inside the circular region of the diffracted wave emanating from the crack tip to Laplace's equation in a semi-infinite strip containing slits. The boundary value problem for the strip is solved by application of the conformal mapping technique, and other elements of the theory of functions of a complex variable.

Expressions for the shear stresses in the vicinities of the crack tips are obtained by means of a limiting process which is analogous to the one worked out by Sih [8] for problems of static equilibrium. An interesting aspect of the results is that the dependence of the stress intensity factors on the angle of crack propagation separates from the dependence on polar angles centered at the moving crack tips.

In a stationary system of polar coordinates, r and $\theta$, two-dimensional antiplane wave motions are governed by the following equation for the out-of-plane displacement $w(r,\theta,t)$.

$$\frac{1}{r}\frac{\partial}{\partial r}\left(r\frac{\partial w}{\partial r}\right)+\frac{1}{r^2}\frac{\partial^2 w}{\partial\theta^2}=\frac{1}{c^2}\frac{\partial^2 w}{\partial t^2} \tag{1}$$

where c is the velocity of transverse waves,

$$c=(\mu/\rho)^{\frac{1}{2}} \tag{2}$$

In Eq. (2) $\mu$ and $\rho$ are the shear modulus and the mass density, respectively. The relevant shear stresses are

$$\tau_{rz}=\mu\partial w/\partial r, \quad \text{and } \tau_{\theta z}=(\mu/r)\partial w/\partial\theta \tag{3a,b}$$

NEAR-TIP FIELDS FOR A RAPIDLY PROPAGATING CRACK

Let us consider a crack which propagates along a rather arbitrary but smooth trajectory as shown in Fig. 1. At time $t = t_1$ the crack tip is located at point O; at time $t = t_1 + \Delta t$ the crack tip is located at point $O_1$. A system of moving polar coordinates $R,\varphi$ is centered at the moving crack tip. The velocity of the crack tip along the trajectory is $v(t)$.

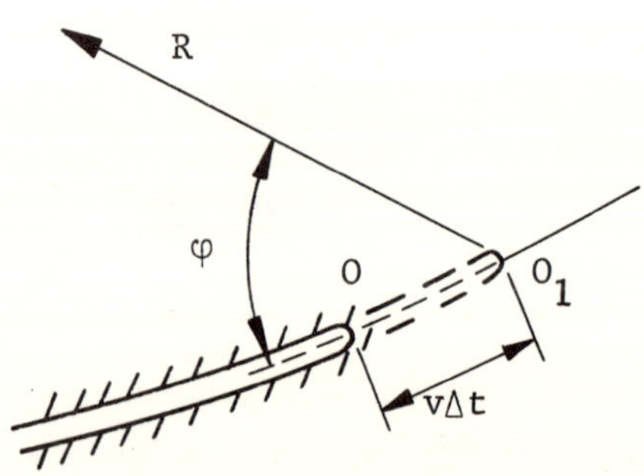


Fig. 1. Propagating crack tip with moving coordinate system.

Let the antiplane displacement be defined in terms of the moving coordinate system, i.e. $w = w(R,\varphi,t)$. The material derivative with respect to time, which is indicated by a superscript dot, then is of the form

$$\dot{w} = \frac{\partial w}{\partial t} + v\cos\varphi \frac{\partial w}{\partial R} + \frac{v}{R}\sin\varphi \frac{\partial w}{\partial \varphi} \tag{4}$$

Also,

$$\begin{aligned}\ddot{w} = {} & \frac{\partial^2 w}{\partial t^2} + \frac{\partial}{\partial t}\left(v\cos\varphi \frac{\partial w}{\partial R}\right) + \frac{\partial}{\partial t}\left(\frac{v}{R}\sin\varphi \frac{\partial w}{\partial \varphi}\right) \\ & + v\cos\varphi \left[\frac{\partial^2 w}{\partial R \partial t} + v\cos\varphi \frac{\partial^2 w}{\partial R^2} - v\sin\varphi\left(\frac{1}{R^2}\frac{\partial w}{\partial \varphi} - \frac{1}{R}\frac{\partial^2 w}{\partial R \partial \varphi}\right)\right. \\ & + \frac{v}{R}\sin\varphi\left[\frac{\partial^2 w}{\partial \varphi \partial t} - v\left(\sin\varphi \frac{\partial w}{\partial R} - \cos\varphi \frac{\partial^2 w}{\partial R \partial \varphi}\right)\right. \\ & + \left.\frac{v}{R}\left(\cos\varphi \frac{\partial w}{\partial \varphi} + \sin\varphi \frac{\partial^2 w}{\partial \varphi^2}\right)\right]\end{aligned} \tag{5}$$

Relative to the moving coordinates the displacement equation of motion may be written

$$\frac{\partial^2 w}{\partial R^2} + \frac{1}{R}\frac{\partial w}{\partial R} + \frac{1}{R^2}\frac{\partial^2 w}{\partial \varphi^2} = \frac{1}{c^2}\ddot{w}, \tag{6}$$

where $\ddot{w}$ is defined by Eq. (5).

In the immediate vicinity of the moving crack tip the displacement is of the general form

$$w(R,\varphi,t) = R^{\frac{1}{2}}\, T(t)\, \Phi_w(\alpha,\varphi) \tag{7}$$

where $\alpha$ is defined by

$$\alpha = v/c \tag{8}$$

Note that in the present considerations the velocity v may be a function of time. Since the shear stress vanishes on the crack surface we have

$$\text{at } \varphi = 0, 2\pi: \quad \partial\Phi_w/\partial\varphi = 0 \tag{9}$$

The antisymmetric nature of the displacement field implies

$$\text{at } \varphi = \pi: \quad \Phi_w = 0 \tag{10}$$

Upon substituting the displacement distribution given by Eq. (7)

into Eq. (6), multiplying the result by $R^{3/2}$, and subsequently considering the limit $R \to 0$ we obtain the following equation for $\Phi_w(\alpha,\varphi)$

$$(1 - \alpha^2\sin^2\varphi)\frac{\partial^2\Phi_w}{\partial\varphi^2} - \frac{\alpha}{2}\sin 2\varphi\frac{\partial\Phi_w}{\partial\varphi} + \frac{1}{2}\left[\frac{1}{2} + \alpha^2\left(\frac{3}{2}\cos^2\varphi - 1\right)\right]\Phi_w = 0 \tag{11}$$

Let us consider the domain $0 \leq \varphi \leq \pi$. It may be checked that in this domain the solution of Eq. (11) which satisfies the conditions given by Eqs. (9) and (10) may be written in the form

$$\Phi_w(\alpha,t) = \Phi_1(\alpha,t)\cos\varphi + (1-\alpha^2)^{\frac{1}{2}}\,\Phi_2(\alpha,t)\sin\varphi \tag{12}$$

where

$$\Phi_1 = \left[\frac{(1-\alpha^2\sin^2\varphi)^{\frac{1}{2}} + \cos\varphi}{1 - \alpha^2\sin^2\varphi}\right]^{\frac{1}{2}} \tag{13}$$

$$\Phi_2 = \left[\frac{(1-\alpha^2\sin^2\varphi)^{\frac{1}{2}} - \cos\varphi}{1-\alpha^2\sin^2\varphi}\right]^{\frac{1}{2}} \tag{14}$$

The stress $\tau_{\varphi z}$ corresponding to the displacement distribution Eq. (7), where $\Phi_w(\alpha,t)$ is defined by Eq. (12), then is easily obtained as

$$\tau_{\varphi z} = \frac{1}{2}T(t)\,R^{-\frac{1}{2}}(1-\alpha^2)^{\frac{1}{2}}\,\Phi(\alpha,t), \tag{15}$$

where

$$\Phi = (1-\alpha^2)^{-\frac{1}{2}}\,\Phi_1\sin\varphi - \Phi_2\cos\varphi \tag{16}$$

For various values of $\alpha$ the function $\Phi(\alpha,\varphi)$ has been plotted versus $\varphi$ in Fig. 2. It should be noted that the maximum value of $\Phi$ moves out of the plane of crack propagation, $\varphi = \pi$, as $\alpha$ increases beyond a certain critical value. For a special case of steady-state elastodynamic crack propagation an analogous result was first shown by Yoffe [1].

Equation (15) is interesting, in that it shows in rather general terms that even for a continuously turning crack the maximum value of the shear stress cannot be maintained in the plane of crack propagation.

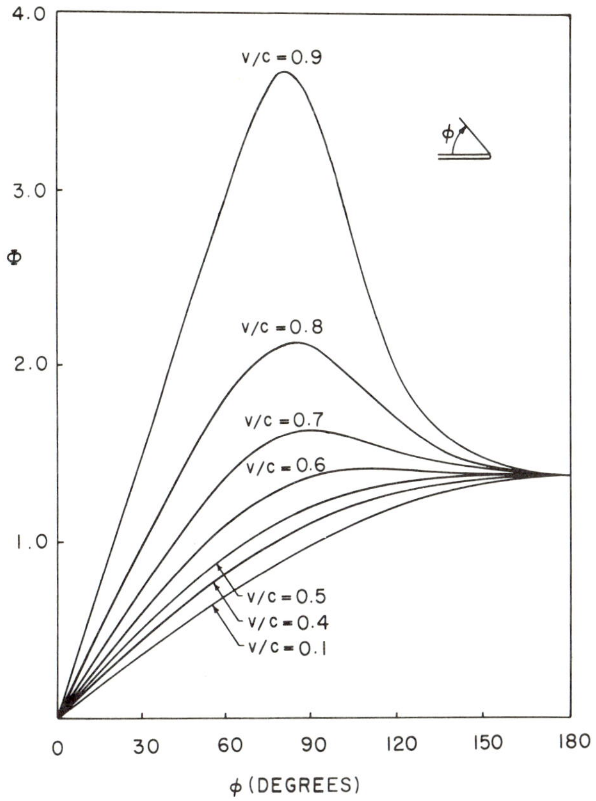


Fig. 2 Variation of $\Phi(\alpha,\varphi)$ with $\varphi$ for various $\alpha$

A PROBLEM OF ELASTODYNAMIC CRACK BIFURCATION

The analysis of elastodynamic stress fields for a crack not propagating in it's own plane is very difficult. In this section an analysis is presented for the two-dimensional case of antiplane strain. We investigate the elastodynamic fields which are generated when two branches emanate from the tip of a semi-infinite crack, upon the sudden application of uniform antiplane shear tractions to the surfaces of the semi-infinite crack. The sudden application of the shear tractions produces two plane waves, and a diffracted cylindrical wave centered at the original crack tip. It is assumed that at the instant the shear tractions are applied, the semi-infinite crack starts to bifurcate with velocities $v_1$ and $v_2$, where both $v_1/c < 1$ and $v_2/c < 1$, and at angles $-\varkappa_1\pi$ and $\varkappa_2\pi$, respectively. Thus at time $t > 0$, the crack tips are located at points D and C, which are defined by $r = v_1 t, \theta = -\varkappa_1\pi$, and $r = v_2 t$, $\theta = \varkappa_2\pi$, respectively. The pattern of wavefronts and the positions of the crack tips at time t are shown in Fig. 3.

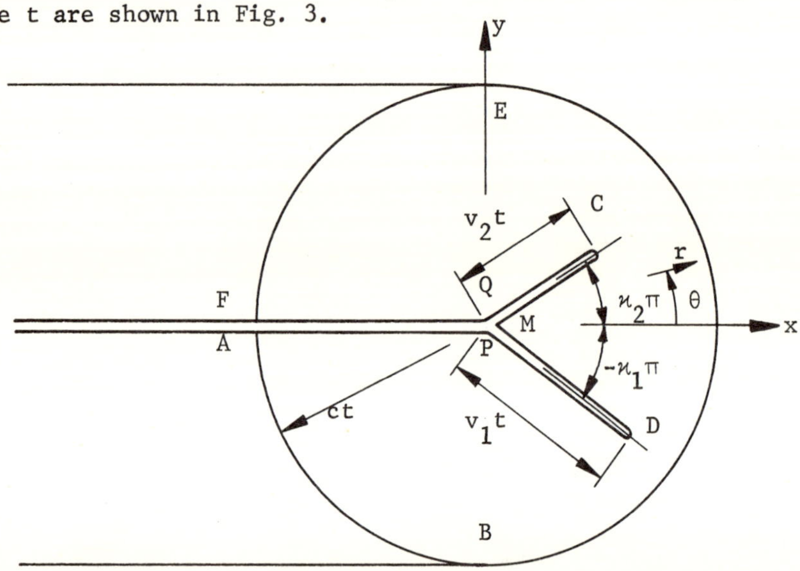


Fig. 3 Pattern of wavefronts and positions of crack tips

The wave motions in the elastic solid are governed by Eq. (1). The relevant shear stresses are given by Eqs. (3a,b). The boundary conditions are

$$\theta = \pm\pi, \ r > 0 \quad \tau_{\theta z} = -\tau_o H(t) \tag{17}$$

$$\theta = -\varkappa_1\pi \pm \epsilon\ , \ r < v_1 t: \quad \tau_{\theta z} = 0 \tag{18}$$

$$\theta = \varkappa_2\pi \pm \epsilon\ , \ r < v_2 t \quad \tau_{\theta z} = 0 \tag{19}$$

In Eq. (17), H(t) is the Heaviside step function. The initial conditions are

$$w(r,\theta,t) = \dot{w}(r,\theta,t) \equiv 0 \quad \text{for } t < 0 \tag{20}$$

It is easily verified that the plane waves generated by the stress distribution (17) are of the form

$$w = \mp(c\tau_o/\mu)\ (t \mp y/c)\ H(t \mp y/c), \tag{21}$$

where the minus and plus signs apply for $y > 0$ and $y < 0$, respectively.

In this paper we seek solutions to Eq. (1) that are self-similar inside the region of the cylindrical wave. Such solutions depend on $r/t$ and $\theta$ rather than on $\theta$ and $r$ and $t$ separately. For a brief discussion of self-similar solutions we refer to the book by Achenbach [9]. For the boundary conditions defined by Eqs. (17) - (19), the particle velocity

$$\dot{w} = \partial w/\partial t \tag{22}$$

is self-similar.

In polar coordinates $r,\theta,t$, the equation for $\dot{w}(r,\theta,t)$ is

$$\frac{1}{r}\frac{\partial}{\partial r}\left(r\frac{\partial \dot{w}}{\partial r}\right) + \frac{1}{r^2}\frac{\partial^2 \dot{w}}{\partial \theta^2} = \frac{1}{c^2}\frac{\partial^2 \dot{w}}{\partial t^2} \tag{23}$$

It is now convenient to introduce the new variable

$$s = r/t \tag{24}$$

whereupon the equation for $\dot{w}(s,\theta)$ follows from Eq. (23) as

$$s^2\left(1 - \frac{s^2}{c^2}\right)\frac{\partial^2 \dot{w}}{\partial s^2} + s\left(1 - \frac{2s^2}{c^2}\right)\frac{\partial \dot{w}}{\partial s} + \frac{\partial^2 \dot{w}}{\partial \theta^2} = 0 \tag{25}$$

Within the region of the cylindrical wave ABE, see Fig. 3, the boundary conditions on $\dot{w}(s,\theta)$ assume the following form for $t>0$:

$$\theta = -\pi, \qquad s \le c: \quad \partial\dot{w}/\partial\theta = 0 \tag{26}$$

$$-\pi \le \theta < -\pi/2, \quad s = c: \quad \dot{w} = c\tau_o/\mu \tag{27}$$

$$-\pi/2<\theta<\pi/2, \qquad s = c: \quad \dot{w} = 0 \tag{28}$$

$$\pi/2 <\theta< \pi, \qquad s = c: \quad \dot{w} = -c\tau_o/\mu \tag{29}$$

$$\theta = \pi, \qquad s \le c: \quad \partial\dot{w}/\partial\theta = 0 \tag{30}$$

$$\theta = -\varkappa_1\pi + \epsilon, \qquad s < v_1: \quad \partial\dot{w}/\partial\theta = 0 \tag{31}$$

$$\theta = \varkappa_2\pi \pm \epsilon, \qquad s < v_2: \quad \partial\dot{w}/\partial\theta = 0 \tag{32}$$

For $s < c$ the following transformation

$$\beta = \cosh^{-1}(c/s), \tag{33}$$

which is known as Chaplygin's transformation, reduces Eq. (25) to Laplace's equation

$$\partial^2\dot{w}/\partial\beta^2 + \partial^2\dot{w}/\partial\theta^2 = 0 \tag{34}$$

The real transformation (33) maps the interior of the circular domain into a semi-infinite strip in the $\theta$-$\beta$ plane, containing two slits, see Fig. 4, where the corresponding points are indicated

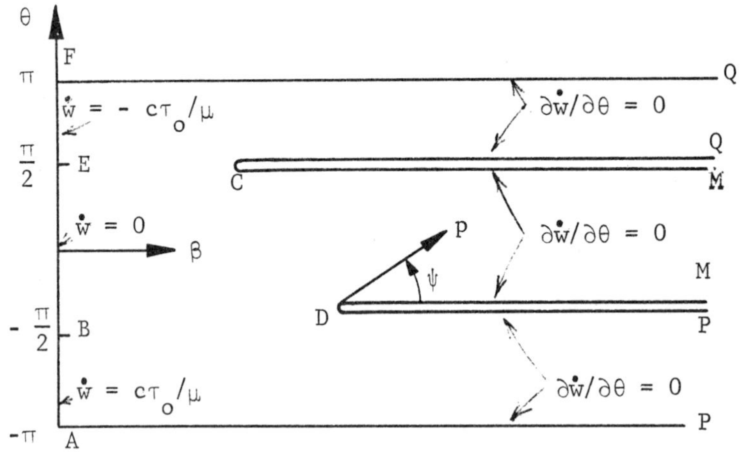


Fig. 4 Domain in the $\theta$-$\beta$ plane

The boundary conditions on $w(\beta,\theta)$ in the $\theta$-$\beta$ plane are indicated in Fig. 4. The harmonic function $\dot{w}$ can be taken as the real part of an analytic function $G(\gamma)$

$$\dot{w} = \mathrm{Re}\, G(\gamma) \tag{35}$$

where $\gamma = \beta + i\theta$. The function $G(\gamma)$ can, in principle, be obtained by conformal mapping techniques. The domain in the $\gamma$-plane can be related to the upper half of the $\zeta$-plane, where $\zeta = \xi + i\,\eta$, by means of the following Schwarz-Christoffel transformation

$$\gamma = \omega(\zeta) = C_1 \int_1^{\zeta} \frac{(u+\xi_D)(u-\varkappa_2)}{(u+\xi_P)(u-\xi_M)(u-\xi_Q)(1-u^2)^{\frac{1}{2}}}\, du + C_2, \tag{36}$$

Here $\xi_D$, $\xi_P$, $\xi_M$ and $\xi_Q$ define the points in the $\zeta$-plane corresponding to D, P, M and Q. The three arbitrary choices in the mapping are the points A, F and C which are mapped into the points $\zeta = -1$, $\zeta = 1$ and $\zeta = \varkappa_2$, respectively. The $\zeta$-plane is shown in Fig. 5, where $\xi_B$ and $\xi_E$ define the points corresponding to B and E in Fig. 4.

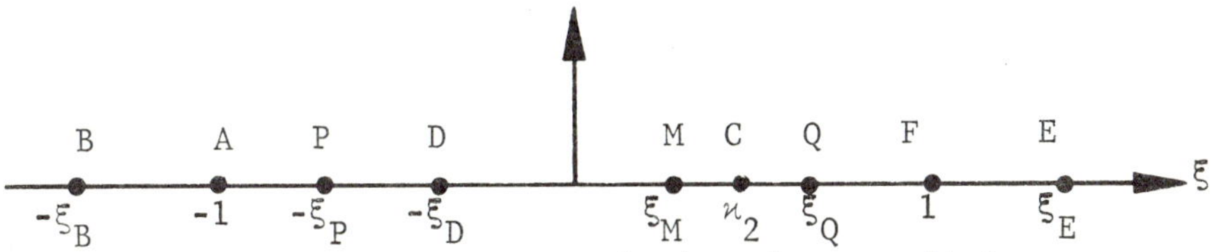


Fig. 5: Points in the $\zeta$-plane. The boundary conditions are: $-\infty < \xi < -\xi_B$ and $\xi_E < \xi < \infty$: $\dot{w} = 0$; $-\xi_B < \xi < -1$: $\dot{w} = c\tau_o/\mu$; $-1 < \xi < 1$: $\partial\dot{w}/\partial\eta = 0$; $1 < \xi < \xi_E$: $\dot{w} = -c\tau_o/\mu$

The integral in Eq. (38) can be evaluated to yield $\omega(\zeta) =$

$$\frac{(\xi_P-\xi_D)(\xi_P+\varkappa_2)}{(\xi_P+\xi_M)(\xi_P+\xi_Q)}\frac{C_1}{(1-\xi_P^2)^{\frac{1}{2}}}\{\ell n[(1-\xi_P^2)^{\frac{1}{2}}(1-\zeta^2)^{\frac{1}{2}}+\zeta\xi_P+1]-\ell n(\zeta+\xi_P)\}$$

$$+\frac{(\xi_M+\xi_D)(\xi_M-\varkappa_2)}{(\xi_M-\xi_Q)(\xi_P+\xi_M)}\frac{C_1}{(1-\xi_M^2)^{\frac{1}{2}}}\{\ell n[(1-\xi_M^2)^{\frac{1}{2}}(1-\zeta^2)^{\frac{1}{2}}-\zeta\xi_M+1]-\ell n(\zeta-\xi_M)\}$$

$$+\frac{(\varkappa_2-\xi_Q)(\xi_Q+\xi_D)}{(\xi_P+\xi_Q)(\xi_M-\xi_Q)}\frac{C_1}{(1-\xi_Q^2)^{\frac{1}{2}}}\{\ell n[(1-\xi_Q^2)^{\frac{1}{2}}(1-\zeta^2)^{\frac{1}{2}}-\zeta\xi_Q+1]-\ell n(\zeta-\xi_Q)\}$$

$$+ C_2. \tag{37}$$

The mapping of the point F gives $C_2 = i\pi$. By examining the changes of the imaginary parts at P, M and Q, three equations are obtained. Two additional equations are obtained by comparing the coordinates of points C and D in the $\gamma$-and $\zeta$-planes. Thus we have a total of five equations from which the constant $C_1$ and the coordinates $\xi_P$, $\xi_Q$, $\xi_M$ and $\xi_D$ can be computed for any given set of values of $\varkappa_1$, $\varkappa_2$, $v/c_1$ and $v/c_2$. The coordinates $\xi_B$ and $\xi_E$ can subsequently be computed from Eq. (37). The procedure for the computation of these points is described for a very similar case of skew crack propagation in Ref [10].

The boundary conditions shown in Fig. 4 are transformed into conditions on the real axis in the $\zeta$-plane. These conditions are shown in Fig. 5. To construct the appropriate analytic function $F(\zeta)$, where

$$\dot{w} = \mathrm{Re}\, F(\zeta) \ , \tag{38}$$

it is convenient to work with

$$f(\zeta) = F'(\zeta) = \frac{\partial \dot{w}}{\partial \xi} - i\,\frac{\partial \dot{w}}{\partial \eta} \ , \tag{39}$$

where the Cauchy-Riemann conditions have been employed. The conditions on the real axis in the $\zeta$-plane now yield

$$\eta = 0 \ , \quad -1 < \xi < 1 \quad : \quad \mathrm{Im}\, f(\zeta) = 0 \tag{40}$$

$$\eta = 0 \ , \quad |\xi| > 1 \quad : \quad \mathrm{Re}\, f(\zeta) = 0 \tag{41}$$

Since there are discontinuities in $\dot{w}$ at $\xi = -\xi_B$ and $\xi = \xi_E$, we expect simple poles at these points in the expression for $f(\zeta)$. The function $f(\zeta)$ satisfying the foregoing requirements is written as

$$f(\zeta) = \frac{i}{(\zeta^2-1)^{\frac{1}{2}}}\left[\frac{A}{\zeta + \xi_B} + \frac{B}{\zeta - \xi_E}\right] , \tag{42}$$

where A and B are real constants. In view of Eqs. (38) and (39), $\dot{w}$ is obtained as

$$\dot{w} = \mathrm{Re}\left\{ \int_{-\infty}^{\zeta} \frac{i}{(u^2-1)^{\frac{1}{2}}}\left[\frac{A}{u + \xi_B} + \frac{B}{u - \xi_E}\right] du + C\right\} . \tag{43}$$

It is not difficult to show that

$$C = 0 \tag{44}$$

$$A = -\,(c\tau_o/\pi\mu)(\xi_B^2 - 1)^{\frac{1}{2}} \tag{45}$$

$$B = (c\tau_o/\pi\mu)(\xi_E^2 - 1)^{\frac{1}{2}} \tag{46}$$

## ELASTODYNAMIC STRESS INTENSITY FACTORS FOR A BIFURCATING CRACK

In the $\theta$-$\beta$ plane shown in Fig. 4, let us consider the polar coordinate system $\rho$, $\psi$, centered at D. It is well-known that the derivatives of a field quantity which is governed by Laplace's equation have square root singularities at the tip of a slit. Thus, as the point D is approached $\partial\dot{w}/\partial p$ and $(1/p)\,\partial\dot{w}/\partial\psi$, or $\dot{\tau}_{pz}$ and $\dot{\tau}_{\psi z}$, show square root singularities. The forms of $\dot{\tau}_{pz}$ and $\dot{\tau}_{\psi z}$ follow, for example, from the results presented by Sih [8] as

$$\dot{\tau}_{pz} - i\,\dot{\tau}_{\psi z} \sim [K/(2p)^{\frac{1}{2}}]\exp(i\psi/2), \tag{47}$$

The intensity factor K can be related to $G(\gamma)$ by the steps outlined by Sih [8]. Since $\gamma-\gamma_D = p\exp(i\psi)$, see Fig. 4, we find

$$\dot{\tau}_{pz} - i\dot{\tau}_{\psi z} \sim [K/\sqrt{2}(\gamma-\gamma_D)^{\frac{1}{2}}]\exp(i\psi) \tag{48}$$

It follows, however, from Eq. (35) that we can write

$$\dot{\tau}_{pz} - i\dot{\tau}_{\psi z} = \mu G'(\gamma)\exp(i\psi), \tag{49}$$

where the prime denotes a derivative with respect to the argument of the function $G(\gamma)$. By comparing Eqs. (48) and (49) it follows that

$$K(\gamma_D) = \mu\sqrt{2}\lim_{\gamma\to\gamma_D}(\gamma-\gamma_D)^{\frac{1}{2}}G'(\gamma) \tag{50}$$

Concomitant with the mapping $\gamma = \omega(\zeta)$, defined by Eq. (37), the expression for the intensity factor K can also be written as

$$K = \mu\sqrt{2}\lim_{\zeta\to\xi_D}\frac{[\omega(\zeta)-\omega(\xi_D)]^{\frac{1}{2}}F'(\zeta)}{\omega'(\zeta)} \tag{51}$$

In general the function $dF/d\zeta$ is continuous at $\zeta = \xi_D$, and $d\omega/d\zeta$ vanishes as $\zeta$ approaches $\xi_D$. To evaluate the limit we employ L'Hospital's rule, to obtain

$$K = \mu F'(\xi_D)/\left[\omega''(\xi_D)\right]^{\frac{1}{2}} \tag{52}$$

By employing Eqs. (36) and (42) the intensity factors are now easily evaluated.

At the crack tips D and C we find

$$K_1 = \frac{\mu[(\xi_Q + \xi_D)(\xi_Q - \varkappa_2)(\xi_P - \xi_D)(\xi_D + \xi_M)(\xi_D + \xi_Q)]^{\frac{1}{2}}}{[(1-\varkappa_2)(\xi_P + \xi_Q)(\xi_Q - \xi_M)(\xi_D + \varkappa_2)]^{\frac{1}{2}}[(1-\xi_D^2)(1-\xi_Q^2)]^{\frac{1}{4}}}$$

$$\times \left[\frac{A}{\xi_B - \xi_D} - \frac{B}{\xi_D + \xi_E}\right], \tag{53}$$

and

$$K_2 = \frac{\mu[(\xi_Q + \xi_D)(\xi_P + \varkappa_2)(\varkappa_2 - \xi_M)]^{\frac{1}{2}}(\xi_Q - \varkappa_2)}{[(1-\varkappa_2)(\xi_P + \xi_Q)(\xi_Q - \xi_M)(\xi_D + \varkappa_2)]^{\frac{1}{2}}[(1-\varkappa_2^2)(1-\xi_Q^2)]^{\frac{1}{4}}}$$

$$\times [A/(\varkappa_2 + \xi_B) - B/(\xi_E - \varkappa_2)], \tag{54}$$

respectively.

To obtain expressions for the shear stress in the vicinity of the crack tip, in the physical plane, we introduce a set of polar coordinates, R, $\varphi$, centered at the crack tip, similarly to the one shown in Fig. 1. The shear stress $\tau_{\varphi z}$ is related to $\tau_{rz}$ and $\tau_{\theta z}$ by

$$\tau_{\varphi z} = -\tau_{rz}\sin\varphi - \tau_{\theta z}\cos\varphi \tag{55}$$

By employing Eqs. (24), (33), and (35), we obtain

$$\dot{\tau}_{rz} = (\partial\beta/\partial r)\mu \mathrm{Re}G'(\gamma) \text{ and } \dot{\tau}_{\theta z} = (\mu/r)\mathrm{Re}[iG'(\gamma)] \tag{56a,b}$$

Eliminating $G'(\gamma)$ by means of Eq. (50), and employing $\gamma - \gamma_D = p \exp(i\psi)$, we can write Eqs. (56a) and (56b) in the following forms for small values of p

$$\dot{\tau}_{rz} = -\frac{K}{(2p)^{\frac{1}{2}}}\frac{ct/r}{(c^2t^2-r^2)^{\frac{1}{2}}}\cos\tfrac{1}{2}\psi \tag{57}$$

and

$$\dot{\tau}_{\theta z} = \frac{K}{(2p)^{\frac{1}{2}}}\frac{1}{r}\sin\tfrac{1}{2}\psi \tag{58}$$

For small values of R/vt, the following relations can easily be derived as

$$\theta \sim \varkappa\pi + R\sin\varphi/vt \tag{59}$$

$$r \sim vt\,[1-(R/vt)\cos\varphi] \tag{60}$$

$$\beta \sim \text{arcosh}\ (c/v) + [(R/vt)\cos\varphi]/(1-v^2/c^2)^{\frac{1}{2}} \tag{61}$$

Expressing p and $\psi$ in terms of $\beta$ and $\theta$, and using Eqs. (59) - (61) we can obtain expressions for p, $\cos(\psi/2)$ and $\sin(\psi/2)$ for small values of R/vt. Substituting these expressions in Eqs. (57), (58) and (55), and subsequently integrating with respect to t, we find for small values of R

$$\tau_{\varphi z} \sim k_T \Phi(v/c,\varphi)/R^{\frac{1}{2}} \tag{62}$$

where

$$k_T = \left(1 - \frac{v^2}{c^2}\right)^{\frac{1}{4}} \left(\frac{t}{v}\right)^{\frac{1}{2}} K \tag{63}$$

In Eq. (63) the constant K is defined either by Eq. (53) or Eq. (54), and $\Phi(v/c,\varphi)$ is defined by Eq. (16).

RESULTS

The expression for the shear stress in the vicinity of a moving crack tip of a bifurcating crack, given by Eq. (62), shows that the dependence of the stress intensity factor on the angle $\varphi$ separates from its dependence on $\varkappa$. The dependence of the stress intensity factor on $\varphi$ is governed by $\Phi(v/c,\varphi)$, while the dependence on $\varkappa$ is governed by $K(v/c,\varkappa)$. Both $\Phi$ and K are functions of the relative speed, v/c, of the moving crack tip. For various values of v/c the function $\Phi(v/c,\varphi)$ is plotted versus $\varphi$ in Fig. 2. To compute the intensity factors $K_1$ and $K_2$ from Eqs. (53) and (54), the constants $\xi_P$, $\xi_Q$, $\xi_M$ and $\xi_D$ must first be computed from a system of four non-linear equations, as described following Eq. (37).

We will present some numerical results for the case of skew crack propagation, i.e., $v_1 = 0$, $v_2 = v$ and $\varkappa_1 = 0$, $\varkappa_2 = \varkappa$. This case was studied in detail in Ref. [10]. The intensity factor K was computed for various values of v/c, as a function of $\varkappa$. The results, which are shown in Fig. 6, indicate that K is a maxiumum for a crack propagating in its own plane until the velocity ratio v/c reaches a critical value, in between v/c = 0.2 and v/c = 0.3. For a higher value of v/c a larger value of K occurs for propagation out of the plane of the original crack. By comparing Figs 2 and 6 it can be noted that a range of v/c exists, for which K would be larger for skew crack propagation, but where the maximum of $\Phi(v/c,\varphi)$ would be in the plane of the propagating branch.

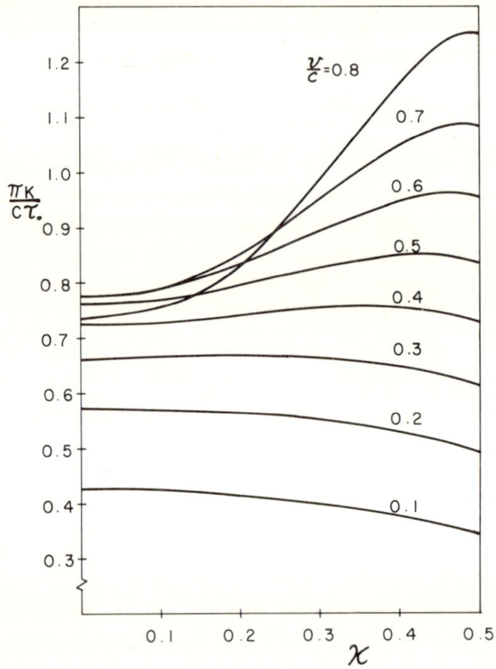


Fig. 6 Factor K as a function of $\varkappa$ for skew crack propagation.

In Fig. 7 the flux of energy into the propagating crack tip is shown versus $\varkappa$, and for various values of v/c. The flux of energy into the propagating crack tip enters in the criterion for brittle fracture.

In summary, we have obtained the angular variation of the singularity in the elastodynamic stress field for a crack tip propagating along a rather arbitrary trajectory. We have also obtained the complete elastodynamic stress intensity factor for a special case of crack bifurcation, and we have presented numerical results for skew crack propagation. An interpretation of these results within the context of a fracture criterion will be presented elsewhere.

ACKNOWLEDGEMENT

The results reported here were obtained in the course of research sponsored by the National Science Foundation under Grant GK 26217 to Northwestern University. The author should like to acknowledge the assistance and cooperation of Dr. V. K. Varatharajulu.

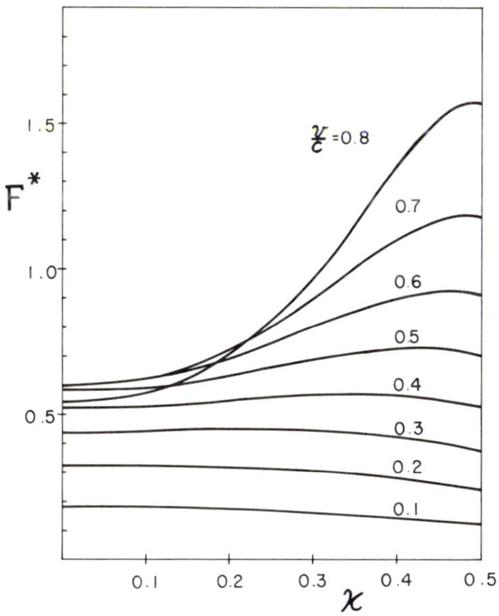


Fig. 7 Dimensionless flux of energy into the propagating crack tip; $F^* = \pi\mu F/2c^2\tau_o^2 t$.

REFERENCES


[1] Yoffe, E. H., The moving Griffith crack, Philosophical Magazine, 42, 739, 1951.

[2] Baker, B. R., Dynamic stresses created by a moving crack, Journal of Applied Mechanics, 29, 449, 1962.

[3] Knein, M., Zur Theorie des Druckversuchs, Abhandlungen Aerodyn. Inst. T. H. Aachen, 7, 62, 1927.

[4] Williams, M. L., Stress singularities resulting from various boundary conditions in angular corners in extension, Journal of Applied Mechanics, 19, 526, 1952.

[5] Karp, S.N., and Karal, F.C., The elastic field behavior in the neighborhood of a crack of arbitrary angle, Communications of Pure & Applied Mathematics, 15, 413, 1962.

[6] Achenbach, J. D. and Bazant, Z.P., Elastodynamic near-tip stress fields for rapidly propagating cracks, to be published.

[7] Achenbach, J. D., Dynamic effects in brittle fracture, in Mechanics Today, Pergamon Press, New York, 1973.

[8] Sih, G.C., Stress distribution near internal crack tips, Journal of Applied Mechanics, 32, 51, 1965.

[9] Achenbach, J.D., Wave Propagation in Elastic Solids, North-Holland/American Elsevier, Amsterdam/New York 1973, 154

[10] Achenbach, J.D., and Varatharajulu, V.K., Skew crack propagation at the diffraction of a transient stress wave, Quarterly of Applied Mathematics, Vol. XXXII, 123, 1974.

# SOME ASPECTS OF DYNAMIC CRACK PROPAGATION : A REVIEW AND SOME GENERALISATIONS

C. Atkinson

Mathematics Department, Imperial College, London S.W.7

## ABSTRACT

The two most widely used crack propagation models the semi-infinite slit and the crack expanding uniformly from zero size are briefly reviewed. Particular attention being paid to the limiting velocity for a crack running along the interface between two different elastic media. The stress analysis for the semi-infinite slit is extended to the case of media with a spatially varying elastic modulus where some special cases are considered in detail.

## INTRODUCTION

This paper is principally concerned with some aspects of dynamic crack propagation i.e. crack propagation including the effects of inertia. We shall briefly review the two most widely used dynamic crack models, the propagating semi-infinite slit and the uniformly expanding crack which grows from zero size. The former has the mathematical advantage that for linear elastic and viscoelastic bodies the stress analysis can be approached via the Wiener-Hopf technique, the solution depending to a large extent on how simply one can do certain factorisations in the complex plane. The second model has the advantage that for the linear elastic case the stress field possesses the property of dynamic similarity which for plane problems gives a dependence on only two variables $x/t$ and $y/t$ rather than the three variables $x$ , $y$ and $t$ . As well as the above models which are concerned with the effect of inertia on a growing crack, a number of authors have considered the diffraction of waves by a crack tip. We shall not discuss detailed diffraction effects here, although for the semi-infinite crack the Wiener-Hopf analysis is much the same whether we consider moving cracks impacted by plane waves or growing under a uniform applied stress. Also although specific reference will be made to some work on expanding crack problems which possess the dynamic similarity property described above our main concern will be with the uniformly moving semi-infinite crack since this model can be more readily applied to more general media. We have already mentioned that the semi-infinite crack model can be applied straightforwardly, at least in principle, to linear viscoelastic media and we shall discuss later its use for crack tip stress analysis in non-homogeneous media i.e. media with a spatially varying elastic modulus.

The plan of the paper is as follows, we begin in Section 1 with a brief review of some recent theoretical work on crack propagation and then give some attention to the question of the limiting velocity of the crack tip i.e. the velocity at which $G$ the energy flow into the crack tip first goes to zero. Strictly speaking the latter remark applies to the truly transient situation (not steady state), however the steady state analysis appears to give the same final result if a different interpretation of $G$ is made in that case. The main result in this section is a demonstration that for a crack moving along the interface between two isotropic elastic media under plane strain conditions the limiting velocity (as defined above) will be the smaller of the Rayleigh wave speeds associated with each of the media separately. This result seems reasonable, particularly since in the anti-plane case it can be shown that the limiting speed should be the smaller wave speed of the two media. Section 2 is mainly concerned with generalising the procedures which have been used for the homogeneous elastic situation to the situation where the elastic moduli vary with position. The simplest case of a semi-infinite crack growing in mode 3 (anti-plane strain) is considered.

## 1. DYNAMIC CRACK PROPAGATION. A BRIEF REVIEW

As indicated in the introduction the main topics of this section are a brief review of recent solutions of some dynamic crack growth problems and some comments on limiting velocities for cracks propagating along the interface between two media. It is worth noting that as far as the limiting velocity is concerned it seems that a steady state analysis (the whole field depending only on $x-vt$ instead of $x$ and $t$ ; with $v$ crack speed), if interpreted in a certain way, will give the same result as a genuinely transient analysis which has the crack starting at a certain time and then beginning to grow. At least this is certainly the case in a homogeneous elastic medium where both situations have been considered. For example Craggs [1] considered the steady state motion of a semi-infinite crack and interpreted the limiting crack speed as that speed which required zero applied load to maintain the motion and obtained the Rayleigh wave speed as the limiting velocity for the plane strain case. The same limiting velocity result is obtained when an energy release rate argument is applied to the more physically realistic crack models of Broberg [2] and Baker [3]. A discussion and comparison of the results of these models when applied to a crack in an infinite elastic medium is given by Cotterell [4] and is reviewed by Rice [5]. The model proposed by Broberg, which is a crack of initially zero size which expands with uniform velocity, has the mathematical advantage that the resulting stress field has the property of dynamic similarity mentioned in the introduction. This seems to have been first exploited in this context by Craggs [6]. Since then a number of authors have solved special problems with this property. More recently Willis [7] has given a general treatment of a class of these self-similar problems which includes most of the known solutions (see [7] for earlier references). Most of these known solutions have involved reducing the problem to something like a Hilbert problem either by a suitable change of variables (e.g. the Chaplygin transformation) or via an integral representation, Willis provides a unifying treatment as well as extending the class of problems treated. The difficulty remains, of course, as to whether one can solve the resulting complex variable problem and this in its turn is connected with whether one can solve the equilibrium elasticity problem which results when one applies a limiting process to the expanding crack. For example, in the limit, crack velocity $V$ tending to zero but $Vt \to a$ a constant, for the Broberg crack one gets the equilibrium problem of a crack of length $2a$. One of the new problems that Willis treats is that of a crack expanding along an interface between two anisotropic elastic media but he is not able to give an exact solution for the general case. Recently Atkinson [8] has given an exact solution to the problem of a crack expanding across an interface between two elastic media rather than along it, under some simplifying assumptions, one of which is an overall anti-plane shear deformation. The main result of this calculation is that the limiting velocity appears to be set by the wave speed of the medium in which the crack tip lies, the presence of the interface merely modifying the value of the energy release rate without changing the velocity at which it goes to zero. One disadvantage with these dynamically similar problems is that if one works out the energy release rate following Atkinson and Eshelby [9] or Broberg [10] one finds that $G$ is proportional to $Vt$, the instantaneous length of the crack. Viewing $G = 2\gamma$ as an equation for the crack growth then leads to an inconsistency unless one has associated with the crack tip a Dugdale type region which is chosen to remove the stress singularity at the crack tip see for example the solution given by Atkinson [11] and used by Embley and Sih [12] to investigate dynamic crack motion with a plastic zone. However, because of the dynamic similarity condition required for the solution of these problems there is a limit to how much one can vary the conditions in this end region and for this purpose a semi-infinite crack analysis is more flexible see for example Glennie and Willis [13]. Notwithstanding the above limitations Bergkvist [14] has used the expanding crack solution to give velocity factors which he has used for an approximate analysis of certain experimental configurations and he claims good agreement with experiment.

The model of Baker [3] was originally that of a semi-infinite crack which suddenly appears and begins to propagate at time $t = 0$ with speed $V$. Because of the semi-infinite geometry the analysis can be extended to more complicated material behaviour, linear viscoelastic media for example (Atkinson and List [15]) and media in which the elastic moduli vary with position see section 2. Freund [16] has used Baker's analysis to treat the case when the crack is subjected to a general loading while Achenbach and co-workers have considered the semi-infinite crack model with transient motion produced by stress waves interacting with the tip of a pre-existing crack. In particular Brock and Achenbach [17] have considered the extension of a crack on an interface between two isotropic elastic media, with different moduli, under the influence of transient shear waves. The analysis is complicated but we expect that a similar analysis plus an energy release rate argument at the crack tip would give a limiting crack speed at the lower of the shear wave speeds of the two media. A similar conclusion can be obtained from the much simpler steady state analysis. The plane-strain steady state analysis of a semi-infinite crack propagating along a welded interface between two different elastic half spaces appears to have been given first by Gol'dshtein [18], he finds for the stresses in front of the crack:

$$\sigma_{y_1} = \frac{Q(1+|\lambda_1|)\, s^{-\frac{1}{2}}}{2\pi \ell^{\frac{1}{2}} |\lambda_1|^{\frac{1}{2}}} \cos\left(\frac{1}{2\pi}\log|\lambda_1|\log\frac{s}{\ell}\right)$$

$$\tau_{xy_1} = \frac{Q(1+|\lambda_1|)}{2\pi\ell^{\frac{1}{2}}|\lambda_1|^{\frac{1}{2}}} \left(\frac{p}{q}\right)^{\frac{1}{2}} s^{-\frac{1}{2}} \sin\left(\frac{1}{2\pi}\log|\lambda_1|\log\frac{s}{\ell}\right) \tag{1.1}$$

where the crack tip is at $x=0$, $y=0$, the crack lies on $y=0$ $x<0$, $y=0$ is the interface between the two half-spaces and the above stresses are evaluated at $x=0+s$, $y=0$, $0<s<<1$. The loading on the crack faces is $\sigma_{yj} = -Q\delta(x+\ell)$, $\tau_{xyj} = 0$ $(x\leq 0$, $y=0)$ $j=1$ and $2$ and there is continuity of stresses and displacements across the interface ahead of the crack. For $V<C_{R1}<C_{R2}$ where $C_{Rj}$ is the Rayleigh wave speed associated with medium $j$, $j=1,2$ then

$$\lambda_1 = \frac{k^2+pq}{k^2-pq} + \frac{2|k|(pq)^{\frac{1}{2}}}{|k^2-pq|} \tag{1.2}$$

where $k=A_1-A_2$ , $p=C_1-C_2$ , $q=B_1+B_2$ and
$A_j=2(1-m_j{}^2)^{\frac{1}{2}}(1-n_j{}^2)^{\frac{1}{2}}-(2-n_j{}^2)/\mu_j R_j$ , $B_j=n_j{}^2(1-n_j{}^2)^{\frac{1}{2}}/\mu_j R_j$ ,
$C_j=n_j{}^2(1-m_j{}^2)^{\frac{1}{2}}/\mu_j R_j$ , $R_j=(2-n_j{}^2)^2-4(1-m_j{}^2)^{\frac{1}{2}}(1-n_j{}^2)^{\frac{1}{2}}$ ,
$m_j=V/a_j$ , $N_j=V/b_j$ $j=1,2$ with $\mu_j$ the shear modulus of medium $j$ and $a_j$ and $b_j$ are the dilatational and transverse wave speeds of medium $j$ . We are interested in the behaviour of $\lambda_1$ as $V$ tends to $C_{R1}$ from below. As $V\to C_{R1}$ $(C_{R1}<C_{R2})$ then $k\to -V^2/2b_1{}^2(2-V^2/b_1{}^2)$ and $(1-m_1{}^2)^{\frac{1}{2}}(1-n_1{}^2)^{\frac{1}{2}}\to(2-n_1{}^2)^2/4+L$ where L is a positive constant which tends to zero as $V$ tends to $C_{R1}$ from below. Substituting in the expression for $\lambda_1$ shows that $\lambda_1$ tends to zero as $V\to C_{R1}$ and hence that the limiting velocity will be $V=C_{R1}$ where $C_{R1}$ is the lower of the Rayleigh wave speeds i.e. $C_{R1}<C_{R2}$ . This result differs from the result given by Willis [19] who from a numerical evaluation with particular elastic constants comes to the conclusion that the limiting speed will be fairly close to the smaller Rayleigh wave speed but slightly greater. The above result suggests that an algebraic solution would have given a result exactly equal to the lower Rayleigh speed. The above result also leads one to suspect that a similar result will occur for a crack propagating between two anisotropic half-spaces. For crack velocities greater than the Rayleigh velocities of each medium there are special situations which occur when the velocity coincides with the surface wave velocities obtained under contact and non-slip conditions at the interface, these situations have been discussed by Gol'dshtein [20].

## 2. GENERALISATIONS : INHOMOGENEOUS ELASTIC MEDIA

One possible application of the semi-infinite crack model is to situations where the elastic moduli vary with position. Interest in such media has arisen recently because of possible applications to soil mechanics and geophysics, a third possibility is when the variation of elastic moduli is caused by a temperature variation. A number of calculations of the effect of line and point sources on a half-space of such an inhomogeneous elastic solid have appeared in the geophysical literature (see for example Scott [21] and related references). No calculations in such a medium, to our knowledge, have appeared concerning the rather more difficult mixed boundary value problems usually encountered in fracture mechanics. As a first step in this direction we restrict ourselves to the simplest crack growth situations one might imagine in such media. Our choice of the crack plane and direction will be such that the usual elastic stress singularities which are encountered in fracture mechanics will also occur here. Although it is not difficult to imagine a situation where a crack might propagate in a stratified medium in a direction perpendicular to the stratification we know, by analogy with the situation of a crack perpendicular to and just touching a bimaterial interface, that the stress singularity will be other than $r^{-\frac{1}{2}}$ and the usual fracture mechanics arguments cannot be applied in a straightforward way. The situations considered below will thus be concerned with cracks running in a direction in which the appropriate elastic modulus is constant, the elastic modulus being a function of $z$, where the z axis is perpendicular to the plane of the crack. We shall further restrict ourselves to situations of anti-plane shear and semi-infinite crack geometries for mathematical convenience.

## 2.1 Analysis

The equations appropriate to anti-plane shear motions of an inhomogeneous medium are given by Scott [22] in the following form

$$\frac{\partial^2 u}{\partial x^2} + \frac{\partial^2 u}{\partial z^2} - a(x,z)u = \frac{\rho(x,y,z)}{\mu_o f(x,z)}\frac{\partial^2 u}{\partial t^2} \tag{2.1}$$

where $\mu$ (the shear modulus) does not depend on $y$ and

$\mu = \mu_o f(x,z)$ with $\mu_o$ a constant,

$$u(x,z,t) = u_y(x,z,t)f^{\frac{1}{2}}(x,z)$$

$$a(x,z) = \frac{1}{4f}\left[\frac{2\partial^2 f}{\partial x^2} + \frac{2\partial^2 f}{\partial z^2} - \left(\frac{1}{f}\frac{\partial f}{\partial z}\right)^2 - \left(\frac{1}{f}\frac{\partial f}{\partial x}\right)^2\right] \tag{2.2}$$

and $\sigma_{yz} = \mu_o f^{\frac{1}{2}}\left(\frac{\partial u}{\partial z} - \frac{1}{2f}\frac{\partial f}{\partial z}u\right)$

In the sequel we will assume that the only non-zero stresses are $\sigma_{xy}$ and $\sigma_{zy}$ and the crack lies in the xy plane and grows in the positive x direction, the only non-zero displacement being $u_y$ the displacement in the y direction and the crack motion is produced by anti-plane shear stresses $\sigma_{yz}$ at infinity or related internal stresses applied to the faces of the crack. As they stand equations (2.1) and (2.2) are quite complicated, further simplifications can be made under the following two assumptions:-
i) $\mu$ and $\rho$ vary in the z direction only and they also have the same variation so that the S-H wave speed is constant,

ii) the variation is such that $a(x,z)$ as given by (2.2) is a constant. This latter property holds for a number of different forms for $f(z)$ e.g. (a) $f(z)$ = constant, $a=0$ ; (b) $f(z) = \exp(2\beta z)$ , $\beta$ a real constant, $a=\beta^2$ ; (c) $f(z) = \cosh^2\beta z$, $a=\beta^2$ ; (d) $f(z) = (z/z_o)^2$ , $z_o$ a real constant, $a=0$ . Other possibilities are given in [22].

## 2.2 Steady state motion with a specified internal stress

We consider the steady state motion of a semi-infinite crack with speed $V$ under a specified internal stress. The full time-dependent problem is rather more complicated although the analysis proceeds in a similar way. Making the substitution $x_1 = (x-Vt)/(1-\alpha V^2)^{\frac{1}{2}}$ for co-ordinates moving with the crack tip equation (2.1) becomes

$$\frac{\partial^2 u}{\partial x_1^{\,2}} + \frac{\partial^2 u}{\partial z^2} - au = 0 \tag{2.3}$$

where it is now assumed that $u$ is a function of only $x_1$ and $z$ . If we consider the half-space $z>0$ (it is assumed that there is symmetry about the axis $z=0$ ) the boundary conditions on $z=0$ will be:-

$$\sigma_{yz} = -\sigma(x_1) \quad x_1 < 0 \quad , \quad u_y = 0 \ , \quad x_1 > 0 \quad . \tag{2.4}$$

In (2.4) it is assumed that the internal loading of the crack is such that $u$ depends on only $x_1$ and $z$ . Transforming equation (2.3) with a Fourier transform over $x_1$ reduces it to

$$\frac{d^2\bar{u}}{dz^2} - \gamma^2\bar{u} = 0 \quad \text{with} \quad \gamma^2 = s^2 + a \tag{2.5}$$

where

$$\bar{u} = \frac{1}{(2\pi)^{\frac{1}{2}}}\int_{-\infty}^{\infty} u e^{isx_1}\,dx_1 \quad .$$

The solution of (2.5) which tends to zero as $z\to\infty$ is

$$\bar{u} = A(s)e^{-\gamma z} \tag{2.6}$$

where $a$ is assumed constant and $\gamma = (s^2+a)^{\frac{1}{2}}$ is defined so that the real part of $\gamma$ is positive.

Also, from the relation between $\sigma_{yz}$ and $u$ given in (2.2) one gets

$$\bar{\sigma}_{yz} = \mu_o f^{\frac{1}{2}}\left(\frac{\partial \bar{u}}{\partial z} - \frac{1}{2f}\frac{\partial f}{\partial z}\bar{u}\right) \tag{2.7}$$

provided $f$ depends only on $z$. If we write

$$u = g(x_1),\ x_1<0,\ z=0+ \quad \text{and} \quad \sigma_{yz} = b(x_1),\ x_1>0,\ z=0+ \tag{2.8}$$

then applying the Fourier transform over $x_1$ to $\sigma_{yz}$ and $u$ gives

$$\bar{\sigma}_{yz} = \bar{\sigma}_-(s) + B_+(s) \quad \text{and} \quad \bar{u} = G_- \tag{2.9}$$

where $z = 0+$ and $\bar{\sigma}_-$ is a known function coming from the condition (2.4). $B_+(s)$ and $G_-(s)$ are both unknown functions defined as

$$B_+(s)=(2\pi)^{-\frac{1}{2}}\int_0^{\infty} b e^{isx_1} dx_1 \ ,\ G_-(s)=(2\pi)^{-\frac{1}{2}}\int_{-\infty}^{0} g e^{isx_1} dx_1 \tag{2.10}$$

From (2.4) and (2.6) with $z=0$ one gets $A(s)=G_-$ and from (2.7) and (2.9) we get

$$G_-(s)H(s) = \bar{\sigma}_- + B_+(s) \tag{2.11}$$

where

$$H(s) = -\mu_o f^{\frac{1}{2}}(0)\left(\gamma + \frac{1}{2f(0)}\left(\frac{\partial f}{\partial z}\right)_0\right) \tag{2.12}$$

Equation (2.11) is of the standard Wiener-Hopf type, the subscripts + and - denoting functions which are regular in certain upper and lower regions of the complex s plane. Strictly speaking the above derivation is for a constant, in the more general situation when a is a function of $z$, a similar equation to (2.12) would be obtained with $\gamma$ in $H(s)$ replaced by the corresponding expression for $(-1/\bar{u})(\partial\bar{u}/\partial z)$ evaluated at $z=0$ in the solution of (2.5).

The first step in the solution of equation (2.11) is to factorise $H(s)$ as $H_+H_-$ where the functions $H_+$ and $H_-$ are regular and non-zero in appropriate regions of the complex s plane. Once this is done (2.11) can be rearranged as

$$J(s) = G_-(s)H_-(s) - C_-(s) = \frac{B_+(s)}{H_+(s)} + C_+(s) \tag{2.13}$$

where we have written

$$\frac{\bar{\sigma}_-}{H_+(s)} = C_+ + C_- \tag{2.14}$$

The subscripts + and - denoting certain regions of the complex s plane as before, what these regions are for each case will be specified later. Theorems exist for the formal completion of the above steps (see for example Noble [23] ), the difficulty in practice is to get sufficiently explicit results to provide useful information.

## 2.3 Results

For convenience we choose the internal stress acting on the crack as:-

$$\sigma(x_1) = (2\pi)^{\frac{1}{2}}\sigma e^{\lambda x_1}, \; x_1 < 0, \; \lambda > 0 \tag{2.15}$$

Then $\bar{\sigma}_- = -\sigma/(is+\lambda)$ and the factorisation of $\bar{\sigma}_-/H_+(s)$ can be made by inspection since the pole $s=i\lambda$ is to lie in the + region, the result is

$$C_-(s) = \frac{-\sigma}{(is+\lambda)(H_+(i\lambda)}$$
$$C_+(s) = \frac{-\sigma}{(is+\lambda)}\left(\frac{1}{H_+(s)} - \frac{1}{H_+(i\lambda)}\right) \tag{2.16}$$

It will be shown later that $H_+(s)$ and $H_-(s)$ behave like $s^{\frac{1}{2}}$ as $s\to\infty$ in their respective regions of regularity and this together with the required behaviour of $G_-(s)$ and $B_+(s)$ as $s\to\infty$ (i.e. the behaviour of the stress and displacement near the crack tip) and Liouville's theorem in (2.13) is sufficient to show that $J(s) = 0$ in this crack problem. As a consequence

$$B_+(s) = -C_+(s)H_+(s) \quad \text{and} \quad G_-(s) = \frac{C_-(s)}{H_-(s)} \tag{2.17}$$

Using the relations (2.10) and the behaviour of $B_+(s)$ and $G_-(s)$ as $s\to\infty$ in the respective half-planes of regularity the following asymptotic results can be obtained for the stresses at the crack tip

$$\sigma_{yz} \sim 2^{\frac{1}{2}} \frac{\sigma(x_1)^{-\frac{1}{2}}}{T}$$

$$u \sim 2^{3/2} \frac{\sigma(-x_1)^{\frac{1}{2}}}{\mu_0 f^{\frac{1}{2}}(0)T} \qquad (2.18)$$

the first expression holding for $x_1>0$ , $z=0$ , the second for $x_1<0$ , $z=0$ where $H_+(i\lambda) = i^{\frac{1}{2}}T$ .

## 2.4 Special Cases

(i) $f(z) = \cosh^2\beta z$ then $(\partial f/\partial z)_0 = 0$ and $a=\beta^2$ ; by inspection from (2.12) we can write

$$H_+(s) = (s + i|\beta|)^{\frac{1}{2}}$$

$$H_-(s) = -\mu_0(s - i|\beta|)^{\frac{1}{2}} \qquad (2.19)$$

so that

$$T = (\lambda + |\beta|)^{\frac{1}{2}} \qquad (2.20)$$

and the + and - regions of the complex s plane are $\mathrm{Im}s>-|\beta|$ , $\mathrm{Im}s<|\beta|$ respectively.

(ii) $f(z) = \exp(2\beta z)$ , $a = \beta^2$ , $\frac{1}{2f(0)}(\frac{\partial f}{\partial z})_0 = \beta$ and from (2.12)

$$H(s) = -\mu_0(\gamma+\beta) \quad , \quad \gamma^2 = s^2 + \beta^2 \quad . \qquad (2.21)$$

The factorisation $H = H_+H_-$ is much more difficult in this case and so we give only the final results:-

a) When $|\beta| \leq \lambda$ ,

$$T = \lambda^{\frac{1}{2}}\exp(\pm \frac{1}{\pi}\int_{\cosh^{-1}(\lambda/|\beta|)}^{\infty} \frac{tdt}{\cosh t}) \qquad (2.22)$$

b) When $|\beta| \geq \lambda$ then when

$$\beta>0, \quad T=\lambda^{\frac{1}{2}}\left[|\beta|/\lambda+(\frac{\beta^2}{\lambda^2}-1)^{\frac{1}{2}}\right]^{\frac{1}{2}}\exp\left(\frac{1}{\pi}\int_0^{\sin^{-1}(\frac{\lambda}{|\beta|})}\frac{xdx}{\sin x}\right)$$

and

$$\beta<0, \quad T=\lambda^{\frac{1}{2}}\left[\frac{|\beta|}{\lambda}+(\frac{\beta^2}{\lambda^2}-1)^{\frac{1}{2}}\right]^{-\frac{1}{2}}\exp\left(-\frac{1}{\pi}\int_0^{\sin^{-1}(\frac{\lambda}{|\beta|})}\frac{xdx}{\sin x}\right). \tag{2.23}$$

The results (2.22) and (2.23) being, of course, the same when $|\beta| = \lambda$, the integrals

$$\int_0^\infty \frac{tdt}{\cosh t} \quad \text{and} \quad \int_0^{\pi/2} \frac{xdx}{\sin x}$$

both being equal to $2G$ where $G = 1-\frac{1}{3^2}+\frac{1}{5^2}-\frac{1}{7^2}\ldots = 0.9159656$ see Dwight [24].

## 2.5 Discussion

To interpret the above results we first of all consider the case $V=0$, then $x_1=x$ and the crack is acted upon by the internal stress $\sigma_{yz} = -\sigma e^{\lambda x}$, $x<0$, $z=0$. For fixed $\lambda$ we can compare the stress intensity factors produced for different $\mu = \mu_0 f(z)$. For example, from (2.20) for $f(z) = \cosh^2\beta z$ we see that the stress intensity factor, and hence the crack extension force, decreases as $|\beta|$ increases, and tends to zero as $|\beta|$ tends to infinity; the case $|\beta| = 0$ corresponding to the homogeneous elastic case with constant modulus $\mu = \mu_0$. This conforms with one's intuition if one imagines the crack growing in a thin strip which has shear modulus $\mu = \mu_0$ (constant) and which is surrounded by material of modulus $\mu > \mu_0$. In the extreme case when $|\beta|\to\infty$ this surrounding material is rigid. A similar conclusion applies in example (ii), $f(z) = \exp(2\beta z)$, when $\beta>0$. When $\beta$ is less than zero the opposite conclusion holds, the stress intensity increasing as $|\beta|$ increases ($\beta$ decreases), the extreme case $\beta\to-\infty$ corresponds to a crack in a thin strip of shear modulus $\mu = \mu_0$, the surrounding regions being free space. The stress intensity factor tending to infinity like $|\beta|^{\frac{1}{2}}$ as $\beta\to-\infty$. A simple comparison can also be made between the results for $f(z) = \cosh^2\beta z$ and for $f(z) = \exp(2\beta z)$. Clearly for $\beta$ large $\cosh^2\beta z$ behaves like $e^{2\beta z}/4$ so we might expect a similar form for $T$ in each case when $\beta$ is large although we expect a larger result for $T$ when $f(z) = \exp(2\beta z)$ since this case is giving the larger constraint in the surrounding medium (larger shear modulus) and thus a smaller stress-intensity factor. This is indeed the case, (2.20) gives $T = \lambda^{\frac{1}{2}}(|\beta|/\lambda)^{\frac{1}{2}}$ for $|\beta|>>\lambda$ and (2.23), gives $T = 2^{\frac{1}{2}}\lambda^{\frac{1}{2}}(|\beta|/\lambda)^{\frac{1}{2}}$ .

When $V \neq 0$ for the steady state problem, to make a suitable comparison we require the loading although depending on $x-Vt$ to be otherwise independent of $V$ . Suppose we take
$\sigma_{yz} = -\sigma e^{\lambda_1(x-Vt)}$ for $x<Vt$ , $z=0$ on the faces of the crack where $\lambda_1$ is a constant independent of $V$ , then formulae (2.18) to (2.23) will still hold with $\lambda$ replaced by $\lambda_1(1-\alpha V^2)^{\frac{1}{2}}$ . Clearly these results are just a first step in any crack tip stress analysis in media with a varying elastic modulus and the results given in (2.22) and (2.23) although non-trivial and holding for the whole range of $\beta$ for the law $\mu(z) = \mu_0\exp(2\beta z)$ have been possible because of the simplifications indicated at the beginning of the section. Even with these simplifications however the corresponding problems for these media are rather more complicated than their homogeneous elastic counterparts. For example even for equilibrium cracks (no-inertia) equation (2.3) has the Helmholtz equation character which means anti-plane crack problems will be like wave diffraction problems or the corresponding transient time-dependent problems in homogeneous elasticity. Specifically for $f(z) = \cosh^2\beta z$ and a finite crack the problem is just like the wave diffraction problem for a finite slit and this generally requires numerical analysis for solutions valid for all $\beta$ . Whereas for $f(z) = \exp(2\beta z)$ , the fact that $(1/2f)\,(\partial f/\partial z)$ is not zero at $z=0$ in (2.7) makes the situation a good deal worse. Further a semi-infinite anti-plane crack on an interface between two media, one with $\mu_1 = \mu_0\cosh^2\beta_1 z$ and a second with $\mu_2 = \mu_0\cosh^2\beta_2 z$ will be just like the transient problem of a semi-infinite anti-plane crack between two homogeneous elastic solids treated in [17] , again only results for small (or large) $\beta_1$ and $\beta_2$ can be obtained in anything like a manageable form. The more realistic $\mu = \mu_0\exp(2\beta z)$ case is appreciably more difficult.

## REFERENCES


1 Craggs, J.W., On the propagation of a crack in an elastic-brittle material, J. Mech. Phys. Solids, 8, 66, 1960.

2 Broberg, K.B., The propagation of a brittle crack, Arkiv för Fysik, 18, 159, 1960.

3 Baker, B.R., Dynamic stresses created by a moving crack, J. Appl. Mechs., 29, 49, 1962.

4 Cotterell, B., On the nature of moving cracks, J. Appl. Mech., 31, 12, 1964.

5 Rice, J.R., Mathematical Analysis in the mechanics of fracture, in Fracture Vol. II, Liebowitz, H., Ed., Academic Press, New York, 1968.

6 Craggs, J.W., Fracture criteria for use in continuum mechanics, in Fracture of Solids, Drucker, D.L. and Gilman, J.J., Eds., Interscience publishers Inc., New York, 1963, 51.

7 Willis, J.R., Self-similar problems in elastodynamics, Phil. Trans. Roy. Soc., 274, 435, 1973.

8 Atkinson, C., On the dynamic stress and displacement field associated with a crack propagating across the interface between two media, Int. J. Eng. Sci. (In the press).

9 Atkinson, C. and Eshelby, J.D., The energy flow into the tip of a moving crack, Int. Jnl. Frac. Mechs., 4, 308, 1968.

10 Broberg, B.R., On the speed of a brittle crack, J. Appl. Mech., 31, 546, 1964.

11 Atkinson, C., A simple model of a relaxed expanding crack, Arkiv fur Fysik, 35, 469, 1968.

12 Embley, G.T. and Sih, G.C., Plastic flow around an expanding crack, Engng. Frac. Mechs., 4, 431, 1972.

13 Glennie, E.B. and Willis, J.R., An examination of the effects of some idealised models of fracture on accelerating cracks, J. Mech. Phys. Solids, 19, 11, 1971.

14 Bergkvist, H., The motion of a brittle crack, J. Mech. Phys. Solids, 21, 229, 1973.

15 Atkinson, C. and List, R.D., A moving crack problem in a viscoelastic solid, Int. J. Engng. Sci., 10, 309, 1972.

16 Freund, L.B., Crack propagation in an elastic solid subjected to general loading, J. Mech. Phys. Solids, 20, 129, 1972.

17 Brock, L.M. and Achenbach, J.D., Extension of an interface flaw under the influence of transient waves, Int. J. Solids and Structures, 9, 53, 1973.

18 Gol'dshtein, R.V., On the steady motion of a crack along a straight line boundary between two joined materials, Inzh. Zh. M.T.T., 5, 93, 1966.

19 Willis, J.R., Fracture mechanics of interfacial cracks, J. Mech. Phys. Solids, 19, 1, 1971.

20 Gol'dshtein, R.V., On surface waves in joined elastic materials and their relation to crack propagation along the junction, P.M.M. (Applied Maths and Mechanics), 31, 468, 1967.

21 Scott, R.A., Transient elastic waves in an inhomogeneous layer, Bull. Scis. Soc. Am., 60, 383, 1970.

22 Scott, R.A., Line loads moving on the surface of an inhomogeneous elastic half space, Appl. Sci. Res., 21, 356, 1969.

23 Noble, B., Methods based on the Wiener-Hopf technique, Pergamon Press, 1958.

24 Dwight, H.B., Tables of integrals and other mathematical data, 4th Edition, Macmillan, 1961, P.216 and 233.

# Section V

# STRUCTURAL ANALYSIS

# FRACTURE MECHANICS AND SHIPS

J.J.W. Nibbering

Ghent University, Belgium/Delft University, The Netherlands

## SUMMARY

The amount of welding in ships is so large that defects can never be completely avoided. These defects may grow as a consequence of the cyclic loading a ship is subjected to. Mostly this does not endanger the situation because often the quality of the material at the tip of an extended defect will be better than that at the original tip.

Thus the problem of fracture control in ships is more a matter of significance of defects in a metallurgical sense than in a geometrical sense. In other words: *location* of defects and cracks is of greater importance than *size*. A small defect at a weld-crossing may be more dangerous than a large crack elsewhere.

Apart from these considerations it should be realized that ships are redundant stiffened plate structures. This has the following consequences:

a. Brittle fractures, once initiated, may propagate to large lengths; 100% safety can only be obtained by using crack-arresting steels.
b. Relatively large slowly developed cracks do hardly impair the strength of a ship as long as the involved reduction in sectional area is not too large. The steels used are so tough that crack length is hardly critical.

It is obvious that limited room is left for usefully applying linear elastic fracture mechanics in commercial ships, except in the field of fatigue crack growth. General yielding fracture mechanics are somewhat more valuable. Then crack opening displacement

(COD) is a sufficiently accurate parameter for static loading conditions. Whether these are sufficiently representative for ship loadings is doubtful. In connection to brittle fracture propagation and arrest there is a need for methods of analysis for welded plate-stiffener assemblies. The input information will mostly be given in terms of Charpy-test results.

## INTRODUCTION

It is not so much a freak of nature, but nevertheless pityful that in the field where fracture has once played a most dramatic role - viz. in merchant shipbuilding - the scope for useful applications of fracture mechanics has remained limited. The main reason is that welding and flame cutting may cause large variations in fracture toughness which are much more important in connection to danger of fracturing than crack length.

A consequence is that a relatively rude fracture mechanics analysis will suffice. For this and other reasons COD is considered to be an appropriate parameter as far as static loading is concerned.

Another consequence is that realistic fracture toughness testing is required which is expensive. Due to that, the other way being relatively simple testing combined with an intelligent evaluation and interpretation of the results is often preferred.

Then the difficulty is, that for less gifted people, or rather people in want of time these interpretations should be translated into simple specifications. But then the advantages of simple testing are often lost. Either unsafe or oversafe structures may be the result. For, the number of steels electrodes and welding processes used in shipyards is really large.

At first sight it may seem that in the above mentioned interpretation of results of simple tests and subsequent translation into specifications, fracture mechanics can be of great help.

Even when it would be true, many difficulties will be met. Some of these will be discussed with the aid of specific examples.

## CRITICAL CRACK LENGTHS

From a fracture-mechanics point of view a ship is a very heterogeneous structure. An extreme case is demonstrated in Figs. (1) and (2). Figure (1) shows differences in grain size in the so-called Heat-Affected Zone (HAZ) of a welded ship plate. Figure (2) gives the corresponding fracture-toughness variations in terms of

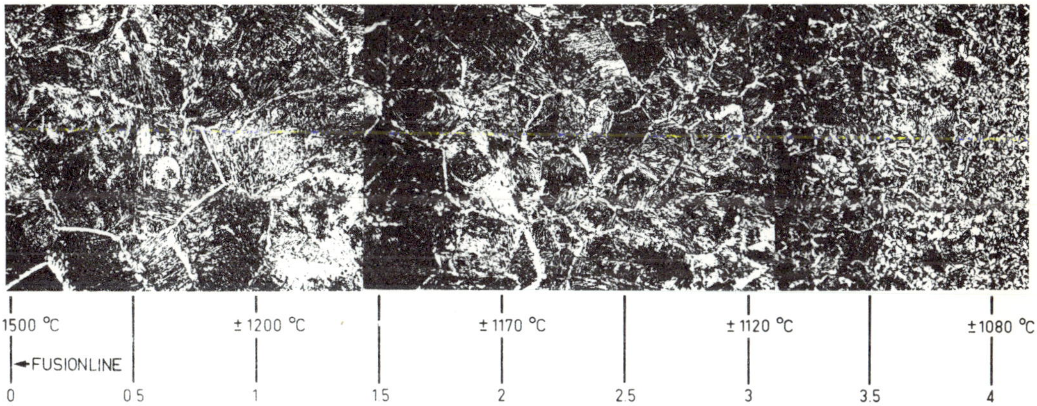


Fig. 1. Extreme grain coarsening in the Heat-Affected Zone of an Electrogas-welded fine grain steel.

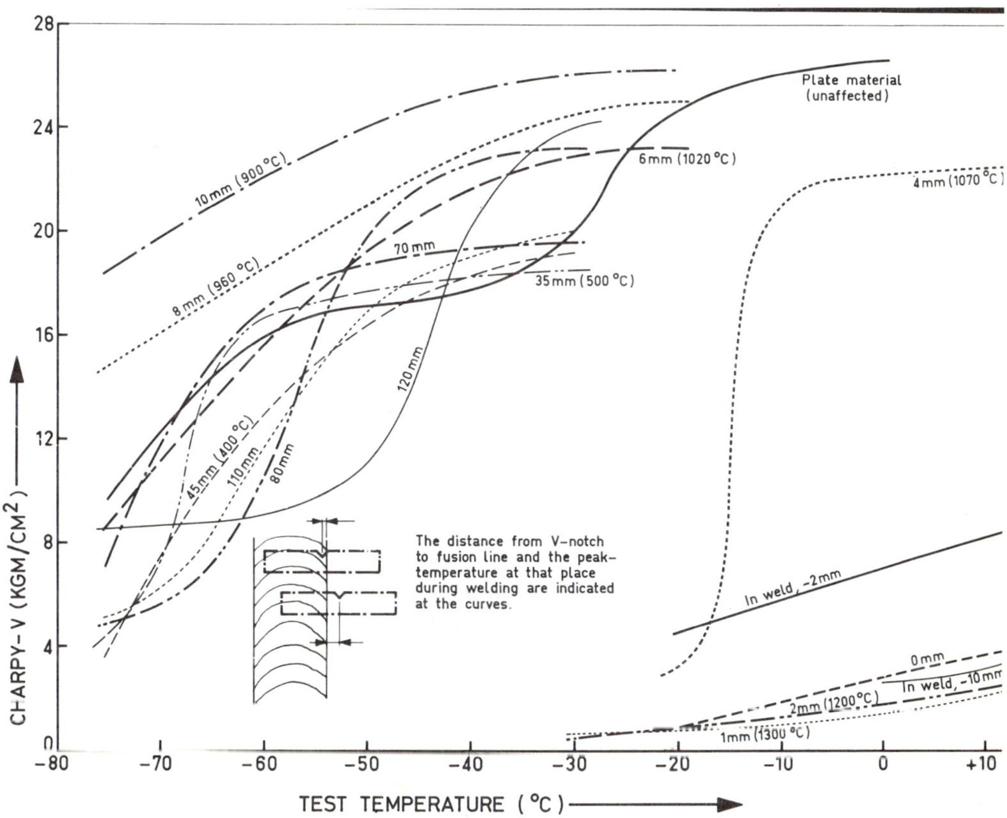


Fig. 2. Charpy-V-values of the HAZ of Fig. (1). (St. 52 + Nb; 34 mm.).

Charpy energy values. They are caused by the high heat-input of the welding process concerned (Electrogas or Electroslag). These processes are used in order to be able to make vertical welds in thick plates in one pass, which is very attractive from a cost point of view. But it has caused that in shipbuilding the problem of brittle fracture is still real. Indeed the competition in shipbuilding is so heavy that the border line between safe and unsafe is often approached when cost reductions can be obtained. Such a situation needs not be alarming except when one does not know precisely where that border line is situated. Unfortunately the latter is often true and - undoubtedly to the disappointment of the audience - the coming of fracture mechanics has not very much improved that state of affairs. Figure (3) may illustrate this. The full lines represent fracture toughnesses at various temperatures as a function of crack tip *position* (for convenience in terms of $K_{Ic}$, although only partly valid).

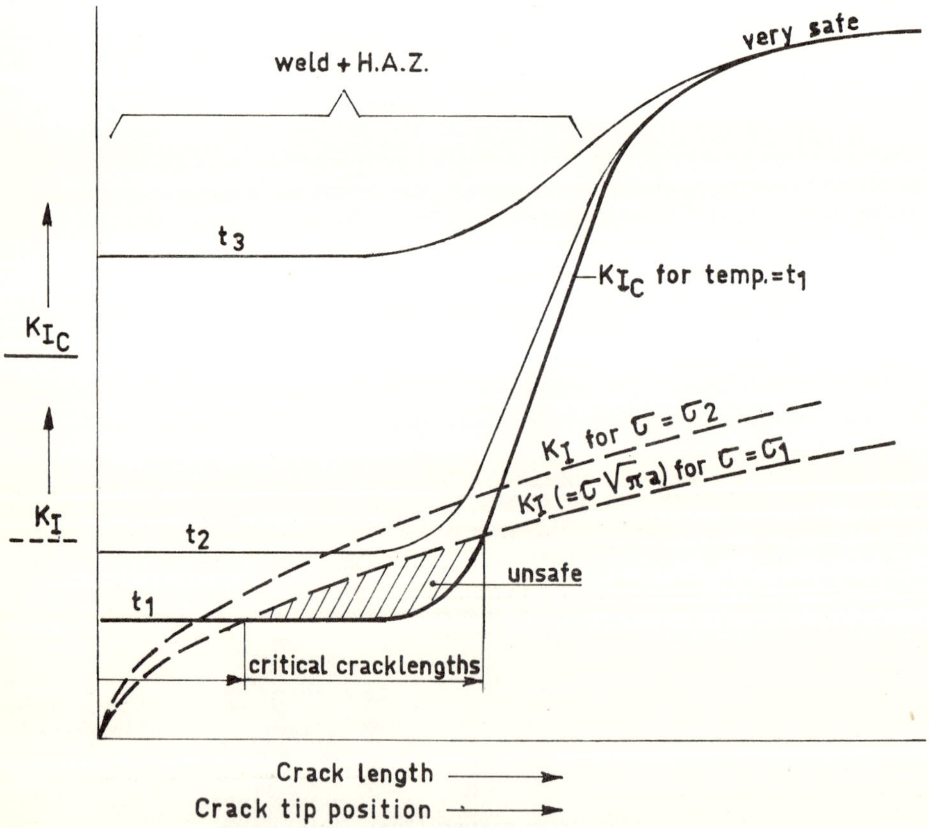


Fig. 3. Relative importance of crack length and crack-tip position in a welded region.

The dotted lines are stress intensities $K_I$ as a function of crack length. It can be seen that the danger of fracture is far more determined by crack tip position than by crack length. It can even be said that a crack of a few mm. in length may be more dangerous than a crack of hundredths of millimeters in length.

The example represents that of a transverse crack started at a defect in a longitudinal weld and grown under the influence of the wave-induced cyclic loading.

The figure gives rise to mention an often overlooked aspect in fracture analysis, viz. the statistical side. Brittle fracture is, fortunately, an extremely rare phenomenon in ships. It can only occur when several unfavourable factors coincide: high loading, low temperature, bad structural detail and *critical defect length*. Critical should now be interpreted in the sense of Fig. (3).

The main point of Fig. (3) remains that *metallurgical* effects throw into the shade the effect of crack size. On the other hand without detrimental influences of high heat-input, or hydrogen or rapid cooling etc. even large cracks would be harmless in ships.

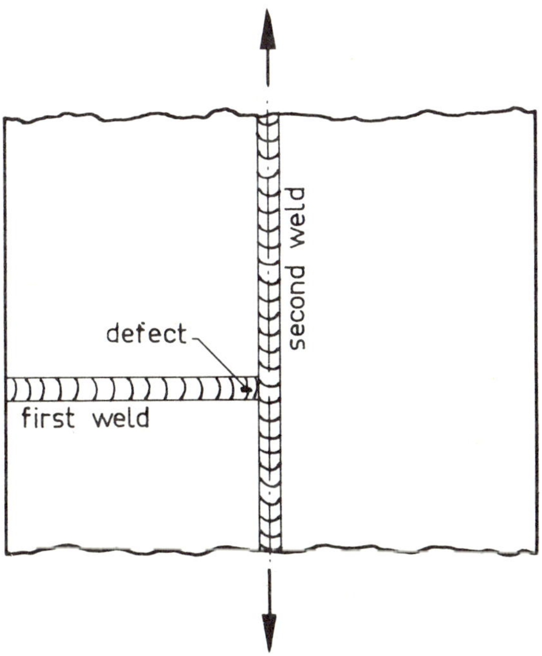


Fig. 4. Classic case of embrittlement at weld crossing.

This is mainly a consequence of the well-known temperature dependency of steel. We still speak in shipbuilding about transition temperatures, knowing very well that for instance under static loading conditions they differ appreciably from those for dynamic loading conditions. For modern shipbuilding steels both transition temperatures are so much lower than winter service temperatures that brittle fractures cannot initiate in as-delivered steel. Only by welding and flame cutting these transition temperatures can be raised to dangerous values.

As for ships most relevant and extremely dangerous case of embrittlement due to welding is possible at weld crossings. (Not only those indicated in Fig. (4) but also crossings of butt welds with stiffener welds). This case corresponds to the classic one developed by Wells [2] on the basis of experimental work by Greene [1]. The defect in the first weld of Fig. (4) may be so severely hot-strained when welding the second one that a rise in transition temperature of some 60°C is quite possible at the tip of the defect. Again position of the original defect is far more critical than defect length.

## CRACK-SAFE DESIGN

From the foregoing section it will be clear that it is difficult to build a ship without any potential starting point for a brittle fracture.

Of course one may avoid high-heat-input welding and one may control non-destructively every weld-crossing and use the best of electrodes etc. But it is expensive and it is doubtful whether it is really necessary. For as said before, most of the ship steels in use nowadays are of high quality. Especially the D- and E-grades used for the extreme fibers of the ship hull are generally so good that in 9 out of 10 cases any brittle crack started in a welded region will be arrested.

Then why not rely upon that property? In fact shipbuilders do so, but they use it rather as a second barrier. In the actual situation this means wise policy for not all steels conforming to D and E grade requirements are capable of arresting brittle fractures. The cause is that the acceptance test used, viz. the Charpy V-notch test gives only poor indications with respect to crack-arrest capability. An example is the steel of Fig. (2), (full thick lines), which is even at -70°C extremely good from a Charpy point of view. Yet its isothermal Robertson crack-arrest temperature is only -25°C.

It remains difficult to understand why Classification Societies do not require real crack-arrest tests for huge structures like ships in which thousands of tons of steel are used.

There is another feature of ship structural design which is responsible for the dual-barrier policy with respect to fracture. One prerequisite for a design based on the crack-arrest properties of steels, is that any brittle crack, once initiated in a welded region, should immediately leave that region under the influence of the residual welding stresses. This indeed has occurred nearly always in the past. But it is doubtful whether it will also always happen in modern ships. In these ships the nominal stresses are much higher than in older ships - partly due to the use of higher strength steels. But the residual stresses have not much changed. The result will be that after some time the high nominal stresses due to seaway will have reduced the residual stresses greatly. Experiments [3] have proved that then in transverse welds fractures may propagate all along the welded region. In case of high heat input welding the situation is even worse because the gradient of the residual stresses is smaller than in case of low heat-input welding. The effect on crack path is shown in Fig. (5).

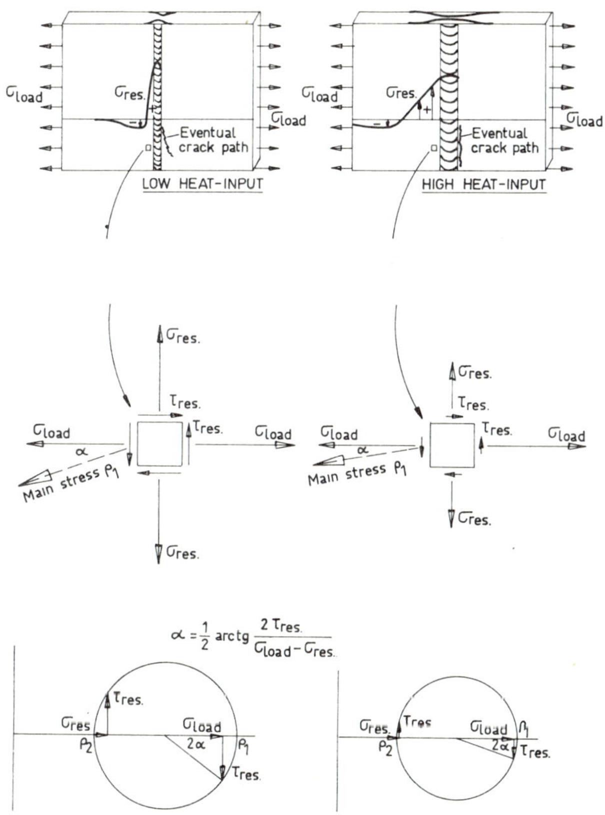


Fig. 5. Residual stress field in case of welding with low and with high heat-input.

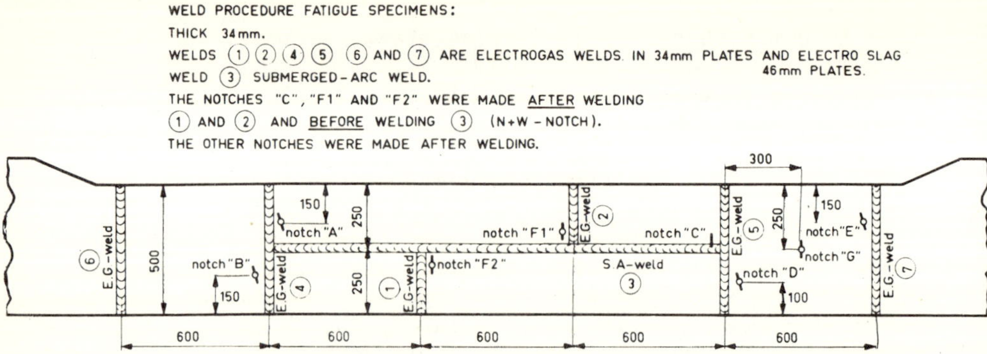


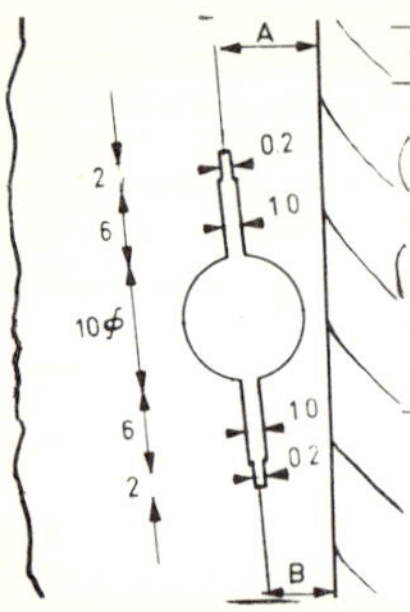


| | A | B |
|---|---|---|
| NOTCH "1A" | 3 | 1 |
| (NOTCH "1B" not reported) | 6 | 4 |
| NOTCH "1D" | 3 | 1 |
| (NOTCH "1E" not reported) | 6 | 6 |
| NOTCH "1F1" | 1 | 1 |
| NOTCH "1F2" | 3 | 3 |

Fig. 6. Specimen fatigue-loaded at reduced temperature in a 1000 tons testing machine.

Bearing in mind the poor properties of the Heat-Affected Zones of the welds concerned, it will be clear that it is worthwhile to carry out experiments in order to investigate whether brittle fractures can really propagate in zones as narrow as a few mm. The specimens tested in the Delft Ship Structures Laboratory were of the type of Fig. (6), [4].

The fractures are shown in Fig. (7). It is indeed astonishing that they remained completely within the 3 mm. wide worst part of the HAZ. A few mm. deviation would have sufficed for arresting, see Fig. (2).

The interesting problem for the audience may be, whether a running brittle crack indeed prefers to propagate in a zone of minimum fracture toughness. Such a thought is more or less supported by a former investigation by van Elst and Verbraak [5]. It was proved that brittle cracks propagate by forming small cracks ahead of the main one and subsequent cracking of the intermediate bridge. It may be that the small cracks easiest develop in the weakest part of the region ahead of the main crack.

## FATIGUE-CRACKS AND BRITTLE FRACTURE

Ships are cyclically loaded structures and as such suffer from fatigue. When critical defects (in the sense of Figs. (3) and (4)) are present in ships, fatigue may aggravate the situation by *sharpening* the defect and further *damageing* of the crack-tip metal.

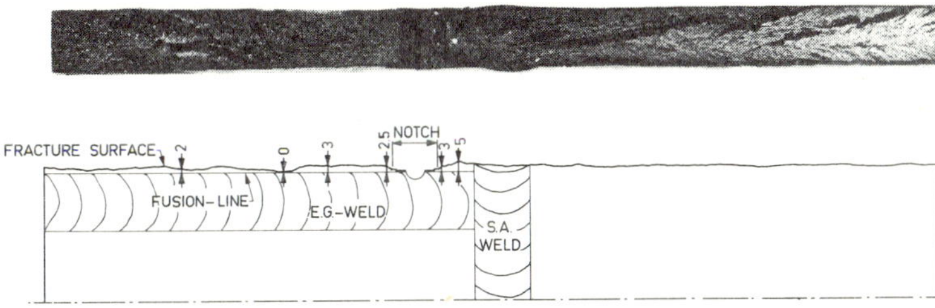


FRACTURE PLANE
SPECIMEN. 2
NOTCH. 2F1

FIG. Nº.

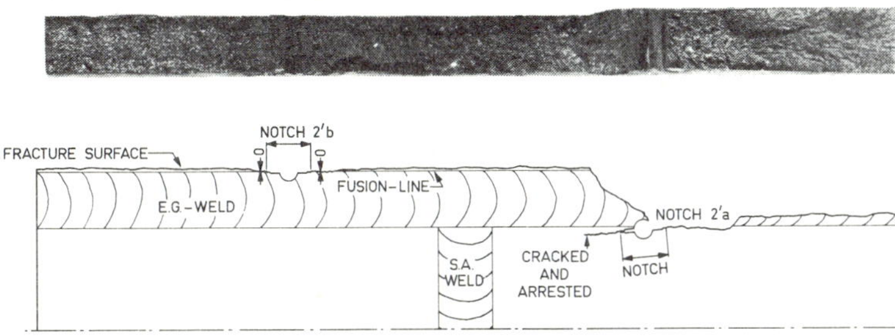


FRACTURE PLANE
SPECIMEN. 2'
NOTCH. 2'a AND 2'b

FIG. Nº.

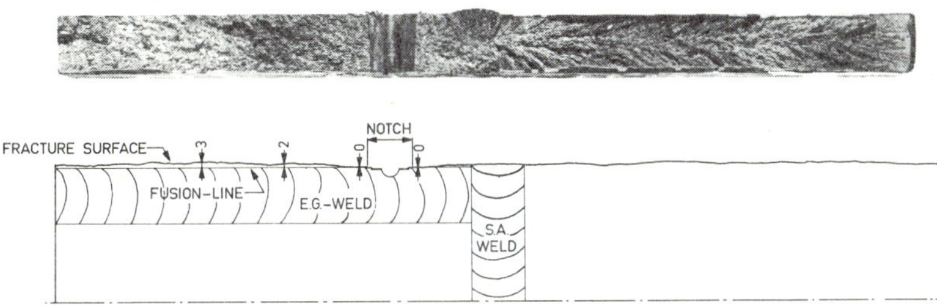


FRACTURE PLANE
SPECIMEN. 12
NOTCH. 2'F2 (12)

FIG. Nº.

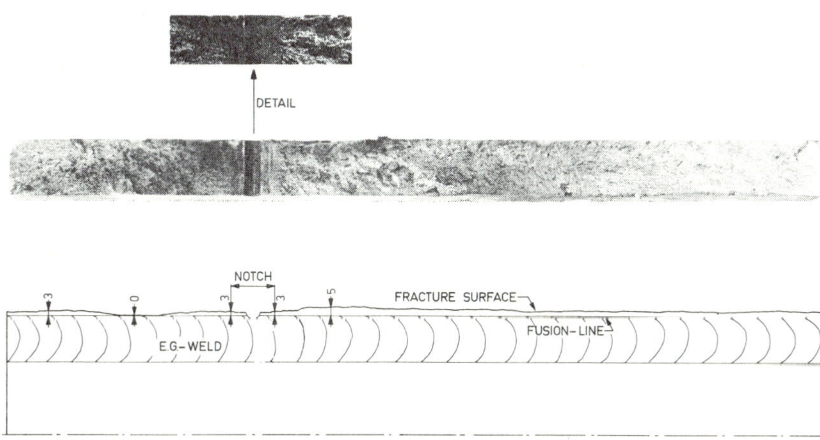


FRACTURE PLANE
SPECIMEN 12'
NOTCH. 2e (12')

FIG. Nº.

But when the - mostly small - critical region has been left, the situation becomes quite safe. In general fatigue cracks do show the same tendency as brittle cracks in connection to crack path. Mostly they deviate from the welded region under the influence of residual welding stresses. But the exceptions are more numerous than in brittle fracturing, because fatigue cracks are often guided along undercuts of butt and fillet welds.

There is another unpleasant factor that adds to the possibility of crack initiation at fatigue cracks. The high sharpness of those cracks makes the crack-tip-metal extremely sensitive to high speed loading, (in case of metals which are highly strain-rate sensitive like ships' steels). The point is illustrated in Fig. (8).

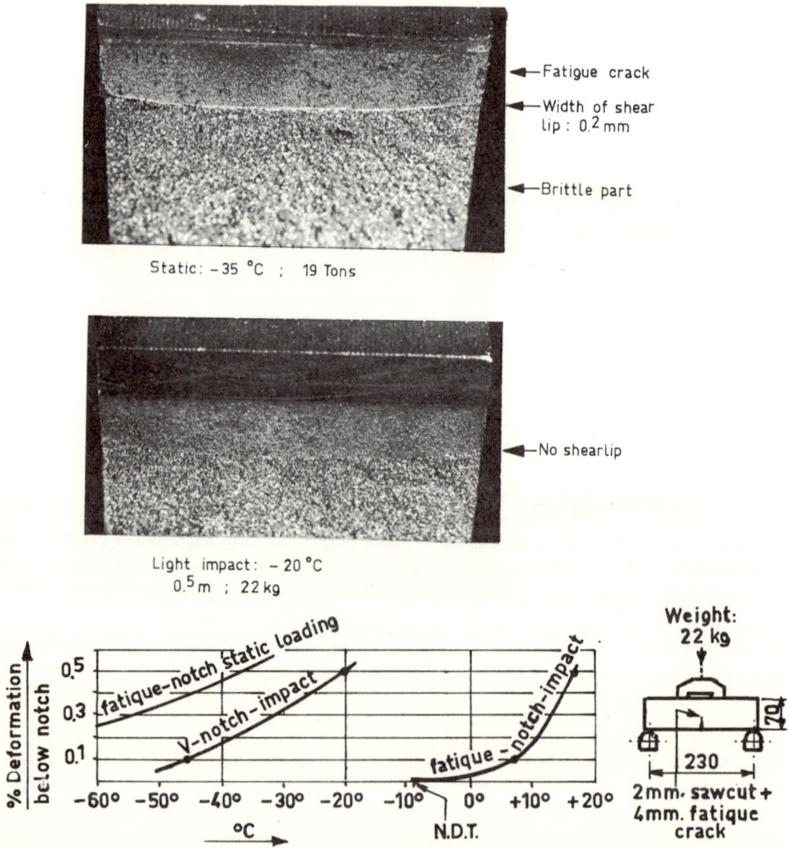


Fig. 8. Difference in cracking behaviour between specimens loaded statically and dynamically. Influence of notch acuity.

From the experiments it could be concluded that strain rates corresponding to about 10.000 Hz. were required for suppressing any plastic deformation at the crack tip below about $-10^{\circ}$C. Frequencies like that cannot be caused by sea-induced loads, not even when the forebottom of a ship is slamming on the sea-surface. The only possibility is that accidental cargo-drops or impacts of rolling hatch-covers against supports and similar phenomena occur close to fatigue cracks. Then brittle fractures may develop, but only when the tip of the fatigue crack is situated in an embrittled part of the structure. Of course the general stress level must also be high, and the temperature low.

## CONCLUSIONS

It will be clear that the possibilities for applying fracture mechanics in merchant ship design are limited. The main causes are: the influence of production methods, the temperature and strain rate sensitivity of steel, the complexity of a ship's structure, the presence of residual stresses and the combination of in-plane and on-plane loadings.

Besides a ship must be able to undergo appreciable plastic deformations in *extreme* conditions. The corresponding loading is sufficiently slow to permit a static approach. Then the COD-parameter is quite adequate and Wells' plastic zone correction [6] sufficiently accurate. It has been experimentally proved by Kanazawa and others [7] that COD is only slightly dependent on crack length. This allows prescribing *one* critical COD-value for a large range of crack lengths, f.i. 0.30 mm.

It has been discussed that static testing cannot represent all possible loads a ship's structure may meet in practice. Fatigue cracks may be particularly dangerous in case of impacts. Apart from that, Fig. (4) gives an example for which 100% safety with respect to crack initiation cannot be guaranteed. Should a fracture start, then the adjacent metal should be capable of arresting a *short* running crack; not a long one.

A test in which that situation has been simulated is Pellini's Drop Weight Test [8]. The same test represents well conditions in which accidental impact loading causes a small crack run. The test unfortunately cannot be applied for any case. For, the brittle crack-starting-weld may favourably influence any underlying HAZ or weld metal to be tested. Moreover a maximum thickness exists. Therefore a test is sought in which the essence of Pellini's test is maintained.

Another possibility is the use of small size specimens like Charpy-types, preferably with extra sharp notches in order to

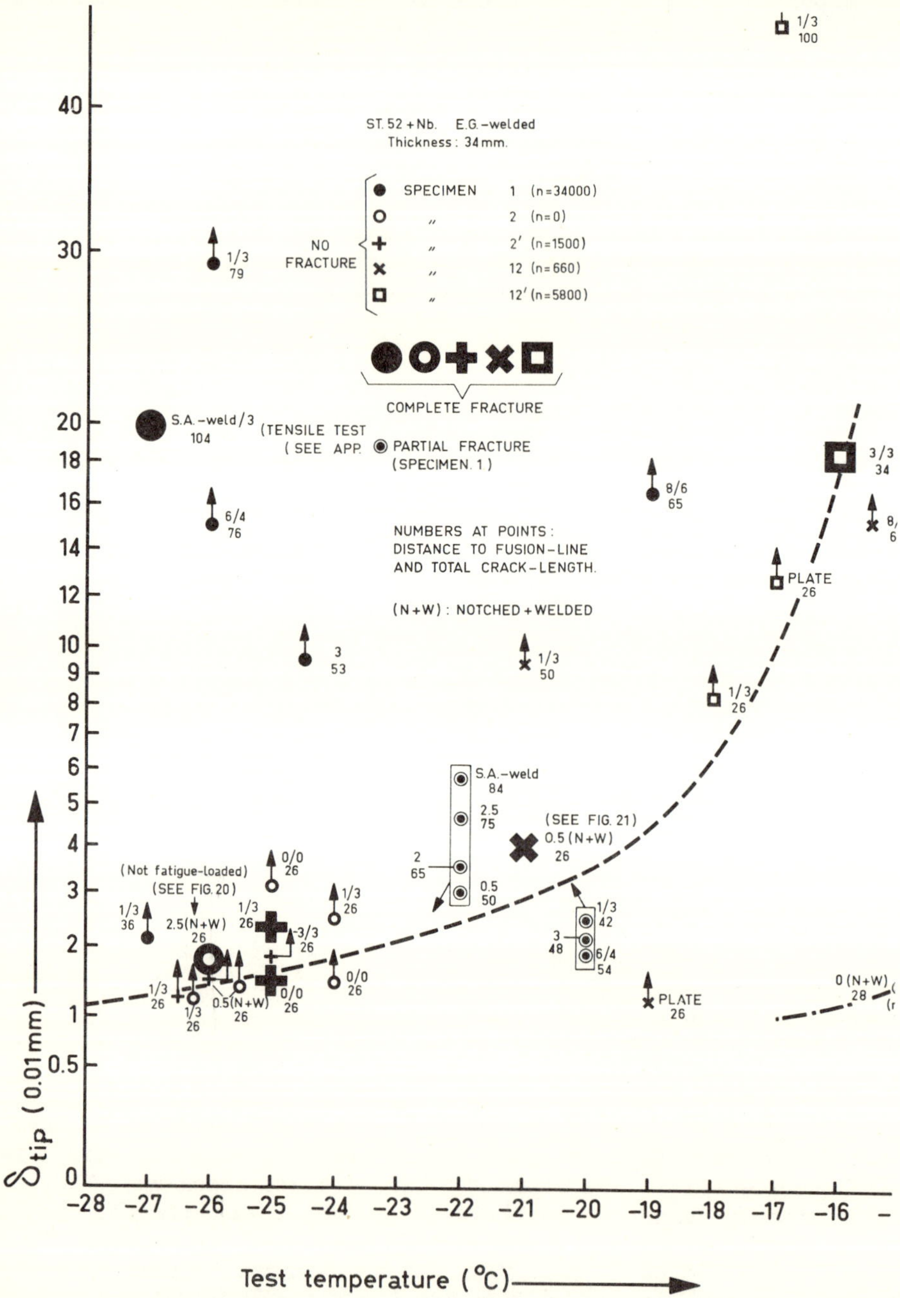


Fig. 9. Deformations at the notch's tips at the moment fracture started.

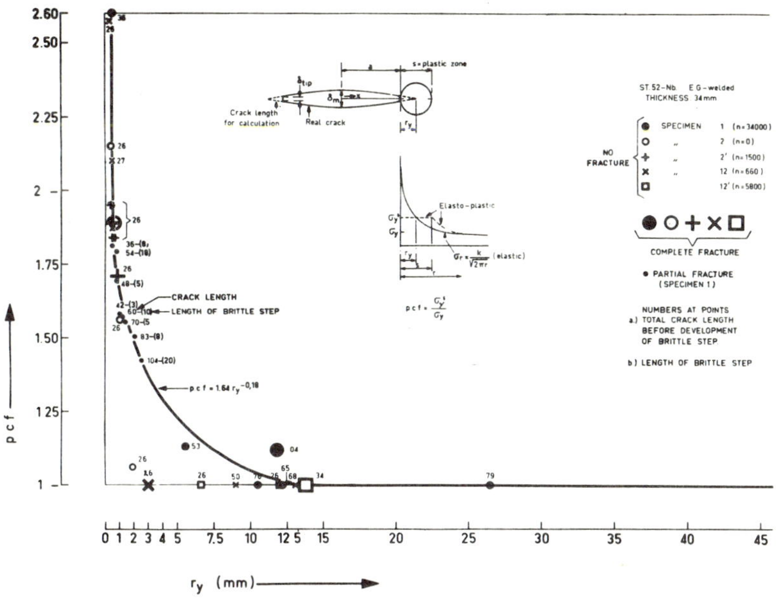


Fig. 10. Relation between plastic constraint factor and $r_y$.
Length of brittle steps compared with plastic zone width.

limit initiation energy. But the fracture surface should be included in the analysis and proper allowance for the difference between bending and axial loading and the reduced size (plate thickness) should be made.

The results of the experiments with the specimens of Fig. (6) illustrate well a number of the foregoing observations. From Fig. (9) it is obvious how well COD serves as a parameter in presenting results of static tests with fatigue-loaded specimens to fracture. Below -20$^{o}$C partial and complete fractures developed practically without prior plastic deformation at the notches' tips. The nominal stresses were significantly below yield point (except of course the result of S.A.-weld/3). The mentioned results were typical plane strain fractures, which could be verified by a simple calculation. (From $\delta_m$, $r_y$ can be calculated and from it the local yield point $\sigma_y'$, see Fig. (10)). It can also be seen that at $r_y$-values as small as half plate thickness, a fully plane stress condition already existed in contrast to the often supposed value $r_y \geqq t$.

The partial fractures which had formed during fatigue loading had a length equal to or greater than $2r_y$. These so-called brittle steps which initiated in the fatigue-embrittled plastic zone material could arrest outside that zone partly as a consequence of the small size of that zone.

Finally in Fig. (9) the importance of crack tip position as compared to crack length is obvious.

REFERENCES


[1] Greene, T.W., Welding Journal, vol. 28, p. 193-S, 1949.

[2] Wells, A.A., Brittle fracture strength of welded steel plates, British Welding Journal, May 1961.

[3] van den Blink, W.P., Dechaene, R., Nibbering, J.J.W., Soete, W., Investigations into the strength of automatically made welds containing notches, I.I.W.-doc. X-507-69.

[4] Nibbering, J.J.W., Lalleman, A.W., Low-cycle fatigue tests at low temperature with E.G.-welded 34 mm. plates, I.I.W.-doc. X-593-70.

[5] van Elst, H.C., Intermittent propagation of brittle fracture in steel, Trans. of Met. Soc. of A.I.M.E., vol. 230, April 1964.

[6] Wells, A.A., Application of fracture mechanics at and beyond general yielding, British Welding Journal, Nov. 1963.

[7] Kanazawa, T., Machida, S., Miyata, T., Hagiwara, Y., A consideration on the variables included in the critical COD value in brittle fracture problems of steel, Int. Conf. Mech. Prop. of Metals, Kyoto, April 1971.

[8] Pellini, W.S., Correlations of brittle fracture service failures with laboratory notch-ductility tests, Welding Journal, vol. 37, no. 9, Sept. 1958.

# ANALYSIS OF ENERGY QUANTITIES FOR FRACTURE UNDER BIAXIAL STRESSES

Samuel H. Smith

Battelle's Columbus Laboratories
505 King Avenue
Columbus, Ohio 43201
The United States of America

## ABSTRACT

This paper describes the fracture crack extension process in terms of analyzed energy release quantities. Biaxial boundary stresses, loads, or displacements imposed on the cracked element and the fracture strain energy release rate were derived for the states of plane strain and plane stress. Elastic strain energy release rate parameters are derived for three conditions of remotely applied in-phase load and displacements which can occur during crack extension. Applications of the analysis to stable crack extension under steady-state and fatigue conditions are discussed. Also the influence of biaxial applied stresses on crack tip plastic deformation is discussed.

## INTRODUCTION

Most engineering structures are subjected to combined loadings which produce biaxial tension or compression stresses on a structural element. In order to perform a crack growth or residual strength analysis of the structural component, the crack extension behavior under biaxial applied conditions is needed.

Recent fracture mechanics technology has been directed at analyses and testing conditions for uniaxial applied stress, load, or displacement conditions. The purpose of this analytical study was to review and expand the strain energy release rate concept of crack extension for biaxial applied stress, load, and displacement conditions.

The three applied boundary conditions which can be applied and cause crack extension were analyzed. These were (1) constant load or stress and changing displacement; (2) constant displacement and changing load; and (3) changing load and displacement, with biaxial boundary conditions considered in each case.

In addition to elastic strain energy release rate considerations, the effect of biaxiality on crack tip plastic zone size was derived. Discussion of the influence of crack tip plastic deformation on fracture energy is presented.

## ANALYTICAL STRESS FUNCTIONS

A literature search has provided two stress functions which can handle the stress analysis of a plate subjected to biaxial remote stress. The two stress functions which were proposed are those of Westergaard [1] and Irwin [2].

The Westergaard stress function, $Z(\xi)$, which solves for the stresses everywhere in an infinite plate subjected to equal biaxial remote stresses, is:

$$Z(\xi) = \sigma \left[\frac{\xi^2}{\xi^2 - a^2}\right] \tag{1}$$

where $\xi$ is the complex variable x + iy.

A modified stress function proposed by Irwin for the near field stress solution is:

$$Z(\xi) = \frac{K}{\sqrt{2\pi(\xi - a)}} \tag{2}$$

where K is the well-known elastic-stress-intensity factor of fracture mechanics.

It seemed appropriate to compare the two stress functions with respect to $\sigma_y$ stress distribution along the line of expected crack extension. Figure (1) shows this comparison for uniaxial loading. The upper curve for the Westergaard solution shows load balance across the net section, whereas the Irwin solution agrees with Westergaard only near the crack tip and drops below $\sigma_y/\sigma = 1.0$ far from the crack. The underlying assumption of linear elastic fracture mechanics is that the strength of the structure is determined by the singularity solution of K at the crack tip. However, this approach seems very unlikely from an energy viewpoint since boundary conditions and the entire structural body contribute to

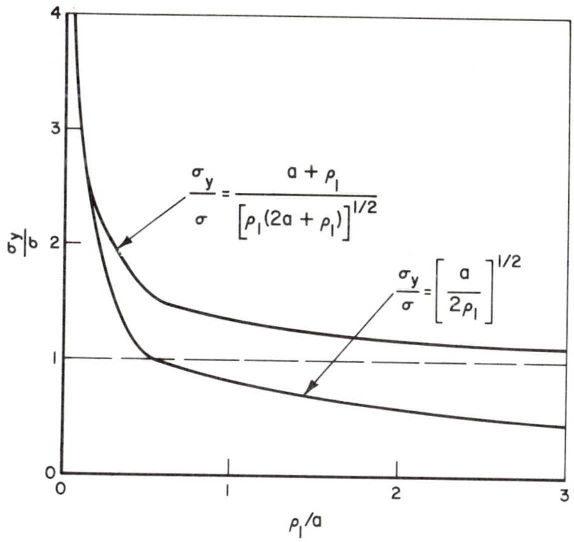


FIGURE 1. DISTRIBUTION OF $\sigma_y$ STRESS ($y = 0$) - UNIAXIAL STRESS

the strain energy that is released during fracture crack extension. Therefore, the Westergaard solution was considered for energy quantity analysis.

## ELASTIC STRAIN ENERGY OF DISTORTION AND VOLUME CHANGE

Timoshenko [3] considered an energy approach to structural strength. The hypothesis was that in a structural member the total energy associated with deformation, $E_T$, is the elastic strain energy due to change in volume, $E_V$, and the elastic strain energy of distortion, $E_D$. These energy quantities were derived by substitution of the derived strain and stress components based on Eq. (1). The techniques of complex variable analysis were used in deriving the strain and stress components. Uniform biaxial stresses are the assumed boundary conditions on a center-cracked element and the energy quantities are per unit volume of material.

For a plane strain-stress state, $E_T$ and $E_D$ are given as follows:

$$E_T = \frac{\sigma^2}{E}\left\{\left[1-\nu+2\nu^2(C-2)\right]\left[\left(\frac{\rho_1}{\rho_1+2a}\right)^{1/2} + \frac{a}{[\rho_1(\rho_1+2a)]^{1/2}}\right]^2 + \right.$$

$$\left. + (5\nu^2-1)(1-C)\left[\left(\frac{\rho_1}{\rho_1+2a}\right)^{1/2} + \frac{a}{[\rho_1(\rho_1+2a)]^{1/2}}\right] + (1-\nu^2)(1-C)^2\right\} \tag{3}$$

$$E_D = \frac{\sigma^2}{3E}\left\{\left(1-3\nu+2\nu^2(C-1)+4\nu^3\right)\left[\left(\frac{\rho_1}{\rho_1+2a}\right)^{1/2}+\frac{a}{[\rho_1(\rho_1+2a)]^{1/2}}\right]^2\right.$$

$$-(1-C)(1-9\nu^2+4\nu^3)\left[\left(\frac{\rho_1}{\rho_1+2a}\right)^{1/2}+\frac{a}{[\rho_1(\rho_1+2a)]^{1/2}}\right] \tag{4}$$

$$\left.+(1-C)^2(1-3\nu^2-\nu^3)\right\}$$

For a plane stress-stress state, $E_T$ and $E_D$ are given as follows:

$$E_T = \frac{\sigma^2}{E}(1-\nu)\left\{\left[\left(\frac{\rho_1}{\rho_1+2a}\right)^{1/2}+\frac{a}{[\rho_1(\rho_1+2a)]^{1/2}}\right]-(1-C)\cdot\right.$$

$$\left.\left[\left(\frac{\rho_1}{\rho_1+2a}\right)^{1/2}+\frac{a}{[\rho_1(\rho_1+2a)]^{1/2}}\right]\right\}+\frac{\sigma^2}{2E}(1-C)^2 \tag{5}$$

$$E_D = \frac{\sigma^2}{3E}(1+\nu)\left\{\left[\left(\frac{\rho_1}{\rho_1+2a}\right)^{1/2}+\frac{a}{[\rho_1(\rho_1+2a)]^{1/2}}\right]^2-(1-C)\cdot\right.$$

$$\left.\left[\left(\frac{\rho_1}{\rho_1+2a}\right)^{1/2}+\frac{a}{[\rho_1(\rho_1+2a)]^{1/2}}\right]+(1-C)^2\right\} \tag{6}$$

The elastic strain energy associated with volume change is merely $E_T - E_D$.

In considering the unbroken ligament of the biaxially stressed plate where the energy density is the largest, the strain energy release rate quantities per unit crack extension, d2a, for plane-strain and plane-stress conditions are as follows:

For plane strain,

$$\frac{dE_T}{d2a} = \frac{\sigma^2}{2E}\left(\frac{t}{w}\right)\left(\frac{2a}{w}\right)\left[[1-\nu+(2C-4)\nu^2]\left(\frac{1}{1-\left(\frac{2a}{w}\right)^2}\right)^2+\right.$$

$$\left.+(5\nu^2-1)(1-C)\left(\frac{1}{1-\left(\frac{2a}{w}\right)^2}\right)^{3/2}\right] \tag{7}$$

$$\frac{dE_D}{d2a} = \frac{\sigma^2}{6E}\left(\frac{t}{w}\right)\left(\frac{2a}{w}\right)\left[\left[1-3\nu+4\nu^3-6(1-C)\nu^2\right]\left(\frac{1}{1-\left(\frac{2a}{w}\right)^2}\right)^2 - (1-9\nu^2+4\nu^3)(1-C)\left(\frac{1}{1-\left(\frac{2a}{w}\right)^2}\right)^{3/2}\right] \tag{8}$$

For plane stress,

$$\frac{dE_T}{d2a} = \frac{\sigma^2}{2E}\left(\frac{t}{w}\right)\left(\frac{2a}{w}\right)\left[\frac{1-\nu}{\left[1-\left(\frac{2a}{w}\right)^2\right]^2} + \frac{(\nu-1)(1-C)}{2\left[1-\left(\frac{2a}{w}\right)^2\right]^{3/2}}\right] \tag{9}$$

$$\frac{dE_D}{d2a} = \frac{\sigma^2}{6E}\left(\frac{t}{w}\right)\left(\frac{2a}{w}\right)(1+\nu)\left[\frac{1}{\left[1-\left(\frac{2a}{w}\right)^2\right]^2} + \left(\frac{1-C}{2}\right)\left(\frac{1}{1-\left(\frac{2a}{w}\right)^2}\right)^{3/2}\right] \tag{10}$$

Figures (2) and (3) show the total strain energy release rate parametrically plotted for various degrees of applied biaxial tension and compression stresses. The plane strain-strain energy release rates peak at 2a/w ratios in a range of 0.3 to 0.5. Plane stress-strain energy release rate behavior, as shown in Fig. (3), shows a steadily increasing release rate with increasing crack size. Biaxial tension, as compared to uniaxial tension, shows an increase in strain energy release with increase in degree of biaxiality; whereas, biaxial compression applied parallel to the crack results in a decrease in strain energy with increase in biaxiality. An expansion of the energy scale in the small crack region for plane stress will show the same effects of biaxiality.

## ELASTIC STRAIN ENERGY RELEASE RATE ANALYSIS

In structural analysis methods, the Law of Conservation of Energy is used to describe the energy equilibrium of the structure; that is, the external work done by boundary forces is equal to the internal strain energy of the structure. However, in the case of a structural member with a crack that is extending, an unbalance of the energy equilibrium occurs. Irwin [2] has defined this unbalance of energy as the elastic energy released during crack extension. This basic relationship is:

$$G = \frac{PdV}{dA} - \frac{dE_S}{dA} \tag{11}$$

where G is elastic strain energy release, PdV/dA is boundary force work, and $dE_S/dA$ is internal strain energy release. All quantities are per unit area of crack extension. This approach to describing

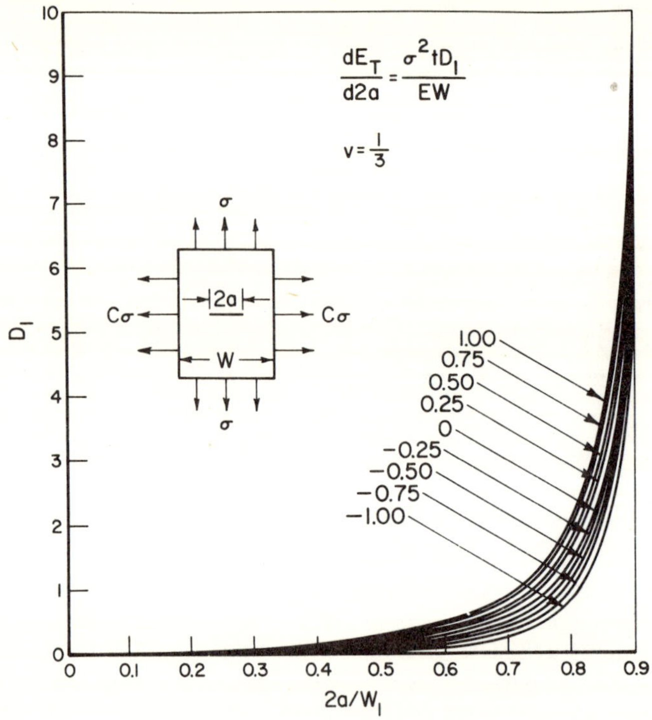


FIGURE 2. PLANE STRESS TOTAL STRAIN ENERGY RELEASE RATE

the fracture crack extension process can be applied to a stable slowly extending crack where the dynamic kinetic energy, thermal energy, and inertia can be neglected.

The application of the elastic strain energy release concept to a structural plate subjected to biaxially applied pin loadings (Fig. (4)) is described in the following sections of this paper. The mechanical compliance approach and Westergaard's displacement equations were used to derive the plane strain, $G_{IS}$, and plane stress, $G_S$, strain energy release quantities. These strain energy release quantities were derived for three boundary situations imposed on the biaxially loaded plate which can occur during the extension of the crack. This continuum mechanics approach is based on results of the author [4].

<u>Constant Load and Changing Displacement</u>. This applied load-displacement situation is applicable to crack growth under steady-state-load application.

Newman [5] has shown by boundary collocation analysis that under applied constant biaxial load or stress the degree of biaxiality influences the elastic stress-intensity factor. Grandt [6],

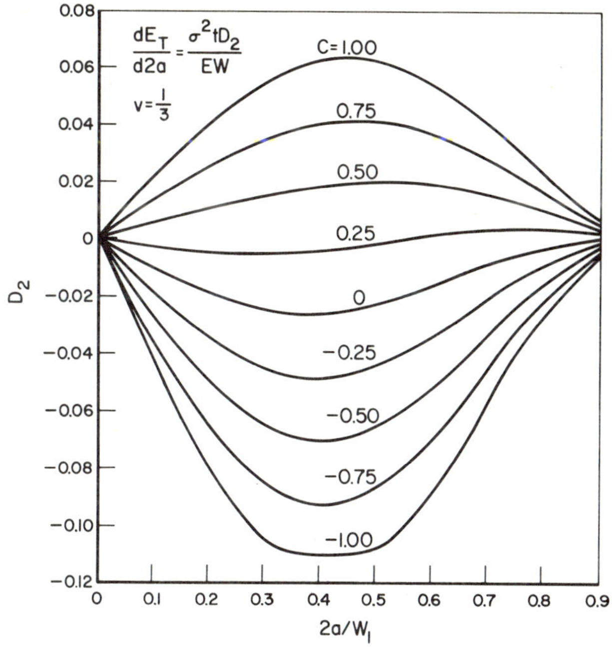


FIGURE 3. PLANE STRAIN TOTAL STRAIN ENERGY RELEASE RATE

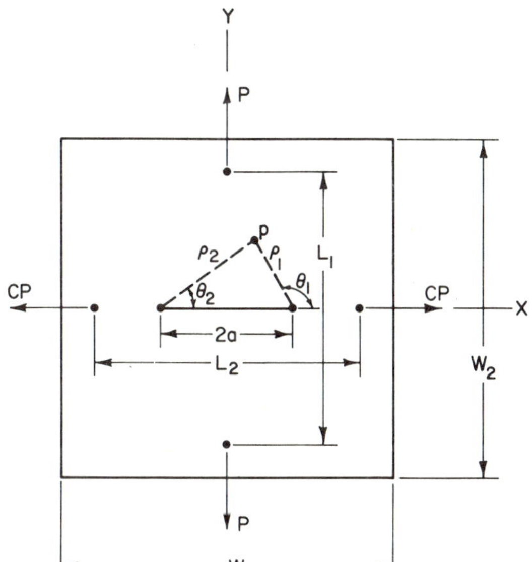


FIGURE 4. PLATE SUBJECTED TO BIAXIAL LOADS

in treating the crack surface as "$\sigma_y$ stress free", has shown similar influences for cracks emanating from open holes and subjected to constant applied stresses. Both of these analytical approaches have shown that a higher stress intensity results from lateral compression stresses applied parallel to the crack as compared to uniaxial tension or biaxial tension-tension stresses.

Paris and Sih [7] in their extensive treatise of the linear elastic fracture problem have shown that in-plane combined loadings can be treated by opening and edge-sliding stress intensity factors, $K_I$ and $K_{II}$.

The mechanical compliance approach shows that for constant applied uniaxial loading the boundary force work and internal strain energy release during crack extension are:

$$\frac{PdV}{dA} = P^2 \left(\frac{d\lambda}{dA}\right) \tag{12}$$

$$\frac{dE_S}{dA} = \frac{P^2}{2} \left(\frac{d\lambda}{dA}\right) \tag{13}$$

where $\lambda$ is the inverse spring constant. Therefore, the G quantity is:

$$G = \frac{P^2}{2} \left(\frac{d\lambda}{dA}\right) = \frac{P}{2} \left(\frac{dV}{dA}\right) \tag{14}$$

Thus, the G value is calculated by differentiating the vertical displacement with respect to the crack area.

The application of a biaxial lateral force will alter the mechanical compliance and thus the G quantity. By applying the original strain energy release approach, the lateral force will effect the internal strain energy release quantity. Equation (14) then becomes:

$$G = \frac{P}{2} \left(\frac{dV}{dA}\right) + CP \left(\frac{dU}{dA}\right) \tag{15}$$

Differentiating the vertical and horizontal displacement equations with respect to crack length, a; the G equations for applied constant load are derived in terms of panel geometry. For plane strain and plane stress conditions, the G equations are respectively:

$$G_{IS} = \frac{2(1+\nu)P^2}{W_1Et^2} \left(\frac{2a}{L_1}\right)\left[(1-\nu)\left(1+\left(\frac{2a}{L_1}\right)^2\right)^{-1/2} - \frac{1}{2}\left(1+\left(\frac{2a}{L_1}\right)^2\right)^{-3/2}\right] - \frac{1}{2}\,\frac{(1-\nu^2)(CP)^2}{W_2Et^2} \tag{16}$$

$$G_S = \frac{2P^2}{W_1Et^2}\left(\frac{2a}{L_1}\right)\left[\left(1+\left(\frac{2a}{L_1}\right)^2\right)^{-1/2} + \left(\frac{1+\nu}{2}\right)\left(1+\left(\frac{2a}{L_1}\right)^2\right)^{-3/2}\right] - \frac{1}{2}\,\frac{(CP)^2}{W_2Et^2} \quad (17)$$

The above equations have been plotted parametrically for two plate geometries. Figure 5 is for a square plate where $L_1 = W_1 = W_2$ whereas, Fig. (6) is for a long rectangular plate with $L_1 = 3W_1$ and $W_2 = L_1$. Biaxial conditions are plotted for C = -1.0, -0.5, 0, +0.5, and +1.0. The $G_{IS}$ and $G_S$ energy release rate quantities are evaluated in terms of $A_1$ and $A_2$ which are parameters that are geometry dependent and vary with crack length.

For the square plate, the $G_{IS}$ value increases to a peak value and then decreases and becomes negative for long crack lengths. The $G_S$ behavior shows a steady increasing value with increase in crack length. A paradox exists for the effect of biaxiality since the $A_1$ and $A_2$ values are equivalent for positive and negative values. However, upon examination of Eqs. (16) and (17), the constant load applied parallel to the crack acts as a constraint during crack extension, but is independent of crack length.

When one considers the stress state that exists for a slowly extending crack, the $G_{IS}$ and $G_S$ behaviors are considered as follows. For small 2a/w, the mode of fracture is predominantly plane strain. As the crack extends, the mode of fracture changes to mixed mode and

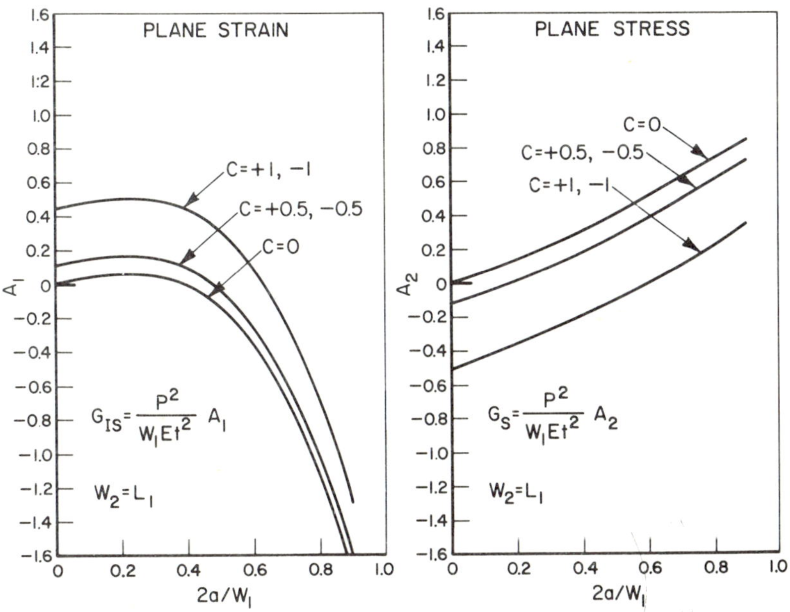


FIGURE 5. $G_{IS}$ AND $G_S$ FOR APPLIED CONSTANT LOAD AND CHANGING DISPLACEMENT, $L_1 = W_1$

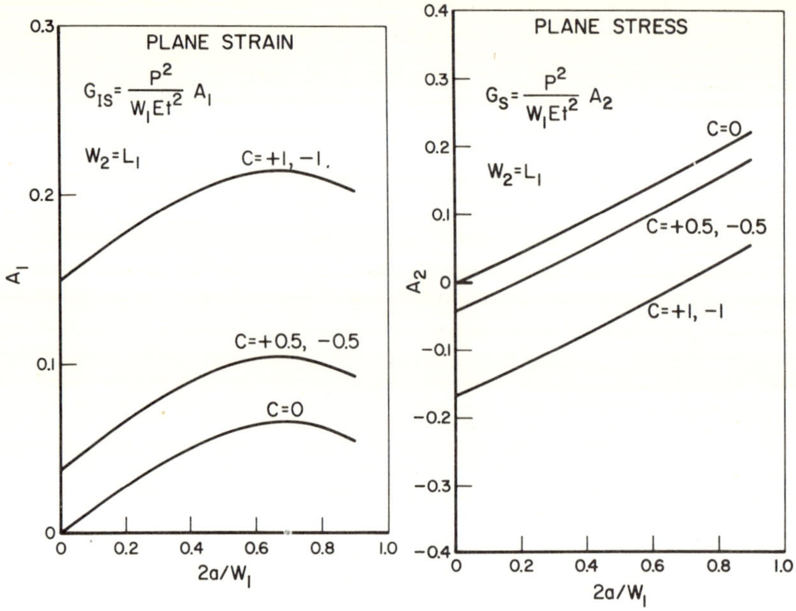


FIGURE 6. $G_{IS}$ AND $G_S$ FOR APPLIED CONSTANT LOAD AND CHANGING DISPLACEMENT, $L_1 = 3W_1$

then predominately plane stress. Therefore, a suggested application of the constant load $G_{IS}$ and $G_S$ quantities is $G_{IS}$ for small $2a/w_1$ and $G_S$ for large $2a/w_1$.

The results for the rectangular plate, shown in Fig. (6), demonstrate the same general trend of behavior as for the square plate. However, the peak value for $G_{IS}$ is shifted to larger $2a/w_1$ and resulting in only positive $A_1$.

Constant Displacement and Changing Load. This applied displacement-load situation is applicable to crack growth under steady state remote displacement application. The elastic strain energy G quantities for this imposed boundary condition are derived similarly to the previous approach. The mechanical compliance approach results in the following strain energy release equation for constant applied vertical displacement:

$$G = -2V\left(\frac{dP}{dA}\right) \tag{18}$$

The constraint of a constant lateral displacement alters the above G equation to:

$$G = -2V\left(\frac{dP}{dA}\right) - 2U\left(\frac{dCP}{dA}\right) \tag{19}$$

Differentiation of the vertical and horizontal displacements and considering plane strain and plane stress conditions results in the following G quantities:

$$G_{IS} = - \left\{ \frac{2(1+\nu)P}{Et^2} \left(\frac{L_1}{W_1}\right) \left[ (1-\nu)\left(1+\left(\frac{2a}{L_1}\right)^2\right)^{1/2} - \frac{1}{2}\left(1+\left(\frac{2a}{L_1}\right)^2\right)^{-1/2} + \right.\right.$$

$$\left.\left. + \frac{\nu}{2}\right]\right\} 2\left(\frac{dP}{d2a}\right) - \frac{2(1-\nu^2)}{W_2 Et} CPa\left(\frac{d(CP)}{d2a}\right) \qquad (20)$$

$$G_S = - \left\{ \frac{2P}{Et^2} \left(\frac{L_1}{W_1}\right) \left[ \left(1+\left(\frac{2a}{L_1}\right)^2\right)^{1/2} - \left(\frac{1-\nu}{2}\right)\left(1+\left(\frac{2a}{L_1}\right)^2\right)^{-1/2} + \right.\right.$$

$$\left.\left. + \frac{\nu}{2}\right]\right\} 2\left(\frac{dP}{d2a}\right) - \frac{2CPa}{W_2 Et^2} \frac{d(CP)}{d2a} \qquad (21)$$

Parametric charts of the above equations are shown in Figs. (7) and (8) for the two plate geometries described in the previous section. The $G_{IS}$ and $G_S$ energy release rate quantities are evaluated in terms of the parameters $B_1$ and $B_2$ for plane strain and plane stress conditions respectively. The $G_{IS}$ and $G_S$ quantities will be positive since in order to maintain the constancy of the displacements during crack extension the applied loading will decrease with crack extension resulting in dP/d2a of a negative slope.

The effects of biaxial applied loading is quite evident for these boundary conditions. As a reference, consider the uniaxial applied loading situation. The application of tensile loading applied parallel to the crack will cause a higher strain energy release rate. Whereas, compression loading will lower the strain energy release rate. In comparing Figs. (7) and (8), the effect of biaxiality is decreased for the longer panel length ratio. Also, the strain energy release quantities increase with increasing panel length.

Bluhm [8] and Heyer [9] have extensively discussed the strain energy release rate behavior with crack size for applied constant uniaxial load or constant deflection (displacement) testing conditions. Bluhm, in his extensive discussion of the G behavior with increasing crack size for several fracture testing specimen configurations, argues that under constant load the condition for possible crack arrest due to a drop in G with increasing crack size is not possible. However, the applied constant displacement boundary condition for edge cracked specimen configurations sets up decreasing G with increasing crack length. Heyer, in his analysis of the crack line wedge loaded specimen, has shown the same results as Bluhm in considering either constant load or constant displacement as applied boundary conditions.

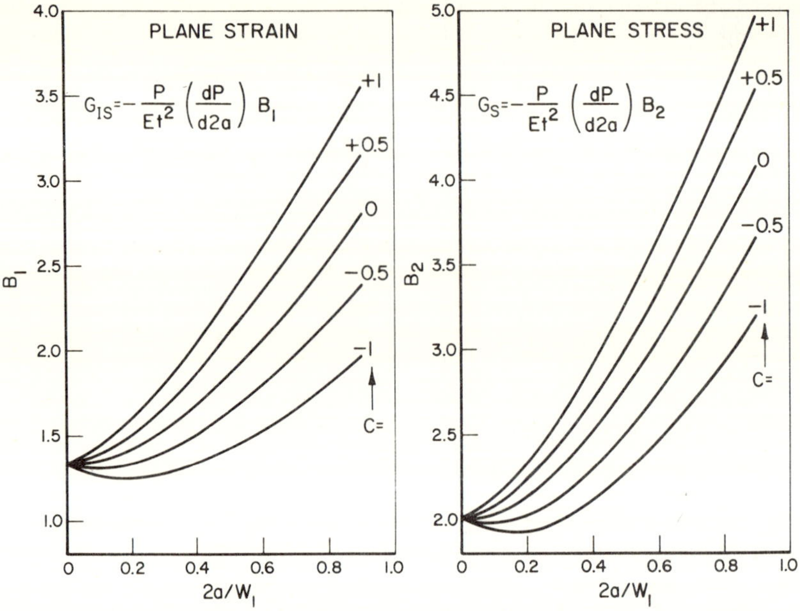


FIGURE 7. $G_{IS}$ AND $G_S$ FOR APPLIED CONSTANT DISPLACEMENT AND CHANGING LOAD, $L_1 = W_1$

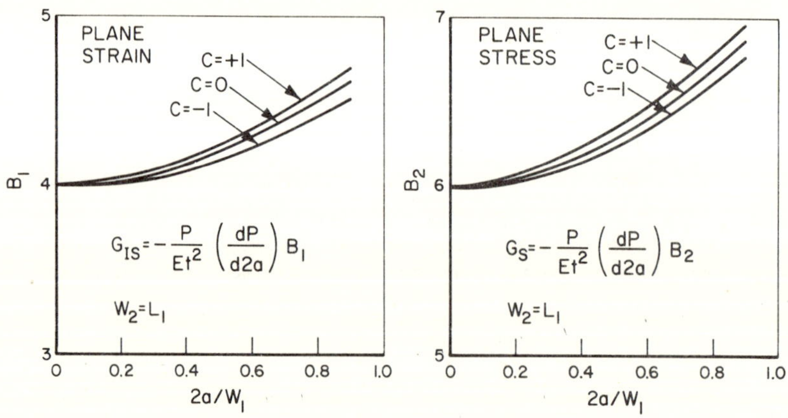


FIGURE 8. $G_{IS}$ AND $G_S$ FOR APPLIED CONSTANT DISPLACEMENT AND CHANGING LOAD, $L_1 = 3W_1$

Changing Load and Changing Displacement. These physical boundary conditions are apropos to crack extension during cyclic fatigue loading.

The mechanical compliance as applied to this boundary results in the following G relationship for uniaxial applied loading:

$$G = \frac{PdV}{dA} - 2V\frac{dP}{dA} \tag{22}$$

A lateral changing loading alters the above equation to:

$$G = \frac{PdV}{dA} - 2V\frac{dP}{dA} - 2U\frac{dCP}{dA} \tag{23}$$

Again, differentiating the displacement equations results in the following plane strain and plane stress $G_{IS}$ and $G_S$ strain energy release quantities:

$$G_{IS} = \frac{4(1+\nu)P}{Et^2}\left\{\left(\frac{2a}{L_1}\right)\left(\frac{P}{W_1}\right)\left[(1-\nu)\left(1+\left(\frac{2a}{L_1}\right)^2\right)^{1/2} + \frac{1}{2}\left(1+\left(\frac{2a}{L_1}\right)^2\right)^{-3/2}\right]\right.$$
$$\left. - \frac{dP}{d2a}\left(\frac{L_1}{W_1}\right)\left[(1-\nu)\left(1+\left(\frac{2a}{L_1}\right)^2\right)^{1/2} - \frac{1}{2}\left(1+\left(\frac{2a}{L_1}\right)^2\right)^{-1/2} + \frac{\nu}{2}\right]\right\} \tag{24}$$
$$- \frac{2(1-\nu^2)CPa}{W_2Et^2}\left(\frac{d(CP)}{d2a}\right)$$

$$G_S = \frac{4P}{Et^2}\left\{\left(\frac{2a}{L_1}\right)\left(\frac{P}{W_1}\right)\left[\left(1+\left(\frac{2a}{L_1}\right)^2\right)^{-1/2} + \left(\frac{1+\nu}{2}\right)\left(1+\left(\frac{2a}{L_1}\right)^2\right)^{-3/2}\right]\right.$$
$$\left. - \frac{dP}{d2a}\left(\frac{L_1}{W_1}\right)\left[\left(1+\left(\frac{2a}{L_1}\right)^2\right)^{1/2} - \left(\frac{1+\nu}{2}\right)\left(1+\left(\frac{2a}{L_1}\right)^2\right)^{-1/2} + \frac{\nu}{2}\right]\right\} \tag{25}$$
$$- \frac{2CPa}{W_2Et^2}\left(\frac{d(CP)}{d2a}\right)$$

Parametric representation of these strain energy quantities are presented in Figs. (9) through (12). The $G_{IS}$ and $G_S$ strain energy quantities are comprised of two terms associated with change of load with crack extension dP/d2a and the applied load P. The influence of biaxiality with respect to lateral tension and compression loading is similar to the constant displacement situation. Parametric charts for the two plate sizes are shown. The longer plate resulting in larger $G_{IS}$ and $G_S$ quantities than the square plate.

Liu [10] and Toor [11] have shown that biaxial in-plane loading influences the fatigue crack growth rate and residual strength of 7075-T6 aluminum alloy. Their analysis of reported data from shear panel testing was based on combined $K_I$ and $K_{II}$ fracture modes based

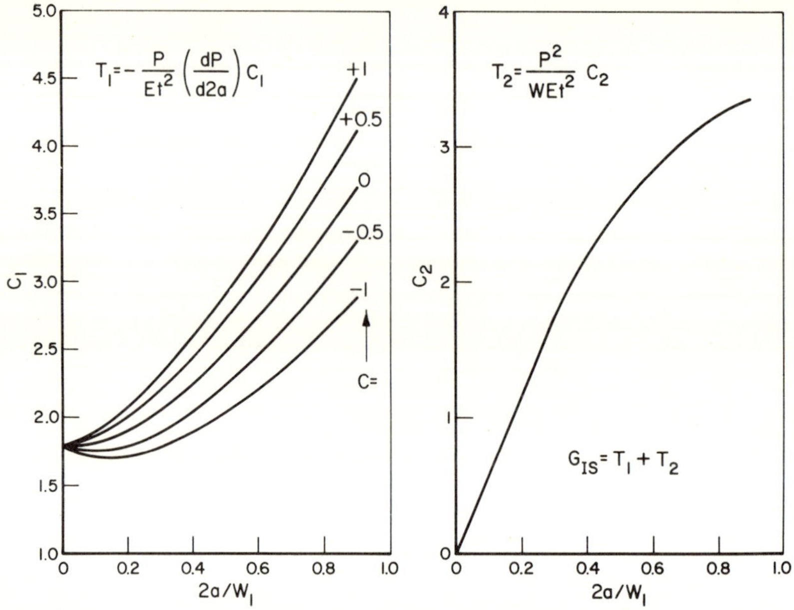


FIGURE 9. $G_{IS}$ FOR CHANGING LOAD AND DISPLACEMENT, $L_1 = W_1$

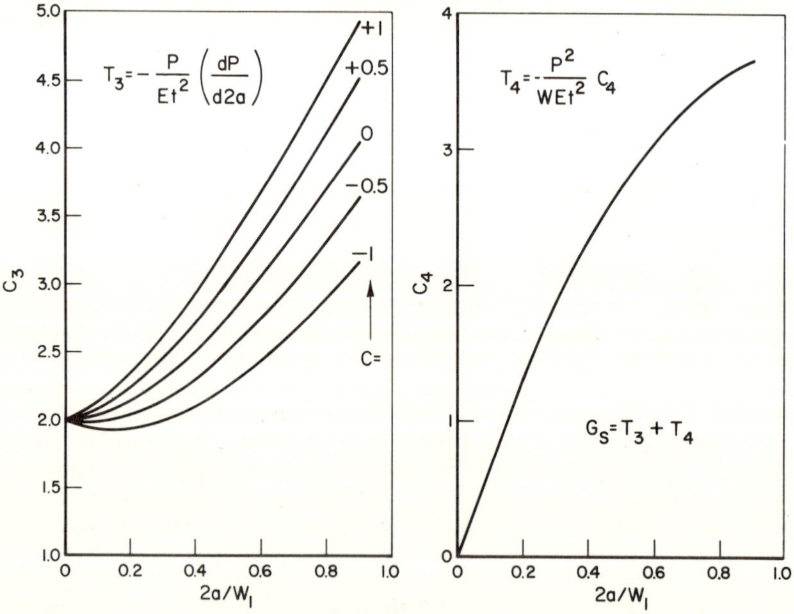


FIGURE 10. $G_S$ FOR CHANGING LOAD AND DISPLACEMENT, $L_1 = W_1$

on the formulations as derived by Paris and Sih [7]. The total strain energy release rate equation for combined modes resulting from in-plane crack surface displacements is:

$$G = \left(\frac{1-\nu^2}{E}\right)\left(K_I{}^2 + K_{II}{}^2\right) \tag{26}$$

However, the applied shear stress condition allows treatment of the problem as an applied biaxial stress problem through resolution with Mohrs circle. Under pure shear loading the maximum and minimum principal stresses are $\sigma$ and $-1.0\,\sigma$.

Figure 13 shows the influence of biaxial applied stresses on fatigue crack growth rate, as compared to uniaxial applied stress. These comparative data as analyzed by Eq. (26) show faster fatigue crack growth rate due to applied compression stresses parallel to the crack. Joshi and Shewchuk [13] and Oh, et al [14] have shown that an increase in degree of biaxial tension-tension stresses results in a slower fatigue crack growth rate.

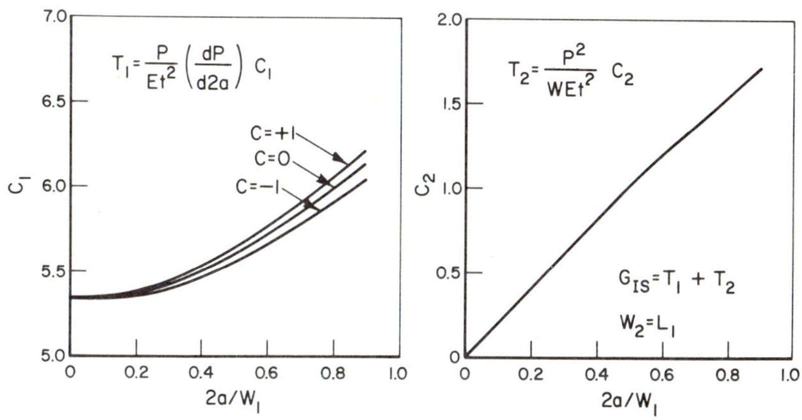


FIGURE 11. $G_{IS}$ FOR CHANGING LOAD AND DISPLACEMENT, $L_1 = 3W_1$

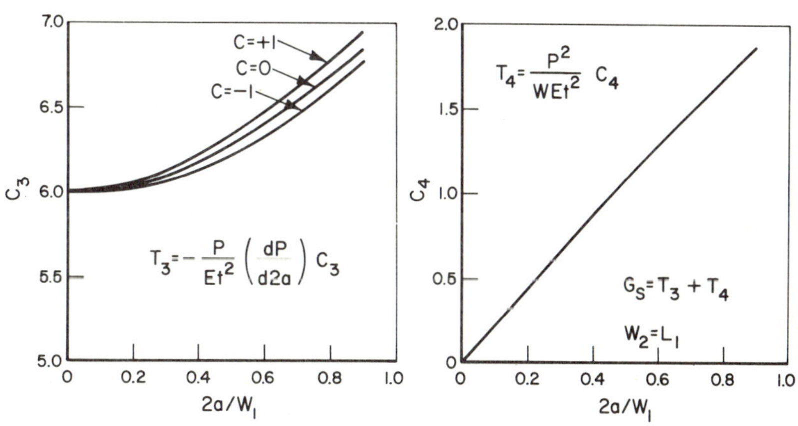


FIGURE 12. $G_S$ FOR CHANGING LOAD AND DISPLACEMENT, $L_1 = 3W_1$

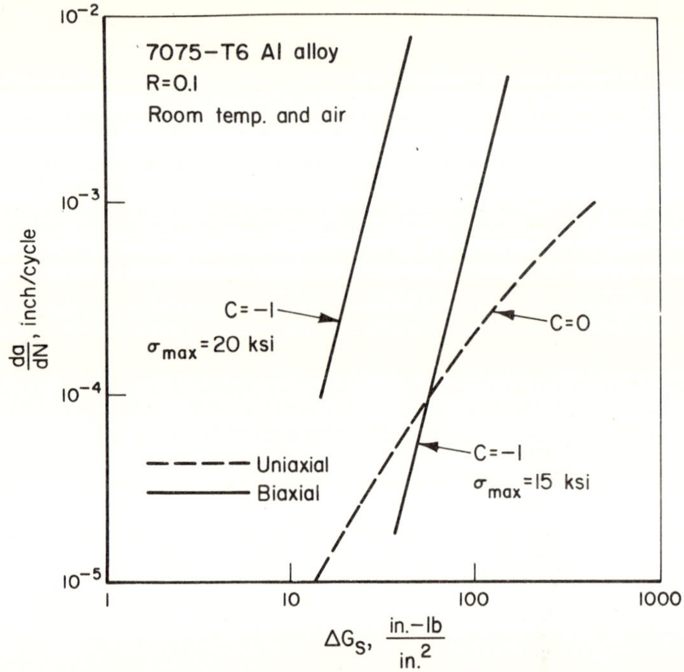


FIGURE 13. FATIGUE CRACK GROWTH RATE VERSUS $\Delta G_S$ FOR BIAXIAL AND UNIAXIAL STRESSES (REF. [11] and [12])

## CRACK TIP PLASTIC DEFORMATION AND INFLUENCE ON ENERGY RELEASE RATE

So far in this paper the analysis has been concerned with the fracture energy problem as analyzed by the theory of elasticity and elastic energy rate analysis. The energy associated with plastic deformation of the plastic zone at a crack tip is influenced by the degree of biaxial applied stresses. The practical structural crack problem considered in this analysis is a wide plate with a colinear array of cracks subjected to biaxial stresses as shown in Fig. 14. Westergaard suggested a stress function which solves this problem.

The elastic-plastic boundary surrounding the crack tip was derived by substituting the elastic stress field equations into a yield criteria. The most widely accepted yield criterion for combined stresses is the von Mises strain energy of distortion and for the states of plane strain and plane stress are given respectively as:

$$\sigma_{ys} = \left[(\nu^2-\nu+1)(\sigma_x^2+\sigma_y^2) + (2\nu^2-2\nu-1)\sigma_x\sigma_y + \tau_{xy}^2\right]^{1/2} \qquad (27)$$

$$\sigma_{ys} = \left[(\sigma_x-\sigma_y)^2 + \sigma_x\sigma_y - 3\tau_{xy}^2\right]^{1/2} \qquad (28)$$

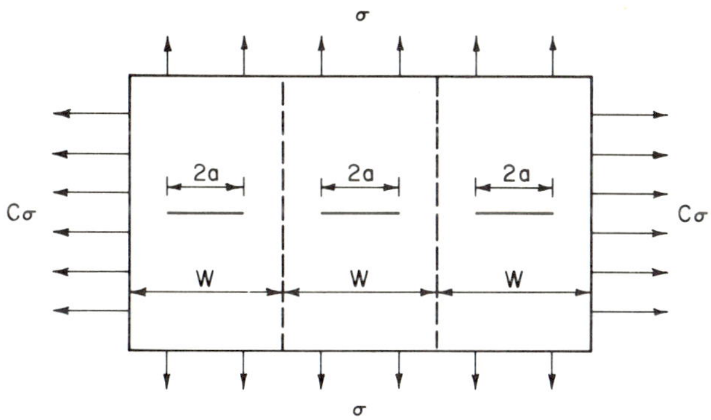


FIGURE 14. COLINEAR ARRAY OF CRACKS IN BIAXIAL STRESSED WIDE PLATE

Figure 15 shows the shape of the elastic-plastic boundary due to various degrees of biaxial applied tensile stresses. It is assumed that the material stress-strain curve is elastic perfectly plastic. This computerized analysis, using transcendental analysis, shows that biaxial applied tension stresses parallel to the crack cause the plastic zone size to decrease when compared to uniaxial stress. These results can be applied to finite width panels of width w since the results for very wide plates are the same as the infinite plate solutions reported in Ref. [17] and [18]. If one assumes a critical plastic zone size as a fracture instability criterion, then an increased remote tension stress with a lateral biaxial component, Cσ, could be applied to produce a higher residual

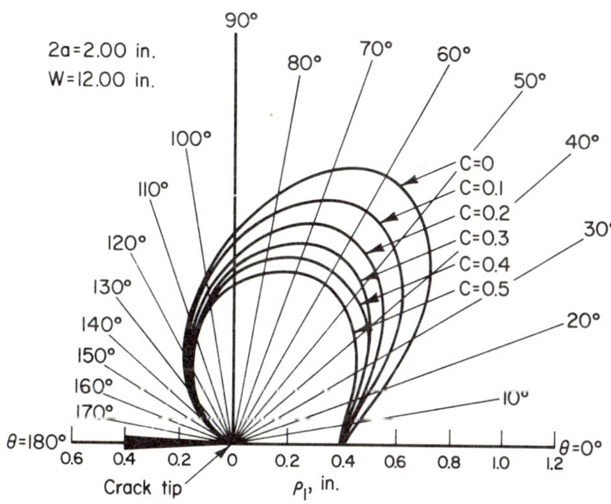


FIGURE 15. VARIATION OF PLASTIC ZONE SIZE WITH BIAXIAL APPLIED STRESSES

strength. However, lateral compression stresses, $-C\sigma$, combined with positive remote longitudinal stress could cause a lower residual strength. This combination of biaxial loading should be of extreme concern to residual strength analysts.

Hilton [15] has shown analytical results very similar to the above analysis. For small scale yielding and material hardening considerations, Hilton showed by plastic intensity factors that lateral biaxial tensile stress components resulted in a reduced plastic zone size, whereas lateral compressive stress components resulted in an increased plastic zone size. Kibler and Roberts [16] in their experiments on biaxial loaded plates of aluminum alloys 6061-T6 and -T4, which are nearly perfectly elastic plastic, demonstrated an increasing residual strength with increasing lateral biaxial tension stresses.

Further examination of this analysis shows that the plastic zone width ahead of the crack tip is constant and independent of the degree of biaxiality. This argument coupled with the independency of crack opening displacement due to biaxial stresses leads to the concept of a necessary energy criterion for fracture. The energy criterion would involve the entire extent of the zone of plastic deformation due to biaxiality and coupling of the elastic and plastic energy quantities.

The elastic-plastic energy quantity has been approached by Kobayashi [19] with the "J-integral" method for the boundary condition of applied uniaxial constant stress. Further extension of the "J-integral" method by Kobayashi, et al, to the elastic plastic analysis, has shown that the plasticity contribution is quite large at moderate to high applied stress levels.

Since the effects of biaxiality are more pronounced on the shape of the plastic zone around a crack tip, the plastic energy release rate would be a predominate energy quantity for considerations in a fracture energy criterion due to biaxial applied stress, load, or displacements.

## CONCLUSIONS

Based on the analytical work and discussions presented in this paper, the following conclusions are made.

(1) Biaxial applied boundary stresses effect the strain energy release rate due to crack extension.

(2) The analysis of elastic strain energy release rate and plastic zone size indicate that longitudinal tension and lateral compression stresses parallel to the crack surface are the most critical conditions of crack instability.

(3) The strain energy release quantities derived for the three load-displacement considerations that can occur during crack extension are applicable to steady-state or cyclic fatigue loadings.

(4) A critical set of experiments under biaxial loading conditions, tension and compression, need to be conducted to establish critical $G_{IS}$ and $G_S$ values and to establish the constancy of material fracture toughness.

(5) Crack extension due to applied constant steady-state displacement needs to be further evaluated.

REFERENCES

[1] Westergaard, H. M., "Bearing pressures and cracks", Journal of Applied Mechanics, 6, June 1939 (A49-A53).

[2] Irwin, G. R., "Analytical aspects of crack stress field problems", University of Illinois, Theoretical and Applied Mechanics Report No. 213, March 1962.

[3] Timoshenko, S. and Goodier, J. N., Theory of elasticity, McGraw Hill Book Company, New York, 1951 (149-151).

[4] Smith, S. H., "Linear elastic fracture mechanics analysis by a continuum approach", Boeing Research Report D6-24974, July 1971.

[5] Newman, J. C., Jr., "An improved method of collocation for the stress analysis of cracked plates with various shaped boundaries", NASA TND-6376, August 1971.

[6] Grandt, Captain A. F., Jr., "A general stress intensity factor solution for thru-cracked fastener holes", International Journal of Fracture (in press).

[7] Paris, P. C. and Sih, G. C., "Stress analysis of cracks", in Fracture Toughness Testing and Its Applications, ASTM STP 381, 1965 (41).

[8] Bluhm, Joseph I., "Fracture arrest", in Fracture, Liebowitz, H., Academic Press, 5, 1969 (2-60).

[9] Heyer, R. H., "Crack growth resistance curves (R-curves -- literature review", in Fracture Toughness Evaluation by R-Curve Method, ASTM STP 527, April 1973 (3-6).

[10] Liu, A. F., "Crack growth and failure of aluminum plate under in-plane shear", AIAA 11th Aerospace Meeting, Washington, D.C., January 1973.

[11] Toor, Pir M., "On fracture mechanics under complex stress", presented at Seventh National Symposium on Fracture Mechanics, University of Maryland, August 1973.

[12] Hudson, C. M., "Effect of stress ratios on fatigue-crack growth in 7075-T6 and 2024-T3 aluminum alloy specimens", NASA TND-5390, August 1969.

[13] Joshi, S. R. and Shewchuk, J., "Fatigue crack propagation in a biaxial stress field", Experimental Mechanics, SESA, December 1970 (529-533).

[14] Oh, K. P. L., Vardar, Ö., and Finnie, I., "Failure of brittle solids under biaxial stresses", International Journal of Fracture, 9, 1973.

[15] Hilton, Peter D., "Plastic intensity factors for cracked plates subjected to biaxial loading", International Journal of Fracture, 9, 2, June 1973 (149-156).

[16] Kibler, J. J. and Roberts, R., "The effect of biaxial stresses on fatigue and fracture", Transactions of ASME, Journal of Engineering for Industry (in press).

[17] Smith, S. H., "Analysis of plane stress plastic zone size at a crack tip using various yielding theories and stress analysis of cracks", Boeing Research Report D6-15644, December 1966.

[18] Smith, S. H., "Analysis of plane strain plastic zone size at a crack tip using various yielding theories and stress analysis of cracks", Boeing Research Report D6-19457, October 1967.

[19] Kobayashi, A. S., Chiu, S. T., and Beeuwkes, R., "A numerical and experimental investigation on the use of J-integral", Engineering Fracture Mechanics, 5, 1973 (pp. 293-305).

## DISCUSSION by L. P. Pook, National Engineering Laboratory, Scotland

Discussions of the effect of biaxiality are in reality discussions of the effect of the coefficient of the second term of the series expansion for the opening mode crack tip stress field. Using Cotterell's notation [1], the coefficient of the first term is $a_1$ and equals $\frac{K_I}{2(\pi a)^{1/2}}$, and is equivalent to the stress intensity factor. Because it is a singularity the first term dominates fracture in many situations, and this is the reason for the success of the stress intensity factor concept in fracture mechanics. The second term is associated with the coefficient $a_2$ and is simply a stress parallel to the crack. Higher terms contain powers of the distance from the crack tip are less important because they vanish as the crack tip is approached.

Cotterell [1] pointed out that a crack tends to be directionally unstable when $a_2$ is positive and stable when it is negative. It is well known that cracks are directionally unstable in double cantilever beam specimens, moderately stable in compact tension specimens, and very stable in centre crack specimens loaded in uniaxial tension. If a material's toughness (in the broad sense of resistance to some type of crack growth) depends on the deflection of a crack by a microstructural feature perpendicular to the direction of crack growth, it follows that test results can depend on the value of $a_2$, and hence the type of specimen used.

In thin sheets reducing (in an arithmetic sense) the stress parallel to the crack increases the likelihood of buckling. Buckling increases the stress intensity factor and consequently fatigue crack growth rates.

An additional result can easily be obtained from the Westergaard solution for an infinite series of collinear cracks. If directional stability is assumed to be constant when $a_2/a_1$ is constant, then the effective biaxiality $B_E$ is given by

$$B_E = 1 + \frac{(B-1)}{\left(\frac{W}{\pi a} \tan \frac{\pi a}{W}\right)^{1/2}}$$

where the biaxiality B is the ratio of applied stresses parallel and perpendicular to the crack and w is the specimen width.

The well known "tangent" equation for a finite width centre crack plate derived from the solution for a series of infinite collinear cracks, becomes seriously inaccurate when the overall crack length exceeds half the plate width. Other results, such as those quoted in the paper, should therefore be regarded as of dubious validity at beyond this limit.

REFERENCE

[1] Cotterell, B., Int. J. Frac. Mech., 2, 31, 526-533, 1966.

## AUTHOR'S CLOSURE

The discussion of Pook has pointed out some crack instability and geometrical aspect considerations of the biaxial stressed fracture problem.

As discussed in the paper, the combined stress situation of applied tension stress which is perpendicular and applied compression stress parallel to the crack surface is a condition for extreme structural instability. The exact instability criterion would hinge on the predominating mode of failure. That is, whether the panel would collapse from fracture or buckling instability. However, this paper has dealt with the fracture energy release rate problem. The biaxial fatigue-crack-growth-rate data presented in Fig. (13) was generated from thick panels loaded in pure shear where in-and-out of plane buckling did not occur.

Certain comments are offered on the discussion of Dr. Pook relative to the validity of the results with respect to the aspect ratio of crack length and panel width. The well known "tangent-function" applied as a finite width correction factor to centrally cracked tension panels does have a limitation of $2a/w \leq 0.5$. The function comes from the Westergaard stress function for a colinear array of cracks where the shear stress field behavior is satisfied for that problem. When applied to a finite width panel with zero shear on the boundary, the limitation of $2a/w \leq 0.5$ is imposed. However, the results presented in the paper for the colinear array of cracks deals with the biaxial stressed crack tip plastic zone solution to aid in understanding the propagation and eventual joining of the cracks and associated plastic zones.

Certain biaxial boundary conditions of stress load and displacement were assumed in the energy release rate analysis. These boundary conditions are assumed to maintain constant during the slow extension of the crack. Therefore, it is recommended that the results be utilized in the small crack or aspect region. That is, for $2a/w \leq 0.35$ where the net-section stresses are at a minimum and the structure remains elastic.

# AN IMPROVED TECHNIQUE FOR RESIDUAL STRENGTH PREDICTION - A MODIFIED CRACK GROWTH RESISTANCE APPROACH

D. P. Wilhem

Northrop Corporation, Aircraft Division, Hawthorne, California 90250 USA

## INTRODUCTION

The implementation of the United States Air Force Damage Tolerance Design Requirements in the form of Military Standard, MIL-A-XXXX, requires that residual strength requirements be met for all military aircraft structures. These requirements which are the result of numerous in-service crack problems and the Air Force's need for improved, dependable equipment represent an additional step beyond those fail-safe operating requirements imposed for civil aircraft by the procuring countries.

To meet the requirements it is necessary to know residual strength not only in the presence of large flaw (crack) conditions but also factors of past and present fatigue history and environmental conditions in order to meet the damage tolerant analysis requirement.

In most cases the residual strength conditions must be met for materials, thicknesses, and crack lengths where failure in other than plane-strain modes are operative. This implies high-toughness, thinner-gage materials where crack extension prior to failure is accompanied by large amounts of slow-stable tear, and gross plastic deformation. In many cases the plastic deformation is not only confined to the cover material (skin) itself, but is also transferred to the support structure (frames, spars, etc.) as well, due to the presence of the crack.

To treat the residual strength prediction problem for other than plane-strain fracture, the basic element of a widely accepted failure criterion has been addressed by numerous investigators.

A comprehensive review of the current state-of-the-art of fracture testing of thin sections has been recently compiled as part of a Task Group activity of the American Society for Testing and Materials, Committee E-24.01 [1]. Similar reviews are underway for the particular application of fracture mechanics to aircraft design [2] and have also addressed the fracture strength problem of fatigue cracked materials [3] through various Panels of AGARD.

The basic conclusion still remains, which is that a widely accepted fracture criteria has not as yet been established for non-plane strain failure.

The available fracture criteria can be classified into three general categories: notch strength analysis, modifications or extensions to critical stress intensity or the $K_c$ approach, and what can be classed as "critical energy" methods. Of these methods the latter two have received much attention and form the basis for several residual strength analysis methods (see e.g., [2] and [4]). However, any extensions of linear elastic fracture mechanics to materials which exhibit large plastic zones and slow-stable crack extension have met with limited success. To this end, interest in crack growth resistance curve approaches to fracture first introduced by Irwin and Kies [5] and modified by Krafft, Sullivan and Boyle [6] and more recently refined by Heyer and McCabe [7,8] has taken resurgence and shows promise of treating the slow-stable tear problem. In recent years the application of Rice's [9] path independent "J" integral has shown promise in treating the large scale yielding problem and inroads have been made into better understanding of the mixed mode (mode here referring to $K_I$, $K_{II}$) problem by Sih [10] and his strain-energy-density factor. Thus the search for a fracture criterion or criteria continues and the problem has become more complex due to a dual requirement that the criterion must apply not only to materials but their use in combination of structural arrangements with associated complex geometries and loading. Since the most critical aircraft structure is analyzed using finite element techniques, it became apparent that a suitable residual strength prediction scheme should utilize information obtained from the normal stress analysis obtained from the finite element model. Thus the basis for the analysis would be finite element procedures with either a critical stress, strain or energy criterion. The crack growth resistance (or R curve) criterion fulfills one requirement of the technique - that is to treat the slow-stable tear problems [2,11]. It does not, however, treat the large scale yielding problem which must be included in any structural residual strength prediction scheme. The J-integral shows promise of meeting this requirement as well as providing a link between the slow-tear and gross-plasticity problem. As a first step, crack growth resistance data have been obtained for a variety of materials indicating the scope of the slow-tear problem. The second step has involved examining a typical structural arrangement in light of an energy criterion, namely the J-integral. Discussion of the requirement of a tie between the two is currently under investigation.

CRACK GROWTH RESISTANCE

Crack growth resistance data have been obtained for the materials in the thicknesses shown in Table 1. Material property data are also given. The crack line wedge loaded (CLWL) specimen geometry is the selected specimen configuration due to its inherent stability and current use to establish accurate measurement of slow stable tear behavior [7,8]. A double compliance technique has been employed to measure the extent of both physical and effective (physical crack length plus plastic zone) crack sizes. Summary plots of the crack growth resistance data are presented in Figs. (1) through (5). Several trends are evident from these data based on physical crack extension, which are:

i. The resistance curve increases with decreasing material thickness, Figs. (1) - (4).

ii. A plateau value of stress intensity can be measured from the CLWL specimen for some materials with the crack oriented in the TL direction.

iii. The initial portion of the R-curve (i.e., at small crack extension) can be similar for two different thicknesses of the same material.

iv. Condition of the initial stress riser (fatigue crack, EDM slot, etc.) does not significantly effect the R-curve trend. This has also been observed by others [13,14] during residual strength testing.

v. Significant changes in crack resistance as measured by $K_R$ can occur with small crack extension during the early life of the crack.

If establishing a material fracture criterion were of prime importance observations i, ii, and iv would be significant. However, in a given aircraft structure the short crack, high stress problems predominate. These problems necessarily lead to small, initial crack extensions with large increases in crack resistance. Thus in a structural application the majority of crack problems concern the early stages of crack extension where the R-curve is rapidly changing and items i, iii, and v are of more importance. This is evident if a tangency condition between the crack driving force and R-curve is used as a fracture criterion [11] for finite width panels. However, in a structural arrangement the crack driving force or K-curve is not a simple function of crack length and stress but is influenced by the substructure due to the load transfer caused by the presence of the crack. Various investigators have studied the parametric effect of idealized structural configuration (i.e., by varying areas of reinforcement, rivet size, spacing, etc.) on the elastic stress intensity [11,15,16] factor. Using these solutions in conjunction with the resistance curve [12-14] have produced reasonable results for simple, structural configurations made of materials which exhibit small amounts of crack tip plasticity (e.g., 7075-T6). However, in tougher materials and for conditions where the crack is in the vicinity of the reinforcement, the load transfer can produce a reinforcement fracture critical problem from one which may have been initially a skin critical case.

If the R-curve concept is to become a universally acceptable failure criterion for plane stress fracture, the crack tip plasticity problem must become a direct part of the analysis.

TABLE 1

| MATERIAL | THICKNESS, B (INCHES) | TRANSVERSE PROPERTIES | | | LONGITUDINAL PROPERTIES | | |
|---|---|---|---|---|---|---|---|
| | | 0.2% OFFSET YIELD, $F_{ty}$ (ksi) | ULTIMATE STRENGTH, $F_{tu}$ (ksi) | MODULUS E (ksix$10^3$) | 0.2% OFFSET YIELD, $F_{ty}$ (ksi) | ULTIMATE STRENGTH, $F_{tu}$ (ksi) | MODULUS, E (ksix$10^3$) |
| 7075-T651 | .195 | 74.97 | 80.74 | 10.30 | 72.77 | 83.42 | 10.27 |
| 7075-T6 | .063 | 77.93 | 81.49 | 10.13 | 70.57 | 80.85 | 10.01 |
| 7075-T751 | .195 | 67.25 | 74.44 | 10.39 | 67.45 | 76.23 | 10.12 |
| 7075-T7 | .063 | 62.60 | 72.61 | 9.92 | 61.58 | 72.30 | 9.74 |
| 2024-T351 | .258 | 47.58 | 71.08 | 10.59 | 53.17 | 72.56 | 10.02 |
| 2024-T3 | .064 | 45.00 | 70.47 | 10.49 | 52.97 | 74.22 | 10.61 |
| Ti-6Al-6V-2Sn | .210 | 144.71 | 153.44 | 14.58 | 156.14 | 163.60 | 16.84 |
| Ti-6Al-6V-2Sn | .063 | 156.57 | 160.92 | 16.66 | 159.44 | 166.78 | 17.08 |
| 9Ni-4Co-.2C | .064 | 136.08 | 162.05 | 26.88 | 147.78 | 169.82 | 28.39 |

## GENERALIZATION OF THE RESIDUAL STRENGTH ANALYSIS

Any effective, accurate, and complete fracture analysis method must include or account for as a minimum requirement the following: crack tip plasticity, crack buckling, slow crack growth, crack-tip geometry, load or strain rate, thermal effects, aggressive environments, finite geometry. As a practical consideration the majority of the parameters listed can be treated through changes in the basic material R-curve although the use of two or more specimen configurations may be necessary. For example, the CLWL specimen

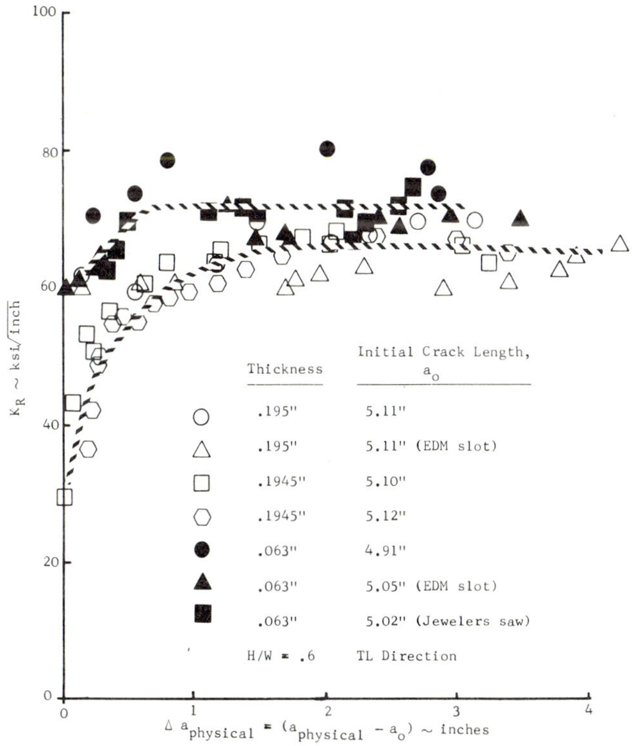


Fig. (1) - 7075-T6 Aluminum crack growth resistance data

could not be used to study load or strain rate effects. Once the resistances curve data were obtained it would then be possible to generalize the R-curve to account for those "environmental" factors which alter its shape. Its use as a fracture criterion could then be accepted as long as stress intensity solutions were available for the structure of interest. But most importantly crack tip plasticity must be contained within an elastic stress field so that linear elastic fracture mechanics can be employed. For the case of structure which can be readily idealized it is a relatively simple matter to use either available closed form solutions or standard finite element techniques to determine the stress intensity factor K, for a series of anticipated in-service crack lengths. These elastic stress intensity solutions can then be used in conjunction with the R-curve to produce reasonable fracture stress predictions [11] as long as linear elastic fracture mechanics governs the stress field behavior. To use this approach could lead to ultra-conservative designs in structural arrangements utilizing materials of high toughness and short crack situations as mentioned previously.

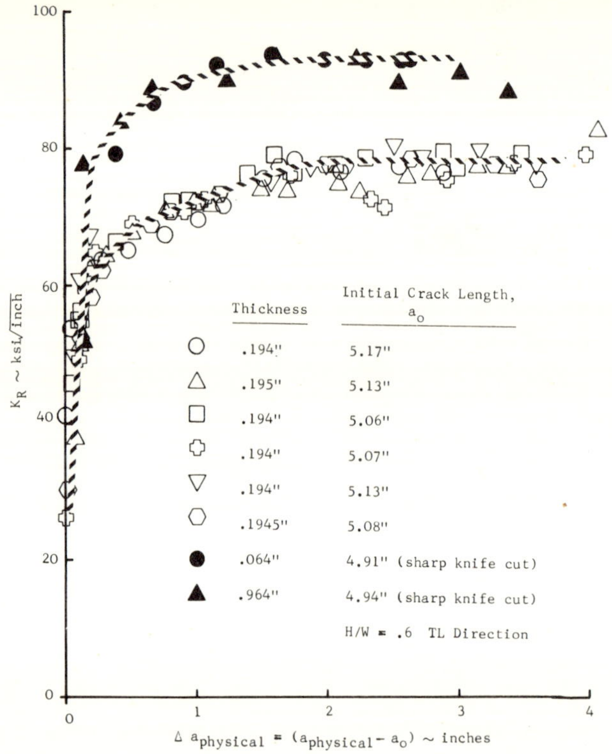


Fig. (2) - 7075-T7 Aluminum crack growth resistance data

It is proposed that a more reasonable approach would be to generalize the R-curve so that the relative merits of each of the other available fracture criterion could be assessed. This can be shown schematically as in Fig. (6) for a reinforced center cracked panel geometry.

As shown in Fig. (6), curves of K, crack opening displacement (COD) and J can be computed for various, applied stress levels and will have the general shape indicated which will be dependent on factors such as stringer area, method of attachment and other structural factors. If linear elastic material behavior prevails at fracture then a well known link exists between K, COD, and J through the critical energy release rate $G_{Ic}$, [17]

$$G_{Ic} = \frac{K_{Ic}^2}{E} = \delta_c \; F_{ty} = J_{Ic} \tag{1}$$

If tangency to the $K_R$, $\delta$ or J curve is employed as a failure criterion the effect of excessive crack tip plasticity (beyond the limit of linear elastic fracture mechanics) will cause not only

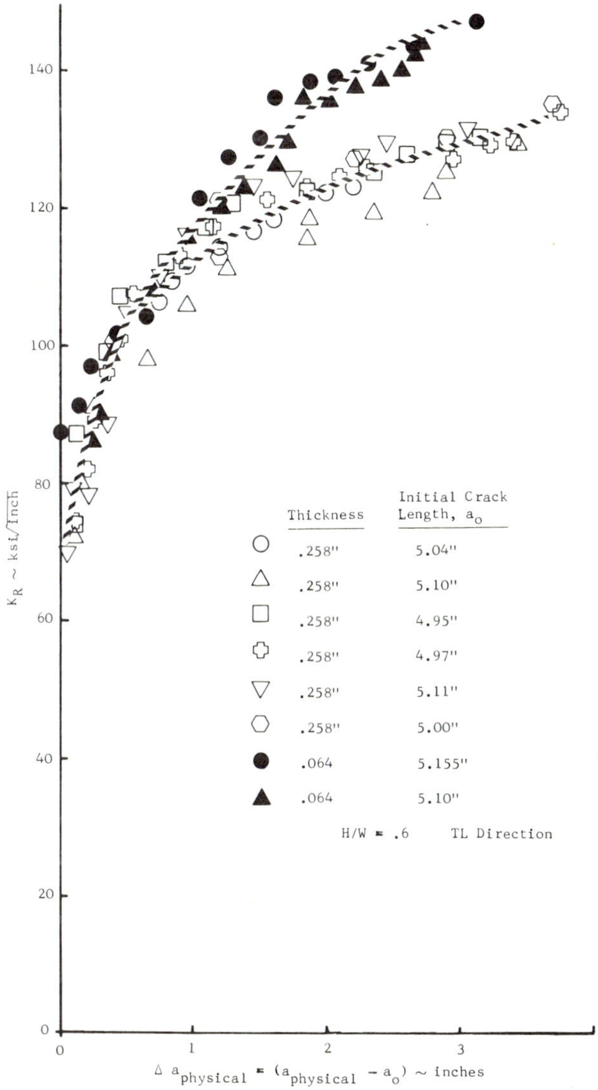


Fig. (3) - 2024-T3 Crack growth resistance data

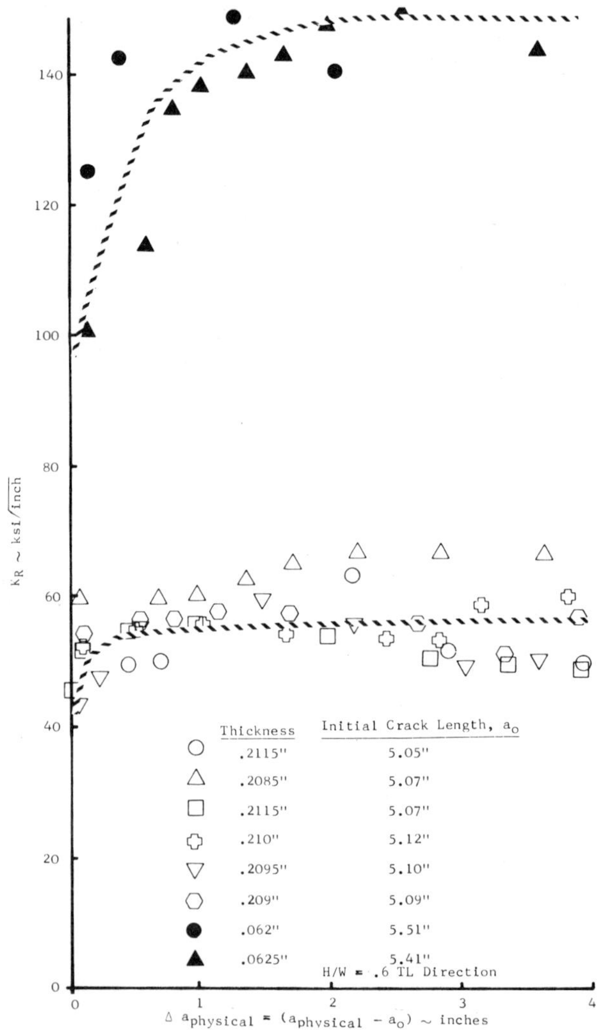


Fig. (4) - Titanium - 6Al-6V-2Sn Crack growth resistance data

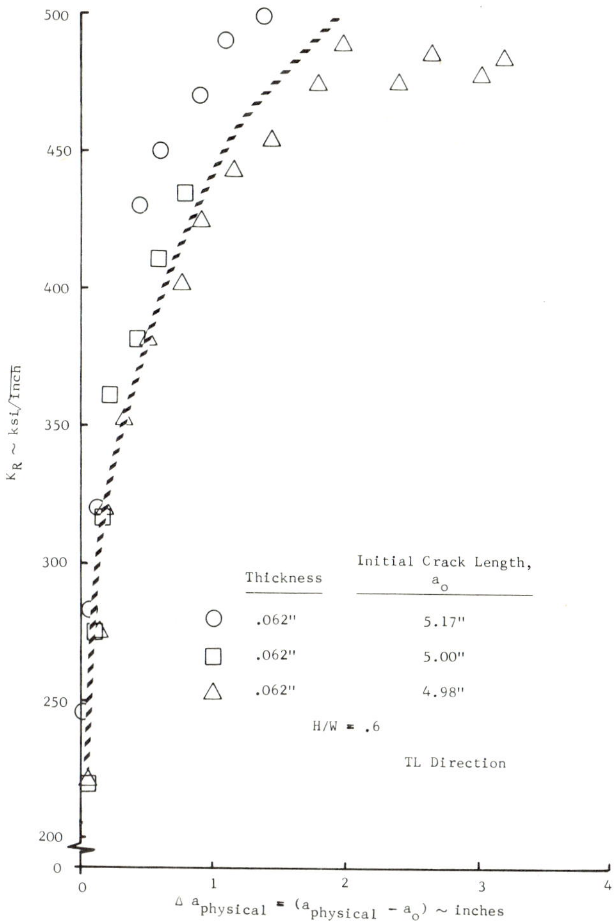


Fig. (5) - 9Ni-4Co-.2C Crack growth resistance data

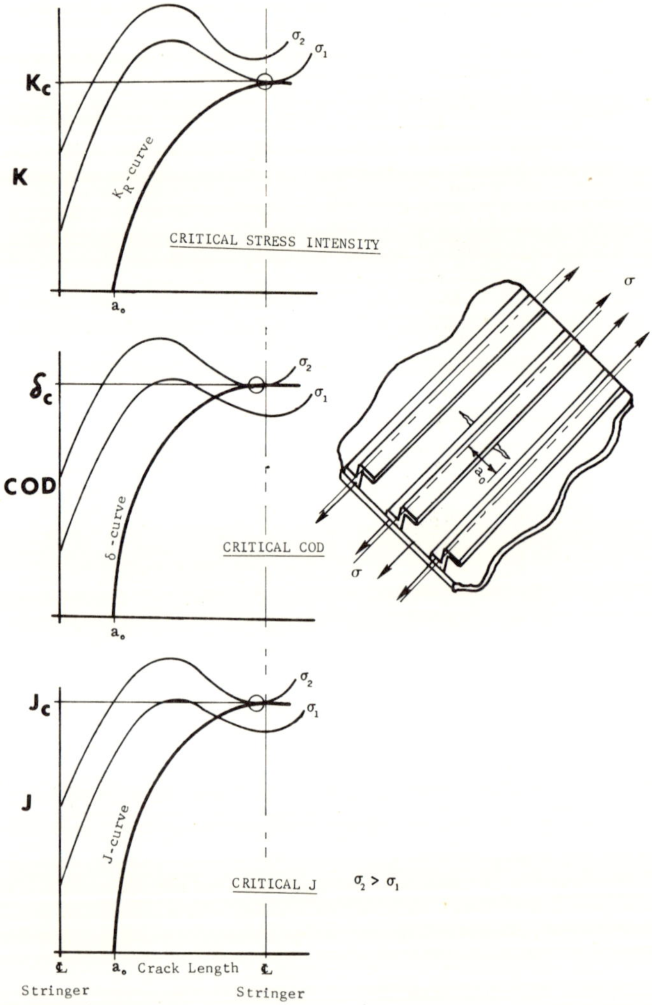


Fig. (6) - Schematic of the available plane stress fracture criteria

an elevation in the failure curves themselves but requires some corrective change to the crack driving force curves. For example, in the stress intensity approach the addition of a plastic zone correction can be used on the crack driving curves in conjunction with the $K_R$ curve developed from compliance techniques. The $K_R$ curves are then based on effective crack size which includes some measure of the extent of physical crack length plus plastic zone. For those materials which exhibit resistance curves of the type shown in Figs. (1) and (2), this approach can produce meaningful residual strength predictions in simple structure. However, the trend in aircraft materials is toward the tougher material, with extensive crack tip plasticity and resistance curves as shown in Figs. (3) and (5). With these materials the use of a linear elastic approach or the plasticity modified linear elastic approach is not warranted and would lead to ultra conservative designs.

The J-integral [9] as a failure criterion has been discussed extensively in recent years [17-24]. Some recent experimental research has indicated its correlation with simple panel geometries [18,20]. The use of finite element techniques, considering elasto-plastic behavior, have shown the path independency of J [20-24] for simple panel geometries. Based on these findings, a modified R-curve approach has been suggested [25] which would be an experimentally determined J-curve as shown in Fig. (6) which would include the effects of slow tear. Recent experimental work by McCabe [26] has indicated that an experimental correspondence exists between $K_R$ and J for a high toughness aluminum alloy. Further experimental studies are underway to verify this correspondence for other materials, and preliminary findings are encouraging.

USE OF THE GENERALIZED RESISTANCE CURVE FOR RESIDUAL STRENGTH PREDICTION

For generalized plane stress conditions Rice's [27] J-integral can be written as [21]

$$J = \int_{\Gamma} \{\int[\sigma_x \, d\varepsilon_x + \tau_{xy} \, d\gamma_{xy} + \sigma_y \, d\varepsilon_y]dy - \vec{T} \cdot \frac{\partial \vec{u}}{\partial x} \, ds\} \tag{2}$$

where Γ is the integration path and ds is an arc positive in the counterclockwise direction along Γ. Using finite element procedure and assuming Prandtl-Reuss material behavior J values can be computed from Eq. (2) for materials deformed into the plastic range [21] where the strain energy density $\int[\ ]dy$ contains equivalent stress and equivalent plastic strain. For conditions of monotonic loading, values of J have been obtained for two crack lengths for the three-wing channel stiffened panel configuration of Fig. (7), but not without large computer run times.

If Dugdale [28] plastic zone behavior is assumed the method of Hayes and Williams [29] which employs the Buekner [30] approach can be used to compute crack opening displacements using finite element techniques for a series of assumed plastic zone lengths at fixed physical crack lengths [25,29]. J-integral values can then be computed from the relationship $J = F_{ty}\delta$ proposed by Rice [27]. Computations of J have been made for the structure of Fig. (7) using this procedure. The results are presented in Fig. (8) along with the assumptions made in the finite element model of the wing channel stiffened panel. As noted, equivalent area rivet model assumptions were used in the original finite element model, subsequently the more realistic shear element (based on rivet deflection data) rivet model proposed by Swift [31] was employed in the model. The difference in rivet modeling can be seen to increase the value of $\sqrt{J}$ by 15% to 20% (depending on applied stress) for a 5.4 inch crack length.

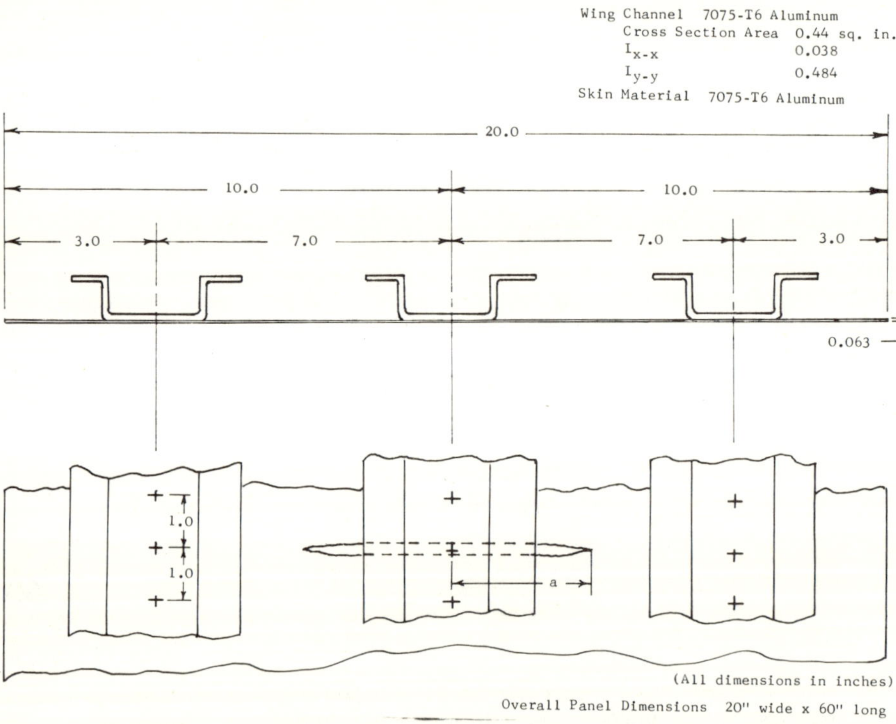


Fig. (7) - Cracked, reinforced panel geometry used in finite element modeling

The results of Fig. (8) indicate the large amount of load transfer which takes place as the crack approaches the edge wing channels and is reflected through a decrease in $\sqrt{J}$. Using these data in conjunction with the proposed material J-curve tangency concept (see Fig. (6)) will provide the failure criterion where both slow stable tear and large crack tip plasticity predominate. With Dugdale zone assumptions and the Hayes/Buekner approach com-

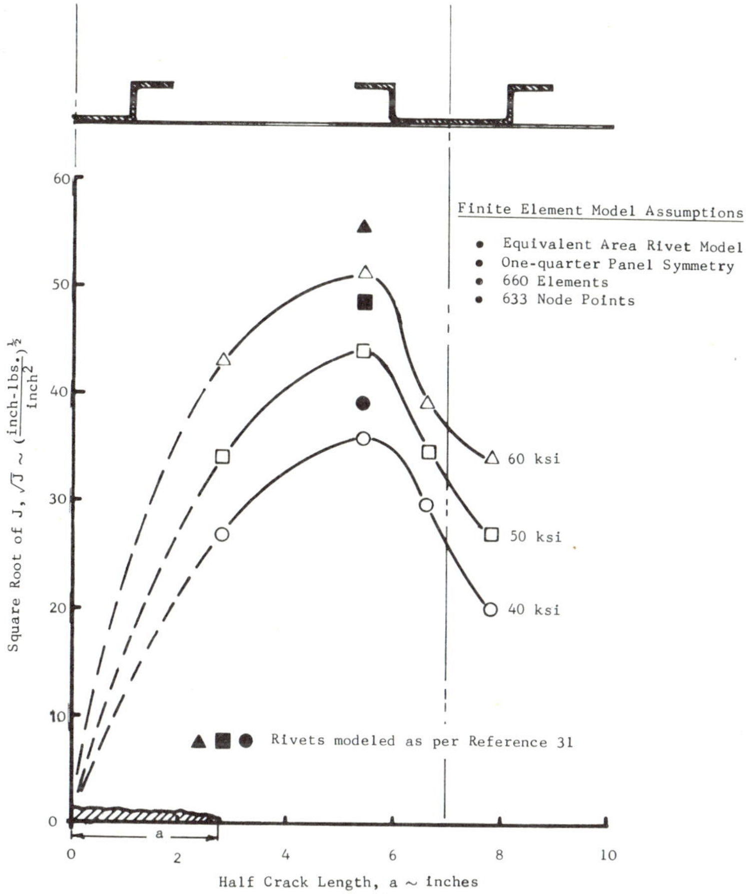


Fig. (8) - Square root of J integral for panel geometry of Fig. (7)

puting time is minimal. Another factor of utmost importance in any structural residual strength scheme is the separation of the skin and stiffener critical fracture case. Using the finite element approach permits monitoring of the stiffener/attachment stress levels so that with a suitable failure criterion for the stiffener one can determine the applied stress-crack length combination which will cause stiffener failure. The details of this failure criterion and its integration into the modified R-curve approach are currently under study.

## CONCLUSIONS

The use of a modified crack growth resistance curve as a structural fracture criterion appears to be a feasible approach to prediction of residual strength of built-up structures.

Depending on the shape of the $K_R$ curve the use of a tangency condition and linear elastic fracture mechanics principles, adjusted for plasticity, may be a sufficient fracture criterion. In the other extreme the modified resistance curve approach or, J curve will not only account for slow stable tear but extensive crack tip plasticity which is the case for most modern aircraft materials.

Additional research on fracture of panels with realistic structural arrangements is currently underway. Analytical predictions of the fracture stress for these panels will be attempted based on the fracture criteria proposed in this study. It is anticipated that the results will aid in producing a better insight into the prediction of residual strength and crack arrest for complex structure manufactured from high toughness material which fails in a predominately plane stress mode.

## REFERENCES


[1] Kaufman, J. G. and Wilhem, D. P., editors, Fracture testing of thin sections - a review of current methods, <u>a task group report of ASTM committee E-24</u> (to be published).

[2] Vlieger, H. and Broek, D., "Residual strength of built-up structures", National Aerospace Laboratory, NLR MP 72029 U, 1972.

[3] Kuhn, P., "Residual strength in the presence of fatigue cracks", paper presented to the Structures and Materials Panel - AGARD, 1967.

[4] Swift, T., "Development of the fail-safe design features of the DC-10", ASTM STP 486, 164, 1971.

[5] Irwin, G. R. and Kies, J. A., "Critical energy rate analysis of fracture strength", Welding Journal Research Suppl., Vol. XIX, No. 4, 193, 1954.

[6] Krafft, J. M., Sullivan, A. M. and Boyle, R. W., "Effect of dimensions on fast fracture instability of notched sheets", Proc. of the Crack Propagation Symposium, College of Aero., Vol. 1, 8, 1961.

[7] Heyer, R. H. and McCabe, D. E., "Plane-stress fracture toughness testing using a crack-line-loaded specimen", Eng. Fract. Mech., Vol. 4, 393, 1972.

[8] Heyer, R. H. and McCabe, D. E., "Crack growth resistance in plane-stress fracture testing", Eng. Fract. Mech., Vol. 4, 413, 1972.

[9] Rice, J. R., "A path independent integral and the approximate analysis of strain concentration by notches and cracks", J. of Appl. Mech., 379, June 1968.

[10] Sih, G. C., "Mechanics of Fracture", Vol. 1, Methods of Analysis and Solutions of Crack Problems, Noordhoff International Publishing, Leyden, 1973, XXI.

[11] Creager, M. and Liu, A. F., "The effect of reinforcements on the slow stable tear and catastrophic failure of thin metal sheet", paper presented to AIAA 9th Aerospace Sciences Meeting, No. 71-113, 1971.

[12] "Fracture toughness evaluation by R-curve methods", American Society for Testing and Materials, STP 527, 1973.

[13] Broek, D., "The residual strength of aluminum alloy sheet specimens containing fatigue cracks or saw cuts", National Aerospace Laboratory, NLR TR 2143, 1965.

[14] Freed, C. N., et al., "Crack-growth resistance characteristics of high-strength sheet alloys", Naval Research Laboratory Report 7374, 1972.

[15] Poe, Jr., C. C., "The effect of riveted and uniformly spaced stringers on the stress intensity factor of a cracked sheet", Proceedings of the Air Force Conference on Fatigue and Fracture of Aircraft Structures and Materials, AFFDL TR 70 144, 1969.

[16] Vlieger, H., "Residual strength of cracked stiffened panels", National Aerospace Laboratory, NLR-TR 71004 U, 1971.

[17] Begley, J. A. and Landes, J. D., "The J-integral as a fracture criterion", ASTM STP 514, 1, 1972.

[18] Landes, J. D. and Begley, J. A., "The effect of specimen geometry on $J_{Ic}$", ASTM STP 514, 24, 1972.

[19] Bucci, R. J., et al, "J-integral estimation procedures", ASTM STP 514, 40, 1972.

[20] Kobayashi, A. S., et al, "A numerical and experimental investigation on the use of J-integral", Eng. Fract. Mech., Vol. 5, 293, 1973.

[21] Hayes, D. J., "Some applications of elastic-plastic analysis to fracture mechanics", Dept. of Mech. Eng., Imperial College, University of London, 1970.

[22] Sumpter, J. D. G. and Turner, C. E., "Fracture analysis in areas of high nominal strain", Proc. of 2nd Inter. Conf. on Pressure Vessel Tech., ASME, Part II: Mat., Fab. & Insp., 1095, 1973.

[23] Sumpter, J. D. G., et al, "Post yield analysis and fracture in notched tension pieces", Third Inter. Conf. on Fract., Munich, I-433, 1973.

[24] Andersson, H., "A finite-element representation of stable crack-growth", J. Mech. and Phys. of Solids, Vol. 21, 337, 1973.

[25] Verette, R. M. and Wilhem, D. P., "Development and evaluation of methods of plane stress fracture analysis", Air Force Report AFFDL TR 73 42, 1973.

[26] McCabe, D. E., "R-curve determination using COS methods", presentation to ASTM Committee E24.01, 1973.

[27] Rice, J. R., "Fracture - An Advanced Treatise", ed. H. Liebowitz, Academic Press, N. Y., 1968.

[28] Dugdale, D. S., "Yielding of steel sheets containing slits", J. of Mech. and Phys. of Solids, Vol. 8, 100, 1960.

[29] Hayes, D. J. and Williams, J. G., "A practical method for determining Dugdale model solutions for cracked bodies of arbitrary shapes", Inter. J. of Fract. Mech., Vol. 8, No. 3, 1972.

[30] Buekner, H. F., "Propagation of cracks and the energy of elastic deformation", Trans. of Amer. Soc. of Mech. Eng., 80E, 1958.

[31] Swift, T. and Wang, D. Y., "Damage tolerant design - analysis methods and test verification of fuselage structure", Proceedings of the Air Force Conference on Fatigue and Fracture of Aircraft Structures and Materials, AFFDL TR 70-144, 653, 1970.

# ON THE PREDICTION OF CATASTROPHIC FAILURES IN PRESSURIZED VESSELS

E. S. Folias

University of Utah, Salt Lake City, Utah 84112, USA

## ABSTRACT

Using a Dugdale-type model, the size of the plastic zone ahead of the crack tip is estimated, and a fracture criterion incorporating geometry and plasticity corrections is suggested. This criterion predicts failures in pressurized vessels of arbitrary shape by knowing only shell geometry, material properties and crack size. A comparison with some of the experimental data in existing literature substantiates the validity of the fracture criterion and its potential use.

## INTRODUCTION

Experience has taught us that when a sharp object penetrates a pressurized balloon two things may occur: (i) if the hole is small the balloon loses pressure; (ii) if the hole is large the balloon explodes. One conjectures, therefore, that for a given pressure there exist a critical flow size beyond which the balloon fails catastrophically and vice versa. In fact, one may further conjecture that the critical pressure is a function of the material properties, the geometry of the structure and the flaw shape and size.

Thin-walled pressure vessels do resemble balloons and like balloons are subject to puncture and explosive loss. For a given material, under a specified stress field due to internal pressure, there will be a crack length in the material which will be self-propagating. Crack lengths less than the critical value will cause leakage but not destruction. However, if the critical length is ever reached, either by penetration or by the growth of a small fatigue crack, an explosion and complete loss of the structure may occur. The subject of eventual concern therefore is to assess analytically the relation between critical pressure and critical crack lengths in sheets which are initially curved.

## GENERAL THEORY

Let us consider a portion of a thin, shallow shell*, of constant thickness $h$, subjected to an internal pressure $q(x,y)$ and containing a through the thickness line crack of length $2c$. In addition, we shall limit our considerations to elastic, isotropic and homogeneous segments of shells which are subjected to small deformations.

The basic variables in the theory of shallow shells are the displacement function $w(x,y)$ in the direction of the z axis and a stress function $F(x,y)$ which represents the stress resultants tangent to the middle surface of the shell. Following Marguerre [2], the coupled differential equations governing $w$ and $F$, with $x$ and $y$ as rectangular Cartesian coordinates of the base plane, are given by

$$\nabla^4 F = Eh\left[2\,\frac{\partial^2 w_0}{\partial x \partial y}\,\frac{\partial^2 w}{\partial x \partial y} - \frac{\partial^2 w_0}{\partial x^2}\,\frac{\partial^2 w}{\partial y^2} - \frac{\partial^2 w_0}{\partial y^2}\,\frac{\partial^2 w}{\partial x^2}\right] \qquad (1)$$

$$D\nabla^4 w = -q - 2\frac{\partial^2 F}{\partial x \partial y}\,\frac{\partial^2 w_0}{\partial x \partial y} + \frac{\partial^2 F}{\partial x^2}\,\frac{\partial^2 w_0}{\partial y^2} + \frac{\partial^2 F}{\partial y^2}\,\frac{\partial^2 w_0}{\partial x^2} \qquad (2)$$

where $w_0(x,y)$ describes the initial shape of the shell in reference to that of a flat plate and $D$ and $E$ the flexural rigidity and Young's modules respectively.

A theoretical attach of the general problem for an arbitrary initial curvature presents formidable mathematical complexities. However, for spherical and cylindrical shells, exact solutions have been obtained in an asymptotic form and the results can be found in references [3] - [11]. For other more complicated shell geometries, the results can be obtained by a proper superposition of the above two solutions [10].

Without going into the mathematical details, the stress distribution around the crack tip for a symmetrical loading is found to be:

---

*According to Ogibalov [1], a shell will be called shallow if the least radius of curvature is greater by one order of magnitude than the linear dimensions, i.e. $L/R \leq 0.1$; and thin if $h/R \leq 0.01$.

*Extensional stresses*--through the thickness:

$$\sigma_x^{(e)} = P^{(e)}\left(\frac{c}{2r}\right)^{1/2}\left(\frac{3}{4}\cos\frac{\theta}{2} + \frac{1}{4}\cos\frac{5\theta}{2}\right) + 0(r^0) \tag{3}$$

$$\sigma_y^{(e)} = P^{(e)}\left(\frac{c}{2r}\right)^{1/2}\left(\frac{5}{4}\cos\frac{\theta}{2} - \frac{1}{4}\cos\frac{5\theta}{2}\right) + 0(r^0) \tag{4}$$

$$\tau_{xy}^{(e)} = P^{(e)}\left(\frac{c}{2r}\right)^{1/2}\left(-\frac{1}{4}\sin\frac{\theta}{2} + \frac{1}{4}\sin\frac{5\theta}{2}\right) + 0(r^0) \tag{5}$$

*Bending stresses*--on the "tension side" of the shell:

$$\sigma_x^{(b)} = P^{(b)}\left(\frac{c}{2r}\right)^{1/2}\left(-\frac{3-3\nu}{4}\cos\frac{\theta}{2} - \frac{1-\nu}{4}\cos\frac{5\theta}{2}\right) + 0(r^0) \tag{6}$$

$$\sigma_y^{(b)} = P^{(b)}\left(\frac{c}{2r}\right)^{1/2}\left(\frac{11+5\nu}{4}\cos\frac{\theta}{2} + \frac{1-\nu}{4}\cos\frac{5\theta}{2}\right) + 0(r^0) \tag{7}$$

$$\tau_{xy}^{(b)} = P^{(b)}\left(\frac{c}{2r}\right)^{1/2}\left(-\frac{7+\nu}{4}\sin\frac{\theta}{2} - \frac{1-\nu}{4}\sin\frac{5\theta}{2}\right) + 0(r^0) \tag{8}$$

where $\nu$ is Poisson's ratio and $(r,\theta)$ are the polar coordinates around the crack tip. The stress coefficients $P^{(e)}$ and $P^{(b)}$ are functions of the crack size, geometry of the shell, material properties and loading characteristics. While their exact behavior can be found in reference [10], for certain simple loadings one may approximate them within a 7% error by the relations* of Table 1,

*The expressions in this table are valid for all $\lambda$.

TABLE 1

| | |
|---|---|
| Long cylinder axial crack 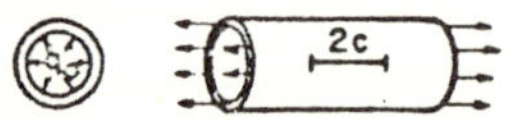 | $p^{(e)} \approx \sqrt{1 + 0.317\lambda^2}\ (q_0R/h)$ <br> $p^{(b)} \approx 0$ |
| Long cylinder peripheral crack 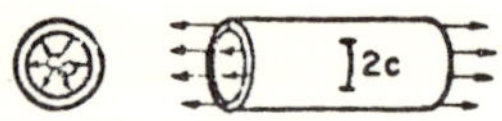 | $p^{(e)} \approx \sqrt{1 + 0.05\lambda^2}\ (q_0R/2h)$ <br> $p^{(b)} \approx 0$ |
| Spherical cap  | $p^{(e)} \approx \sqrt{1 + 0.466\lambda^2}\ (q_0R/2h)$ <br> $p^{(b)} \approx 0$ |

where for convenience we have defined $R$ to be the radius of curvature and

$$\lambda \equiv \left(\frac{Ehc^4}{R^2D}\right)^{1/4} \equiv \{12(1-\nu^2)\}^{1/4}\left\{\frac{R}{h}\right\}^{1/2}\frac{c}{R} \equiv \{12(1-\nu^2)\}^{1/4}\,\frac{c}{(Rh)^{1/2}} \quad (9)$$

It must be emphasized that classical bending theory has been used in deducing the foregoing results. Hence it is inherent that only the Kirchhoff equivalent shear free condition is satisfied along the crack, and not the vanishing of both individual shearing stresses. While outside the local region the stress distribution should be accurate, one might expect the same type of discrepancy to exist near the crack point as that found by Knowles and Wang [12] in comparing Kirchhoff and Reissner bending results for a flat plate. In this case the order of the stress singularity remained unchanged but the angular distribution around the crack changed so as to precisely be the same as that due to solely extensional loading.

Recently, Sih and Hagendorf [13] investigated this matter further by deriving an improved theory of shallow shells which incorporates the effect of a transverse shear deformation. As expected, their results showed that classic theory cannot adequately predict the exact angular dependence of the bending stresses in the vicinity of a crack. However, in general, these bending stresses are so small when compared to the extensional stresses that can be neglected. On the other hand, for very long cracks such contributions become significant and consequently may no longer be neglected. Unfortunately, in such cases bulging effects become extremely important and any theory, whether classic or shear, is inadequate.

## ELASTIC BRITTLE FRACTURE

From the foregoing discussion it becomes apparent that initially curved sheets containing through cracks present a reduced resistance to fracture initiation. Consequently, the presence of a crack in the walls of a pressure vessel can severely reduce the strength of the structure and can cause sudden failure at nominal tensile stresses less than the material yield strength. To ensure, therefore, the integrity of a structure, the designer must be cognizant of the relation that exists between fracture load, material properties, flow shape and size and structural geometry.

The principal task, however, of Fracture Mechanics is precisely the prediction of failure due to the presence of sharp discontinuities. Specifically, the approach is based on a corollary of the first law of thermodynamics which was first applied to the phenomenon of fracture by Griffith [14]. His hypothesis was that the total energy of a cracked system subjected to external loading remains constant as the crack extends an infinitesimal distance. That is

$$\frac{\partial U_{system}}{\partial c} = 0. \tag{10}$$

Following the work of Griffith, the total energy of the system is given by

$$U_{system} = U_{loading} + U_{surface} + U_{strain}, \tag{11}$$

where the increase in strain energy due to the presence of the crack may be calculated by considering the discontinuity in the v-displacement across the crack. In particular

$$U_{strain} = -\frac{1}{2}\int_{-c}^{c} \sigma_y(x,0)(v^+ - v^-)dx, \tag{12}$$

which in view of the results of ref. [3] now becomes

$$U_{strain} = -\frac{\pi}{E}c^2\left\{\left[P^{(e)}\right]^2 + \frac{2}{3}(1+\nu)^2\left[P^{(b)}\right]^2\right\}. \tag{13}$$

Thus for crack instability

$$4\gamma c = - \frac{\partial U_{strain}}{\partial c} , \tag{14}$$

where $\gamma$ is the surface energy per unit area, or

$$\begin{aligned} 4\gamma c = \frac{2\pi c}{E} \left\{ \left[ P^{(e)} \right]^2 + \frac{2}{3} (1 + \nu)^2 \left[ P^{(b)} \right]^2 \right\} + \\ + 2 \frac{\pi}{E} c^2 \left\{ P^{(e)} \frac{dP^{(e)}}{dc} + \frac{2}{3} (1 + \nu)^2 P^{(b)} \frac{dP^{(b)}}{dc} \right\} \end{aligned} \tag{15}$$

The reader should notice that as $R \to \infty$ the elastic fracture criterion for flat plates is recovered.

## PLASTICITY CORRECTION

Due to the presence of high stresses in the vicinity of the tip of a crack, when the appropriate yield criterion is satisfied then localized plastic deformation occurs and a plastic zone size is created. This phenomenon effectively increases the crack length and therefore must be accounted for. Following Dugdale [15], the size of the plastic zone (see fig. 1) is determined by the relation

$$\frac{c}{c_e} = \cos \left( \frac{\pi \sigma}{2\sigma_y} \right) \tag{16}$$

where $2c_e$ is the effective crack length and $\sigma_y$ the yield stress of the material. This relation, however, applies only to a perfect elasticplastic non-strain-hardening material. Subsequently, McClintock [16] has suggested that a strain-hardening material may be approximated by an ideally, plastic one if a stress higher than the yield stress and lower than the ultimate stress is chosen. In particular, the author suggests the value

$$\sigma^* = \frac{1}{2} \left\{ \sigma_y + \frac{\sigma_y + \sigma_u}{2} \right\}. \tag{17}$$

Correcting, therefore, the Griffith-Irwin equation so as to include plasticity and geometry effects one has

$$P^{(e)} = \frac{2\sigma^*}{\pi} \cos^{-1} \left[ \exp \left( - \frac{\pi K^2}{8\sigma^{*2} c} \right) \right] . \tag{18}$$

where K represents the fracture toughness of the material. It is interesting to note that one may also treat $\sigma^*$ and K as floating constants to be determined from scaled-down models.

EXPERIMENTAL VERIFICATION

In general, a theory is inadequate unless there exists experimental evidence to substantiate its validity and its potential use. Therefore, in the following we compare our results with some of the experimental data existing in literature.

1. R. P. Sopher, A. L. Lowe, D. C. Martin, and P. J. Rieppel, Evaluation of weld joint flaws on initiating points of brittle fracture, Welding J. (Nov. 1959), p. 4415. The results of tests on 9-ft-diameter, 3/4-in.-thick spheres with full thickness cracks are given.

   Material: ABS-B Steel

   $\sigma_y$: 30.7 ksi, $\sigma^* = 37.9$ ksi

   $\sigma_u$: 59.4 ksi, $K = 102.0$ ksi $\sqrt{\text{in}}$

   The results are plotted in fig 2. The agreement is very good.

2. R. B. Anderson and T. L. Sullivan, Fracture mechanics of through-cracked cylindrical pressure vessels. NASA TN D-3252. This paper gives results of tests on 6-in.-diameter, 0.060-in.-thick cylinders with full thickness cracks.

   Material: 2014-T6Al

   $\sigma_y$: 90.5 ksi, $\sigma^* = 91.9$ ksi

   $\sigma_u$: 93.9 ksi, $K = 51.6$ ksi $\sqrt{\text{in}}$

   The results are plotted in fig. 3. The agreement is good.

3. W. H. Irvine, A. Quirk and E. Bevitt, Fast fracture of pressure vessels: An appraisal of theoretical and experimental aspects and application to operational safety. J. Brit. Nucl. Energy Soc. (Jan. 1964). The results of test on 5-ft-diameter, 1-in.-thick, cylindrical vessels with through-cracks are given.

   Material: 0.36% C Steel

   $\sigma_y$: 33 ksi, $\sigma^* = 40$ ksi

   $\sigma_u$: 61 ksi, $K = 179$ ksi $\sqrt{\text{in}}$

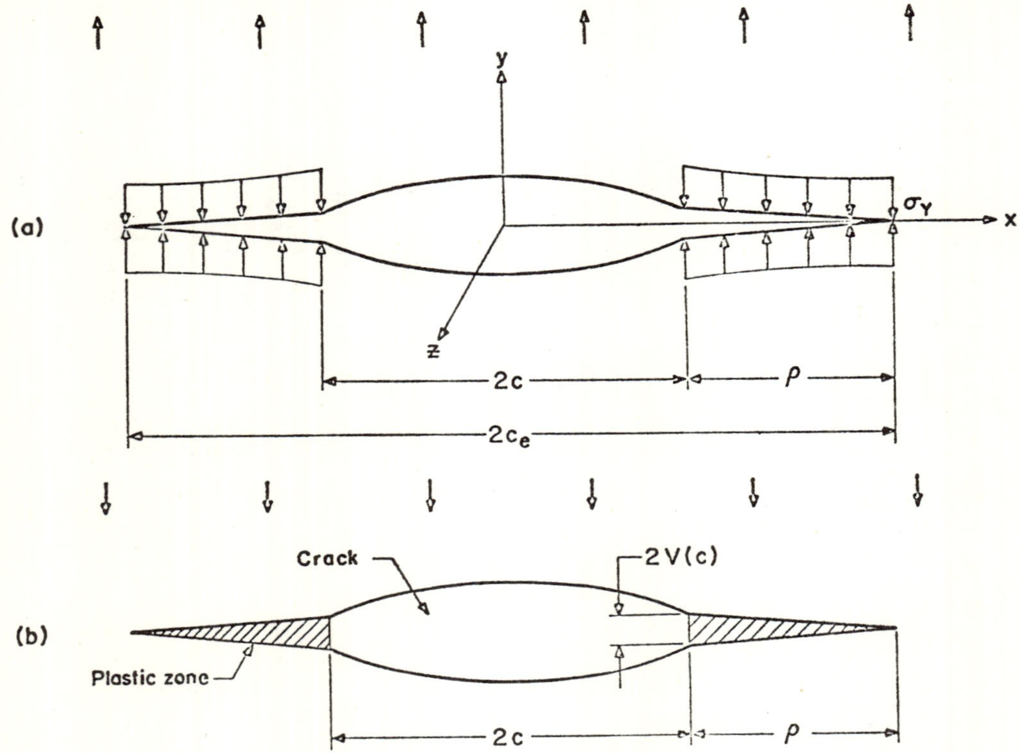


Fig. 1. (a) Internal stress distribution used in the Dugdale model of elastic-plastic deformation near a crack of length 2c under plane-stress tensile loading. (b) Displacements 2V associated with crack opening.

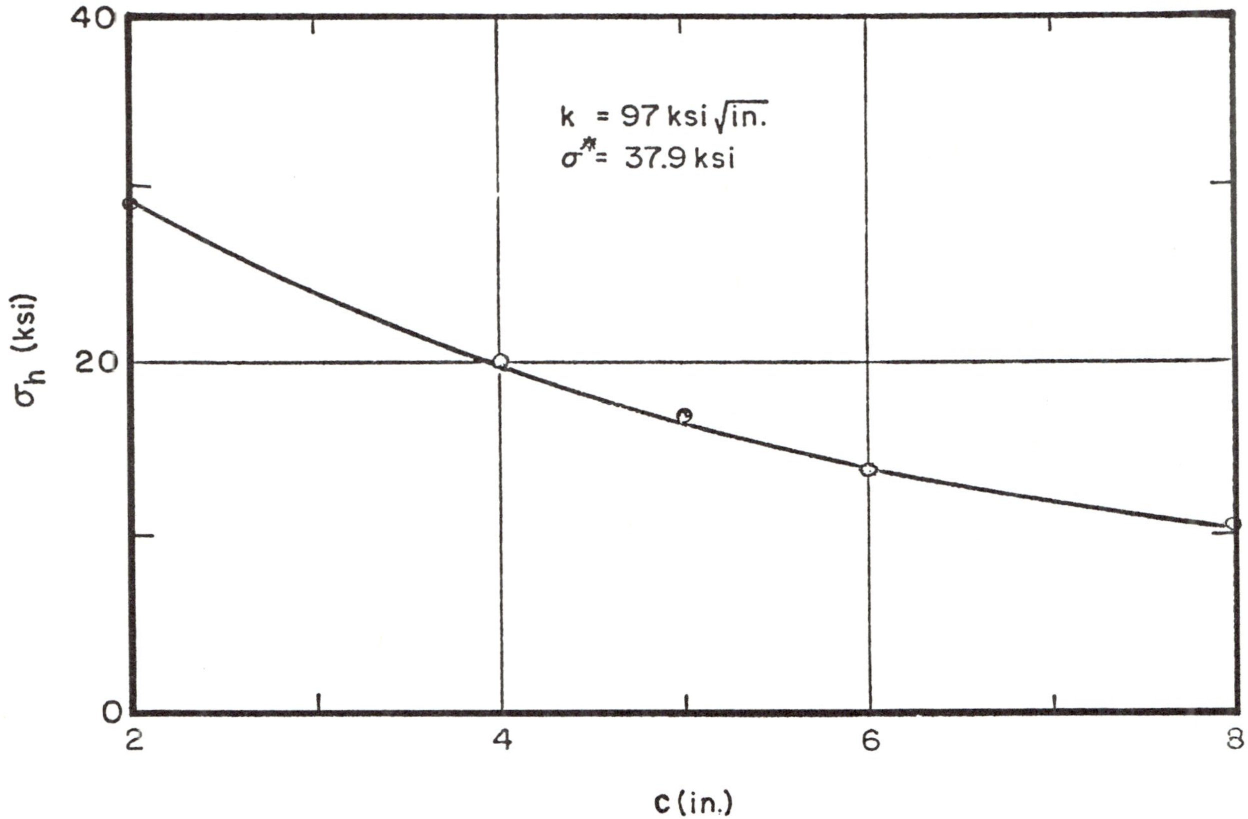


Fig. 2. Comparison between theory and experiment for ABS-B steel spherical vessels.

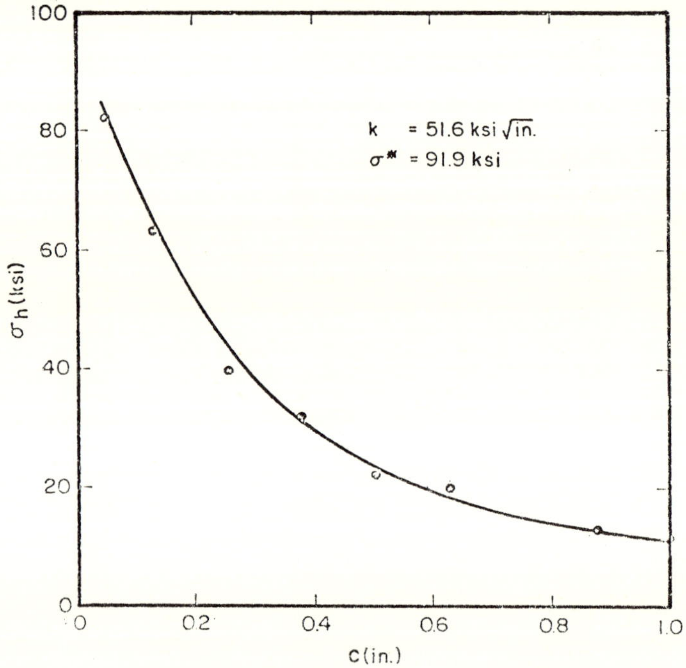


Fig. 3. Comparison between theory and experiment for 2014-T6Al cylindrical vessels at -423°F.

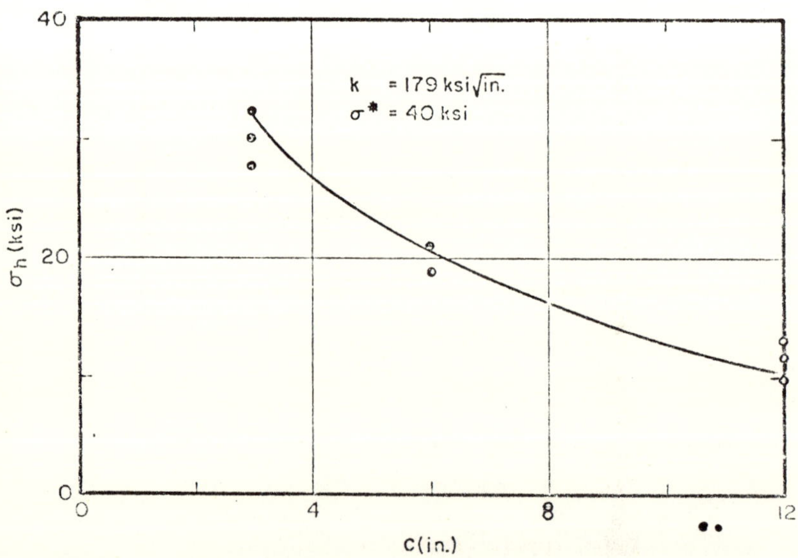


Fig. 4. Comparison between theory and experiment for 0.36% C steel cylindrical vessels.

The results are plotted in fig. 4. The agreement is fairly good.

Comparison with other experimental data available in the literature leads to equally good agreements.

## CONCLUSIONS

The close agreement between the theoretically predicted fracture strengths and the experimental data suggests that eq. (18) may be used to predict failures in pressurized vessels knowing only the structural geometry, the crack length, the ultimate and yield stresses, and the fracture toughness of the material.

An interesting point, however, should be stressed. It seems from the above examples that for very small crack lengths the cosine angle is greater than $\pi/2$ and the criterion, therefore, fails. This failure is due to the fact that the material, at least local to the crack, undergoes considerable strain-hardening and as a result the fracture hoop stress is higher than $\sigma^*$. At the present time there are no adequate criteria to handle this problem.

Recently, using a slightly different model than that of Dugdale's, the author was able to derive* a more accurate fracture criterion which for very small crack lengths shows better agreement with experimental observations. On the other hand, for larger crack lengths the predicted results are very close to those of eq. (18).

As an example, for a long cylindrical pressure vessel containing an axial crack the more accurate fracture criterion gives the hoop stress $\sigma_h$ as

$$\sigma_h = \sigma^* \left\{ \frac{1 + 0.317\lambda^2 \left(\frac{1-\alpha}{\alpha}\right)^2}{1 + 0.317\lambda^2 \left(\frac{1}{\alpha}\right)^2} (1 - \alpha) \right\}^{1/2} \tag{19}$$

where $\alpha$ is the ratio of the actual to the effective crack length, ie

$$\alpha = \frac{c}{c_e} = e^{-\frac{\pi K^2}{8\sigma^{*2} c}} \tag{20}$$

Although eq. (19) will yield more accurate results, it does require some rather lengthy calculations.

*For details see ref. [17].

In addition to the external applied loads, pressure vessels are also subjected to vibrations. Consequently, an investigation was carried out ref. [18] in order to assist analytically what effect, if any, do vibrations have on the mechanism of fracture. The analysis has shown that in general transverse vibrations reduce the stress intensity factor. However, when the forcing frequency approaches the natural frequency of the uncracked shell, the stress intensity factor increases without bound. This phenomenon coupled with the usual $1/\sqrt{r}$ singular behavior, causes the pressure vessel to fail at nominal values even lower than the yield stress.

REFERENCES

[1] Ogibalov, Dynamics and strength of shells, translated from Russian, published for NASA and NSF by the Israel Program for Scientific Translations.

[2] Marguerre, K., Zur Theorie der Gekrummten Platte Grosser Formanderung, Proceedings of the 5th International Congress of Applied Mechanics, 93-101, (1938).

[3] Folias, E. S., Journal of Mathematics and Physics, Vol. 44, No. 2, June 1965, pp. 164-176.

[4] Erdogan, F. and Kibler, International Journal of Fracture Mechanics, 5, 3 (September 1969).

[5] Folias, E. S., International Journal of Fracture Mechanics, 1, 2, 104-113 (1965); ARL 64-174, Aerospace Research Laboratories, United States Air Force, Dayton, Ohio, October 1964.

[6] Folias, E. S., International Journal of Fracture Mechanics, 3, 1, 1-11 (1967).

[7] Copley, L. G. and Sanders, J. L., International Journal of Fracture Mechanics, 5, 2 (June 1969).

[8] Duncan, M. E. and Sanders., J. L., International Journal of Fracture Mechanics, 5, 2 (June 1969).

[9] Folias, E. S., International Journal of Fracture Mechanics, 1, 1, 20-46 (1965).

[10] Folias, E. S., On the effect of initial curvature on cracked flat sheets, International Journal of Fracture Mechanics, Vol. 5, No. 4, pp. 327-346, December 1969.

[11] Fung, Y. C. and Sechler, E. E., Thin Shell Structures, Prentice-Hall, Chapter 21, January 1974.

[12] Knowles, J. K. and Wang, N. M., On the bending of an elastic plate containing a crack, Journal of Mathematics and Physics, Vol. 39, pp. 223-236, 1960.

[13] Sih, G. C. and Hagendorf, H. C., A new theory of spherical shells with cracks, Thin Shell Structures, edited by Y. C. Fung and E. E. Sechler, Prentice-Hall, pp. 519-545, 1974.

[14] Griffith, A. A., The theory of rupture, Proceedings of the 1st International Congress of Applied Mechanics, pp. 55-63, Delft, 1924.

[15] Dugdale, D. S., Journal of Mechanics and Physics of Solids, Vol. 8, p. 100, 1960.

[16] McClintock, F. A., On notch sensitivity, Welding Journal of Research Supplement, Vol. 26, pp. 202-208.

[17] Folias, E. S., Estimating plastic zone sizes ahead of a crack, to appear in International Journal of Fracture Mechanics, 1974.

[18] Do, S. H. and Folias, E. S., On the steady-state transfers vibrations of a cracked spherical shell, International Journal of Fracture Mechanics, Vol. 7, No. 1, March 1971.

## DISCUSSION by D. Radenkovic, Ecole Polytechnique, France

The solutions of Marguerre's equations presented in this communication will prove certainly very useful in the applications where membrane forces are predominant - the experimental results quoted by the author show this clearly. However, I would like to point out that on the vicinity of the crack tip the work of shearing forces cannot be neglected as a rule; in order to take it into account, on the calculations concerning through-cracks on plates and shells, it is necessary to use a theory of Reissner's type. In this way, by a local analysis, Radenkovic and Bergez [1,2] were able to define the five intensity factors which, in the general case, resume at the crack tip, the shell geometry and the loading. Once these factors are clearly defined, the solution of the global problem, which permits to find their values, can be obtained by numerical methods.

REFERENCES

[1] Bergez, D. and Radenkovic, D., "La caracterization des", etc.

[2] Bergez, D. and Radenkovic, D., "On the definition of stress-intensity factors in cracked plates and shells", Proc. 2nd Int. Conf. on Pressure Vessel Technology, San Antonio, Texas, pp. 1089-1096, 1973.

## AUTHOR'S CLOSURE

Yes, this inadequacy was also pointed out by the author in the paper. I would furthermore like to add that in the case of spherical shells, Sih and Hagendorf upon deriving a higher order theory, they were able to explain precisely this very point. Their results were reported at the Symposium on Thin Shell Structures, 1972. (See author's paper for reference).

# THE EFFECTS OF FASTENER FLEXIBILITY AND STIFFENER GEOMETRY ON THE STRESS INTENSITY IN STIFFENED CRACKED SHEET

T. Swift

McDonnell Douglas Corporation, Douglas Aircraft Division,
Long Beach, California 90846

## ABSTRACT

The fracture analysis of stiffened aircraft structure using closed form solutions, wherever possible, promises a more economical means of evaluating configurations for damage tolerance than finite element methods. This paper presents a closed form solution for the analysis of stiffened cracked panels including the effects of fastener flexibility and stiffener bending. One-bay cracks and two-bay cracks with center stiffener both intact and broken are included for a typical fuselage configuration. It is shown that considerable error in crack tip stress intensity and stiffener stress concentration can result by assuming compatibility of strain between stiffener and sheet without accounting for fastener flexibility. The effects of stiffener bending and stiffener material distribution are shown. In addition, the effects of fastener yielding on crack tip stress intensity and stiffener stress concentration are illustrated with test panel correlation.

## NOMENCLATURE

$a$ = half crack length, cm

$d$ = rivet diameter, cm

$E$ = Young's modulus, $MN/m^2$

$K$ = crack tip stress intensity, $MN/m^{3/2}$

$\sigma$ = gross applied stress, $MN/m^2$

$\sigma_s$ = stiffener stress, $MN/m^2$

$\nu$ = Poisson's ratio

Other terms are defined within the text.

## INTRODUCTION

Damage tolerance analysis depends on the determination of stress intensity factors and stiffener stress concentration factors. Finite element analysis methods have been used in the past for fracture analysis of stiffened structure [1,2]; however, this type of analysis requires considerable computer running time due to the inversion of large matrices. In the early development of a transport aircraft structure, a large number of parametric and trade studies is usually required to determine the most efficient structural configuration. Due to the expense of large finite element analyses, the cost of parametric configuration studies for damage tolerance becomes prohibitive. Closed form solutions, wherever possible, promise a more economical means of evaluating the many configurations necessary during the development of a large aircraft.

Many valuable contributions have been made to the development of closed form solutions for cracked stiffened structure [3]. The solution which has been more useful to the author during parametric studies is that of Poe [4]. The present paper illustrates the effects of refining some of the basic assumptions made in previous analysis of riveted stiffened sheet structure. Some of these refinements are as follows:

Not assuming compatibility of strain between the sheet and stiffener by considering fastener nonlinear flexibility.

Accounting for increased stiffener flexibility by considering stiffener bending.

Not confining the number of effective fasteners to those close to the crack.

Including the effects of a biaxial stress field.

## DAMAGE TYPES

The three types of damage considered in this paper, illustrated by Fig. (1), are: (a) one-bay skin crack, (b) two-bay skin crack with center stiffener intact, (c) two-bay skin crack with center stiffener broken. The degree of damage for which large commercial aircraft are now being designed is that of type (c). The present work is confined to plates of infinite width with three stiffeners considered for types (b) and (c) damage. Other authors [4] have considered cracks propagating across more than one stiffener. It is, however, pointed out that the structure must be able to withstand limit load conditions at any damage size within the selected tolerance. Structure designed for damage greater than type (c) would incur a severe weight penalty. Cracks extending beyond two bays with type (b) damage are also unrealistic since continuity of the center stiffener under limit load conditions with two bays of skin cracked is extremely difficult due to the heavy load concentration in the center stiffener. For cracks within two bays, additional outboard stiffeners will have little effect on the results. In view of these facts the three-stiffener analysis is considered justified.

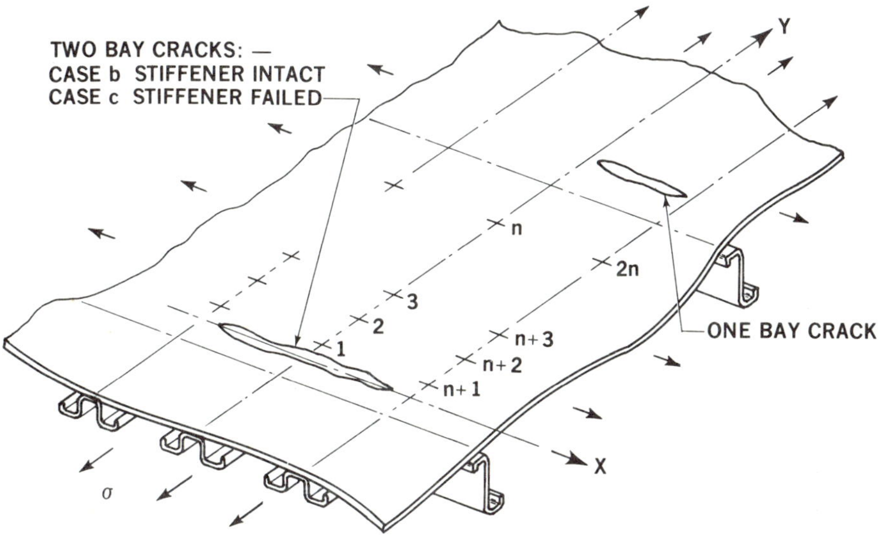


Fig. (1) - Damage types

ANALYSIS

In an unstiffened cracked panel all of the load given up by the cracked material is transferred to the sheet ahead of the crack. When the panel is stiffened, however, some of this load is transferred into the stiffeners through the fastener system thus reducing the stress ahead of the crack tip. The proportion of load redistributed into the intact stiffeners as the crack propagates depends on the stiffener geometry and the stiffness of its attachment to the sheet. If the fasteners are infinitely flexible then no load will be transferred from the cracked sheet; on the other hand, if the fasteners are infinitely rigid, a large portion of the load given up by the crack will be transferred to the stiffeners, thus reducing the stress intensity at the crack tip. The following analysis does not assume compatibility of strain between the stiffener outer fibers and the sheet, as is the usual case, but compatibility of deflection of the sheet with each fastener shear face allowing for fastener slip or deflection. In addition, the effects of increased stiffener outer fiber extension due to simple bending are included.

Space limitations make it impossible to include complete development of the compatibility equations involved, and so this paper will be confined to discussion of the philosophy, statement of the equations and illustration of some of the effects of fastener flexibility and stiffener bending.

## SHEET DEFLECTIONS

Sheet deflections due to overall stress and concentrated fastener forces are obtained using similar methods to those proposed by Poe [4]. Considering the complex Airy stress function approach of Westergaard [5], the deflection anywhere in a biaxially loaded unstiffened cracked panel can be expressed as

$$v = [2 \,\mathrm{Im}\, \bar{Z} - (1+\nu) y \,\mathrm{Re}\, Z]/E \tag{1}$$

where Z is a function of a complex variable given by Eq. (2).

$$Z = \sigma z/[z^2-a^2]^{1/2} \tag{2}$$

$z = x+iy$, $\bar{Z}$ is the first integral of Z, Im and Re are the imaginary and real parts of Z respectively. Deflection due to uniaxial loading can be obtained by applying a compressive stress in a direction parallel to the crack. The deflection in a cracked panel due to fastener forces can be determined by first considering the stress and deflection in an uncracked panel using a development of Eq. (77) of [6]. The resulting stress, along the x-axis, in the plane of the crack due to fastener forces over a distance equal to the crack length, is cancelled by applying a stress of opposite sign. Deflections due to this condition can be obtained by substituting into Eq. (1) a complex stress function proposed by Irwin [7], given by Eq. (3), for a system of concentrated forces applied to the crack face (see Fig. (2)).

$$Z = \frac{2Qa}{\pi t(z^2-b^2)} \left[\frac{1-(b/a)^2}{1-(b/z)^2}\right]^{1/2} \tag{3}$$

Total deflection due to the distribution of stress is obtained by substituting the concentrated forces by a stress acting on an increment of crack face area and integrating over the crack length.

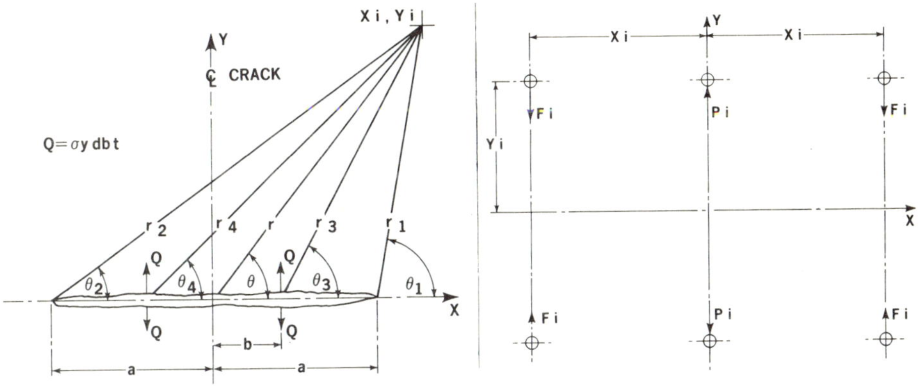


Fig. (2) - Coordinate system and fastener load convention

## STIFFENER EXTENSION

The stiffener extensions considered are those due to direct loads and bending from fastener loads, direct applied stress and fastener deflections. For this particular study, stiffener bending distribution is obtained using continuous beam theory with reactions provided by frames or ribs. For the center broken stiffener case the loads applied to the center stiffener are shown in Fig. (3). Stiffener bending is assumed to damp out over two frame spacings. Bending loads applied to the outer, intact stiffeners are also reacted by frames in a similar manner. Stiffener extensions considered are those in the y direction at the outer fiber adjacent to the sheet.

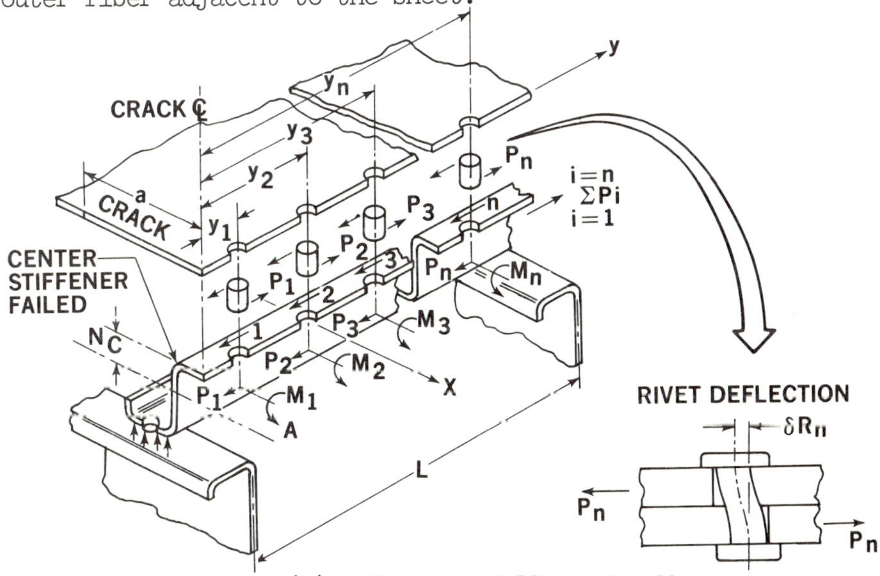


Fig. (3) - Center stiffener bending

## CENTER STIFFENER BROKEN CASE

Because the center stiffener is assumed broken, the reference point for extension of this stiffener is made at the nth fastener (Fig. (1)). The total stiffener extension relative to the x-axis at the ith fastener is given by Eq. (4).

$$\delta sc_i = v_n - \delta D_i - \delta M_i - \delta R_i + \delta R_n \tag{4}$$

where $v_n$ is the skin deflection at the nth fastener, $\delta D_i$ is the stiffener extension due to direct fastener loads, $\delta M_i$ is the stiffener outer fiber extension due to bending, $\delta R_i$ and $\delta R_n$ are the ith and nth fastener deflections, respectively. Total outer stiffener extension relative to the x-axis at the ith fastener is given by Eq. (5).

$$\delta so_i = \delta D_i + \delta M_i + \delta R_i + \delta G_i \tag{5}$$

where the first three terms are as described above and $\delta G_i$ is the stiffener extension at the ith fastener due to overall stress.

Compatibility at the center broken stiffener is given by Eq. (6).

$$\sum_{j=n+1}^{j=2n} F_j[v_2(x_n,y_n,x_j,x_j) - \frac{(1+\nu)y_j}{2\pi^2 Et}\int_0^a \alpha(x_j,y_j,b)\ \varepsilon(x_n,y_n,b)db]$$

$$- \sum_{j=1}^{j=n} P_j[v_3(x_n,y_n,y_j) - \frac{(1+\nu)y_j}{2\pi^2 Et}\int_0^a \beta(y_j,b)\ \varepsilon(x_n,y_n,b)db]$$

$$- \sum_{j=n+1}^{j=2n} F_j[v_2(x_i,y_i,x_j,y_j) - \frac{(1+\nu)y_j}{2\pi^2 Et}\int_0^a \alpha(x_j,y_j,b)\ \times$$

$$\times\ \varepsilon(x_i,y_i,b)db]$$

$$+ \sum_{j=1}^{j=n} P_j[v_3(x_i,y_i,y_j) - \frac{(1+\nu)y_j}{2\pi^2 Et}\int_0^a \beta(y_j,b)\ \varepsilon(x_i,y_i,b)db]$$

$$+ \delta R_n - \delta D_i - \delta M_i - \delta R_i = \sigma[v_1(x_i,y_i) - v_1(x_n,y_n)] \tag{6}$$

where i is considered from 1 to n-1

$$\delta D_i = [\sum_{j=1}^{j=i} P_j(y_n-y_i) + \sum_{j=i+1}^{j=n-1} P_j(y_n-y_j)]/AE \quad (7)$$

$$\delta M_i = \frac{c}{EI}[\sum_{i=i}^{i=n-1} Ma_i(y_{i+1}-y_i)] \quad (8)$$

$$Ma_i = \sum_{j=1}^{j=i} cP_j - \frac{c}{L}[\sum_{j=1}^{j=n} P_j(\frac{5}{4} - \frac{3y_j}{4L^2})][\frac{y_{i+1}+y_i}{2}] \quad (9)$$

The nth equation is obtained from center stiffener equilibrium under fastener loads and applied stress.

Equation (10) is the final equation for compatibility at the outer intact stiffener. E is Young's modulus, a is half crack length, ν Poisson's ratio, t sheet thickness, L frame spacing and c is the distance from cg of the stiffener to inside skin line.

$$- \sum_{j=n+1}^{j=2n} F_j[v_2(x_i,y_i,x_j,y_j) - \frac{(1+\nu)y_j}{2\pi^2Et}\int_0^a \alpha(x_j,y_j,b) \times$$

$$\times \varepsilon(x_i,y_i,b)db]$$

$$+ \sum_{j=i}^{j=n} P_j[v_3(x_i,y_i,y_j) - \frac{(1+\nu)y_j}{2\pi^2Et}\int_0^a \beta(y_j,b)\ \varepsilon(x_i,y_i,b)db]$$

$$+ \delta D_i + \delta M_i + \delta R_i = \sigma[v_1(x_i,y_i) - y_i/E] \quad (10)$$

where i is considered from n+1 to 2n

$$\delta D_i = [\sum_{j=n+1}^{j=1} F_j y_j + y_i \sum_{j=i+1}^{j=2n} F_j]/AE \quad (11)$$

$$\delta M_i = \frac{c}{EI}\sum_{i=n+1}^{i=i} Ma_i(y_i-y_{i-1}) \quad (12)$$

$$Ma_i = \sum_{j=i}^{j=2n} cF_j - \frac{3c}{2L^3}[\sum_{j=n+1}^{j=2n} F_j(2Ly_j-y_j^2)] \times$$

$$\times [L - \frac{y_{i-1}+y_i}{2}] \quad (13)$$

Terms used in these compatibility equations are described by Eqs. (14) to (19) with coordinate system and fastener load convention illustrated in Fig. (2).

$$v_1(x_i,y_i) = \{2\sqrt{r_1 r_2}\,\sin\left(\frac{\theta_1+\theta_2}{2}\right) - \frac{(1+\nu)y_i r}{\sqrt{r_1 r_2}} \times$$

$$\times\left[\cos\left(\theta - \frac{\theta_1+\theta_2}{2}\right)\right] + \nu y_i\}/E \tag{14}$$

$$v_2(x_i,y_i,x_j,y_j) = \frac{(1+\nu)(3-\nu)}{16\pi Et}\left((Xa+1)\mathrm{Ln}\left[\frac{(Xa+1)^2+Ya^2}{(Xa+1)^2+Yb^2}\right]\right.$$

$$- (Xa-1)\mathrm{Ln}\left[\frac{(Xa-1)^2+Ya^2}{(Xa-1)^2+Yb^2}\right] + (Xb+1)\mathrm{Ln}\left[\frac{(Xb+1)^2+Ya^2}{(Xb+1)^2+Yb^2}\right]$$

$$- (Xb-1)\mathrm{Ln}\left[\frac{(Xb-1)^2+Ya^2}{(Xb-1)^2+Yb^2}\right] + \frac{4(1-\nu)}{(3-\nu)}\{Ya\tan^{-1}\left[\frac{2Ya}{Ya^2+Xa^2-1}\right]$$

$$+ Ya\tan^{-1}\left[\frac{2Ya}{Ya^2+Xb^2-1}\right] - Yb\tan^{-1}\left[\frac{2Yb}{Yb^2+Xb^2-1}\right]$$

$$\left.- Yb\tan^{-1}\left[\frac{2Yb}{Yb^2+Xa^2-1}\right]\}\right) \tag{15}$$

$$v_3(x_i,y_i,y_j) = \frac{(1+\nu)(3-\nu)}{16\pi Et}\left(\left(\frac{2x_i}{d}+1\right)\mathrm{Ln}\left[\frac{\{(2x_i/d)+1\}^2+Ya^2}{\{(2x_i/d)+1\}^2+Yb^2}\right]\right.$$

$$- \left(\frac{2x_i}{d}-1\right)\mathrm{Ln}\left[\frac{\{(2x_i/d)-1\}^2+Ya^2}{\{(2x_i/d)-1\}^2+Yb^2}\right] + \frac{4(1-\nu)}{(3-\nu)} \times$$

$$\left.\times\{Ya\tan^{-1}\left[\frac{2Ya}{Ya^2+(2x_i/d)^2-1}\right] - Yb\tan^{-1}\left[\frac{2Yb}{Yb^2+(2x_i/d)^2-1}\right]\}\right) \tag{16}$$

$$\alpha(x_j,y_j,b) = \frac{(3+\nu)}{(1+\nu)}\left[\frac{1}{(b-x_j)^2+y_j^2} + \frac{1}{(b+x_j)^2+y_j^2}\right]$$

$$- \frac{2(b-x_j)^2}{[(b-x_j)^2+y_j^2]^2} - \frac{2(b+x_j)^2}{[(b+x_j)^2+y_j^2]^2} \tag{17}$$

$$\beta(y_j,b) = \frac{(3+\nu)}{(1+\nu)}\left[\frac{1}{b^2+y_j^2}\right] - \frac{2b^2}{(b^2+y_j^2)^2} \tag{18}$$

$$\varepsilon(x_i,y_i,b) = \mathrm{Ln}\left[\frac{(a^2-b^2)+(a^2-b^2)^{1/2}(BC+AD)+r_1r_2}{(a^2-b^2)-(a^2-b^2)^{1/2}(BC+AD)+r_1r_2}\right.$$

$$- \frac{y_i(1+\nu)(a^2-b^2)^{1/2}}{r_1r_2r_3^2r_4^2}\{(x_i^2-b^2-y_i^2)[x_i(AC-BD) + y_i(BC+AD)]$$

$$- 2x_iy_i\,[x_i(BC+AD) - y_i(AC-BD)]\} \tag{19}$$

$$Xa = \frac{2}{d}(x_i-x_j),\ Xb = \frac{2}{d}(x_i+x_j),\ A = \sqrt{r_1+x_i-a},\ B = \sqrt{r_1-x_i+a}$$

$$Ya = \frac{2}{d}(y_i-y_j),\ Yb = \frac{2}{d}(y_i+y_j),\ C = \sqrt{r_2+x_i+a},\ D = \sqrt{r_2-x_i-a}$$

CENTER STIFFENER INTACT CASE

Total stiffener extension at the ith fastener for both center and outer intact stiffeners is given by Eq. (5). Equation (20) is the final equation for compatibility at the center stiffener.

$$- \sum_{j=n+1}^{j=2n} F_j[v_2(x_i,y_i,x_j,y_j) - \frac{(1+\nu)y_j}{2\pi^2Et}\int_0^a \alpha(x_j,y_j,b)\ \times$$

$$\times\ \varepsilon(x_i,y_i,b)db]$$

$$- \sum_{j=1}^{j=n} P_j[v_3(x_i,y_i,y_j) - \frac{(1+\nu)y_j}{2\pi^2Et}\int_0^a \beta(y_j,b)\ \varepsilon(x_i,y_i,b)db]$$

$$+ \delta D_i + \delta M_i + \delta R_i = \sigma[v_1(x_i,y_i) - y_i/E] \tag{20}$$

where i is considered from 1 to n. The terms $v_1$, $v_2$, $v_3$, $\alpha$, $\beta$

and $\varepsilon$ are given by Eqs. (14) - (19), respectively. The terms $\delta D_i$ and $\delta M_i$ are given by Eqs. (21) and (22).

$$\delta D_i = [\sum_{j=1}^{j=i} P_j y_j + y_i \sum_{j=i+1}^{j=n} P_j]/AE \tag{21}$$

$$\delta M_i = \frac{c}{EI} \sum_{i=1}^{i=i} Ma_i (y_i - y_{i-1}) \tag{22}$$

$$Ma_i = \sum_{j=i}^{j=n} cP_j - [\frac{3c}{2L^3} \sum_{j=1}^{j=n} P_j (2Ly_j - y_j^2)][L - \frac{y_{i-1}+y_i}{2}] \tag{23}$$

Compatibility at the outer stiffener is given by Eq. (10) but with the sign of $P_j$ changed from positive to negative.

## ONE-BAY CRACK CASE

For the one-bay crack case with the crack propagating along the x-axis, centered between the stiffeners, compatibility is given by Eq. (24).

$$- \sum_{j=1}^{j=n} F_j [v_2(x_i, y_i, x_j, y_j) - \frac{(1+\nu) y_j}{2\pi^2 Et} \int_0^a \alpha(x_j, y_j, b) \times$$

$$\times \varepsilon(x_i, y_i, b) db] + \delta D_i + \delta M_i + \delta R_i = \sigma[v_1(x_i, y_i) - y_i/E] \tag{24}$$

$$\delta D_i = [\sum_{j=1}^{j=i} F_j y_j + y_i \sum_{j=i+1}^{j=n} F_j]/AE \tag{25}$$

$$\delta M_i = \frac{c}{EI} \sum_{i=n}^{i=i} Ma_i (y_i - y_{i-1}) \tag{26}$$

$$Ma_i = \sum_{j=i}^{j=n} cF_j - [\frac{3c}{2L^3} \sum_{j=1}^{j=n} F_j (2Ly_j - y_j^2)][L - \frac{y_{i-1}+y_i}{2}] \tag{27}$$

## FORMULATION OF EQUATIONS

Equations (6), (10), (20) and (24) have been programmed in FORTRAN language for the IBM computer. The integration on the left-hand side of these equations is performed numerically. Figure (4) illustrates a 2n x 2n matrix of equations for the solution of case (c). The value n is the number of rivets in each stiffener as illustrated by Fig. (1). The matrix is inverted within the program to solve for fastener loads. Crack tip stress intensity factors in the cracked sheet are obtained from Eq. (28).

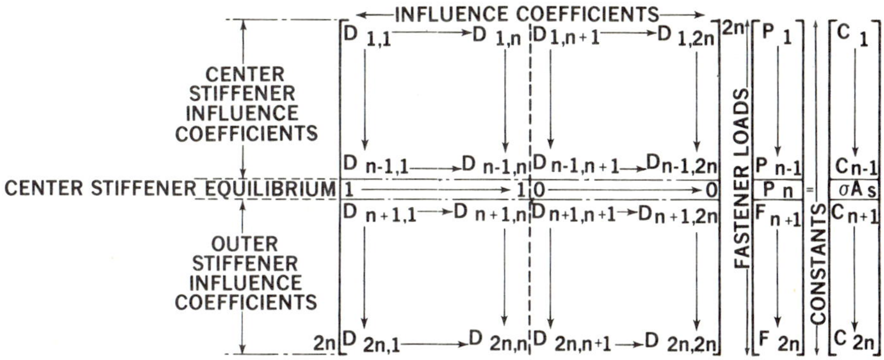


Fig. (4) - Compatibility matrix

$$K = \sigma\sqrt{\pi a} + \sum_{i=1}^{i=n} P_i[K(a,y_i)] - \sum_{i=n+1}^{i=2n} F_i[K(a,x_i,y_i)] \qquad (28)$$

The values $K(a,y_i)$ and $K(a,x_i y_i)$ are the stress intensity factors due to unit values of fastener loads and are obtained from [8]. Equation (28) is also used for case (b) with the sign of the second term changed from positive to negative. Case (a) is obtained by elimination of the second term in Eq. (28).

FASTENER FLEXIBILITY

Local fastener flexibility has been obtained from tests on single shear lap splice specimens tested back-to-back to eliminate joint rotation as illustrated by Fig. (26) of [2]. After subtraction of the sheet extension, the local fastener flexibility can be approximated by Eq. (29).

$$\delta R/P = [A + B(d/t_1 + d/t_2]/Ed \qquad (29)$$

where $\delta R$ is rivet deflection, P is applied load, E is modulus of the sheet material, d is rivet diameter, $t_1$ and $t_2$ are thicknesses of joined sheets. A is 5.0 for aluminum rivets and 1.666 for steel fasteners. B is 0.8 for aliminum rivets and 0.86 for steel fasteners.

## PARAMETRIC STUDIES

The prime purpose of this investigation was to improve the efficiency of analysis by cutting down computer time. Computing time can be minimized by reducing the problem size (number of fasteners considered) to a reasonable level while still maintaining the desired degree of accuracy. Reference [3] had assumed one fastener at each side of the crack while [4] found five fasteners to be appropriate. With the added refinements of fastener flexibility and stiffener bending it was decided to modify the programs to allow for a variable number of fasteners and to find the point at which accuracy was not improved by increasing the problem size.

All of the parametric analyses shown are for one typical fuselage configuration with skin thickness 1.8 mm, hat section stiffeners spaced 20.32 cm apart with area 3.53 $cm^2$, inertia of 3.180 $cm^4$ and distance from skin to centroid of 1.364 cm.

Figure (5) shows the results of analysis on a panel containing a one-bay crack. The figure shows no variation for a half-crack length of 1.27 cm, but a reasonable variation for a half-crack length of 13.97 cm. The solution has stabilized at about n = 15. Figure (6) shows the results of a similar analysis for a two-bay crack case with a broken central stiffener. For this case, however, considerable variation takes place for both short and long crack lengths, but again the solution has stabilized at about n = 15. In view of this it would seem that no advantage can be gained by increasing the problem size beyond 15 fasteners on each side of the crack in each stiffener. Fifteen fasteners were therefore used.

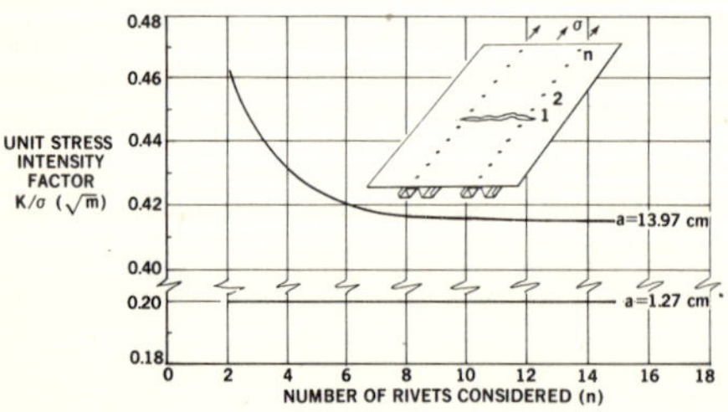


Fig. (5) - Effect of number of rivets considered - case (a)

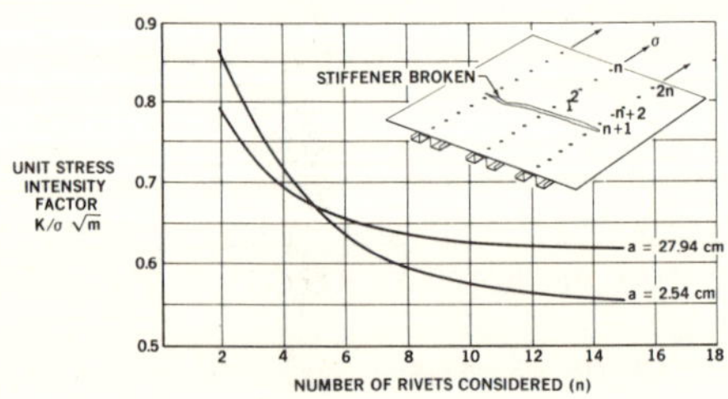


Fig. (6) - Effect of number of rivets considered - case (c)

Figure (7) shows the effects of fastener flexibility and stiffener bending for case (a). The effects of variation of lateral frame spacing L are shown by Fig. (8). Figures (9) - (12) show the effects of fastener flexibility and stiffener bending for cases (b) and (c), respectively. The effects of frame spacing L for cases (b) and (c) are shown on Figs. (13) and (14). Figure (15) shows the percent difference between rigid fasteners - no bending and flexible fastener - with bending assumptions for the three cases considered. Figures (7) - (15) are for rivet spacing 3.175 cm with a crack running between rivets. Figure (16) shows the difference between a crack running through a fastener versus one running between fasteners for case (c). Figures (17) and (18) illustrate the effects of biaxial stress for cases (b) and (c), respectively. The effects of rivet spacing for case (c) are illustrated by Fig. (19). Figure (20) shows the effect of stiffener material distribution on crack tip stress intensity for case (c). Stiffeners are spaced 20.32 cm apart and have an area of 5.16 cm$^2$, skin thickness is 2.54 mm and rivet spacing 2.54 cm.

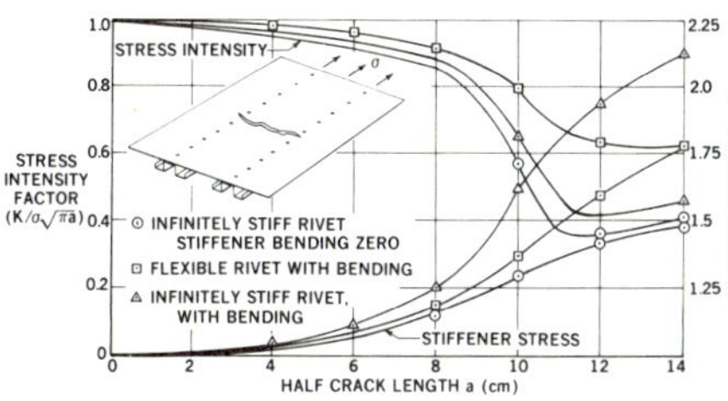


Fig. (7) - Stress intensity and stiffener stress - case (a)

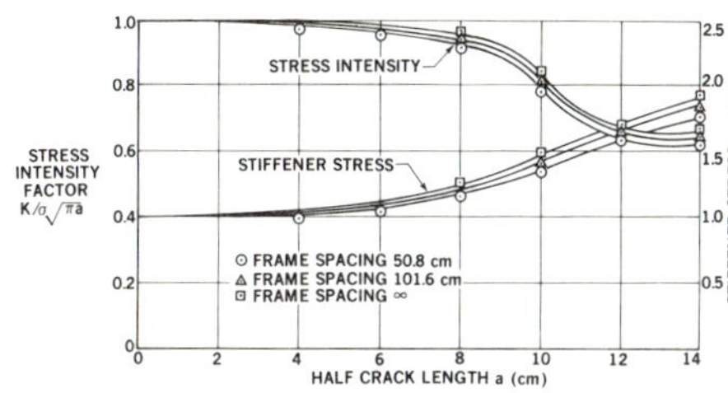


Fig. (8) - Effect of frame spacing - case (a)

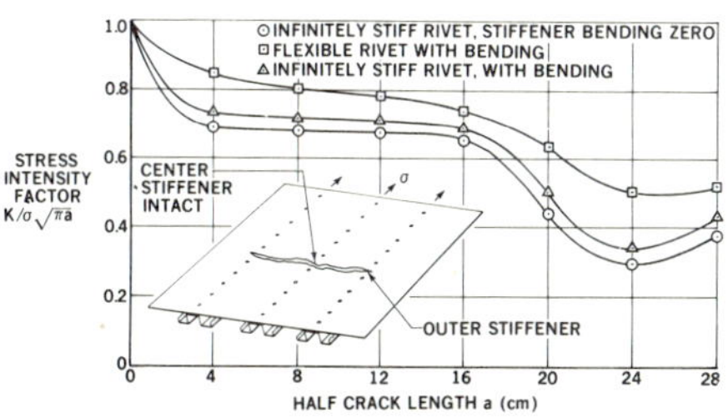


Fig. (9) - Stress intensity - case (b)

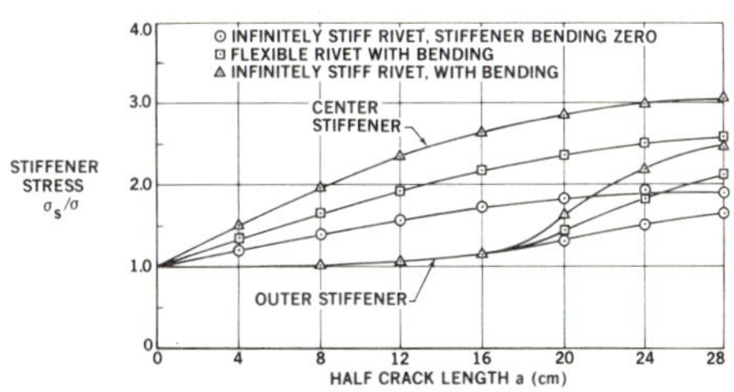


Fig. (10) - Stiffener stress - case (b)

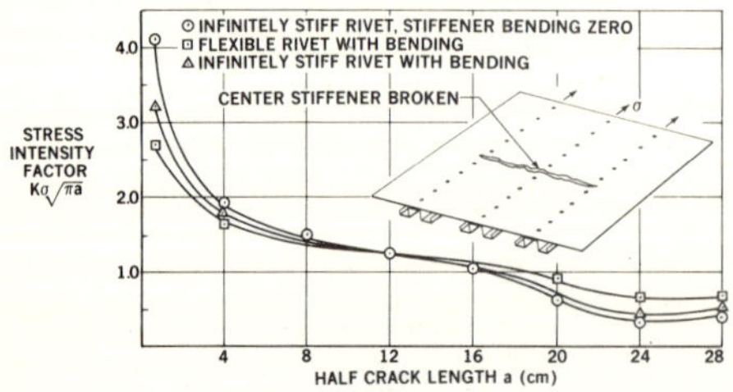


Fig. (11) - Stress intensity - case (c)

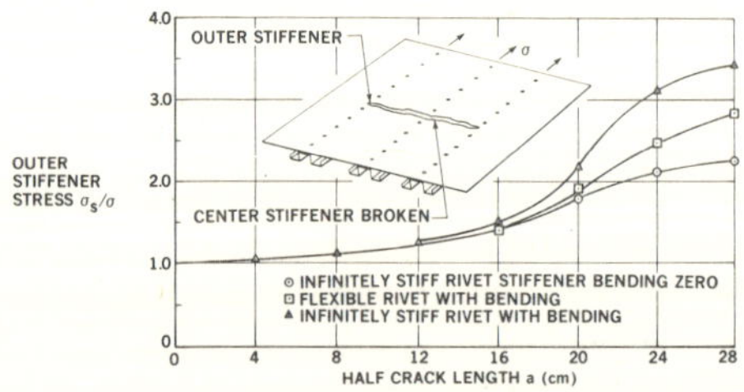


Fig. (12) - Stiffener stress - case (c)

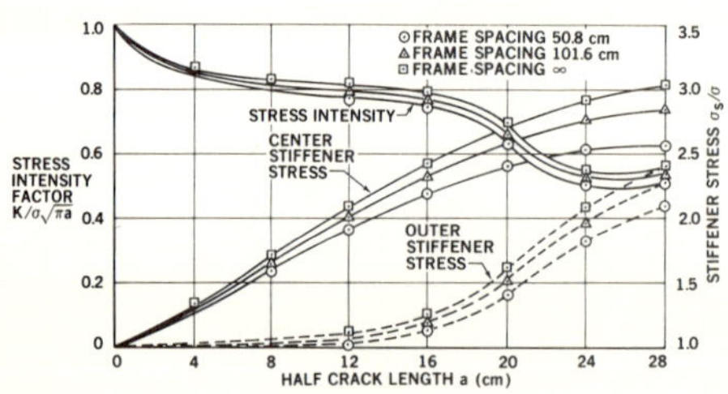


Fig. (13) - Effect of frame spacing - case (b)

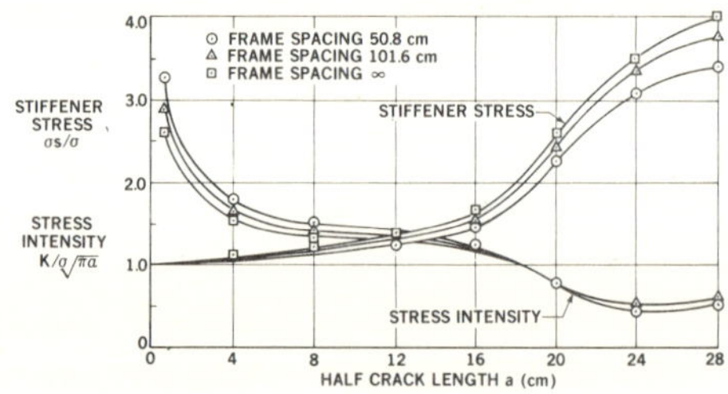


Fig. (14) - Effect of frame spacing - case (c)

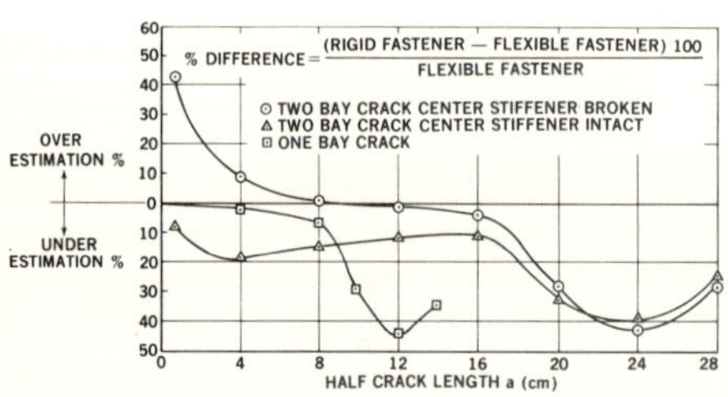


Fig. (15) - Difference between rigid and flexible fastener systems

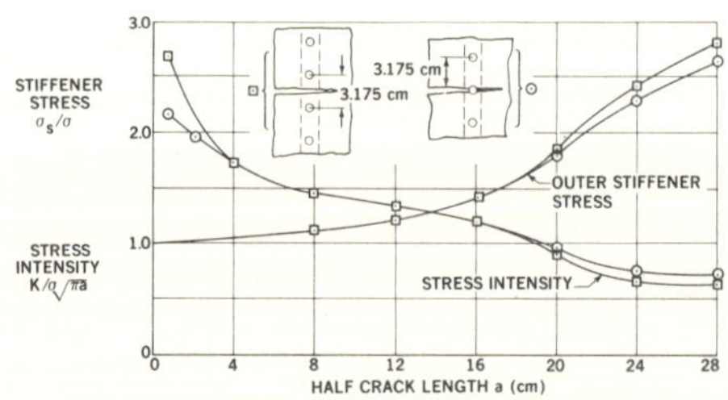


Fig. (16) - Comparison - crack toward versus crack between rivets - case (c)

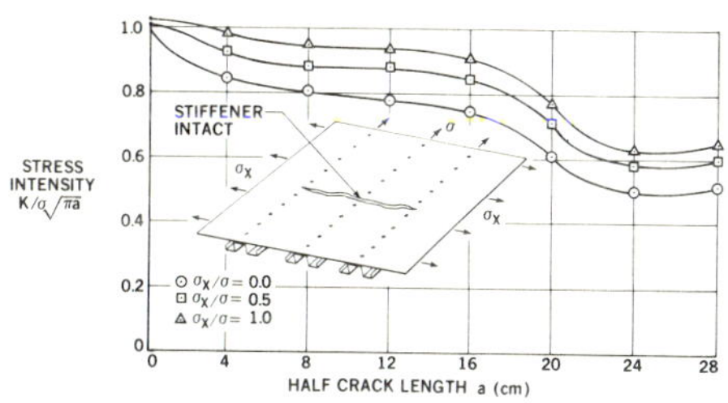


Fig. (17) - Effects of biaxial stress on stress intensity - case (b)

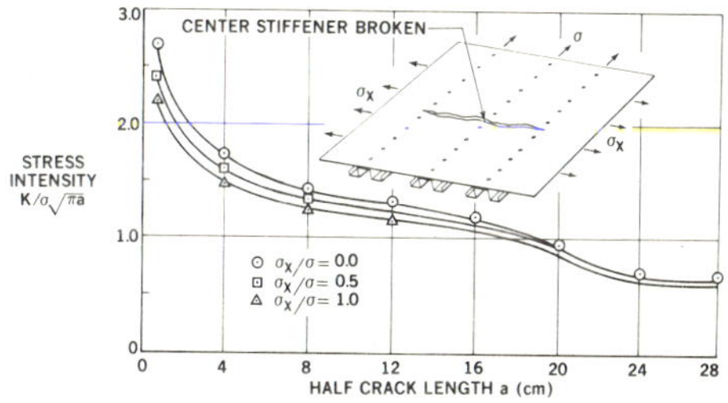


Fig. (18) - Effects of biaxial stress on stress intensity - case (c)

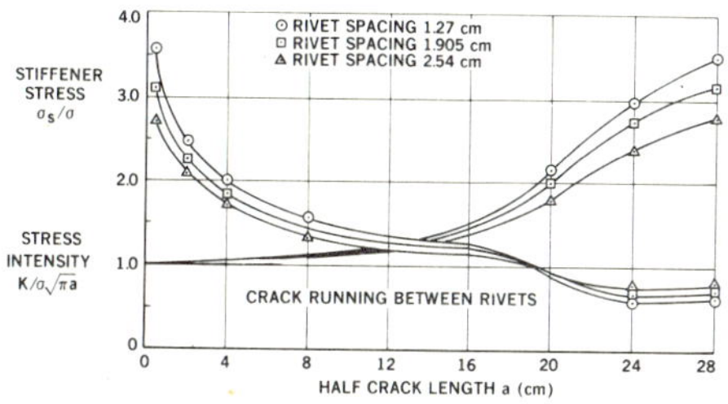


Fig. (19) - Effect of rivet spacing - case (c)

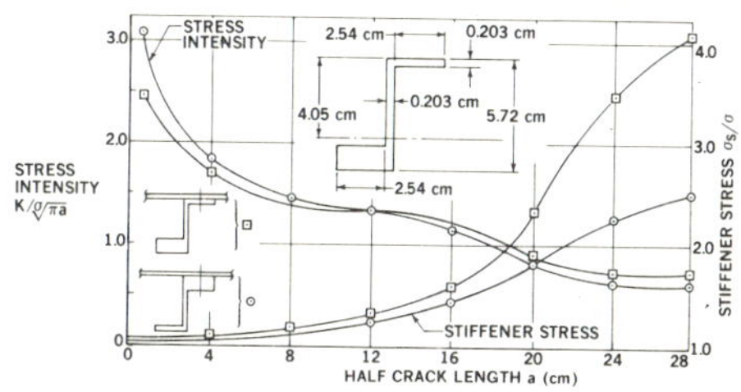


Fig. (20) - Effect of stiffener material distribution - case (c)

## FASTENER YIELDING

Fastener yielding plays an extremely important role in residual strength calculations of stiffened structure particularly when cracks are long. Elastic analysis, assuming either rigid or flexible fasteners, will always predict extremely high loads in those fasteners adjacent to the crack in the crack stopping outer stiffeners. Figure (21) shows a stiffened panel with 2024 T3 skin and 7075 T6 extended stiffeners, riveted with NAS 1097 DD6 rivets at 3.175 cm spacing. Geometrical configurations of stiffeners were identical to those previously described for the parametric analysis. Failure of the panel took place at 275 $MN/m^2$ gross stress with a half-crack length 25.1 cm and center stiffener saw cut. Inspection of the failed parts revealed that fasteners over the entire length of both crack stopping outer stiffeners had sheared as illustrated in Fig. (21). Failure of the stiffeners had taken place approximately 12.07 cm from the crack at a section reduced in area by fastener holes attaching the stiffeners to the frames. Residual strength based on elastic analysis is illustrated in Fig. (22) using $K_c$ of 2.11.4 $MN/m^{3/2}$ for sheet material [1]. Ultimate strength of the stiffener material was 583.3 $MN/m^2$ from coupon tests of the failed parts and single shear rivet allowable of 5560 N from tests using exact material thicknesses. It can be seen that strength is underestimated both from a fastener and stiffener standpoint. The analysis was modified to account for rivet yielding using an elastic-plastic model shown in Fig. (23). The slope of the plastic line was chosen to give equivalent energy while passing through the failure point. This method was chosen rather than an equation simulated curve so as to reduce the number of iterations required to converge the solution. Figure (24) shows the effects of fastener yielding on crack tip stress intensity, stiffener stress and first rivet load for a half-crack length of 25.1 cm. The predicted first rivet load at failure was 5783 N while simple splice tests gave 5560 N (4 percent high). Predicted crack tip stress intensity at failure was 212 $MN/m^{3/2}$ while $K_c$ from [1] gave 211.4 $MN/m^{3/2}$. Predicted stiffener stress at failure of the panel was 512.3 $MN/m^2$ under the crack indicating that failure would not take place under the crack, which was substantiated by the test. The elastic-plastic analysis predicts that failure would be precipitated by fastener failure within 4 percent of the test result, whereas the elastic analysis was off considerably, especially for the fastener load prediction as indicated by Fig. (22). The effect of fastener yielding on residual strength is illustrated on Fig. (22).

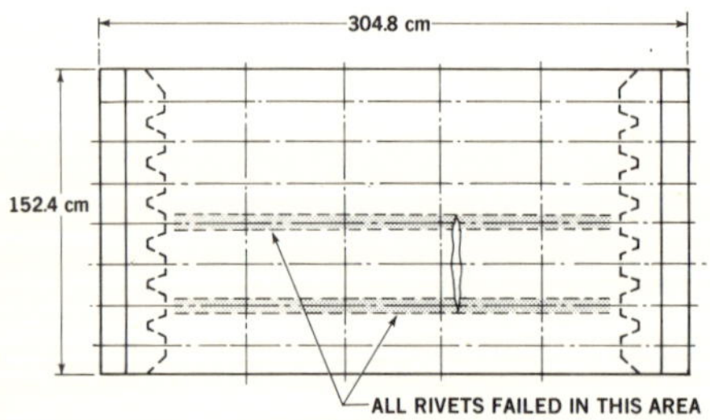


Fig. (21) - Rivet failure in stiffened test panel containing a crack

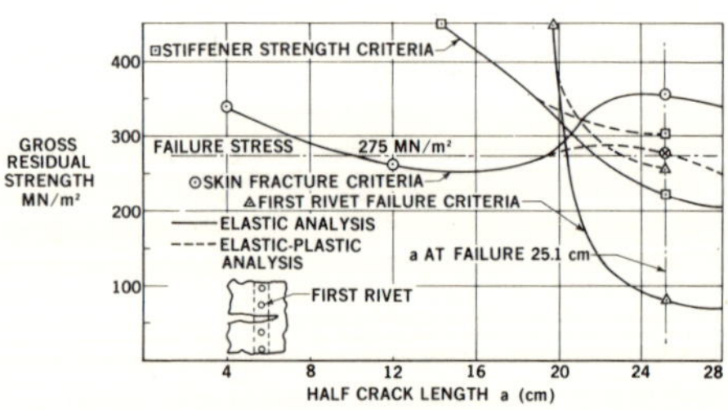


Fig. (22) - Residual strength diagram

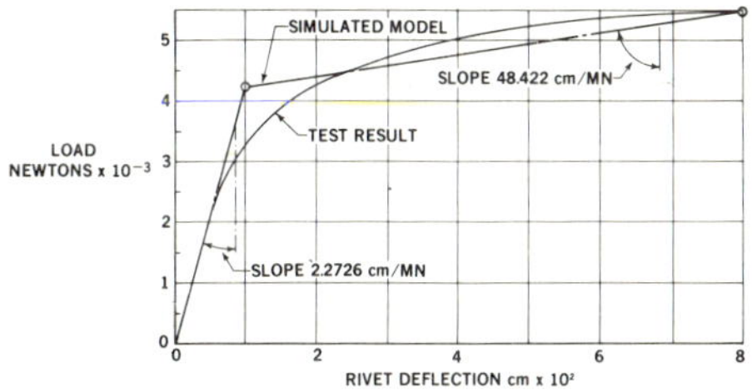


Fig. (23) - Elastic plastic model for rivet deflection

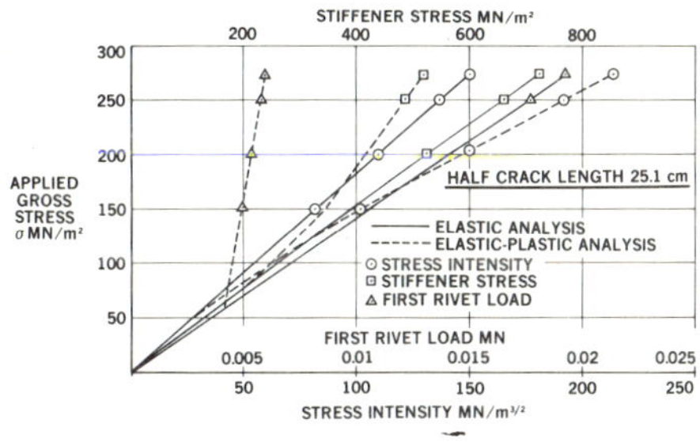


Fig. (24) - Effects of fastener yielding on stress intensity and stiffener stress

ACKNOWLEDGEMENT

The author wishes to express his appreciation to R. E. Darling and S. F. Y. Gee for their valuable assistance.

REFERENCES

[1] Swift, T. and Wang, D. Y., "Damage tolerance design-analysis methods and test verification of fuselage structure", AFFDL TR 70-144, 1970.

[2] Swift, T., "Development of fail-safe design features of the DC-10", Damage Tolerance in Aircraft Structures, ASTM STP 486, 1971

[3] Romualdi, J. P., Frasier, J. T. and Irwin, G. R., "Crack extension force near a riveted stiffener", Naval Research Lab., NRL Report 4956, 1957.

[4] Poe, C. C., Jr., "Stress intensity factor for a cracked sheet with riveted and uniformly spaced stringers", NASA TR R-358, 1971.

[5] Westergaard, H. M., "Bearing pressures and cracks", J. Appl. Mech., Trans. ASME, 1939.

[6] Timoshenko, S. and Goodier, J. N., "Theory of elasticity", Second Edition, McGraw-Hill Book Company, Inc., 1951.

[7] Irwin, G. R., "Analysis of stresses and strains near the end of a crack traversing a plate", J. Appl. Mech., Trans. ASME, September 1957.

[8] Paris, P. C., "Application of Muskhelishvili's method to the analysis of crack tip stress intensity factors for plane problems", Part III, Lehigh University, 1960.

# IMPORTANT ASPECTS OF CRACK GROWTH PREDICTION FOR AIRCRAFT STRUCTURAL APPLICATIONS

Howard A. Wood

Air Force Flight Dynamics Laboratory (FBE)
Wright Patterson AFB, Ohio

## 1. INTRODUCTION

Structural safety for military aircraft is achieved by assuring that flaws, cracks and other types of damage left undetected during inspection, will not grow to critical size during the operational life of the structure. The success of this endeavor depends upon the capability to characterize damage in terms of inspection capability, and to make predictions of damage growth in rather complex materials and structures under the combined influence of variable amplitude cyclic loadings, chemical and thermal enviroment.

In the past, sole reliance upon analysis for predicting growth behavior in structures has not been possible because of the complex load interaction and environmental effects for which sufficent understanding has not existed and for which modeling has only been moderately successful. Thus, heavy emphasis upon experimental simulation has been necessary. Increased need for crack growth prediction methods has resulted because of new requirements (1) in which designs must meet a specified safe crack growth life. In such cases, a large number of candidate materials and structural arrangements are involved and testing of each option may not be economically feasible.

The objective of this paper is to present and discuss the more influential factors that affect total crack growth life of small initial flaws (Figure 1). The discussion will attempt to highlight the relative importance of each factor and the manner in which each is currently accounted for in designing and analyzing crack growth in aircraft applications.

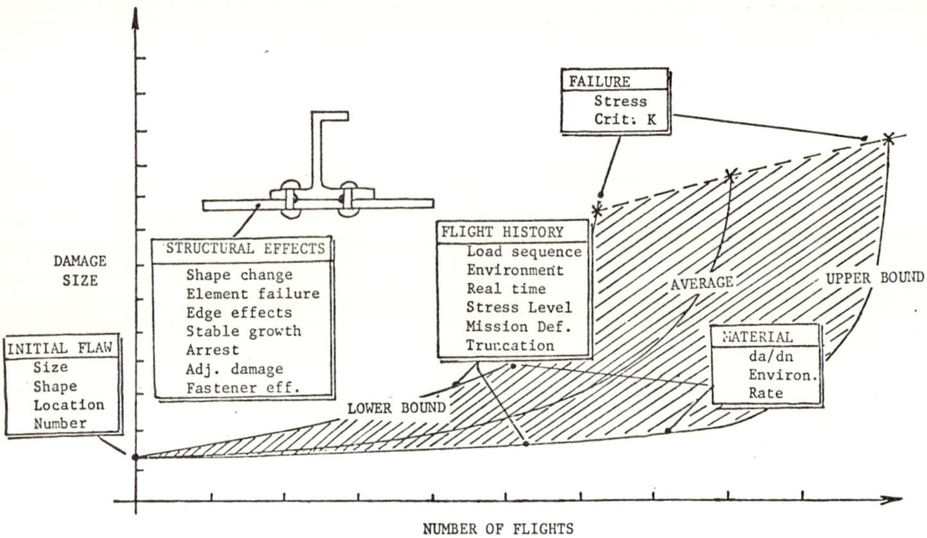


FIGURE 1 - FACTORS THAT INFLUENCE CRACK PROPAGATION LIFE

The reader will note that some rather simple approximations have been reasonably successful in accounting for complex phenomena such as delay due to load interaction effects. In other cases rather arbitrary assumptions are required such as flaw shape change, flaw growth direction and the influence of free edges, structural discontinuities, etc. In presenting these topics it is hoped that some direction and guidance will be given to future researchers.

## 2. THE NEED FOR GROWTH PREDICTION CAPABILITY

The reasons for conducting growth analyses may be due to criteria or specifications (1) such in the case of new designs, or by known or suspected sources of fleet cracking at some point during the service history for older aircraft. For each case safe crack growth is required as the primary means to protect the safety of fleet operation until damage can be detected and repaired. Until recently, structural safety for new USAF aircraft was associated with the successful demonstration of a full scale cyclic test to four times the design lifetime with the anticipation that any service induced damage in the fleet would be detected. prior to failure. The test article was cycled to failure, repaired and tested until the basic structure achieved the four lifetime goal. It is now recognized that the full scale cyclic test alone cannot insure that the fleet will be safe from catastrophic failure. Rather, the fatigue test can uncover design deficiencies or "hot spots" and be a precursor of future in-service economic problems resulting from wide general cracking due to poor detail design. In this sense, the fatigue test article serves a major role in the demonstration of "structural durability".

New aircraft are being designed to comply with specific safe crack growth and residual strength criteria (1). It is recognized that a high probability exists that new structure can be flawed during manufacture. The requirements stipulate that damage will be assumed to exist in each element of new structure in a conservative fashion (i.e. critical orientation highest stress location, etc.). The design must successfully contain the growth of this damage under service conditions for a specified life segment. If scheduled in-service inspections will not be relied upon specifically, or if the part cannot or will not be inspected during service, then safety is totally dependent upon a safe crack growth life equivalent to the expected service life. If the part is in-service inspectable, safety may be assured by designing such that the specified in-service detectable damage (usually greater than the initial specified size) will not grow critical before the next opportunity to discover the damage. These requirements are schematically represented in Figures 2 and 3 where it is noted that a structure can be qualified under one of two general categories of damage tolerance and satisfy the USAF requirements. The "Slow Crack Growth" category (Fig 2) represents structure in which the damage is always sub-critical and stable throughout the entire service life and the residual strength is never allowed to decrease below design limit load. As implied by the name, "Fail Safe" structure (Fig 3) is designed such that damage may be contained and/or temporarily arrested in the failure of a load path or the arrest of a rapidly running crack. The residual strength following this condition may be less than limit if detection is obvious and the inspection is frequent (i.e., on the order of several flights).

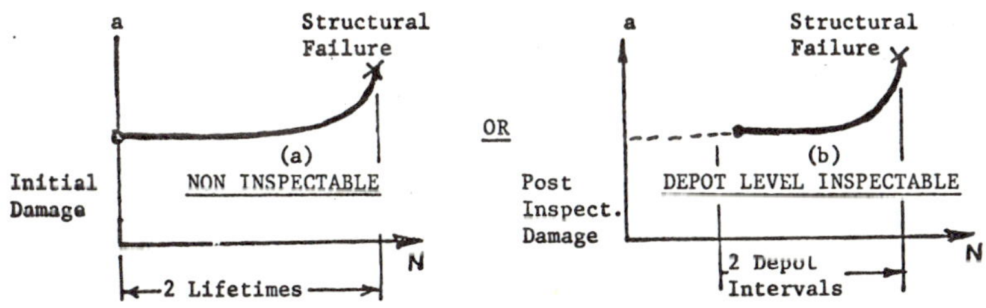


FIGURE 2 - SAFE CRACK GROWTH REQUIREMENTS(1) FOR NEW STRUCTURE QUALIFIED AS SLOW CRACK GROWTH

It should be emphasized that while the "Fail Safe" concept appears to offer a larger degree of safety protection, it is the intent of the new criteria that structure qualified to either category be equally safe.

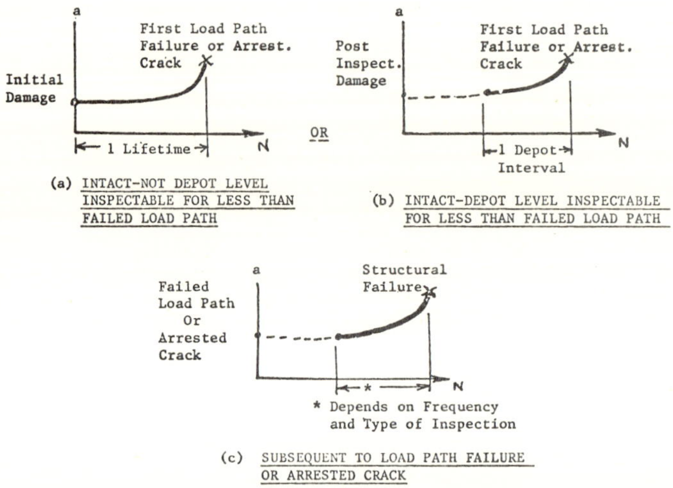


FIGURE 3. SAFE CRACK GROWTH REQUIREMENTS FOR NEW STRUCTURE QUALIFIED AS FAIL SAFE

As will be indicated in later discussion the "Fail Safe" structure is offered relief in the criteria through smaller initial flaw sizes and shorter periods of required safe crack growth. In order to prohibit the early failure of fail safe load paths, there is a safe crack growth design requirement for "intact" structure as indicated by Fig 3 (a) and 3 (b). It is important also to point out that the design conditions for inspectable and non-inspectable are totally independent (i.e., an either/or condition), and the damage assumed initially for the depot level inspectable case is based only on inspection capability and not on the size that initial manufacturing damage would be when the first depot inspection is made.

## 3. INITIAL DAMAGE

As indicated in Fig (1) size, shape, location and quantity are the factors which must be addressed initially in making the growth prediction. In older aircraft damage may either be discovered during scheduled or special inspections or assumed to exist because of a fleet accident to one or more aircraft. In large fleets it may be difficult to establish crack size population because of the wide variation in individual aircraft usage. Conventional NDI methods and structural proof testing are often adequate when suspected damage locations are relatively small.

In cases of suspected wide area cracking or where the safe increment of growth is extremely small it may be advantageous to deliberately tear down an airframe for the express purpose of assessing the flaw population.

In the new criteria (1) primary initial damage is intended to represent demonstrated capability for in-plant inspection. The specified sizes and locations, Figure 4, represent typical values obtained through USAF experiences. Assumed slow crack growth values may be less than specified if a valid demonstration program (90% probable, 95% confidence) is conducted by the contractor. The "Fail Safe" values are deliberately smaller in recognition of the additional level of safety afforded this concept.

Post inspection damage (Fig 5) represents typical depot level inspection capability with NDI aid but without part removal. These sizes are assumed to exist in each structure whose safety depends on the safe growth of damage between depot level inspection.

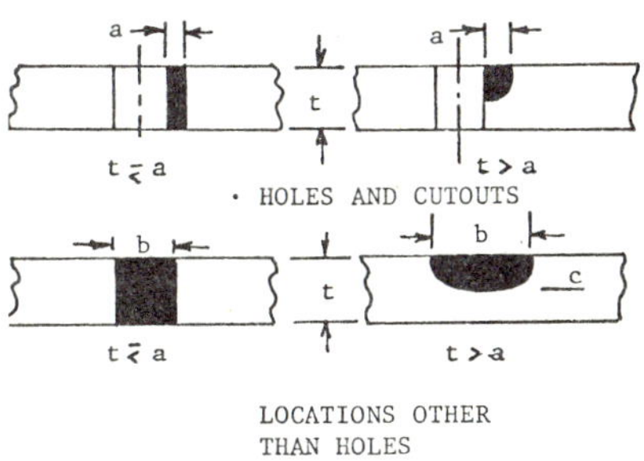


* PRIMARY DAMAGE ASSUMED AT MOST CRITICAL LOCATION IN STRUCTURE

| PRIMARY DAMAGE* - INCHES | | |
|---|---|---|
| | SLOW CRACK GROWTH | FAIL SAFE |
| a | 0.050 | 0.020 |
| b | 0.250 | 0.100 |
| c | 0.125 | 0.050 |
| CONTINUING DAMAGE, ADJACENT STRUCTURE DAMAGE - INCHES | | |
| 0.005 Corner Crack Assumed at Each Fastener Hole Surface Flaw With b=0.02, c=0.01 Assumed Present at Any Other Location | | |

FIGURE 4 - INITIAL FLAW SIZE ASSUMPTIONS (1)

CONDITION: PROOF TEST-DAMAGE SIZE FOR MAX. CRITICAL TOUGHNESS

CONDITION: PART REMOVED FOR INSPECTION WITH MANUFACTURING QUALITY TECHNIQUES - DAMAGE SAME AS INITIAL

CONDITION: PENETRANT, MAG. PART., ULTRASONICS, BUT NO PART REMOVAL

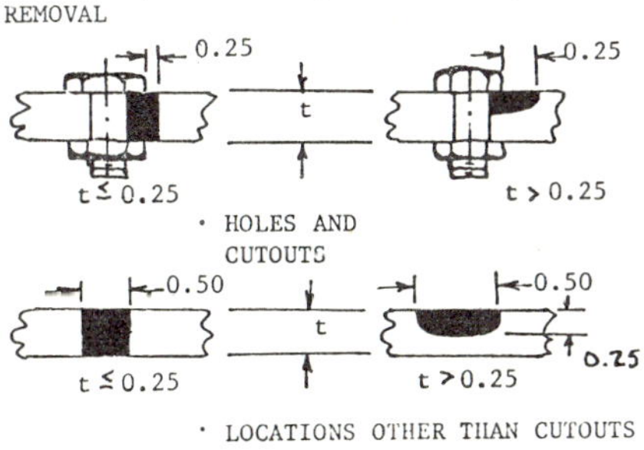


FIGURE 5 -- POST INSPECTION DAMAGE SIZE ASSUMPTIONS (1)

## 4. SAFETY GOALS

As indicated in Figures 2 and 3, the safety goal for life is a function of the category, the inspection technique and the frequency of inspection. An arbitrary factor is assigned to each safe crack growth period to account for variations in the flight history as well as the possibility of a missed inspection. It is noted that this factor is less for "Fail Safe" than "Slow Crack Growth" type structure in an attempt to provide an equal likelihood of success for each category. Safety through assured residual strength is also provided (1), however this topic will not be addressed in this discussion.

While it is possible to meet safety goals without relying on specific inspections, safety is obviously enhanced by inspections and inspectable structure. It is USAF experience that a depot type inspection occurs on the average at about one fourth the design lifetime.

## 5. FLIGHT LOAD HISTORY

A major first order effect on growth can be attributed to the manner in which the actual expected flight load history is simulated (7), (8), (22). This is particularly important in test as well as analysis. As indicated in Figure 6, statistically derived loads must be combined with the expected mission description, including sequence, cyclic frequency, period of sustained load, environment and temperature to produce a sequence and format to use in the analysis.

Short cuts are almost always made (for economic reasons) in flight load simulation. A common error is induced when "typical", "severe" and "mild" discriptions of usage are defined strictly in accordance with frequence of loading, i.e., the more frequent the occurrence, the more severe, the less frequent the less severe. The presumption is made that loads contribute to life degradation in a linear and cumulative fashion. While it is well known (7), (8), (11-21) that periodic high load can produce beneficial residual effects, the lack of a general acceptance of the advanced analysis techniques for both fatigue (23) and crack growth (13-19) has resulted in this practice of defining severity to continue.

Combining positive and negative peak loadings in a flight sequence has been reported to have a significant effect on the growth life (7), (8), (12) and at the current time, crack propagation models cannot account for these differences.

The procedure of "blocking" an entire portion of the life into arbitiary sequences of equal amplitude loads, while practiced rather univerally in the past, has for the most part be abandoned in favor of a multi cycle flight by flight simulation (Fig. 6) in which both frequence and sequence are preserved. With the

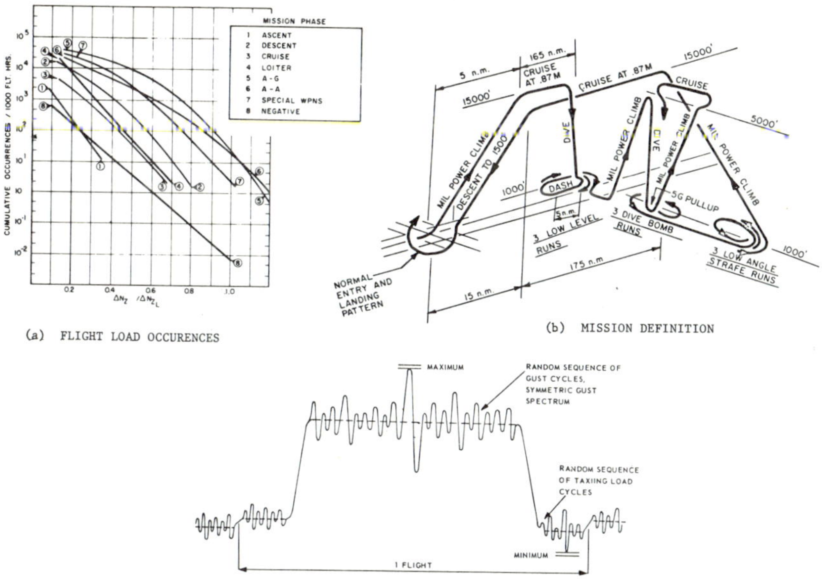

(a) FLIGHT LOAD OCCURENCES

(b) MISSION DEFINITION

(c) STRESS HISTORY SIMULATION

FIGURE 6. DETERMINATION AND SIMULATION OF FLIGHT STRESS HISTORY

introduction of high speed computers and advanced testing systems, simulation by "blocking" is no longer required. Even within flight by flight simulation, variations are observed when flight length, sequence, inclusion of compressive loads, and real time effects are varied (7), (8), (11), (12).

Load truncation to save test and analysis time can seriously influence crack growth life. Judgement concerning truncation can only be made with analyses which consider residual stress and load interaction effects on crack growth.

While by no means exhaustive, this discussion is intended to point out the significance of usage variation and flight simulation on crack growth. In tactical aircraft these effects can alter life by factors of 2-10 in many cases and thus cannot be taken lightly.

## 6. ENVIRONMENT AND CONSTANT AMPLITUDE GROWTH RATE DATA

Although the effects of aggressive chemical environment on crack growth have been well documented for most popular airframe materials, the effects of real time, enviroment and flight loading have not been extensively studied. It is common to analyse situations by using constant amplitude data generated under average cyclic frequencies in the most conservative media expected and to ignore sustained load growth. Figure 7 includes the effect of several environments on D6ac steel. The average growth rate is seen to be significantly influenced and presumably the crack growth life under variable amplitude loading would be shortened accordingly.

It is difficult to find many factors that don't contribute in some way to altering fatigue crack growth rate. Among these are humidity, chemical and thermal environment, yield strength and toughness, grain size and direction, stress level and load ratio R. Any process or heat treatment which alters these factors may be expected to alter the growth rate. Obviously some of these factors are of secondary or even little importance where others are first order factors. An extreme amount of carefully produced test results are required to evaluate data trends for many of these factors and to determine their statistical significance. As a result, little is actually known about many materials currently in use today and variability is generally accepted as normal scatter.

A contributing factor to variability in constant amplitude data may be the lack of standard procedures for obtaining and reporting data. These differences may result from test technique, specimen type, stress level etc. It sufficies to state that the proper characterization and description of da/dn is of first order importance in assembling a growth rate prediction technique.

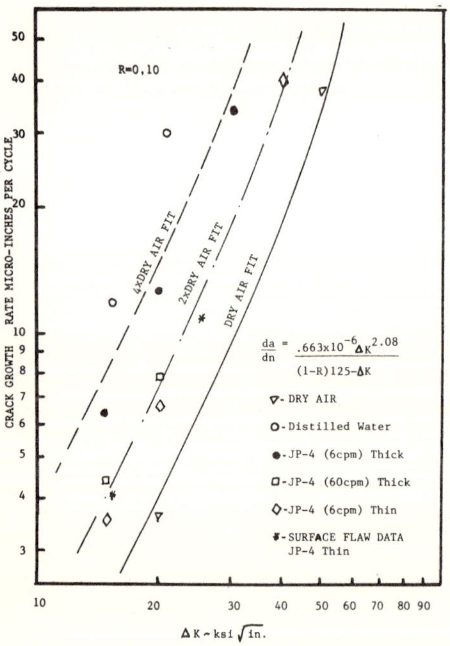


FIGURE 7 - EFFECT OF ENVIRONMENT ON CRACK GROWTH D6ac STEEL

Extensive data is required to establish definite trends of cyclic crack growth and care must be exercised to accurately describe the data over the range of interest. Empirical growth rate laws should be derived and used with extreme care. Small errors particular in the lower ranges of $\Delta K$ can significantly affect the accuracy of prediction.

## 7. STRUCTURAL EFFECTS

Compliance with the safe crack growth design requirements (1) presents many difficulties when applied to real structure. One of the most important of these involves the assumptions which must be made regarding the growth patterns of the assumed initial flaws, the likelihood that the growth will stop due to free edges, and that in multi-element structure, one or more of the individual elements will fail prior to the failure of the major load path. All these effects of course are dependent on the particular geometry, the location of the initial damage and the stress condition. As an example of the complexity of the situation consider the following case (Figure 8). A spanwise splice is indicated and is considered to be designed for the slow crack growth category. Although the patterns of the primary growth may be different than illustrated other modes of growth would be as complex. It is assumed that the critical damage size in the skin is not reached until the stiffener fails. The analysis is started by assuming primary damage equal to an .050" through the thickness crack in both the skin and spar as indicated in Fig 8a. Assumptions must be made everytime the crack stops at an adjacent hole, a free edge and when the stiffener fails. Each time this occurs, an .005" crack is assumed to be present in such a location that the analysis can be continued. According to Ref. 1 the .005" crack is considered to be present initially and thus is growing along with the primary damage (Fig 4.). The life of the structure is significantly affected by the assumptions necessary to conduct the analysis. Among the conditions which are required to complete the analysis are:

- Growth of Multiple Flaws
- Effects of Flaw Shape Change
- Growth Mechanics of Very Small Flaws at Fastener Holes
- Failure Criteria For a Complex Shaped Member
- Possible Redistribution of Load (Geometry Dependent Effects).

While it is common to consider all filled fastener holes as open holes the effects of load transfer and joint clamp-up can alter the life prediction (15). In addition the introduction of beneficial features such as tapered pins and the numerous hole expansion processes has demonstrated that growth of relatively large flaws can be significantly retarded, and to ignore such effects could be overly conservative. In Ref. 1 relief on safe growth requirements is offered for those cases where the benefical process and its means of control are clearly demonstrated. In such cases the initial damage sizes are reduced to .005".

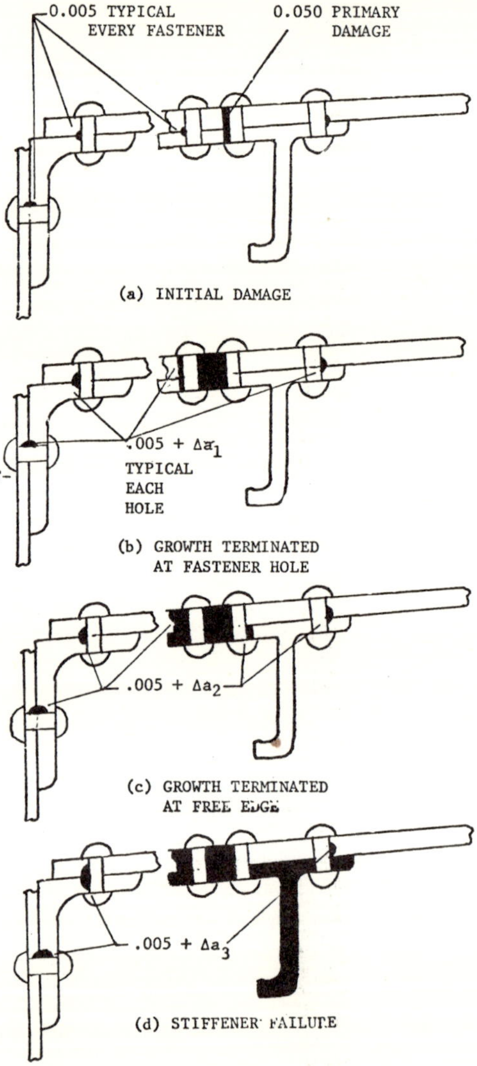


FIGURE 8 GROWTH PATTERN FOR ASSUMED DAMAGE IN BUILT-UP STRUCTURE

## 8. PREDICTION METHODS TO ACCOUNT FOR VARIABLE AMPLITUDE STRESS HISTORIES

It is generally well known that load interaction in a variable amplitude stress history, such as encountered by an aircraft, can cause an apparent delay or retardation over the expected rate of growth if each load amplitude is incremented in a linear fashion without regard of the influence of one level on the other (Fig. 9). To completely neglect this phenomenon will in general yield safe predictions, but mostly they will be too conservative and therefore unacceptable particularly when designing for specific safe crack growth requirements such as Ref 1.

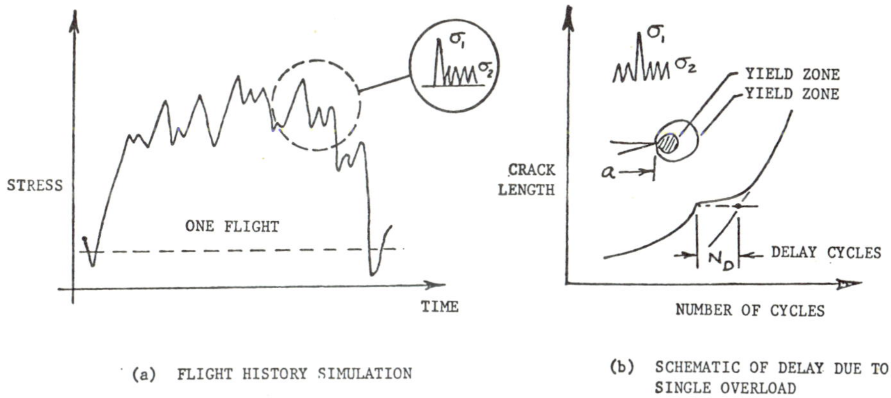


FIGURE 9. PREDICTION OF CRACK GROWTH UNDER VARIABLE AMPLITUDE SPRECTUM LOADING USING EFFECTS OF SINGLE AND MULTIPLE OVERLOADS

At the present time considerable research is being conducted to investigate the quantative effects of overloads(and underloads) on subsequent crack growth behavior, Ref. 12, 13, 14, 16, for example. It is hoped that these studies and the associated delay models which they will support can be utilized to predict complex spectrum effects by simplifying the stress history into a series of single or multiple overload occurrences, Fig. 9. As will be evident from the following discussion, there has been fair success in the past using this approach and some simple first order models of the delay process have been developed.

Wheeler (17) envisioned delay of growth following an overload to be proportional to the relative sizes of the approximate yield zones caused by the overload and the subsequent loading (Fig. 9). Retardation is maximum following the overload, with recovery to "steady state" conditions occuring when the extent of both zones are coincident. The basic model requires empirical derivation of the exponent m from a single or series of spectrum tests. The uniquence of a particular "m" to one material and/or to one spectrum limits the general usefulness of the model in predicting of growth behavior of a particular structure under an entirely different sequence of loads. In recent studies, Gray, (13), has investigated the Wheeler model and has devised a means to compute "m". Some results of these studies will be presented later in this report. The reader is referred to the author's paper (18) in which the historical development of crack growth delay models is reviewed.

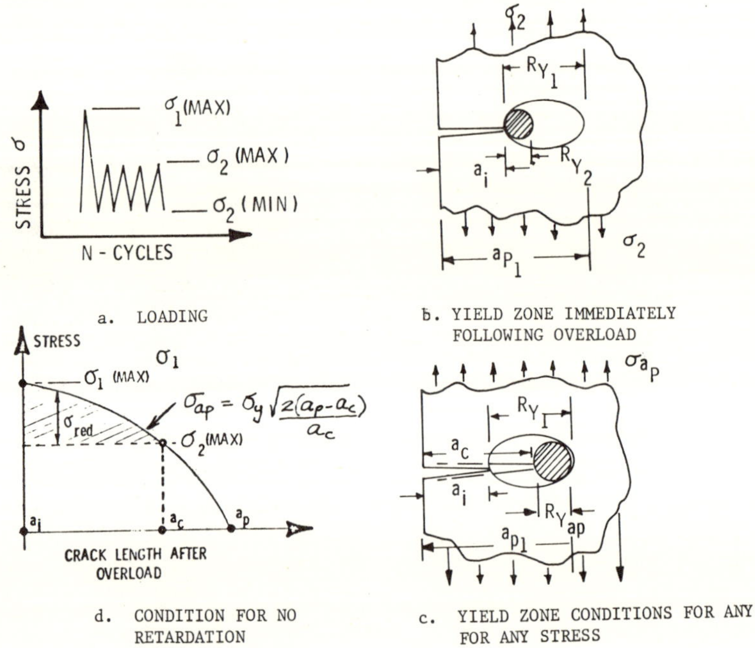


a. LOADING

b. YIELD ZONE IMMEDIATELY FOLLOWING OVERLOAD

d. CONDITION FOR NO RETARDATION

c. YIELD ZONE CONDITIONS FOR ANY FOR ANY STRESS

$$\sigma_{ap} = \sigma_y \sqrt{\frac{2(a_p - a_c)}{a_c}}$$

FIGURE 10. THE WILLENBORG MODEL STRESS FORMAT

## 8.1 The Willenborg Model

One of the more important spin-offs of Wheeler's model was proposed by Willenborg, et al (19), in an attempt to generalize the yield zone concept, allow for a wider use, and not depend upon any empirical factors, except those used to describe the constant amplitude growth rate data.

The Willenborg model was developed under the following assumptions.

a. Retarded growth occurs when the maximum cyclic stress is reduced.

b. Retardation is proportional to the amount that the maximum stress is reduced.

c. The length of retarded growth is that zone caused by the overload (i.e., the approximate yield zone $R_y$ caused by the overload).

d. A new condition is produced each time a load equal to or greater than the original overload is applied or when the yield zone of the subsequent loading is coincident with that produced by the overload.

Figure (10) illustrates the yield zone conditions for the Willenborg model for a two level loading spectrum. Crack growth under loading $\sigma_2$ is always retarded as long as growth remains within the zone of plasticity caused by the overload, $\sigma_1(\max)$. For the latter condition, it is presumed that the "current" crack length, $a_c$, plus the length of "current" yield zone, $R_{yc}$, is less than the value of $a_p$, that is the extent of the yield zone caused by the overload. Irwin's yield zone model is used in the model where

$$R_y = \text{yield zone size} = \frac{K^2}{C(\sigma_y)^2} \qquad (1)$$

$C = 2\pi$ for plane stress conditions

$C = 4\sqrt{2\pi}$ for plane strain conditions

In the example the plane stress assumption is illustrated. Although a single two level stress spectrum is considered here, assume that a third stress level $\sigma_3 = \sigma_{ap}$ (less than $\sigma_1$) occurs following the last cycle of $\sigma_2$, and assume also that growth has not completely progressed through the yield zone caused by the overload, $\sigma_1$. For the previously established conditions of retardation, it can be presumed that retardation will be terminated when the value of $\sigma_{ap}$ is large enough ($\sigma_{ap} \leq \sigma_1$), and the current crack length, $a_c$, is of such extent that the condition exists:

$$a_c + R_{yap} = a_{pl} \qquad (2)$$

where: $R_{yap}$ = yield zone caused by general stress $\sigma_{ap}$ at crack length $a_c$,

$$R_{yap} = \frac{(K_{ap})^2}{2(\sigma_y)^2} = \tfrac{1}{2}(\sigma_{ap}/\sigma_y)^2 \qquad (2a)$$

$\sigma_y$ = material yield strength

Therefore, the value of general stress $\sigma_{ap}$ is as follows:

$$\sigma_{ap} = \sigma_y\sqrt{2(a_p - a_c)/\, a_c} \quad \text{(plane stress)} \qquad (3)$$

The stress, $\sigma_{ap}$ may be termed the generalized "no retardation stress". If $\sigma_2 \geq \sigma_{ap}$, no retardation will occur and the growth is back under "Steady State Conditions". Figure 10 (d) includes a plot of the function $\sigma_{ap}$ following the overload. Note that it is bounded by $\sigma_{ap} = \sigma_1$ and $\sigma_{ap} = 0$.

Since retardation is presumed to be proportioned to the difference between the overload and subsequent loading, and the effect decreases with increase in the growth of the crack following overload, a local residual stress condition may be thought of as being present, where the amount of residual stress is equal to:

$$\sigma_{red} = \sigma_{ap} - \sigma_2 \text{ (at any crack length)} \qquad (4)$$

This is illustrated as the shaded portion of Figure 10 (d).

The residual stress $\sigma_{red}$ is subracted from both the maximum and minimum values of the subsequent load $\sigma_2$, to effect calculation of the retarded growth rate. Retardation takes place in three distinct modes.

(1) $R = \sigma_{min}/\sigma_{max}$ is reduced

(2) $\Delta K$ is reduced, $R = 0$

(3) $\Delta K$ and R both equal zero when $\dfrac{\sigma_1}{\sigma_2} \geq 2.0$

8.2 Model Limitations

The Willenborg model has some limitations which are unattractive.

1. The condition for complete retardation at overload ratios $\geq$ 2.0 appears to be unreal based on data of Hillberry (20) and others. Growth has been observed for overload ratios in excess of 2.5.

2. The model simply ignores the compression portion of the loading. Compression has been observed to have a variable effect depending upon how and when it is introduced.

3. No accounting is made of the difference between single and multiple occurrences of $\sigma_1$, the overload. There have been some data to indicate that there is a difference in behavior between single and multiple overloads (21).

Of these, the shutoff condition appears to be most serious. Occurrences of multiple overloads in flight simulation spectra are generally rare, and the effect appears to be of some what secondary interest. Effects of compression are not well understood and much additional research is required.

8.3 Willenborg Model - Stress Intensity Format

In a recent publication, Ref. 16, Gallagher derived the

Willenborg model in a generalized stress intensity format in order to give a somewhat different association to the effect of the overload. Referring to Figure 11 the equivalent "no retardation" stress intensity factor $K_{ap}$, is illustrated for the conditions $\sigma_{ap}$ at crack length $a_c$. (See also Figure 10 and Equation 3).

$$K_{ap} = K_{1(max)} \sqrt{1 - \Delta a / R_{y_1}} \tag{5}$$

where $K_{1(max)}$ = The Stress Intensity at the point of overload $\sigma_1$,

$\Delta a$ = the incremental growth following the overload = $a_c - a_i$

$R_{y_1}$ = the yield zone associated with the overload $\sigma_1$

No retardation will occur if $K_2 \geq K_{ap}$ and $K_{ap}$ will vary with crack length as does $\sigma_{ap}$ in Figure 10.

As envisioned by Gallagher growth is controlled by the "local" conditions at the crack tip, that is residual stresses or forces, producing residual local values of stress intensity, K. Thus the "local" residual K is

$$K_{red} = K_{ap} - K_2 \tag{6}$$

Growth is therefore controlled by the local effective stress intensity conditions

$$K_{eff} \text{ (max and min)} = K_2 \text{ (max and min)} - K_{red}$$

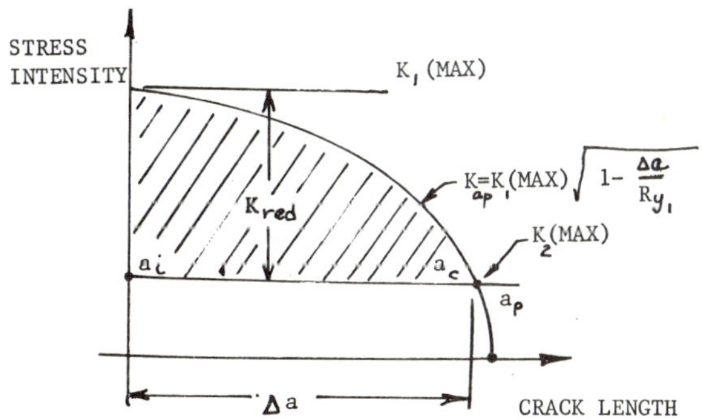


FIGURE 11. WILLENBORG MODEL DERIVED IN STRESS INTENSITY FORMAT

One reason for expressing the Willenborg model in this format is that it allowed Gallagher to more conveniently correlate the results of a series of overload tests being conducted under constant K conditions. A more direct reason, however, was that it allowed a generalization of the overload ratio shut off anomaly associated with this model. Gallagher's reasoning is as follows:

$$K_{red} = K_{ap} - K_2 \tag{7}$$
$$= K_1\sqrt{1 - \Delta a/R_{y_1}} - K_2 \text{ (FOR R = 0).}$$

Immediately following the overload, $\Delta a$ is relatively small and

$$K_{red} = K_1 - K_2 \ . \tag{8}$$

For the crack to grow under these conditions, the maximum effective stress intensity, $K_{eff}$, must be greater than the threshold $K_{th}$ (for a constant $K_{min}$), thus,

$$K_{eff} = \left(K_1 - K_{red}\right) \geq K_{th}, \tag{9}$$

where $K_{th}$ is the low $\Delta K$ threshold observed in constant amplitude loadings. By introducting a proportionality factor

$$\phi = \frac{K_{red}}{K_{red} \text{ (due to Willenborg)}}, \tag{10}$$

for maximum retardation let

$$K_{eff} = K_{th} = K_2 - \phi\ (K_{ap} - K_2) \tag{11}$$

Therefore:

$$\phi = \frac{K_2 - K_{th}}{K_1 - K_2} \tag{12}$$

Thus by knowing $K_{th}$ for a given material and R conditions, $K_2$ and the ratio of $K_1/K_2$ which actually produces shut-off, $\phi$ can be calculated and used in the model. The effectiveness of this generalized model in correlating single overload and spectrum results in currently being investigated.

8.4 Application of the Willenborg Model

The Willenborg model has been used with fairly good success to account for the first order behavior of the load interaction phenomena despite its obvious anamolies and simplification. Many of these results were presented by the author in the AGARD Paper (18), and will not be repeated here. Engle and Rudd, (14) have summarized these cases in a single correlation plot, Fig 12. All types of spectra are represented here, including simple, single and multiple overload conditions as well as blocked and flight-by-flight spectra for aluminum, titanium and steel.

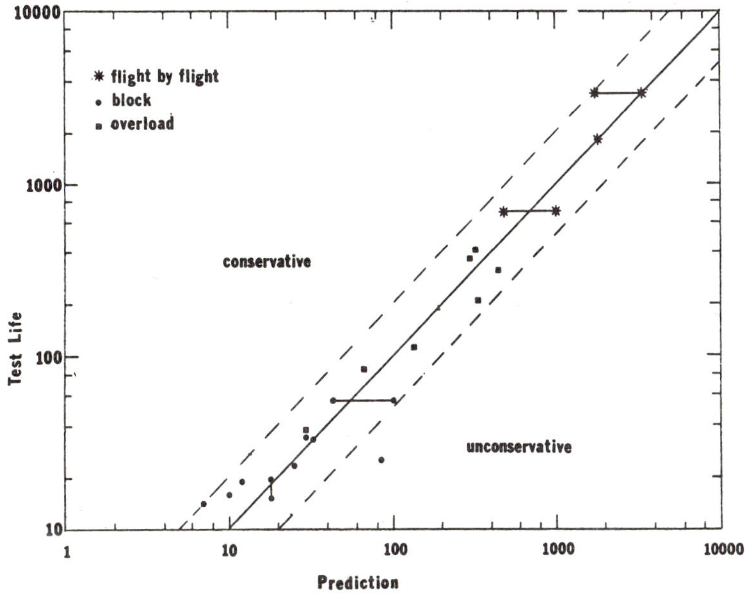


GURE 12. SUMMARY OF CORRELATION STUDY
WILLENBORG MODEL (14)

8.5 Modified Wheeler Model

In Reference 13, Gray and Gallagher present a modified version of the Wheeler Model which allows predictions to be made without reliance on data fitting to derive "m" and without limiting the usefulness of the model to a specific material and a specific set of loading conditions. The modification makes use of the known overload shut-off ratio $K_1/K_2 = S$, observed by Probst and Hillberry (20) and the effective local stress intensity factor following the overload. The exponent "m" is calculated for each condition following the overload utilizing S and the value of observed steady state constant amplitude threshold $\Delta K_{th}$. Figure 13 summarizes the use of the model in correlating several tests of single overloads taken from the literature.

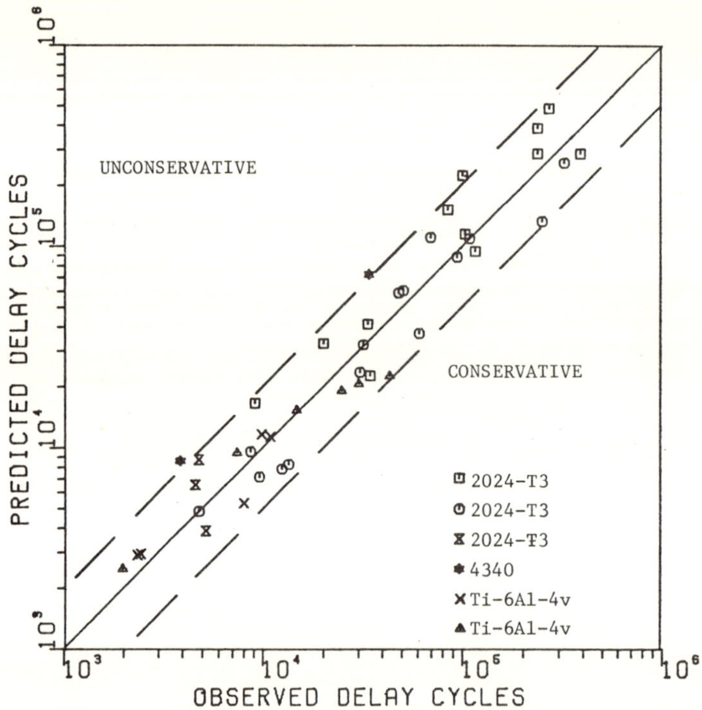


FIGURE 13. CORRELATION STUDY RESULTS - MODIFIED WHEELER MODEL. (13)

## 8.6 Summary - Delay Models

Further research is certainly warranted to both understand and to model growth rate behavior under simple and complex load histories. Among the more important topics are:

a. Systematic study of compressive loading effects on crack growth.

b. A resolution of fact concerning "delayed retardation".

c. A resolution regarding the observations and theories associated with crack closure and the influence on growth rate.

d. Environmental and time associated delay effects such as relaxation of beneficial residual compressive zones.

e. Models based on other than simple yield zone assumptions.

f. Standarization of test and reporting methods for constant amplitude data.

It has been shown that some pretty simple models been successful in predicting a very complex phenomena. The range of correlation between test and prediction is of the same order as observed in scatter for constant amplitude crack growth.

In using currently delay models in design analysis, it is strongly urged that verification tests be performed to validate the range of model applicability for the particular spectra being used. This is especially true for blocked type spectra where the occurrence of a few layers of load can profoundly overestimate the amount of retardation actually present. The use with spectra arbitrarily ordered in ascending and decending fashion should also be avoided.

## 9. CONCLUSIONS

New damage tolerance requirements have created the need for a capability to predict crack growth in very complex structures. The application of such criteria to new design is bound to cause concern among potential contractors because of this complexity. This paper has attempted to bring to light some of the important aspects of the complex process of predicting structural crack growth. Time and experience will test the true value and effectiveness of the criteria.

## REFERENCES


1. MIL-A-83444, Damage Tolerance Design Requirements for Aircraft Structures, USAF, May 1974

2. Tiffany, C.F., "The Air Force's Changing Philosophy on Structural Safety, Durability, and Life Management", presentation to The AIAA, SAE Conference on Structures and Dynamics, Las Vegas, Nevada, April 1974

3. MIL-STD-1530 USAF Structural Integrity Program, Sept, 1972

4. Hinders, U.A., "F-111 Design Experience - Use of High Strength Steel", AIAA 2nd Aircraft Design and Operations Meeting, July 1970

5. Wood, H.A., "Fracture Control Procedures for Aircraft Structural Integrity, AFFDL-TR-71-89, July 1971

6. Wood, H.A., "The Use of Fracture Mechanics Principles in the Design and Analysis of Damage Tolerant Aircraft Structures", AGARDograph No. 176 on Fracture Mechanics of Aircraft Structures, Edited by H. Liebowitz, November 1973

7. Schijve, J., "The Accumulation of Fatigue Damage in Aircraft Materials and Structures", AGARDograph No. 157 (1972)

8. Schijve, J. & Broek, D., et al "Fatigue Tests with Random and Programmed Load Sequences With and Without Ground Air-Ground Cycles a Comparative Study on Full Scale Wing Center Sections", AFFDL-TR-66-143, 1966

9. Smith, Porter, Engstrom, "Fatigue Crack Propagation Behavior and Residual Strength of Bonded Strap Reinforced, Laminated and Sandwich Panels", AFFDL-TR-70-144

10. Poe, C.C., "Fatigue Crack Propagation in Stiffened Panels", ASTM, STP 486, 1971

11. Broek, D., "Fail Safe Design Concepts and Fatigue Crack Propagation in Real Structures, the Prediction of Fatigue Crack Propagation" AGARDograph No. 176, 1973

12. Wei, R.P., Shih, T.T., "Delay in Fatigue Crack Growth" International Journal of Fracture, Vol. 10, No. 1, March 1974

13. Gray, T.D.& Gallagher, J.P., "Predicting Fatigue Crack Retardation Following a Single Overload Using a Modified Wheeler Model" Air Force Flight Dynamics Laboratory, WPAFB, Ohio, submitted to National Symposium on Fracture Mechanics, 1974

14. Engle, R.M. & Rudd, J.L., "Analysis of Crack Propagation Under Variable Amplitude Loading Using the Willenborg Retardation Model", presented at the AIAA Structures and Dynamics Conference, Las Vegas, Nevada, USA, April 1974 Paper No. 74-369

15. Hall, L.R., Shah, R.C. and Engstrom, W.L., "Fracture and Fatigue Crack Growth Behavior of Surface Flaws and Flaws Originating at Fastener Holes", AFFDL-TR-74-47, (to be published. Work conducted by The Boeing Company under Contract F33615-72-C-1740, May 1974

16. Gallagher, J.P. & Hughes, T.F., "The Influence of Yield Strength on Crack Growth Rate Delay in 4340 Steel", Air Force Flight Dynamics Laboratory, AFFDL-TR-74-27, March 1974

17. Wheeler, O.E., "Spectrum Loading and Crack Growth", Trans. ASME, Journal of Basic Engineering, Vol. 94, Series D, No. 1 March 1972

18. Wood, H.A., "A Summary of Crack Growth Prediction Techniques" AGARD Lecture Series No. 62, July 1973

19. Willenborg, Engle, Wood, "A Crack Growth Retardation Model Using an Effective Stress Concept, Air Force Flight Dynamics Laboratory, AFFDL-TM-FBR-71-1, January 1971

20. Probst, E.P & Hillberry, B.M., "Fatigue Crack Delay and Arrest Due to Single Peak Tensile Overloads" presented to the AIAA Dynamics Specialists Conference, Williamsburg, Va., March 19-20, 1973

21. Trebules, V.W. Jr, Roberts, R. and Hertzberg, R.W., "Effects of Multiple Overloads on Fatigue Crack Propagation in 2024-T3 Aluminum Alloy", ASTM STP 536, 1973

22. Wood, H.A., "History of Loading of Aircraft and Examples of Aircraft Failure", AGARDograph No. 176, Fracture Mechanics of Aircraft Structures - 1974

23. AGARD Lecture Series No. 62, "Fatigue Life Prediction for Aircraft Structures and Materials", June 1973

# Section VI

# STRESS ANALYSIS AND ANALYTICAL MODELING

# A CRITICAL SURVEY OF MODELS THAT ACCOUNT FOR NON-LINEARITY OF MATERIAL BEHAVIOUR AT A CRACK TIP

E. Smith

Department of Metallurgy, University of Manchester, Manchester, United Kingdom.

## ABSTRACT

The key features of continuum-type models that account for non-linearity of material behaviour at a crack tip are critically reviewed.

## INTRODUCTION

Griffith's classic crack model consists of an infinite isotropic elastic solid obeying the laws of infinitesimal linear elasticity, subject to the applied tensile stress $p_{22} = \sigma$ and containing the two-dimensional crack $-c < x_1 < c$, $x_2 = 0$, this loading state usually being referred to as mode I deformation. He obtained [1] the crack extension condition by considering the variation, with crack length, of the sum of the energy contributions due to the solid's elasticity, the potential associated with the applied stress and the energy of the crack faces, arguing that a crack extends if the total energy decreases as a consequence of extension; the appropriate condition is

$$\sigma = \left[ \frac{4\mu\gamma}{\pi(1-v)c} \right]^{\frac{1}{2}} \tag{1}$$

where $\mu$ is the shear modulus, $v$ is Poisson's ratio and $\gamma$ is the surface energy. For a prescribed applied stress $\sigma$ and surface energy $\gamma$, eqn. (1) gives a critical crack length $2c = 8\mu\gamma/\pi(1-v)\sigma^2$; if the crack length is less than this value, according to the Griffith theory it should contract, whereas if it exceeds the critical value it should extend and the Griffith theory consequently gives the critical condition for unstable fracture. The crack extension condition is the same for mode II deformation where the applied stress is $p_{12} = \sigma$, while for mode III deformation where the applied stress is $p_{23} = \sigma$, it is given by eqn. (1) with $(1-\nu)$ replaced by unity.

A most important characteristic of the Griffith model is the assumption that the entire solid is in the form of an elastic continuum, obeying the laws of infinitesimal linear elasticity. However the theory associated with the model predicts infinite stresses and strains in the immediate vicinity of a crack tip, and there is therefore an inconsistency in using the model to describe the crack tip region, which is of course vitally important with regard to the crack extension process. Because of this inconsistency it is essential to explore the situation near a crack tip in greater detail than the Griffith model allows, and this paper's general objective is to critically survey approaches that have been made to account for non-linearity of material behaviour, and discuss the relation between the associated models and the original Griffith model. Consideration is restricted to models where non-linearity is confined to material lying along the crack plane, and in order to simplify the mathematical analysis, the paper concentrates on mode III deformation, although the conclusions should be equally applicable to modes I and II loading states.

## THE INCORPORATION OF NON-LINEARITY OF MATERIAL BEHAVIOUR INTO A CONTINUUM-TYPE CRACK MODEL

The simplest way, first suggested by Elliott [2] , of incorporating non-linearity of material behaviour into a continuum-type model is to cut an infinite solid along the plane $x_2 = 0$ and separate the two semi-infinite blocks $x_2 > 0$ and $x_2 < 0$ so that their surfaces become respectively $x_2 = a/2$ and $x_2 = -a/2$. The externally applied stress is $p_{23} = \sigma$ and forces act between the faces $x_2 = \pm a/2$ of the blocks to produce a stress $p_{23} = P(w)$ on these faces, where $w = u_3(x_1, a/2) = -u_3(x_1, -a/2)$ and $2w$ is the relative displacement of the faces in the $x_3$ direction; non-linearity arises by prescribing various forms of $P(w)$ and further realism is introduced into the model because the non-linearity occurs within a region of finite thickness a, although this realism is idealised in that the region is confined to the crack plane (Fig. (1)). The equation expressing the equilibrium of the blocks is

$$\frac{\mu}{\pi} \int_{-\infty}^{\infty} \frac{dw}{d\lambda_1} \frac{d\lambda_1}{(\lambda_1 - x_1)} + \sigma - P(w) = 0 \qquad (2)$$

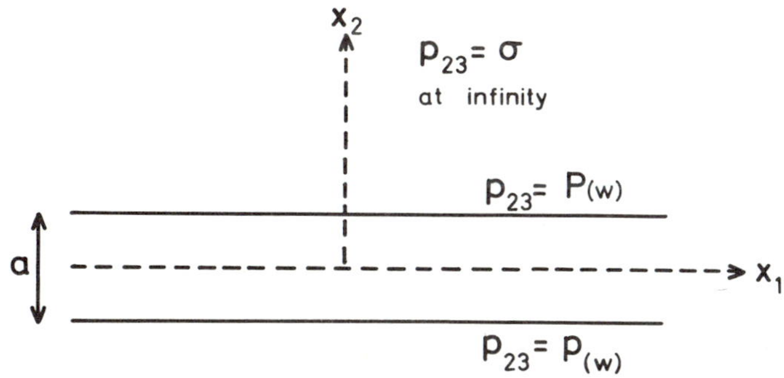


FIGURE 1. The general model for the incorporation of non-linearity of material behaviour within a strip of thickness a. P(w) is a prescribed function of w, where 2w is the relative displacement of the surfaces $x_2 = \pm a/2$.

for all $x_1$, with P(w) having the general form shown in fig. (2), being zero when w = 0 and linearly dependent on w for small w, rising to a maximum $p_m$ at some critical value $w = w_m$ and then falling gradually to zero as $w \to \infty$, the latter region corresponding to the crack; this will be true (Bilby and Eshelby [3]), when describing various fracture processes, i.e. whether P(w) represents the forces associated with atomic bonds that link separating crystal planes or the forces associated with a series of internal necks formed when a sequence of small cracks ahead of a large crack successively link up with the latter. Equation (2) is a non-linear singular integral equation similar to that associated with the Peierls-Nabarro dislocation model [4,5], except that P(w) is not periodic with respect to w; not surprisingly, therefore, the consequential mathematical difficulties render a complete analytical discussion of equ. (2) a formidable problem. It is possible, of course, to solve the equation numerically and Anderson and Bergkvist [6] have adopted this approach for a specific form of P(w). However, in the interests of generality, it is clearly desirable to investigate equ. (2) from an analytical viewpoint, and a specific purpose of this paper is to assess the various attempts that have been made to find solutions. These approaches may be viewed in one of two ways: either as modifications to the general model already outlined in this section, or alternatively as approximate procedures for solving the governing singular integral equ. (2).

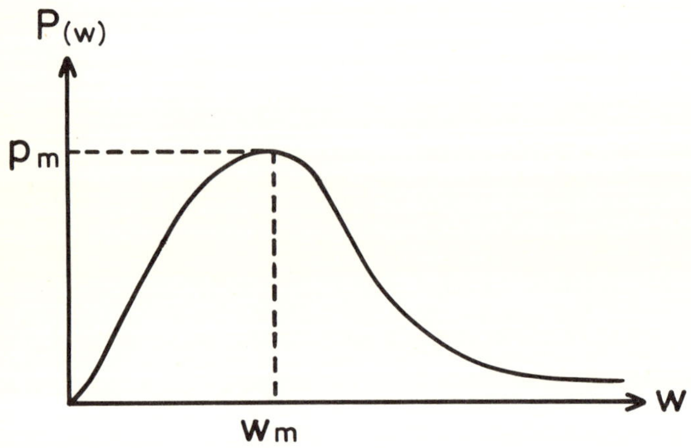


FIGURE 2. The general form of P(w); 2w is the relative displacement of the surfaces of the region of non-linear material.

## ELLIOTT'S SPECIFIC NON-LINEAR CRACK MODEL

Elliott's specific approach [2] starts with the model of a two-dimensional crack $-c < x_1 < c$, $x_2 = 0$ in an infinite isotropic continuum obeying the laws of infinitesimal linear elasticity, and subject to the applied stress $p_{23} = \sigma$. The stresses and displacement on the plane $x_2 = a/2$ in the immediate vicinity of a crack tip (Fig. (3)).

$$\left.\begin{aligned} p_{13} &= -\sigma\sqrt{\frac{c}{a}}\ \sqrt{\sin\theta}\ \sin\left(\frac{\theta}{2}\right) \\ p_{23} &= \sigma\sqrt{\frac{c}{a}}\ \sqrt{\sin\theta}\ \cos\left(\frac{\theta}{2}\right) \\ u_{23} &= \frac{a\sigma}{\mu}\sqrt{\frac{c}{a}}\ \frac{1}{\sqrt{\sin\theta}}\ \sin\left(\frac{\theta}{2}\right) \end{aligned}\right\} \tag{3}$$

are used [7] to discuss the situation where two semi-infinite elastic blocks, originally bounded by the faces $x_2 = \pm a/2$ interact across these faces. Elimination of $\theta$ gives the relation between

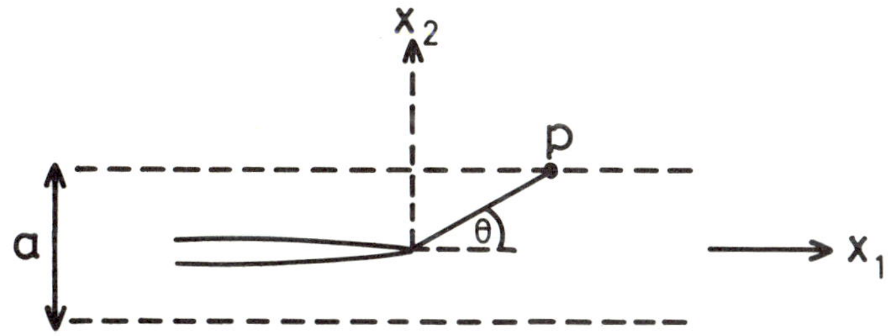


FIGURE 3. The co-ordinate system for representing the stresses and displacement (relations (3) in the immediate vicinity of a crack tip.

the stress $p_{23} \equiv p$ along the plane $x_2 = a/2$ and the displacement $u_3 \equiv w$ as

$$\frac{p}{\mu} = \frac{2(w/a)}{1 + m^4(w/a)^4} \tag{4}$$

where $m = (\mu/\sigma)\sqrt{2a/c}$. Thus for the stress-displacement law (force law) given by relation (4), the stress $p_{23}$ and displacement $u_3$ along the plane $x_2 = a/2$ in the immediate vicinity of the crack tip are given by

$$\begin{aligned} p_{23} &= \frac{\mu}{m}\sqrt{2\sin\theta}\cos\left(\frac{\theta}{2}\right) \\ u_3 &= \frac{\sqrt{2a}}{m}\frac{1}{\sqrt{\sin\theta}}\sin\left(\frac{\theta}{2}\right) \end{aligned} \tag{5}$$

It must be emphasized, however, that these expressions are applicable only to the specific situation where $\sigma\sqrt{c} = \mu\sqrt{2a}/m$; if $\sigma$ or $c$ is increased so that $\sigma\sqrt{c}$ exceeds $\mu\sqrt{2a}/m$ the crack will extend unstably, while if they are reduced so that $\sigma\sqrt{c}$ is less than $\mu\sqrt{2a}/m$ the crack will contract. Elliott's model and the associated results therefore refer only to the instability situation and there is thus a close similarity with the Griffith model. The critical condition may be expressed in terms of the surface energy $\gamma$ appropriate to the force law (4), i.e.

$$\gamma = \int p_{23}(x_1)\,d\{u_3(x_1)\} \tag{6}$$

whereupon $\gamma = \pi a\mu/2m^2$ using relations (5); the critical condition then becomes $\sigma\sqrt{c} = \sqrt{4\mu\gamma/\pi}$ which is precisely the Griffith condition (eqn. (1)) for mode III deformation. The result $\gamma = \pi a\mu/2m^2$ may, of course, also be obtained directly from the force law (4) noting that

$$\gamma = \int_0^\infty p(w)\,dw \tag{7}$$

This equivalence between the crack extension criteria associated with the Griffith and Elliott models also follows [7] by use of the path independent J integral (Eshelby [8], Rice [9]). Thus for any contour Γ that surrounds a crack tip and starts and finishes on respectively the lower and upper faces, the integral

$$J = \int_\Gamma \left\{ \tfrac{1}{2}p_{ij}e_{ij}dx_2 - p_{in}n_j \frac{\partial u_i}{\partial x_j} ds \right\} \tag{8}$$

is independent of the contour; the conventional notation is used for the displacement, strain and stress components, $\delta s$ is an element of arc length and $n_j$ are the direction cosines of the outward normal to the contour. If Γ is a circle of radius r whose centre is the crack tip and $r \to 0$, relations (3) show that J has the value $\pi\sigma^2c/2\mu$. Moreover if Γ is the contour shown in Fig.(4), it again being assumed that Γ is in the immediate vicinity of the

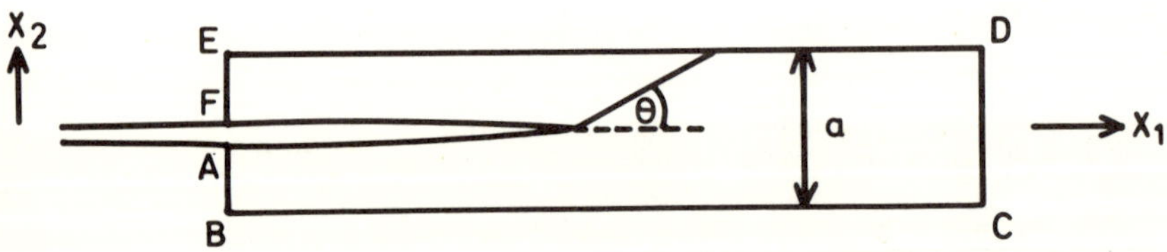


FIGURE 4. The particular contour Γ used to demonstrate, with the aid of the path independent J integral, the equivalence between the crack extension criteria associated with the Griffith and Elliott crack models.

crack tip, the stress and displacement expressions show that, for the limiting case where a/c is small, J reduces to

$$J = 2 \int_0^{\pi} p_{23}(\theta) \frac{du_3(\theta)}{d\theta} d\theta \tag{9}$$

where $p_{23}$ and $u_3$ refer to values along the plane $x_2 = a/2$; since J has the value $\pi\sigma^2 c/2\mu$, it immediately follows that

$$\frac{\pi\sigma^2 c}{2\mu} = 2 \int_0^{\pi} p_{23}(\theta) \frac{du_3}{d\theta} d\theta \tag{10}$$

and the equivalence is again demonstrated.

## BARENBLATT'S APPROACH TO THE NON-LINEARITY PROBLEM

Barenblatt's approach [10] starts with the model (Fig.5) of

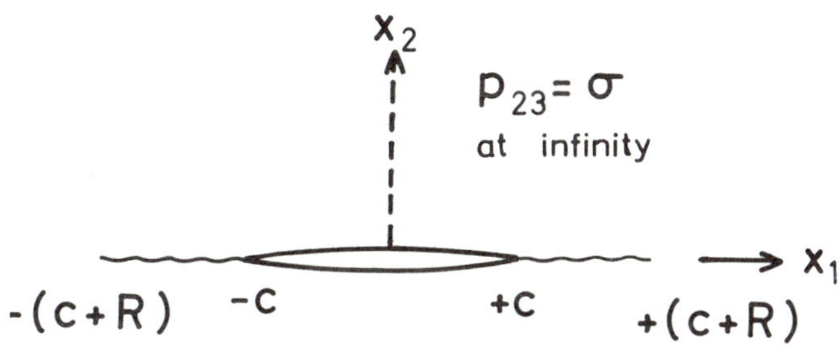


FIGURE 5. Barenblatt's cohesive zone model of a crack. In the intervals $c < | x_1 | < c + R$, $x_2 = 0$, there are cohesive stresses $p(x_1)$ which generate a displacement discontinuity $2v(x_1)$.

a two-dimensional crack $-c < x_1 < c$, $x_2 = 0$ in an infinite isotropic continuum obeying the laws of infinitesimal linear elasticity, and subject to the applied stress $p_{23} = \sigma$. The material along the plane $x_2 = 0$ ahead of the crack tip $x_1 = c$, $x_2 = 0$, is cut as far as $x_1 = C + R$, $x_2 = 0$, the cut faces interacting through cohesive stresses that act in the region $c < x_1 < c + R$, $x_2 = 0$, which is referred to as the cohesive zone (in the symmetric situation a similar procedure is adopted at the other tip). These stresses are represented by $p(x_1)$ a $p_{23}$ component of stress and $v(x_1)$ is the $u_3$ component of displacement of the face of the upper semi-infinite block $x_2 > 0$, $-v(x_1)$ being the corresponding displacement of the face of the lower semi-infinite block $x_2 < 0$; the cohesive zone size is assumed to be small compared with the crack size and $dv/dx_1 = 0$ at the cohesive zone tip $x_1 = c + R$, $x_2 = 0$, so that the stresses are finite. The most realistic way of applying this model is to calculate the stress $p_{23}$ and the displacement $u_3$ along the plane $x_2 = a/2$ for a prescribed $p(x_1)$ or $v(x_1)$ and determine the resulting stress-displacement law; it is therefore possible, in principle, to obtain a wide range of force laws describing the interaction between the planes $x_2 = \pm a/2$, each particular law corresponding to a prescribed variation of $p(x_1)$ or $v(x_1)$ and therefore a specific law relating p and v in the cohesive zone. In this way the original Elliott model, which was discussed in the preceding section and is restricted to the force laws represented by relation (4), is generalized [11]. Elliott's specific model is clearly a special case that arises when the cohesive zone force law is such that $p = p_c$ for $0 < v < v_c$ and $p = 0$ for $v > v_c$, with $p_c$ and $v_c$ being allowed to approach respectively infinity and zero such that $p_c v_c$ remains constant. However this procedure is almost always simplified by ignoring the differences between the displacements and stresses on the planes $x_2 = a/2$ and $x_2 = 0$ and the associated analysis is then relatively easy. This is because equ. (2) is replaced by the equation

$$\frac{\mu}{\pi} \int_{-(C+R)}^{(C+R)} \frac{dw}{d\lambda_1} \frac{d\lambda_1}{(\lambda_1 - x_1)} + \sigma - p(x_1) = 0, \text{ for } |x_1| < C+R, \qquad (11)$$

which is familiar in the theory of slit cracks in an elastic continuum obeying the laws of infinitesimal linear elasticity.

The previous section showed how the Griffith and Elliott models are related, particularly as regards their associated crack extension criteria; Barenblatt's model may also be related to these models in a similar manner [11]. Thus by considering the J integral for any contour Γ that surrounds the cohesive zone and starts and finishes on respectively the lower and upper crack forces, if Γ contracts on to the cohesive zone, it follows [9] from (8) that

$$J = 2 \int_0^{v_c} p(v) dv \qquad (12)$$

where p is the $p_{23}$ stress component and v is the $u_3$ component of displacement of the face of the upper semi-infinite block $x_2 > 0$ within the cohesive zone, and $v_c$ is the value of v at the crack tip. It immediately follows from (9) and (10) that

$$\frac{\pi\sigma^2 c}{2\mu} = 2\int_0^{v_c} p(v)\,dv = 2\int p_{23}\,d(u_3) \tag{13}$$

where $p_{23}$ and $u_3$ are the local stress and displacement along $x_2 = a/2$. The crack extension criterion for the generalized Elliott model is therefore the same as that for the cohesive zone model whose force law generates the generalized Elliott model force law, and is the classic Griffith condition with $\gamma$ replaced by respectively the integrals of $p_{23}$ with respect to $u_3$ and p with respect to v. It must be emphasized that the generalized Elliott model force law is not the same as the generating cohesive zone force law; for example, where the starting model is a sharp crack which, as indicated previously, is a special case of the cohesive zone model for which the cohesive zone force law is $p = p_c$ for $0 < v < v_c$ and $p = 0$ for $v > v_c$ with $p_c \to \infty$, $v_c \to 0$ such that $p_c v_c$ remains constant, the corresponding force law for the Elliott model is given by relation (4). It should also be noted that for a given Elliott model force law, the crack extension criterion may be determined directly without using the generating cohesive zone model, although the displacement and stress distribution near the crack tip cannot be obtained without solving the governing singular integral equation and, as indicated previously, this is in general an extremely difficult procedure. In this context, a major advantage of discussing the crack tip region by means of a generating cohesive zone model is that the displacement and stress for the corresponding Elliott model may, in principle, be determined.

## FURTHER CONSIDERATION OF THE GOVERNING SINGULAR INTEGRAL EQUATION

Equation (2) governs the behaviour of the general non-linear crack model, and the complete expressions for the stress and displacement along $x_2 = a/2$ for the original Elliott model and also the generalized model arising from an appropriate cohesive zone model automatically satisfy this equation. However the discussion of Elliott's original model used expressions for stress and displacement in the immediate vicinity of a sharp crack tip, and these satisfy the simplified singular integral equation

$$\frac{1}{\pi}\int_{-\infty}^{\infty} \frac{dw}{d\lambda_1}\,\frac{d\lambda_1}{(\lambda_1 - x_1)} = \frac{P(w)}{\mu} \quad \text{for all } x_1, \tag{14}$$

where w is the displacement of the face $x_2 = a/2$ of the upper block and P(w) describes the interaction between the faces of the two semi-infinite blocks; equ. (14) is in effect the local representation of equ. (2). For the original Elliott model, the equation becomes

$$\frac{1}{\pi}\int_{-\infty}^{\infty}\frac{dw}{d\lambda_1}\ \frac{d\lambda_1}{(\lambda_1-x_1)} = \frac{2(w/a)}{1+m^4(w/a)^4} \tag{15}$$

with the solution (see relations (3))

$$\left.\begin{aligned}
\frac{w}{a} &= \frac{1}{m}\left[\left(1+\frac{4\lambda_1^{\ 2}}{a^2}\right)^{\frac{1}{2}} - \frac{2\lambda_1}{a}\right]^{\frac{1}{2}} \\
\frac{dw}{d\lambda_1} &= -\frac{1}{m}\left[\left(1+\frac{4\lambda_1^{\ 2}}{a^2}\right)^{\frac{1}{2}} - \frac{2\lambda_1}{a}\right]^{\frac{1}{2}}\Bigg/\left[1+\frac{4\lambda_1^{\ 2}}{a^2}\right]^{\frac{1}{2}} \\
\frac{p}{\mu} &= \frac{1}{m}\left[\left(1+\frac{4\lambda_1^{\ 2}}{a^2}\right)^{\frac{1}{2}} + \frac{2\lambda_1}{a}\right]^{\frac{1}{2}}\Bigg/\left[1+\frac{4\lambda_1^{\ 2}}{a^2}\right]^{\frac{1}{2}}
\end{aligned}\right\} \tag{16}$$

which may also be shown to be correct by direct substitution in (15) and appropriate contour integration. It is valid only for the particular situation where $m = (\mu/\sigma)\sqrt{2a/c}$ when

$$\left.\begin{aligned}
w &= \frac{\sigma}{2\mu}\sqrt{\frac{c}{2x_1}}\ a \qquad &\text{as } x_1 \to \infty \\
w &= \frac{\sigma}{\mu}\sqrt{-2cx_1} \qquad &\text{as } x_1 \to -\infty
\end{aligned}\right\} \tag{17}$$

with the condition (see relation (6))

$$\frac{\pi\sigma^2 c}{4\mu} = \int_0^{\infty} P(w)\,dw \tag{18}$$

This state of affairs is not peculiar to the force law given by relation (4). The displacement and stress near the crack tip for any force law are given by the solution of the non-linear singular integral equation (14) subject to the boundary conditions (17), though recognizing that the solution applies only to the critical situation given by relation (18). Equation (14) is far easier to solve than the more general equ. (2), the simplification arising as a consequence of reframing the original equation around the crack tip neighbourhood. The arguments leading to the simplified equation are non-rigorous and the precise conditions for which it is valid remain to be explored in detail. However the conditions are likely to be somewhat similar to those for which the cohesive zone model is applicable; i.e. for the case where $p = p_c$ for $0 < v < v_c$ and $p = 0$ when $p > p_c$, $v_c\mu/cp_c$ should be small so that the cohesive zone size is small compared with the crack size.

So far this paper has considered only one set of solutions to equ. (14), i.e. that corresponding to the original Elliott model for which the force law $P(w)$ is given by equ. (4), although, as indicated earlier, other solutions may be obtained by appropriate use of the cohesive zone model. However, such a procedure is inverse in that stress and displacement variations are determined and associated force laws obtained by elimination of the relevant space variable. It is obviously more satisfying and straightforward to prescribe $P(w)$ in the first instance and solve (14) directly, taking into account the appropriate boundary conditions; another specific objective of this paper is to focus attention on the possibility of obtaining such solutions since (14) is far easier to discuss than the original equ. (2). A careful search of the literature for other fields should reveal solutions to (14), albeit limited in their range of applicability, for specific forms of $P(w)$.

As an illustration, consider the specific force law for which

$$\left.\begin{aligned} P(w) &= 2\mu w/a && \text{for } 0 < w < w_c \\ P(w) &= 0 && \text{for } w > w_c \end{aligned}\right\} \qquad (19)$$

using a co-ordinate system such that

$$\left.\begin{aligned} P(w) &= 0 && \text{for } -\infty < x_1 < 0 \\ P(w) &= 2\mu w/a && \text{for } 0 < x_1 < \infty \end{aligned}\right\} \qquad (20)$$

Although $P(w)$ is linear with respect to $w$ for $0 < w < w_c$, this law falls into the non-linear category since there is a cut-off at a finite value of $w$. For this law, the governing equ.(14) may be transformed into an equation that arises in connection with Spence's discussion [13] of the lift coefficient of a thin jet-flapped wing. Detailed consideration of this problem will be published elsewhere, but a particular result is that at the critical stage of crack instability,

$$\left.\begin{aligned} \frac{w}{w_c} &= 1 + \frac{2x_1}{\pi a}\left[\log_e\left(\frac{2x_1}{\pi a}\right) + \gamma_o - 2\right] \quad \text{for } \frac{2x_1}{a} < 1 \\ \frac{w}{w_c} &= \left(\frac{2\pi x_1}{a}\right)^{\frac{1}{2}} + \tfrac{1}{2}\left(\frac{2\pi x_1}{a}\right)^{-\frac{3}{2}}\left[\log_e\left(\frac{8x_1}{a}\right) + \gamma_o - 1\right] \quad \text{for } \frac{2x_1}{a} > 1 \end{aligned}\right\} \qquad (21)$$

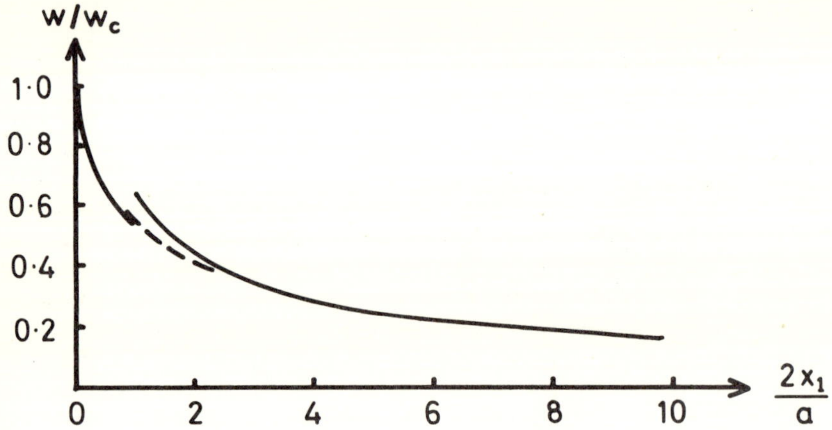


FIGURE 6. The displacement w ahead of the crack tip when the force law is given by relation (19); this result has been obtained using Spence's discussion [13] of the lift coefficient of a thin jet-flapped wing.

neglecting higher terms, where $\gamma_o$ is Euler's constant. This displacement ahead of the crack tip is shown in Fig. (6), the solid lines denoting the limiting forms for small and large $x_1$ while the broken line is faired between them. This result is presented as an illustrative example to reinforce the view that other discussions of equ. (14) are likely to be found elsewhere in the literature.

## DISCUSSION

The preceding sections have discussed methods of accounting for non-linearity of material behaviour in the vicinity of a crack tip, consideration being restricted to the case where the non-linearity is confined to a strip that is coplanar with the crack. Thus Elliott's original model has been reviewed in detail, and it has been shown how this can be generalized using Barenblatt's cohesive zone model; moreover the crack extension criteria for these models have been related to each other and also with the classic Griffith condition. In a sense, these models may be regarded as approximate methods of solving the governing singular integral equ. (2) or its simplified version (14). With respect to the latter, it has been suggested that a careful literature survey could reveal appropriate solutions for given forms of P(w) and a particular example was given in the preceding section.

Of course, one must question the usefulness of the search for solutions of the governing singular integral equation for a given force law, particularly as the crack extension criterion (18) can be determined without the necessity of knowing the stress and displacement variations in the vicinity of a crack tip. The justification is that many physical situations, such as the degradation of material at the tip of a propagating crack in steel plate where it is appropriate to use the type of model considered in this paper, are so complex that it is often difficult to specify the force law with any precision, and a detailed understanding of the fracture process depends on fitting together various fragmentary pieces of experimental evidence; in such situations it is essential to know the displacement and stress variations produced by various force laws.

The paper has concentrated on mode III deformation in the interests of mathematical simplicity; however the conclusions are, in general, equally applicable for modes I and II and the referenced papers consider these modes in detail. It should also be emphasized that the models are of the continuum type even though they account for non-linearity of material behaviour and since the non-linearity can describe various fracture processes, their usefulness is wide-ranging. However care must be exercised in those situations where the associated theory predicts large stress variations over distances of the order of atomic dimensions. In such cases, for example where the non-linearity represents the rupturing of atomic bonds, it is more appropriate to use a model that allows for the atomic structure's discreteness. Recently there have been significant advances in this direction [14,15], following the initial investigation of Goodier and Kanninen [16]; thus consideration has been given to the effect of the atomic structure's discreteness on cleavage crack extension, using models in which the atoms in the separating crystal planes are allowed to be discrete, while the remainder of the material is in the form of an elastic continuum.

REFERENCES

[1] Griffith, A.A., The phenomena of rupture and flow in solids, Phil.Trans.Roy.Soc., A 221, 163, 1921.

[2] Elliott, H.A., An analysis of the conditions for rupture due to Griffith cracks, Proc.Phys.Soc., 59, 208, 1947.

[3] Bilby, B.A., and Eshelby, J.D., Dislocations and the theory of fracture, Fracture, Volume 1, Ed. H. Liebowitz, Academic Press Inc., New York, 1968, 99.

[4] Peierls, R.E., The size of a dislocation, _Proc.Phys.Soc._, 52, 34, 1940.

[5] Nabarro, F.R.N., The stress to move a dislocation, _Proc.Phys.Soc._, 59, 256, 1947.

[6] Andersson, H., and Bergkvist, H., Analysis of a non-linear crack model, _Jnl.Mech.Phys.Solids_, 18, 1, 1970.

[7] Smith, E., A comparison of the crack extension criteria associated with the Griffith and Elliott crack models, _Int.Jnl.Fracture_, to be published.

[8] Eshelby, J.D., The force on an elastic singularity, _Phil. Trans.Roy.Soc._, A 244, 87, 1951.

[9] Rice, J.R., Mathematical analyses in the mechanics of fracture, _Fracture_, Volume 2, Ed. H. Liebowitz, Academic Press Inc., New York, 1968, 192.

[10] Barenblatt, G.I., Mathematical theory of equilibrium cracks, _Advances in Appl.Mech._, 2, 55, 1962.

[11] Smith, E., A generalization of Elliott's model of a crack tip, to be published.

[12] Smith, E., The structure in the vicinity of a crack tip: a general theory based on the cohesive zone model, _Engineering Fracture Mechanics_, to be published.

[13] Spence, D.A., The lift coefficient of a thin jet-flapped wing, _Proc.Roy.Soc._, A 261, 97, 1961.

[14] Thomson, R., Hsieh, C., and Rana, V., Lattice trapping of fracture cracks, _Jnl.App.Phys._, 42, 3154, 1971.

[15] Smith, E., The effect of the discreteness of the atomic structure on cleavage crack extension in brittle crystalline materials, _Jnl.Applied Physics_, to be published.

[16] Goodier, J.N., Mathematical theory of equilibrium cracks, _Fracture_, Volume 2, Ed. H. Liebowitz, Academic Press Inc., New York, 1968, 1.

## DISCUSSION by G. Caglioti, Nucleare Politechnico, Italy

I would like to call attention to Professor Smith that the one-dimensional crack model of brittle fracture is related to the Peierls stress of dislocation as introduced by Frenkel and Kontorova. This problem has also been studied at the Politechnico of Milano. The spring constants in Professor Smith's model can be obtained from a knowledge of the dispersion relations of the normal modes of vibration of a given metal by means of inelastic neutron scattering.

For instance, a model can be devised for the evaluation of the surface energy $\gamma$ for brittle fracture based on the local sublimation along the fracture plane. Referring to Fig. (1), the vertical lines represent a set of crystal planes, say (100) planes of a pcc crystal. Fracture occurs when the interplanar springs are loaded such that the additional displacement from the equilibrium position required to separate two blocks of crystal corresponds to the atomic mean square displacement at sublimation, $u_s$. This process is described by an equation of the type:

$$2\gamma = \frac{1}{2}\,(K_1 + 2K_2 + 3K_3 + \text{---})\,(\overline{2u_s})^2 n$$

where $K_1$, $K_2$, etc., are the interplanar force constants referred to the neighboring crystal planes of order 1, 2, etc., and n is the number density of atoms on the plane of fracture. The quantities $K_1$, $K_2$, -- are deduced from the Foreman-Lomer expansion of the square of frequencies of the normal modes of vibration in terms of the wave vectors, and $u_s$ can be derived from the Lindemann melting formula plus some elementary energy considerations.

The predicted values of $\gamma$ are in general agreement with the experimental data. Reference can be made to a number of recent papers [1-3] concerning $\gamma$, and the Peierls stress as computed from the Frenkel-Kontorova-Hobart model and the origin of acoustic signals from moving dislocations.

REFERENCE

[1] Caglioti, G., Rizzi, G. and Bilells, J. C., J. Appl. Phys., 42, 4271, 1971.

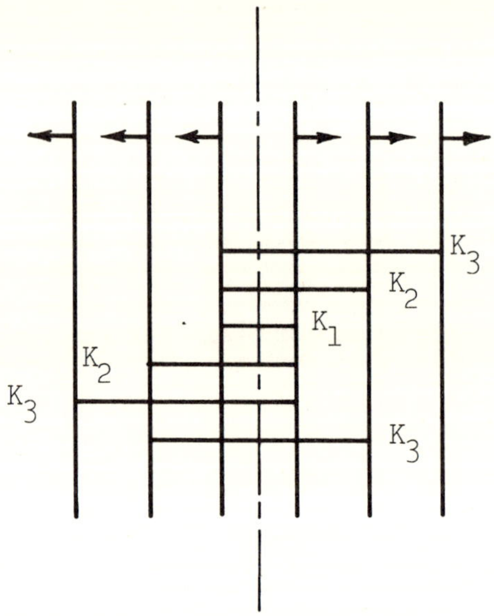


Fig. (1) - Crystal planes

## AUTHOR'S CLOSURE

No comments.

# FRACTURE OF COMPOSITE MATERIALS

F. Erdogan

Lehigh University, Bethlehem, Pennsylvania 18015, U.S.A.

## ABSTRACT

The paper deals with the method of analysis and a related fracture criterion for the problem of a crack running into and crossing a bi-material interface. The general problem arises in fracture studies of composite materials containing a dominant flaw in the neighborhood of a boundary separating two constituents with different mechanical properties. The method is described by applying it to a specific problem. Sample results are given from its application to various crack-second phase interaction problems.

## INTRODUCTION

To a considerable extent, the fracture strength of composite materials depends on the size, the shape, and the orientation of flaws in the nonhomogeneous medium. Thus in fracture studies of composite materials very often the related mechanics problem is that of a dominant crack located near a bi-material interface in an arbitrary orientation. In this analytical phase of the study the general objective is the evaluation of a "critical load factor" which represents the severity of the applied loads and which corresponds to the conjectured mode of fracture. The remaining phases of the fracture study consist of the experimental determination of the corresponding "characteristic material parameter" representing the fracture strength of the material, and the development or selection and application of an appropriate fracture criterion.

Because of the obvious practical importance of the problem, within the past two decades a rather vast amount of effort has been devoted to the strength studies in composites following both the so-called fracture mechanics approach dealing with the initiation and propagation of rupture from a dominant flaw and the statistical approach dealing with the bulk strength of the material. Even to attempt any kind of a review of the general topic would be outside the scope of this paper. Here we will only consider an isolated aspect of the subject concerning the problem of a crack running into and crossing a bi-material interface. An important peculiarity of this problem lies in the fact that for the crack terminating at the interface the power of the singularity is no longer -1/2 and hence, if the crack propagates further, the stress state around the crack tip will not remain self-similar. Two questions which need to be studied in this connection are therefore the behavior of the stress state around the crack tip terminating at or crossing the boundary, and the development or elucidation of a relevant fracture criterion. These two questions will be discussed in the following sections.

## INTEGRAL EQUATIONS AND THE FUNDAMENTAL FUNCTIONS

The general three-dimensional problem of a crack with arbitrary shape and orientation which is located near an interface in the composite medium is clearly intractable. However, the main features of the problem can be studied by idealizing the materials, the geometry, and the loading conditions. Thus, in this study it is assumed that the "local" problem may be represented by two bonded isotropic linearly elastic half planes containing through cracks which are oriented perpendicular to the interface and are subjected to normal tractions (Fig. 1). The problem is then one of plane strain or generalized plane stress. Using Mellin transforms and the polar coordinates shown in Fig. (1), the mixed boundary value problem may be formulated in a rather straightforward manner. Let the known tractions on the crack surfaces be

$$\tau_{1\theta\theta}(r,\pi) = -p_1(r)\,, \quad \tau_{1r\theta}(r,\pi) = 0\,, \quad (a_1 < r < b_1)\,,$$

$$\tau_{2\theta\theta}(r,0) = -p_2(r)\,, \quad \tau_{2r\theta}(r,0) = 0\,, \quad (a_2 < r < b_2)\,. \quad (1.a\text{-}d)$$

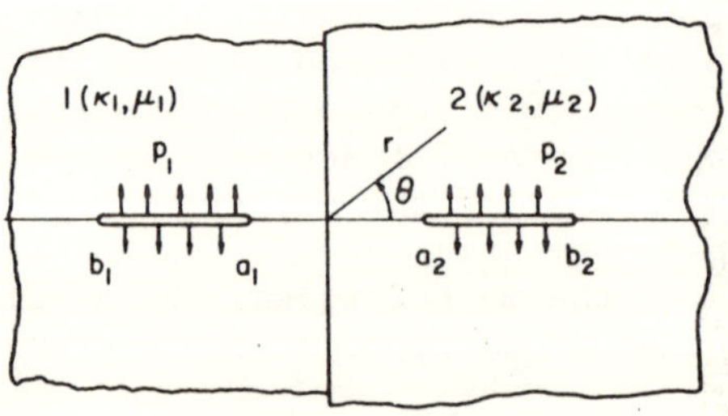


Fig. 1. Cracks in bonded half planes.

Referring to [1] and [2] for details, in terms of the unknown functions

$$f_1(r) = -2 \frac{\partial}{\partial r} u_{1\theta}(r,\pi-0) , \qquad f_2(r) = 2 \frac{\partial}{\partial r} u_{2\theta}(r,+0) \tag{2.a,b}$$

the integral equations of the problem may be obtained as

$$\frac{1}{\pi}\int_{a_1}^{b_1} [\frac{1}{s-r} + k_{11}(r,s)]f_1(s)ds + \frac{1}{\pi}\int_{a_2}^{b_2} k_{12}(r,s)f_2(s)ds$$

$$= - \frac{1+\kappa_1}{2\mu_1} p_1(r) , \qquad (a_1 < r < b_1) ,$$

$$\frac{1}{\pi}\int_{a_1}^{b_1} k_{21}(r,s)f_1(s)ds + \frac{1}{\pi}\int_{a_2}^{b_2} [\frac{1}{s-r} + k_{22}(r,s)]f_2(s)ds$$

$$= - \frac{1+\kappa_2}{\mu_2} p_2(r) , \qquad (a_2 < r < b_2) \tag{3.a,b}$$

where

$$k_{11}(r,s) = \sum_{k=1}^{3} \frac{c_{1k}r^{k-1}}{(s+r)^k} , \qquad k_{22}(r,s) = \sum_{k=1}^{3} \frac{c_{2k}r^{k-1}}{(s+r)^k} ,$$

$$k_{12}(r,s) = \sum_{1}^{2} \frac{d_{1k}r^{k-1}}{(s+r)^k} , \qquad k_{21}(r,s) = \sum_{1}^{2} \frac{d_{2k}r^{k-1}}{(s+r)^k} ; \tag{4.a-d}$$

$$c_{11} = \frac{1}{2} - \frac{m_1(1+\kappa_1)}{2(m_1+\kappa_2)} - \frac{3(1-m_1)}{2(1+m_1\kappa_1)} , \qquad c_{12} = \frac{6(1-m_1)}{1+m_1\kappa_1} ,$$

$$c_{13} = - \frac{4(1-m_1)}{1+m_1\kappa_1} ; \qquad c_{21} = \frac{1}{2} - \frac{m_2(1+\kappa_2)}{2(m_2+\kappa_1)} - \frac{3(1-m_2)}{2(1+m_2\kappa_2)} ,$$

$$c_{22} = \frac{6(1-m_2)}{1+m_2\kappa_2} , \qquad c_{23} = - \frac{4(1-m_2)}{1+m_2\kappa_2} ;$$

$$d_{11} = \frac{3(1+\kappa_1)}{2(m_2+\kappa_1)} - \frac{1+\kappa_1}{2(1+m_2\kappa_2)} , \qquad d_{12} = \frac{1+\kappa_1}{1+m_2\kappa_2} - \frac{1+\kappa_1}{m_2+\kappa_1}$$

$$d_{21} = \frac{3(1+\kappa_2)}{2(m_1+\kappa_2)} - \frac{1+\kappa_2}{2(1+m_1\kappa_1)} , \qquad d_{22} = \frac{1+\kappa_2}{1+m_1\kappa_1} - \frac{1+\kappa_2}{m_1+\kappa_2} ;$$

$$m_1 = \mu_2/\mu_1 , \qquad m_2 = \mu_1/\mu_2 . \tag{5}$$

In Eq. (5) $\mu_1$ and $\mu_2$ are the shear moduli, $\kappa_i = 3-4\nu_i$ for plane strain and $\kappa_i = (3-\nu_i)/(1+\nu_i)$ for the generalized plane stress, $(i = 1,2)$, $\nu_i$ being the Poisson's ratio. Equations (3) have to be solved under the following single-valuedness conditions:

$$\int_{a_j}^{b_j} f_j(r)dr = 0 , \qquad (j = 1,2) . \tag{6}$$

In order to solve the system of singular integral equations (3) first the fundamental functions of the system must be determined. Thus following [3] we let

$$f_j(s) = g_j(s)w_j(s) = \frac{g_j(s)}{(s-a_j)^{\alpha_j}(b_j-s)^{\beta_j}} , \qquad 0 < \mathrm{Re}(\alpha_j,\beta_j) < 1 \qquad (j = 1,2) \tag{7}$$

where $g_j(s)$ satisfies a Hölder condition in $a_j \leq s \leq b_j$ and is non-zero at the ends $a_j$ and $b_j$, and the fundamental function $w_j(s)$ is any definite branch which varies continuously in $a_j < s < b_j$, $(j = 1,2)$. Consider now the following sectionally holomorphic functions

$$G_j(z) = \frac{1}{\pi}\int_{a_j}^{b_j} \frac{f_j(s)}{s-z}\, ds = \frac{1}{\pi}\int_{a_j}^{b_j} \frac{g_j(s)e^{\pi i\beta_j}ds}{(s-z)(s-a_j)^{\alpha_j}(s-b_j)^{\beta_j}} , \qquad (j = 1,2) , \tag{8}$$

where $z = r+i\rho$. From the asymptotic analysis around the end points of the cuts we obtain

$$G_j(z) = \frac{g_j(a_j)e^{\pi i\alpha_j}}{(b_j-a_j)^{\beta_j}\sin\pi\alpha_j}\,\frac{1}{(z-a_j)^{\alpha_j}} - \frac{g_j(b_j)}{(b_j-a_j)^{\alpha_j}\sin\pi\beta_j}\,\frac{1}{(z-b_j)^{\beta_j}} + G_{oj}(z) , \qquad (j = 1,2) \tag{9}$$

where $G_{oj}$ is bounded everywhere except possibly at the ends, $e_{jk}$, near which

$$|G_{oj}(z)| < \frac{C_{jk}}{|z-e_{jk}|^{p_{jk}}} , \qquad (j,k = 1,2) \tag{10}$$

$$e_{j1} = a_j , \quad p_{j1} < \mathrm{Re}(\alpha_j) ; \qquad e_{j2} = b_j , \quad p_{j2} < \mathrm{Re}(\beta_j) ,$$

$C_{jk}$ and $p_{jk}$ being real constants.

Using now the Plemelj formula [3]

$$\frac{1}{\pi}\int_{o}^{b_j} \frac{f_j(s)}{s-r}\, ds = \frac{1}{2}\,[G_j^{+}(r) + G_j^{-}(r)]\,, \qquad (a_j < r < b_j) \tag{11}$$

from Eq. (9) the terms with the Cauchy singularity in Eq. (3) may be expressed as

$$\frac{1}{\pi}\int_{a_j}^{b_j} \frac{f_j(s)}{s-r}\, ds = \frac{g_j(a_j)\cot\pi\alpha_j}{(b_j-a_j)^{\beta_j}(r-a_j)^{\alpha_j}} - \frac{g_j(b_j)\cot\pi\beta_j}{(b_j-a_j)^{\alpha_j}(b_j-r)^{\beta_j}} + G_{1j}(r)\,, \qquad (a_j < r < b_j,\ j = 1,2) \tag{12}$$

where the behavior of $G_{1j}(r)$ in the closed interval $[a_j,b_j]$ is similar to that of $G_{oj}$ $(j = 1,2)$. The remaining terms in the integral equations (3) may be obtained directly from Eq. (9) through differentiation. From Eqs. (3), (4) and (8) it is seen that in these terms $z = -r$ which is outside the cut $a_j < r < b_j$, $(j = 1,2)$. As r varies in the interval $(a_j,b_j)$, z will vary in the interval $(-b_j,-a_j)$ in which $G_j(z)$ is holomorphic. Thus, from Eqs. (8) and (9) the asymptotic behavior of these integrals (for small $a_j$ and r) may be obtained as

$$\frac{1}{\pi}\int_{a_j}^{b_j} \frac{f_j(s)}{s+r}\, ds = G_j(-r) = \frac{g_j(a_j)}{(b_j-a_j)^{\beta_j}\sin\pi\alpha_j}\,\frac{1}{(r+a_j)^{\alpha_j}} + G_{2j}(r)\,,$$

$$\frac{1}{\pi}\int_{a_j}^{b_j} \frac{rf_j(s)}{(s+r)^2}\, ds = -r\frac{d}{dr}\,G_j(-r) = \frac{\alpha g_j(a_j)}{(b_j-a_j)^{\beta_j}\sin\pi\alpha_j}\,\frac{1}{(r+a_j)^{\alpha_j}} + G_{3j}(r)\,,$$

$$\frac{1}{\pi}\int_{a_j}^{b_j} \frac{r^2 f_j(s)}{(s+r)^3}\, ds = \frac{r^2}{2}\,\frac{d^2}{dr^2}\,G_j(-r) = \frac{\alpha_j(\alpha_j+1)g_j(a_j)}{2(b_j-a_j)^{\beta_j}\sin\pi\alpha_j}\,\frac{1}{(r+a_j)^{\alpha_j}} + G_{4j}(r)\,,$$

$$(a_j < r < b_j,\ j = 1,2) \tag{13.a-c}$$

where the behavior of $G_{kj}(r)$ in $a_j \le r \le b_j$ will be similar to that of $G_{oj}$, $(j = 1,2;\ k = 2,3,4)$.

Consider now the following possible crack geometries:

(a) $0 < a_j < b_j$ , $(j = 1,2)$.

In this case the integrals given in Eq. (13) and hence the terms containing the kernels $k_{ij}(r,s)$ in Eq. (3) are bounded. Thus, substituting Eq. (12) into Eq. (3) and noting that $g_j(a_j)$ and $g_j(b_j)$, $(j = 1,2)$ are finite and nonzero, we obtain

$$\cot\pi\alpha_j = 0 , \qquad \cot\pi\beta_j = 0 , \qquad (j = 1,2) \tag{14}$$

giving $\alpha_j = 1/2 = \beta_j$ as the admissible values. This is the well-known result for a crack imbedded in a homogeneous medium.

(b) $0 = a_1 < b_1$ , $0 < a_2 < b_2$ .

If $a_1 = 0$, or $a_2 = 0$, or $a_1 = a_2 = 0$, from Eq. (4) it is seen that some or all the kernels $k_{ij}$, $(i,j = 1,2)$ become unbounded as s and r both approach zero. Consequently, in addition to the simple Cauchy kernels $(s-r)^{-1}$ these kernels will also have an influence on the singular behavior of the solution at the end point(s) $r = a_j$. In this case the integral equations are said to have "generalized Cauchy kernels". For $a_1 = 0$, $a_2 > 0$ note that the terms containing the kernels $k_{12}$, $k_{21}$, and $k_{22}$ in Eq. (3) are bounded, and that containing $k_{11}$ has a singularity of the form $r^{-\alpha_1}$. Thus, letting $a_1 = 0$, substituting from Eqs. (12), (13) and (4) into Eq. (3), and again observing that $g_j(s)$ is finite and nonzero at the end points, we obtain

$$2\cos\pi\alpha_1 + (D_1 + D_2) - 4D_1(1 + \alpha_1)^2 = 0 , \tag{15}$$

$$D_1 = (1 - m_1)/(1 + m_1\kappa_1) , \qquad D_2 = (\kappa_2 - m_1\kappa_1)/(\kappa_2 + m_1) , \tag{16}$$

$$\cot\pi\beta_1 = 0 , \qquad \cot\pi\alpha_2 = 0 , \qquad \cot\pi\beta_2 = 0 . \tag{17}$$

It can be shown that for all practical material combinations the characteristic equation (15) has only one root in the strip $0 < \mathrm{Re}(\alpha_1) < 1$, and this root is always real.

(c) $0 = a_1 < b_1$ , $0 = a_2 < b_2$ .

This is the case of a crack crossing the interface in which the problem has only three irregular points, namely 0, $b_1$, and $b_2$. If $\alpha$ is the power of the singularity at the irregular point $r = 0$, then Eqs. (3) - (13) are still valid with $\alpha_1 = \alpha = \alpha_2$ and $a_1 = 0 = a_2$. From Eqs. (4), (12) and (13) it is clear that all the terms on the left-hand side of Eq. (3) will have a singularity of the form $r^{-\alpha}$. Thus, if we substitute from Eqs. (4), (12), and (13) into Eq. (3), and multiply both sides first by $r^{\alpha}$ and let $r \to 0$, then Eq. (3.a) by $(b_1-r)^{\beta_1}$ and let $r \to b_1$, and Eq. (3.b) by $(b_2-r)^{\beta_2}$ and let $r \to b_2$ we obtain

$$(\cos\pi\alpha + c_{11} + \alpha c_{12} + \frac{\alpha(\alpha+1)}{2} c_{13}) \frac{g_1(0)}{b_1^{\beta_1}\sin\pi\alpha}$$

$$+ (d_{11} + \alpha d_{12}) \frac{g_2(0)}{b_2^{\beta_2}\sin\pi\alpha} = 0 ,$$

$$(d_{21} + \alpha d_{22}) \frac{g_1(0)}{b_1^{\beta_1}\sin\pi\alpha}$$

$$+ (\cos\pi\alpha + c_{21} + \alpha c_{22} + \frac{\alpha(\alpha+1)}{2} c_{23}) \frac{g_2(0)}{b_2^{\beta_2}\sin\pi\alpha} = 0 ,$$

$$\frac{g_1(b_1)}{b_1^{\alpha}} \cot\pi\beta_1 = 0 , \qquad \frac{g_2(b_2)}{b_2^{\alpha}} \cot\pi\beta_2 = 0 . \qquad (18.a\text{-}d)$$

Noting that $g_j(0)$ and $g_j(b_j)$, $(j = 1,2)$ are finite and nonzero, from Eq. (18) we obtain

$$\cot\pi\beta_1 = 0 , \qquad \cot\pi\beta_2 = 0 , \qquad (19)$$

$$[(K_1-K_2) \cos^2 \frac{\pi\alpha}{2} - K_1(\alpha-1)^2]^2$$

$$+ K_3^2 \sin^2 \frac{\pi\alpha}{2} \cos^2 \frac{\pi\alpha}{2} - K_2^2(\alpha-1)^2 = 0 , \qquad (20)$$

$$K_1 = 2(m_1-1) , \qquad K_2 = m_1(1+\kappa_1) - (1+\kappa_2) ,$$

$$K_3 = m_1(1+\kappa_1) + (1+\kappa_2) ,$$

where Eq. (20) is the coefficient determinant of Eqs. (18.a) and (18.b). From Eq. (18) it is clear that $g_1(0)$ and $g_2(0)$ are not independent and are related by

$$g_2(0) = - g_1(0) \sqrt{b_2/b_1} \frac{\cos\pi\alpha + c_{11} + \alpha c_{12} + \frac{\alpha(\alpha+1)}{2} c_{13}}{d_{11} + \alpha d_{12}} . \qquad (21)$$

In the special case of homogeneous medium the acceptable root becomes $\alpha = 0$ giving $g_2(0)/\sqrt{b_2} = -g_1(0)/\sqrt{b_1}$. Considering Eqs. (2) and (7), this means that, as expected, at $r = 0$ the derivative of the crack surface displacement becomes finite and continuous.

In the case of through crack, the two conditions given by Eq. (6) which are necessary to obtain a unique solution of the integral equations are replaced by Eq. (21) and the following single-valuedness condition:

$$\int_0^{b_1} f_1(s)ds = \int_0^{b_2} f_2(s)ds . \qquad (22)$$

STRESS INTENSITY FACTORS

The left-hand sides of the integral equations (3) represent the stress components $\tau_{1\theta\theta}(r,\pi)$ and $\tau_{2\theta\theta}(r,0)$ on the outside as well as the inside of the cuts $(a_j,b_j)$. Thus, in the case of $a_1>0$, $a_2>0$, noting that the terms involving $k_{ij}$ are bounded and outside the cuts, $G_j(z)$ defined by Eq. (8) is holomorphic, $G_j(r)$ directly gives the asymptotic values of $\tau_{j\theta\theta}$ on the real axis, from which the stress intensity factors may readily be evaluated. For example, at the end $r=b_1$ from Eqs. (8), (9) and (3) the stress intensity factor $k(b_1)$ may be defined and evaluated as follows:

$$k(b_1) = \lim_{r\to b_1}\sqrt{2(r-b_1)}\;\tau_{1\theta\theta}(r,\pi) = \frac{2\mu_1}{1+\kappa_1}\lim_{r\to b_1}\sqrt{2(r-b_1)}\;G_1(r)$$

$$= -\frac{2\mu_1}{1+\kappa_1}\frac{\sqrt{2}\,g_1(b_1)}{\sqrt{b_1-a_1}} = -\frac{2\mu_1}{1+\kappa_1}\lim_{r\to b_1}\sqrt{2(b_1-r)}\;f_1(r)\,. \tag{23}$$

Similarly

$$k(a_1) = \lim_{r\to a_1}\sqrt{2(a_1-r)}\;\tau_{1\theta\theta}(r,\pi) = \lim_{r\to a_1}\frac{2\mu_1}{1+\kappa_1}\sqrt{2(r-a_1)}\;f_1(r)\,,$$

$$k(a_2) = \lim_{r\to a_2}\sqrt{2(a_2-r)}\;\tau_{2\theta\theta}(r,0) = \lim_{r\to a_2}\frac{2\mu_2}{1+\kappa_2}\sqrt{2(r-a_2)}\;f_2(r)\,,$$

$$k(b_2) = \lim_{r\to b_2}\sqrt{2(r-b_2)}\;\tau_{2\theta\theta}(r,0) = -\lim_{r\to b_2}\frac{2\mu_2}{1+\kappa_2}\sqrt{2(b_2-r)}\;f_2(r)\,. \tag{24}$$

In the second typical case where one crack tip is at the interface (e.g., $b_1>a_1=0$, $b_2>a_2>0$), at $r=0$ the stress intensity factor will be defined by

$$k(0) = \lim_{r\to 0}\sqrt{2}\;r^{\alpha_1}\;\tau_{2\theta\theta}(r,0)\,, \tag{25}$$

where $\alpha_1$ is obtained from Eq. (15) and $\tau_{2\theta\theta}(r,0)$, $(0<r<a_2,\; b_2<r)$ is given by Eq. (3.b). Since $a_1=0$ and $0<a_2<b_2$, it is clear that in Eq. (3.b) only the first term on the left-hand side will contribute to the stress singularity. Thus, substituting from Eqs. (13), (4.d), and (3.b) into Eq. (25) and ignoring the bounded terms, we obtain

$$k(0) = \lim_{r\to 0}\sqrt{2}\;r^{\alpha_1}\frac{2\mu_2}{1+\kappa_2}\frac{1}{\pi}\int_0^{b_1}\sum_{k=1}^{2}\frac{d_{2k}r^{k-1}}{(s+r)^k}\,f_1(s)ds$$

$$= \sqrt{2}\;\mu_1^*\;g_1(0)/\sqrt{b_1} = \lim_{r\to 0}\sqrt{2}\;\mu_1^*\;r^{\alpha_1}\;f_1(r)\,, \tag{26}$$

$$\mu_1^* = m_1\mu_1 \frac{(3-2\alpha_1)(1+m_1\kappa_1) - (1-2\alpha_1)(m_1+\kappa_2)}{(1+m_1\kappa_1)(m_1+\kappa_2)\sin\pi\alpha_1} .$$

For the remaining three crack tips $r = (b_1, a_2, b_2)$ the stress intensity factors will again be given by Eqs. (23) and (24). In terms of the stress intensity factor $k_1(0)$ the asymptotic values of the stresses around the crack tip $r = 0$ (in both materials) may be expressed as

$$\tau_{ij}(r,\theta) = \frac{k_1(0)}{\sqrt{2}\, r^{\alpha_1}} f_{ij}(\theta) + O(r^{\alpha_1'}) ,$$

$$(0 \leq \theta < \pi,\ (i,j) = (r,\theta),\ \mathrm{Re}(\alpha_1') > 0). \qquad (27)$$

where the functions $f_{ij}$ giving the angular distribution of the stresses are obtained from the original Mellin transform solution and are given in [1].

For the crack crossing the interface, even though the singular behavior of the stresses at $r = 0$ can again be expressed in terms of a single constant, here, because of their direct relevance to the possible debonding phenomenon, the stress intensity factors will be defined as follows:

$$k_\theta = \lim_{r\to 0} r^\alpha \tau_{j\theta\theta}(r,\tfrac{\pi}{2}) , \qquad k_r = \lim_{r\to 0} r^\alpha \tau_{jr\theta}(r,\tfrac{\pi}{2}) ,$$

$$(j = 1 \text{ or } 2). \qquad (28)$$

From the original formulation of the problem all the field quantities may be expressed as definite integrals with the functions $f_1$ and $f_2$ defined by Eq. (2) as the densities [1,2]. In particular, the stress components shown in Eq. (28) are found to be

$$\tau_{1\theta\theta}(r,\tfrac{\pi}{2}) = \tau_{2\theta\theta}(r,\tfrac{\pi}{2}) = \frac{1}{\pi}\int_0^{b_1} [\frac{A_1 s}{s^2+r^2} + \frac{A_2 s(3r^2-s^2)}{(s^2+r^2)^2}]\, f_1(s)ds$$

$$+ \frac{1}{\pi}\int_0^{b_2} [\frac{A_3 s}{s^2+r^2} + \frac{A_4 s(3r^2-s^2)}{(s^2+r^2)^2}]\, f_2(s)ds ,$$

$$\tau_{1r\theta}(r,\tfrac{\pi}{2}) = \tau_{2r\theta}(r,\tfrac{\pi}{2}) = \frac{1}{\pi}\int_0^{b_1} [\frac{A_1 r}{s^2+r^2} - \frac{A_2 r(r^2-3s^2)}{(s^2+r^2)^2}]\, f_1(s)ds$$

$$+ \frac{1}{\pi}\int_0^{b_2} [-\frac{A_3 r}{s^2+r^2} + \frac{A_4 r(r^2-3s^2)}{(s^2+r^2)^2}]\, f_2(s)ds , \qquad (29.a,b)$$

$$A_1 = \frac{\mu_2}{m_1+\kappa_2} ,\quad A_2 = -\frac{\mu_2}{1+m_1\kappa_1} ,\quad A_3 = -\frac{\mu_1}{m_2+\kappa_1} ,\quad A_4 = \frac{\mu_1}{1+m_2\kappa_2} . \quad (30)$$

Since the kernels in Eq. (29) are of the form

$$(s \mp ir)^{-1}, \quad ir(s \mp ir)^{-2},$$

it is clear that the asymptotic values of the integrals in Eq. (29) may be obtained in terms of $G_j(z)$ (and its derivative) given by Eqs. (8) and (9) by letting $z = \mp ir$. Thus, carrying out the necessary manipulations, we obtain

$$k_\theta = \frac{1}{2\sin\frac{\pi\alpha}{2}}\left\{\frac{g_1(0)}{\sqrt{b_1}}\,[A_1 - (1-2\alpha)A_2] - \frac{g_2(0)}{\sqrt{b_2}}\,[A_3 - (1-2\alpha)A_4]\right\},$$

$$k_r = \frac{1}{2\cos\frac{\pi\alpha}{2}}\left\{\frac{g_1(0)}{\sqrt{b_1}}\,[A_1 + (1-2\alpha)A_2] + \frac{g_2(0)}{\sqrt{b_2}}\,[A_3 + (1-2\alpha)A_4]\right\}, \tag{31}$$

where the constants $g_1(0)$ and $g_2(0)$ are linearly related through Eq. (21) and the constants $A_i$ $(i = 1,..,4)$ are given by Eq. (30).

## FRACTURE CRITERION

In composite materials for fracture initiation and propagation from a crack front which is fully imbedded in one homogeneous phase, the conventional fracture theories and related fracture criteria (such as those based on various forms of energy balance) are still applicable. It may be noted that regardless of the generality of the underlying principles of most of these theories, as they are understood and applied in practice, it is tacitly required that as the crack propagates the stress and deformation fields around the crack front remain "self-similar", meaning that the crack must remain in its own plane as it propagates, the crack front advancing in an autonomous fashion. Thus, as may frequently be the case in composite materials, if the stress state around the leading edge of the crack is of the mixed-mode type or if the crack intersects a phase boundary, these theories may either require drastic modifications or may not be applicable at all. In such cases for brittle and most quasi-brittle materials a simple criterion based on the notion of "maximum cleavage stress" or some variation of it may provide an acceptable working tool. In introducing this criterion some modification may be necessary for problems in which, instead of a standard $r^{-1/2}$ singularity, the stresses have an arbitrary power singularity $r^{-\alpha}$, $(0 < \mathrm{Re}(\alpha) < 1)$.

In real solids, due to the difficulty in realistically modeling the material and carrying out the related mathematics, in a small toroidal region enclosing the leading edge of the crack the response of the medium cannot be accurately determined. Let $\delta_p$, measured from the crack front and called the "characteristic process zone size", be a representative dimension describing this region. $\delta_p$ may depend on the microstructure and continuum properties of the material as well as on the environmental conditions. In brittle and quasi-brittle solids it may be assumed that $\delta_p$ is sufficiently small so that the response of the medium outside the process zone is not significantly affected by what takes place inside $\delta_p$. This means that for most applications, an elastic analysis may be sufficient. Thus, a simple fracture criterion may be stated as

> the fracture propagation will take place in the direction $\theta = \theta_c$ for which the cleavage stress $\sigma_{\theta\theta}(\delta_p,\theta_c)$ is maximum and when

$$\sigma_{\theta\theta}(\delta_p,\theta_c) = \sigma_c \, , \tag{32}$$

where $\sigma_c$ is constant for a given material under given environmental conditions. Equation (32) is valid if the "weak link" is a particular homogeneous medium surrounding the crack tip. However if the singular point is at a bi-material interface and if the weak link is the adhesion along the interface, then $\sigma_c$ represents the adhesive strength of the particular joint.

## SOLUTION AND SOME SAMPLE RESULTS

The system of singular integral equations (3) subject to the auxiliary conditions given by Eq. (6) or Eqs. (21) and (22) may be solved by taking advantage of the fact that the fundamental functions of the system $w_j(s)$ defined by Eq. (7) are weights of Jacobi polynomials $P_n^{(\beta_j,\alpha_j)}(s)$, and by using the related Gauss-Jacobi integration formulas. The details of the numerical technique may be found in [4] and [5]. In this section we will give only some sample numerical results.

Table 1 shows the stress intensity factors (normalized with respect to $p_1\sqrt{b_1/2}$) for a crack terminating at the interface for material pairs Aluminum-Epoxy and Boron-Epoxy [1]. An example giving the stress intensity factors for a crack crossing the interface is shown in Table 2 [2]. The limiting values 0 and $\mp\infty$ shown in the table by an arrow as the crack tip approaches the interface are due to the discontinuous change in the power of singularity at $r = 0$.

Table 1. Stress intensity factor for a crack terminating at the interface (Fig. 1: $a_1 = 0$, $a_2 = b_2$).

| $\mu_2/\mu_1$ | $\alpha_1$ | $k(0)$ | $k(b_1)$ | $\mu_2/\mu_1$ | $\alpha_1$ | $k(0)$ | $k(b\ )$ |
|---|---|---|---|---|---|---|---|
| 23.077 | 0.3381 | 2.7845 | 0.8827 | 138.46 | 0.3234 | 3.0723 | 0.8751 |
| 0.0433 | 0.8248 | 0.0700 | 1.3552 | 0.0072 | 0.9258 | 0.0078 | 1.5102 |

Table 2. The stress intensity factors for through crack, $a_1 = a_2 = 0$. Material 1: aluminum, Material 2: epoxy; $2\ell = b_1 + b_2 = 2$ = constant, $c = (b_2 - b_1)/2$ variable, ($\beta_1 = \beta_2 = 0.5$, $\alpha = 0.273692$).

| $c/\ell$ | $k(b_1)/p_1\sqrt{\ell}$ | $k(b_2)/p_2\sqrt{\ell}$ | $k_\theta/p_1\ell^\alpha$ | $k_r/p\ \ell^\alpha$ |
|---|---|---|---|---|
| -1.00 | 1.3552 | $\to\infty$ | $\to -\infty$ | $\to\infty$ |
| -0.95 | 1.40333 | 5.69903 | -0.34124 | 0.12472 |
| -0.75 | 1.37532 | 2.76781 | -0.11960 | 0.04371 |
| -0.50 | 1.29929 | 1.87137 | -0.07423 | 0.02713 |
| -0.25 | 1.20448 | 1.43765 | -0.05310 | 0.01941 |
| 0.00 | 1.09312 | 1.17874 | -0.03839 | 0.01403 |
| 0.25 | 0.96213 | 1.01590 | -0.02575 | 0.00941 |
| 0.50 | 0.80281 | 0.91834 | -0.01293 | 0.00473 |
| 0.75 | 0.59068 | 0.87340 | 0.00313 | -0.00114 |
| 0.95 | 0.29854 | 0.87519 | 0.02903 | -0.01061 |
| 1.00 | $\to 0$ | 0.8827 | $\to\infty$ | $\to -\infty$ |

Figure 2 shows some results for a crack oriented at an arbitrary angle $\theta_0$ with respect to the interface. Here the plane is loaded parallel to the interface and away from the crack region,

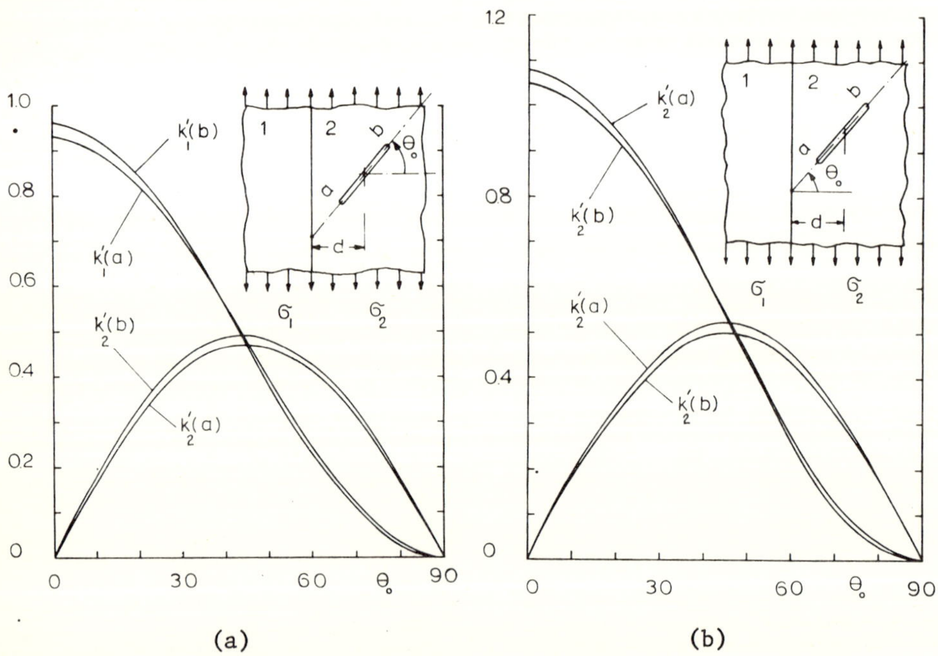


Fig. 2. Stress intensity factors for an arbitrarily oriented crack, $d = 2a_o = b - a$ = constant, (a): $E_1/E_2 = 22.22$, $\nu_1 = 0.3$, $\nu_2 = 0.35$, (b): $E_1/E_2 = 0.045$, $\nu_1 = 0.35$, $\nu_2 = 0.3$.

in the uncracked plane $(1-\nu_1^2)\sigma_1/E_1 = (1-\nu_2^2)\sigma_2/E_2$, $k_j'(c_m) = k_j(c_m)/\sigma_2\sqrt{a_o}$, $(j = 1,2,\ c_1 = a,\ c_2 = b)$, $k_1$ is the symmetric and $k_2$ is the antisymmetric component of the stress intensity factor.

Some results for radial cracks in an elastic matrix stiffened by a circular inclusion described in Fig. 3a are given in Tables 3 - 6. All these results are for plane strain where $\kappa_i = 3-4\nu_i$, $(i=1,2)$, $\nu_i$ being the Poisson's ratio. Note that $k_x$ and $k_{xy}$ given in Table 6 are identical to $k_\theta$ and $k_r$ which are defined by Eq. (28) at the point of intersection of the crack and the interface.

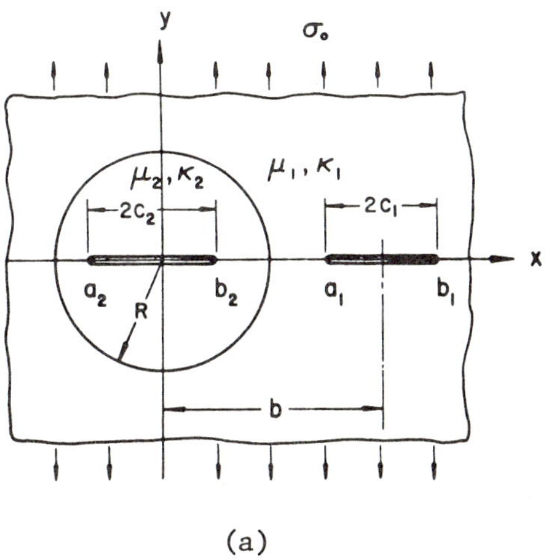


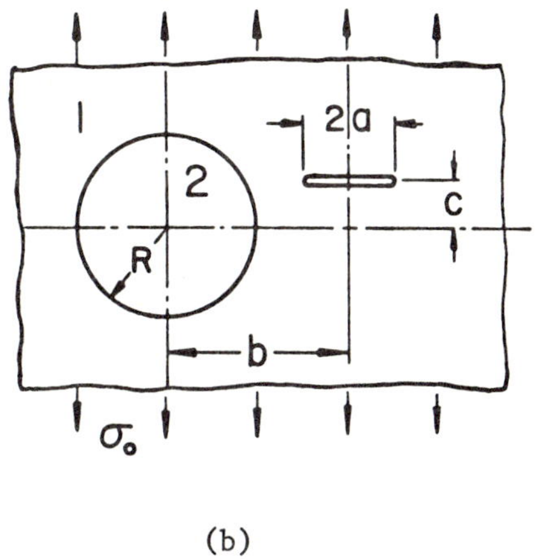


Fig. 3. Geometry of cracks in a matrix containing a circular inclusion.

Table 3. The effect of modulus ratio on the stress intensity factors for a crack terminating at the interface ($a_1 = R$, $b_1/R = 2$, $\kappa_1 = \kappa_2 = 1.8$, $c_1 = (b_1-a_1)/2$).

| $\frac{\mu_2}{\mu_1}$ | $\alpha$ | $\frac{k(b_1)}{\sigma_o\sqrt{c_1}}$ | $\frac{k(R)}{\sigma_o c_1^{\alpha}}$ |
|---|---|---|---|
| 0 | | 1.940 | |
| 0.05 | 0.81730 | 1.615 | 1.053 |
| 1/3 | 0.62049 | 1.229 | 0.5836 |
| 1.0 | 0.5 | 1.000 | 1.000 |
| 3.0 | 0.40074 | 0.8610 | 1.299 |
| 10.0 | 0.33277 | 0.7969 | 1.389 |
| 23.0 | 0.30959 | 0.7796 | 1.375 |
| 100 | 0.29387 | 0.7691 | 1.345 |
| 300 | 0.28883 | 0.7667 | 1.348 |

Table 4. Stress intensity factor for a crack in inclusion ($a_2 = -b_2$, $a_1 = b_1$, $\kappa_1 = \kappa_2 = 1.8$).

| $\frac{b_2}{R}$ | $k(b_2)/\sigma_o\sqrt{b_2}$ | |
|---|---|---|
| | $\frac{\mu_2}{\mu_1} = 3$ | $\frac{\mu_2}{\mu_1} = \frac{1}{3}$ |
| → 0+ | 1.275 | 0.6105 |
| 0.1 | 1.283 | 0.6046 |
| 0.25 | 1.332 | 0.5796 |
| 0.50 | 1.491 | 0.5144 |
| 0.90 | 2.062 | 0.3900 |

Table 5. Stress intensity factor for a completely cracked inclusion ($b_2 = R = -a_2$, $a_1 = b_1$, $k'(R) = k(R)/(\sigma_o R^{\beta_2})$).

| $\frac{\mu_2}{\mu_1}$ | $\kappa_1 = 2.2$, $\kappa_2 = 1.8$ | | $\kappa_1 = 1.8$, $\kappa_2 = 2.2$ | |
|---|---|---|---|---|
| | $\beta_2$ | $k'(R)$ | $\beta_2$ | $k'(R)$ |
| 0.2 | 0.3809 | 0.7848 | 0.3203 | 1.046 |
| 0.6 | 0.4703 | 0.9456 | 0.4212 | 1.174 |
| 1.0 | 0.5199 | 0.9209 | 0.4772 | 1.107 |
| 2.0 | 0.5919 | 0.8165 | 0.5569 | 0.9465 |
| 5.0 | 0.6912 | 0.6194 | 0.6638 | 0.6940 |

Table 6. The stress intensity factors for a crack crossing the interface ($\kappa_1 = 1.6$, $\kappa_2 = 1.8$, $\mu_2/\mu_1 = 23.077$, $\alpha = 0.27326$, $c = (b_1 - a_2)/2$).

| $\frac{a_2}{R}$ | $\frac{b_1}{R}$ | $\frac{k(b_1)}{\sigma_o\sqrt{c}}$ | $\frac{k(a_2)}{\sigma_o\sqrt{c}}$ | $\frac{k_x(R)}{\sigma_o c^\alpha}$ | $\frac{k_{xy}(R)}{\sigma_o c^\alpha}$ |
|---|---|---|---|---|---|
| 0 | 1.1 | 0.548 | 3.564 | -0.847 | 0.170 |
| 0 | 1.5 | 0.570 | 3.756 | -0.282 | 0.0565 |
| 0 | 1.9 | 0.672 | 3.838 | -0.0835 | 0.0167 |
| 0 | 3.0 | 0.811 | 3.996 | 0.219 | -0.0440 |
| -0.9 | 1.5 | 0.920 | 5.600 | -1.102 | 0.221 |
| -0.5 | 1.5 | 0.757 | 5.003 | -0.753 | 0.151 |
| 0.1 | 1.5 | 0.547 | 3.481 | -0.209 | 0.0420 |
| 0.5 | 1.5 | 0.510 | 2.465 | 0.0171 | -0.00343 |
| 0.9 | 1.5 | 0.572 | 1.383 | 0.212 | -0.0426 |
| 0.1 | 1.1 | 0.487 | 3.377 | -0.730 | 0.146 |
| 0.3 | 1.3 | 0.425 | 2.978 | -0.229 | 0.0459 |
| 0.7 | 1.7 | 0.619 | 1.961 | 0.211 | -0.0424 |
| 0.9 | 1.9 | 0.731 | 1.401 | 0.412 | -0.0826 |

Some results regarding an arbitrarily located crack in the matrix shown in Fig. 3b are given in Table 7. The two cases considered are the circular hole, $\mu_2 = 0$, and an epoxy matrix containing an aluminum inclusion. The dimensionless stress intensity factors given in the table are normalized with respect to $\sigma_o\sqrt{a}$, and in $k_{ij}$ the first subscript $i = 1,2$ refers to the first and the second mode, and the second subscript $j = 1,2$ refers to the right and the left crack tip, respectively.

Table 7. Stress intensity factors for a crack perpendicular to the external load ($R = 2a$, $b = 3a$, $\kappa_1 = 1.6$, $\kappa_2 = 1.8$).

| $\frac{c}{a}$ | $\mu_2 = 0$ | | | | $\mu_2 = 23.077\ \mu_1$ | | | |
|---|---|---|---|---|---|---|---|---|
| | $k_{11}$ | $k_{12}$ | $k_{21}$ | $k_{22}$ | $k_{11}$ | $k_{12}$ | $k_{21}$ | $k_{22}$ |
| 0 | 1.940 | $\to \infty$ | $\to 0$ | $\to 0$ | | $\to 0$ | $\to 0$ | $\to 0$ |
| 0.3 | 1.607 | 4.267 | 0.072 | -0.391 | 0.784 | 0.225 | -0.004 | 0.072 |
| 0.5 | 1.544 | 3.070 | 0.113 | -0.412 | 0.792 | 0.341 | -0.006 | 0.101 |
| 1.0 | 1.430 | 1.969 | 0.177 | -0.316 | 0.817 | 0.613 | -0.005 | 0.057 |
| 1.5 | 1.357 | 1.552 | 0.184 | -0.241 | 0.839 | 0.763 | 0.008 | -0.007 |
| 2 | 1.299 | 1.337 | 0.150 | -0.243 | 0.860 | 0.845 | 0.034 | -0.021 |
| 3 | 1.189 | 1.083 | 0.052 | -0.282 | 0.905 | 0.953 | 0.089 | -0.001 |
| 4 | 1.091 | 0.957 | 0.011 | -0.261 | 0.951 | 1.014 | 0.117 | 0.002 |
| 8 | 0.955 | 0.911 | -0.014 | -0.114 | 1.020 | 1.043 | 0.088 | -0.026 |
| $\infty$ | 1.0 | 1.0 | 0 | 0 | 1.0 | 1.0 | 0 | 0 |

More extensive results for the problems described in this paper and the results for a fiber or filament reinforced matrix containing a penny-shaped crack will appear elsewhere.

## REFERENCES


[1] Cook, T. S. and Erdogan, F., Stresses in bonded materials with a crack perpendicular to the interface, Int. J. Engng. Sci., 10, 677, 1972.

[2] Erdogan, F. and Biricikoglu, V., Two bonded half planes with a crack going through the interface, Int. J. Engng. Sci., 11, 745, 1973.

[3] Muskhelishvili, N. I., Singular Integral Equations, P. Noordhoff N. V., Groningen - Holland, 1953.

[4] Erdogan, F., Gupta, G. D., and Cook, T. S., Numerical solution of singular integral equations, Methods of Analysis and Solutions of Crack Problems, G. C. Sih, ed., Noordhoff International Publishing, Leyden, 1973.

[5] Erdogan, F. and Gupta, G. D., On the numerical solution of singular integral equations, Quarterly of Appl. Math., 29, 525, 1972.

## DISCUSSION by C. Atkinson, Imperial College, UK

In connection with Professor Erdogan's numerical results for cracks (or inclusions) approaching or passing through an interface between two media we should like to mention some analytical results we've obtained recently. These analytical results are for situations where the crack is a small distance $\varepsilon$ from the boundary or has passed through the boundary and emerges a distance $\varepsilon$ into the second medium. These results show in a precise way how the stress intensity factor varies in regions where the numerical approach to the problem becomes more difficult. For brevity, we indicate here only one of the simplest results, other results and the method of solution are outlined in a forthcoming paper in Int. J. Eng. Sci.

The example we consider here is a crack of length $(1-\varepsilon)$ approaching a free surface $x=0$, so that one tip of the crack is at position $(\varepsilon,0)$ and the other tip at $(1,0)$ the problem being one of plane strain with a uniform tension $\sigma_{yy} = P$ applied at infinity. For this situation the dominant terms for the singular stresses at the crack tips can be shown to be:

$$\sigma_{yy} \sim = \frac{P\delta^{-1/2}(0.793043)}{\varepsilon^{1/2}\log\varepsilon} \text{ at } x=\varepsilon-\delta,\ y=0,\ \delta<<\varepsilon$$

and

$$\sigma_{yy} \sim P\delta^{-1/2}(0.793043)(1+\frac{1}{\log\varepsilon}) \text{ at } x = 1+\delta,\ y=0$$

It should be noted that $2^{1/2}(0.7930\text{--}) = 1.12$ and the stress at $x = 1+\delta$ is the usual stress singularity associated with the problem of a crack in a half plane with a correction for small $\varepsilon$ that tends to zero like $(\log\varepsilon)^{-1}$. Similarly at $x=\varepsilon$, the coefficient of the stress singularity tends to infinity like $(\varepsilon^{1/2}\log\varepsilon)^{-1}$. For the bimaterial case, the results are more complicated and depend on powers of $\varepsilon$ exponents of which are related to the exponents in the eigensolutions associated with the problem of a crack against the interface.

## AUTHOR'S CLOSURE

No comments.

# RECENT DEVELOPMENTS IN FRACTURE OF FILAMENTARY COMPOSITES

S. W. Tsai and H. T. Hahn

Nonmetallic Materials Division, Air Force Materials Laboratory, Wright-Patterson Air Force Base, Ohio 45433 USA

## ABSTRACT

Survey of available experimental results indicates that the applicability of LEFM approach is by no means universal and depends on the type of composites and the loading and specimen configuration. The limiting factors associated with LEFM should be addressed and broadened to accommodate the inherent difference between homogeneous materials and composites such as anisotropy, heterogeneity and the lack of actual through cracks in composites.

This paper discusses several variations of LEFM in conjunction with the unique fracture behavior of noncoplanar crack extension and $K_I$-$K_{II}$ coupling in unidirectional composites. Most laminated composites show notch sensitivity, the degree of which reduces with the extent of subcritical damage before final failure. It seems necessary to understand the growth of this subcritical damage in order to assure satisfactory structural performance of composites.

## INTRODUCTION

Assured performance of filamentary composites as structural elements depends on thorough knowledge of their behavior in various environments. Thermomechanical properties, such as stiffness, heat conduction, and thermal expansion are describable in terms of average material constants by treating the composite as a homogeneous, generally anisotropic body. The classical plate theory gives a good approximation to the exact elasticity when the region of interest lies a few laminate thicknesses away from the stress-free boundaries [1]. It is not surprising then to see recent efforts directed toward the application of the two-dimensional linear elastic fracture mechanics (LEFM) to composites.

Unidirectional and laminated composites have a wide range of characteristic dimensions that covers the diameter of filaments to their length which can be as large as the specimen dimension. Furthermore, there is enormous difference in properties between reinforcing fibers and matrix. Since fracture is very sensitive to local properties in the small region surrounding the crack tip, characteristic dimensions should be much smaller than the size of this crack tip region in order for LEFM to be applied. Consequently, there has been cast a serious doubt on the applicability of the method developed for homogeneous materials to heterogeneous materials like composites. It is the purpose of this survey to cover recent analytical and experimental observations related to fracture of filamentary composites.

## ELEMENTS OF LINEAR ELASTIC FRACTURE MECHANICS (LEFM)

Griffith's fracture criterion is based on the principle of energy balance during crack extension. The true surface energy density $\gamma$ introduced in the criterion was later replaced by the effective surface energy density $\gamma_e$ which includes, besides $\gamma$, all the energy dissipated in the material damage due to crack extension. When the rate of release of potential energy accompanying the crack extension is $G_c/2$, the energy balance principle requires that

$$G_c/2 = \gamma_e \tag{1}$$

In practice, the critical energy release rate $G_c$ is taken to be the value of the energy release rate G at the load that makes the crack extend and G is determined independently of the actual crack extension by artificially extending the crack. Thus, it is implicitly assumed that the damage accompanying the actual crack extension is restricted to a thin surface layer and this has negligible effect on the elastic field.

Although there are various methods for the determination of G, two methods will serve our present purpose. First, the compliance technique is based on the gross behavior of a notched specimen:

$$G_{gc} = - \lim_{\Delta a \to 0} \frac{1}{\Delta a} \int_0^{u_c} \Delta P \, du = \frac{{P_c}^2}{2} \frac{dS}{da} , \tag{2}$$

where u is the displacement of the applied load P and a is the crack length. The subscript g indicates that the method is "global," i.e., the method does not depend on the local stress analysis around the crack tip. The second equality follows in the case of linear load-displacement relations with S being the compliance.

The second method requires the knowledge of the displacements $u_i$ and stresses $\sigma_{ji}$ around the crack tip:

$$G_{\ell c} = -\lim_{\Delta a \to 0} \frac{1}{\Delta a} \int_{\Delta A} \int_{u_i^b}^{u_i^a} \sigma_{ji}\, du_i\, dA_j \quad , \tag{3}$$

where $\Delta A$ is the newly created surface. The superscripts a and b respectively denote after and before crack extension and $dA_j$ is the vector area. The subscript $\ell$, which stands for "local," is used to distinguish the above equation from $G_g$. Again, if the material is linear, the stresses and displacements in the vicinity of the crack tip are approximated by

$$\sigma_{ji}(r,\theta) = \frac{1}{\sqrt{2\pi r}} \sum_{A=I}^{III} K_A f_{Aji}(\theta) \quad ,$$

$$u_i(r,\theta) = \sqrt{\frac{2r}{\pi}} \sum_{A=I}^{III} K_A g_{Ai}(\theta) \quad , \tag{4}$$

$$j = 1, 2; \quad i = 1, 2, 3 \quad .$$

Here the coordinate axes $x_1$ and $x_2$ are chosen to be parallel with and perpendicular to the crack, respectively, and r and $\theta$ are the cylindrical coordinates. For an infinite isotropic or anisotropic plate, the stress intensity factors $K_A$ are independent of elastic constants. The constants are included in the functions $f_{Aji}(\theta)$ and $g_{Ai}(\theta)$. When the plate is finite, $K_A$ depend on the boundary conditions [2,3].

A relation between $K_A$ and the energy release rate can be obtained from Eqs. (3) and (4):

$$G_\ell = \frac{1}{2} \sum_{A=I}^{III} \sum_{B=I}^{III} M_{AB} K_A K_B \quad , \quad M_{AB} = f_{A2i}(0) g_{Bi}(\pi) \quad . \tag{5}$$

For monoclinic materials with $x_3$=0 as the plane of symmetry, the matrix $M_{AB}$ is given in [4].*

Noting that the stress field in the immediate vicinity of crack tip plays a crucial role in the Griffith's criterion, Irwin proposed to use the stress intensity factors as variables describing fracture. When a small region of finite cohesive forces is assumed to exist at the crack tip, this theory has been shown to be equivalent to Griffith's. This foundation of the fracture theory has been applied to composites.

## APPLICATION OF LEFM TO FRACTURE OF COMPOSITES

Although through-the-thickness cracks are not likely to occur in composites in actual applications, the question is beyond the scope of this review. Thus, flaws to be considered in the following discussion are sharp, through cracks and circular holes.

According to LEFM, the critical stress intensity factor $K_c$ is independent of crack length and $G_{gc}$ should be equal to $G_{\ell c}$. Therefore, applicability of LEFM is usually ascertained by checking those conditions and the mode of fracture.

Data available so far seems to be grouped into three categories: (1) crack perpendicular to fibers in unidirectional composites under tensile loading; (2) crack parallel with fibers in unidirectional composites; and (3) crack aligned with one of material symmetry axes in laminated composites under tensile loading.

### Unidirectional Laminae

When the crack is not parallel with fibers, it is usually observed to extend not in its plane but along the fibers even under a normal tensile loading. An attempt in [5] is related to the generalization of Griffith's criterion as follows:

$$G(\sigma,\theta_c) = 2\gamma_e(\theta_c) \quad , \quad \frac{\partial}{\partial\theta}(G-2\gamma_e)\bigg|_{\theta=\theta_c} = 0 \quad , \tag{6}$$

where $\sigma$ is a parameter representative of external load. The second equation is the condition to determine the direction of crack extension which is chosen in such a way that $(G-2\gamma_e)$ becomes a maximum.

---

*Note that $A_{16}$ and $A_{26}$ are missing in [4]. Also, the exponent (1/2) in $M_{33}$ should read 3/2.

The same splitting problem may be investigated in terms of stress intensity factors as in [6]. Here the condition for crack extension is expressed by

$$\bar{\sigma}_{\theta\theta}(\sigma,\theta_c) = K_c(\theta_c) \quad , \quad \frac{\partial}{\partial\theta}(\bar{\sigma}_{\theta\theta}-K_c)\bigg|_{\theta=\theta_c} = 0 \quad , \tag{7}$$

where

$$\bar{\sigma}_{\theta\theta}(\sigma,\theta) = \lim_{r\to 0}\left(\sqrt{2\pi r}\quad \sigma_{\theta\theta}(\sigma,r,\theta)\right) \quad . \tag{8}$$

For a graphite/epoxy unidirectional composite [7] with a central crack perpendicular to the fibers, the ratio $\bar{\sigma}_{\theta\theta}(90°)/\bar{\sigma}_{\theta\theta}(0°)$ under a tensile loading was 0.119. However, the measured ratio $K_c(90°)/K_c(0°)$ was only 0.048. Therefore, the composite will be expected to split along the fibers before any appreciable crack extension.

The criterion (7) neglects any influence of shear. In fact, the shear stress has about twice as high a concentration as the transverse stress. However, the strengths have the ratios,

$$X_T/X_L = 0.051 \quad , \quad S/X_L = 0.066 \quad , \tag{9}$$

where $X_L$, $X_T$, and S are the longitudinal, transverse, and shear strength, respectively. Thus, the maximum stress criterion will predict that the splitting occurs first in shear.

Based on the results for a narrow elliptical crack, Kelly [8] pointed out that in carbon fiber composites shear failure is more likely to occur along the fibers at the crack tip. This point of view was then taken in [9] to explain the notch insensitivity of surface untreated carbon fiber/epoxy composite. However, the same conclusion results based on the stress field of a sharp crack.

It is a well established observation that crack oriented parallel with the fibers in the unidirectional composites propagates along the fiber direction. Wu [10] showed that the critical stress intensity factors $K_{IC}$, $K_{IIC}$ remained almost independent of the crack length for fiber glass reinforced Scotchply, thus arguing for the applicability of LEFM to this case (see also [11, 12,13]).

Under the combined loading, however, there is no uniform agreement on the theory. While the maximum tensile stress theory (7) was successfully applied to determine the failure load and the direction of crack propagation in isotropic, homogeneous materials [6], it fails to account for the interaction between $K_I$ and $K_{II}$ at fracture, Fig. (1).

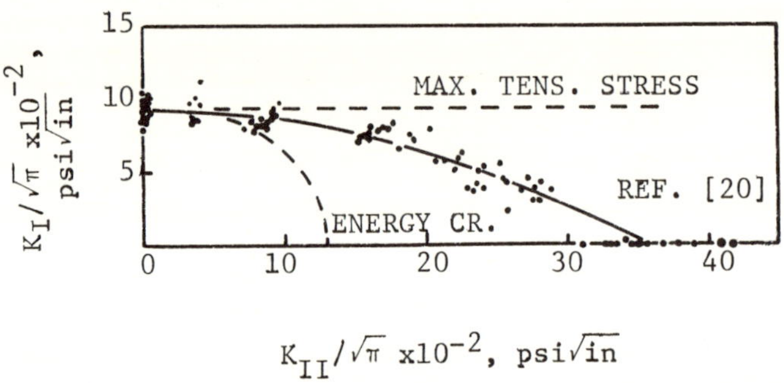


Fig. (1). Interaction of $K_I$ and $K_{II}$ for Scotchply.

For coplanar crack extension in an orthotropic plate, the energy criterion dictates that fracture is described by an ellipse in the $K_I$-$K_{II}$ plane whose major and minor axes are related by

$$K_{IIC}/K_{IC} = (E_1/E_2)^{1/4} , \tag{10}$$

where $E_1$ and $E_2$ are elastic moduli. This relation is shown in Fig. (1) together with the experimental results for glass/epoxy composite (Scotchply) [10].

There is a wide discrepancy between the theories described above and experimental data. While this difference was reconfirmed in [13] for graphite/epoxy composite, Ref. [14] reported contradicting results for glass reinforced plastics which seemed to show the validity of the energy criterion although the data does not include a measurement of $K_{IIC}$. Since most composites exhibit high nonlinearity in shear [15], this in turn may reduce stress concentration [16], resulting in higher values of $K_{IIC}$.

In the vicinity of crack tip, the assumptions of linear elasticity do not hold in reality. Assuming that outside this core region the stress field is given by the linear elasticity solution with the original crack geometry, it may be assumed that the crack extends when the circumferential stress or strain at the core boundary reaches a critical value [17,18].

This concept may be generalized further by introducing an appropriate failure criterion instead of using the circumferential stress alone. Elastic energy density may be suggested as the required criterion, cf. the energy density factor proposed by Sih [19]. Wu [20] chose a second order tensor polynomial failure criterion for unnotched composite [21]. Using the experimental results for $K_{IC}$ and $K_{IIC}$ for glass/epoxy composite, he further showed that the core region can be considered a circle of some radius $r_0$ with the center at the crack tip. His theoretical prediction is shown in Fig. (1).

The agreement between the theory and experiment is remarkable. However, the theory predicts that the failure occurs first at $\theta_c = \pm 60°$ under symmetric loading and at $\theta_c = 8°$ under skew-symmetric loading while the crack is macroscopically observed to propagate in its own plane. By studying the post-fracture path, it was found that the crack propagated not in a smooth plane but by skipping, leaving behind unbroken fibers which bridge the opposite crack surfaces. This behavior was not due to fiber misalignment alone. Moreover, the way those fibers were attached to the top and bottom surfaces of the crack was random under symmetric loading but in the preferred direction under skew-symmetric loading. Thus it seemed that even if the first failure was predicted to be out of the crack plane, the presence of weak matrix layer prevented the crack from going in the predicted direction.

An experimental check on the equality $G_{gc} = G_{\ell c}$ is given in [12]. Values of $G_{gc}$ were uniformly higher than those of $G_{\ell c}$ for a carbon/carbon composite with filaments running at angles $\pm 8°$-$10°$ to the crack. However, the opposite trend was observed for a carbon felt substrate/carbon matrix composite. The specimens were tested in the three-point bend.

Whereas all the approaches discussed above are macroscopic, there are several microscopic approaches proposed. Like in LEFM, those are again grouped into energy method and stress method.

In the first method attention is given to mechanisms of energy dissipation associated with crack extension. Each source of energy dissipation is attributed to a simplified mode of failure. For example, energy is released upon fracture of fibers debonded from the matrix. If fibers are still embedded in the matrix, then frictional work is required to pull out broken fibers from the matrix. Also to be included is the plastic energy dissipation in the matrix. Further details can be found in [22] where analytical models associated with those failure modes were discussed. It should be kept in mind that those models are intended for unidirectional composites with crack perpendicular to fibers.

The second method postulates that the crack extends when the stress in the fiber adjacent to the crack tip reaches the fiber strength [23]. Statistical scatter of fiber strength has also been accounted for [23,24]. Shear lag type of analysis [25] shows that the stress concentration factor is proportional to the square root of the number of broken fibers n when n is large. Then the result is similar to what is expected from LEFM. In addition, there is some question as to the length over which stress concentration is supposed to be dominant. On the whole, it is not yet clear how those theories will compare with experimental results.

In summary, we have seen various analytical models proposed. Each model seems to serve its intended purpose. However, there are not enough experimental data that can guarantee extension of a particular model to more general circumstances.

Laminated Composites

Fracture of laminated composites poses much more difficulty because the problem frequently becomes no more two-dimensional but three-dimensional, let alone heterogeneous. Depending on the type of material system and lay-up, there is an extensive variety of failure modes which seem to defy any simple characterization that the two-dimensional LEFM is capable of. As such, the current approach to fracture of laminates should be regarded as a methodology rather than as an attempt for a detailed explanation of fracture behavior.

Irwin's idea of incorporating the effect of plastic zone into fracture toughness of quasi-brittle material was taken up in [26] and extended to explain the hole size effect. The damage zone of size C in [26] is more like an inherent crack, because the unnotched strength X is related to the fracture toughness $K_c$ by

$$X = K_c/\sqrt{\pi C} \quad , \tag{11}$$

By comparison the plastic zone size $r_y$ is related to the tensile yield strength $\sigma_y$ by

$$\sigma_y = K_c/\sqrt{\pi 2 r_y} \quad . \tag{12}$$

For a hole, the damage zone is treated like a crack emanating from the hole boundary. The reported value of C ranged from 0.028 in. to 0.05 in. depending on the hole size for $[0/\pm45]_S$

graphite/epoxy; and was 0.0671 in. for center cracks in $[0/90]_S$ graphite/epoxy. For boron/epoxy laminates C was in the range of 0.1~0.14 in. for $[0/\pm45/0]_S$ lay-up; and 0.05~0.2 in. for $[0/\pm45/0/90]_S$ lay-up [27].

It is reported in [11] that the damage zone is not artificial but real. By observing the damage zone size near a hole, a fairly accurate strength prediction was possible. However, considering the scatter, this may be fortuitous.

Commonly observed reduction of stress concentration at final failure further indicates that there occurs stress relaxation around the hole boundary. For an infinite plate the stress concentration factor $Q_\infty$ is independent of the hole diameter d but depends on the elastic constants. When the plate has a finite width w, the stress concentration factor $Q_g$ based on the gross stress is shown in Fig. (2). The ratio $Q_g/Q_\infty$ is also shown for the isotropic case because it is known to be fairly insensitive to orthotropy [28]. The experimental data is for a $[0/\pm45/0/\overline{90}]_S$

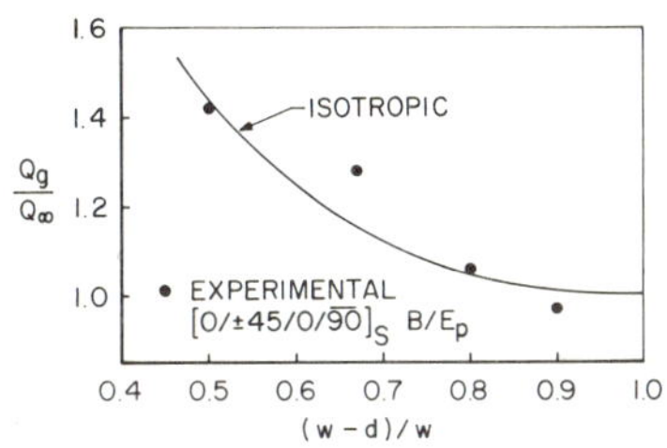


Fig. (2). Stress Concentration for Isotropic Plate and $[0/\pm45/0/\overline{90}]_S$ Boron/Epoxy Laminate within Elastic Range.

boron/epoxy composite [29]. It is important to note that the data was obtained when the boundary was in the elastic range. Above a certain load level the boundary strain was no longer proportional to the applied stress. Thus it is only natural that the stress concentration $Q_{exp}$ based on the failure stresses is lower than the theoretical one, as shown in Fig. (3) [29,30]. Further discussion on the hole size effect is given in [31].

Perhaps the classical plate theory is not sufficient for the stress concentration analysis, as the isochromatic fringe pattern changes with the stacking sequence of $[0_2/\pm45]$ boron/epoxy laminate [30]. However, it is interesting to note that the stacking sequence has no discernible effect on the ultimate failure stress, including [0/90] and [0/±45/90] laminates.

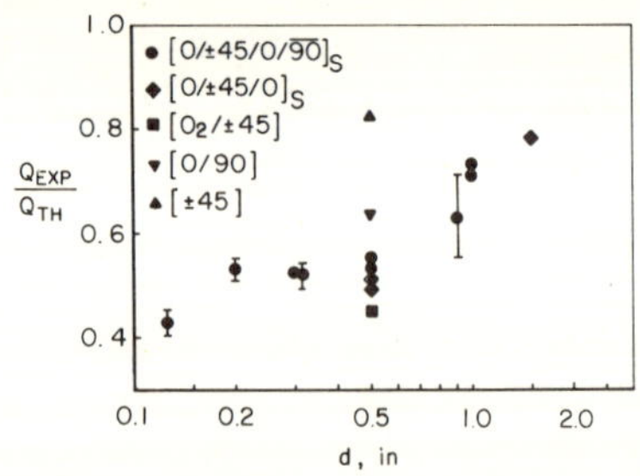


Fig. (3). Experimental Stress Concentration Factors at Failure.

Notch sensitivity of strength is displayed by plotting the ratio of the notched strength to unnotched strength against the ratio of the flaw size to the panel width. Available data are shown in Figs. (4)-(6) for some typical laminates. Here the

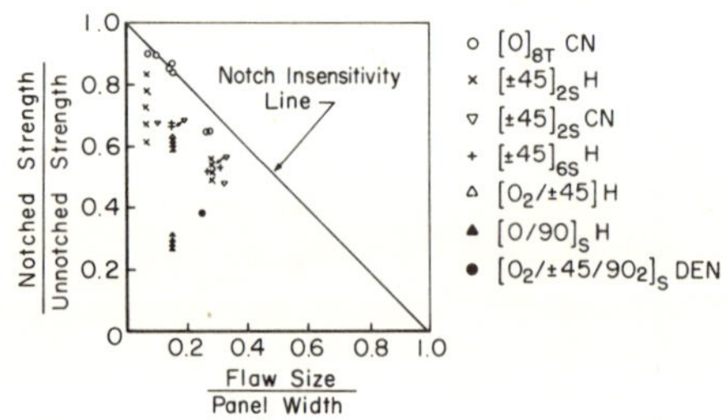


Fig. (4). Notch Sensitivity of Boron/Epoxy Laminates.

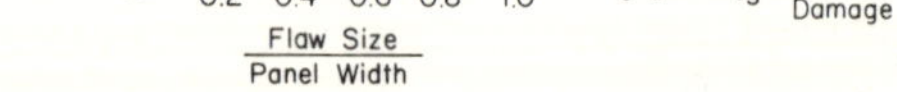


Fig. (5). Notch Sensitivity of Graphite/Epoxy Laminates.

symbols H, CN, and DEN denote hole, center notch, and double edge notch, respectively. In Fig. (5), the first six laminates are intermediate-strength graphite/epoxy and the remaining four high-strength graphite/epoxy. Data in Figs. (4) and (5) are taken from [32]. Figure (6) shows the results for S-glass/epoxy laminates [33]. The type of flaw is a hole with artificially introduced cracks. It is seen that most of the laminates studied show some degree of notch sensitivity.

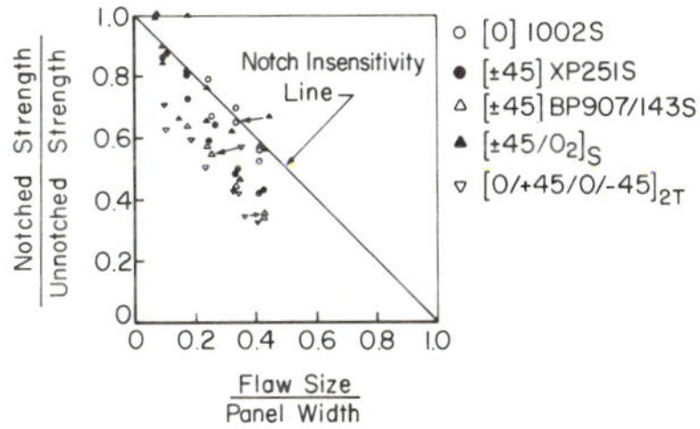


Fig. (6). Notch Sensitivity of S-Glass/Epoxy Laminates.

As a first check on the applicability of LEFM calculated $K_c$ values are plotted in Fig. (7) against crack lengths. All the data are for intermediate-strength graphite/epoxy unless otherwise indicated. Note that the isotropic finite width correction factors have been used except for the data from [2]. The symbol SEN denotes single edge notch and B indicates that the test is the three-point bend.

If the material is elastic up to failure, then the ratio F/X for a laminate with a hole should be

$$F/X = 1/qQ_\infty \quad , \quad q = Q_g/Q_\infty \tag{13}$$

and the strength of a laminate with a sharp crack of length 2a predicted by LEFM becomes

$$\frac{F}{X} = \frac{K_c}{X\sqrt{w}} \; \frac{1}{Y\sqrt{\pi a/w}} \quad , \tag{14}$$

where Y is the finite width correction factor.

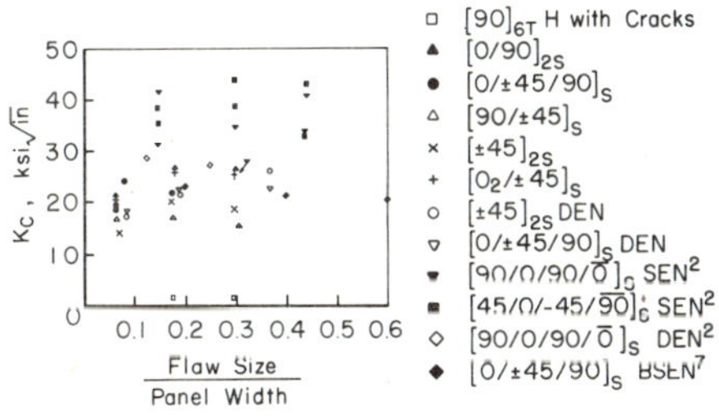


Fig. (7). Fracture Toughness Values versus Ratio of Flaw Size to Panel Width for Graphite/Epoxy Laminates.

Theoretical difference among various types of notches is shown in Fig. (8). The data used are for $[90/0/90/\bar{0}]_S$ graphite/epoxy composite reported in [2]. Isotropic values were used for q and for Y of both SEN and DEN. The result for CN was taken from [2]. Although exact shapes of the curves in Fig. (8) depend on the material properties, they show the trend we could expect from the linear elastic analysis.

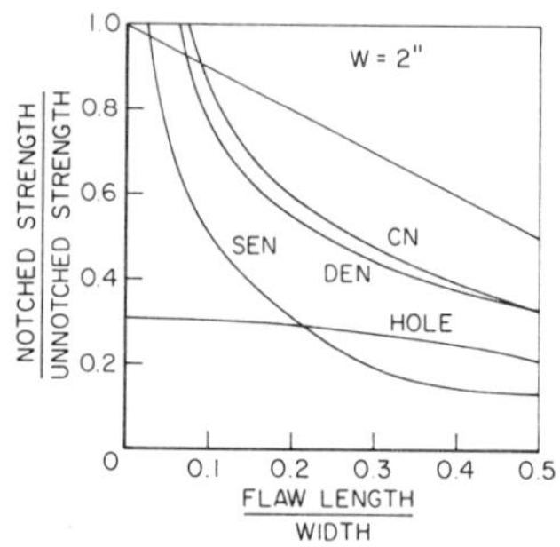


Fig. (8). Theoretical Notch Sensitivity of $[90/0/90/\bar{0}]_S$ Graphite/Epoxy Laminate.

Inspection of fracture modes suggests that the final failure of laminate is usually preceded by subcritical damage in the form of matrix crazing, splitting, and local delamination [2,33,34]. The severity of this subcritical damage depends on the material system and lay-up configuration. For example, glass/epoxy laminates seem to be most susceptible to the subcritical damage [33] and angle-ply graphite/epoxy laminates were found to behave in a rather nonbrittle fashion when the angle exceeded 45° [2].

As the external load is increased, subcritical cracks grow along the fibers from the crack tips rather independently in each layer. This independent failure is constrained by the neighboring layers until a compatible mode of deformation is found via delamination, splitting or fiber failure, or a combination of those. In [±45°] S-glass/epoxy laminates, the subcritical damage zone was observed to grow even as far as to the boundary without major fiber break [33]. It is also possible for the damage zone to be restricted to a thin band along a fiber direction [2].

When the subcritical damage is localized due to a strong bond, further increase of load will induce macroscopic failure; i.e., failure of the laminate as a whole. It is in this case that one may hope for the applicability of LEFM.

In summary, we have seen some experimental evidence that stress relaxation occurs at the hole boundary before the final failure. Consequently, the observed stress concentration is much lower than the theoretical value. Available data on notched strengths show that most composites are affected by the presence of flaws. Although limited data on $K_c$ may give some insight into the applicability of LEFM, it is hardly appropriate to draw any general conclusion, considering the complexity of fracture mode. Judgment should be made on an individual laminate basis.

## CONCLUDING REMARKS

Although a simple conclusion cannot be derived from the existing experimental results, we may sum up the observations as follows.

When a load is applied to a laminate with a crack, a region of subcritical damage appears and grows around the crack tip. If the dimension of this region at final failure is small enough not to affect linear elasticity of gross behavior, it seems plausible to expect the concepts of LEFM to apply. Still, the problem reduces to the following questions: (1) What are the variables that can describe the state of this tip region? (2) What are the critical values of those variables that dictate ultimate failure?

Efforts so far have been directed toward measuring the opening mode $K_c$ values for coplanar crack extension. In reality crack may not extend in the prescribed manner depending on its orientation with fibers and the type of loading. This question remains to be answered.

Perhaps the most difficult situation arises when diffuse subcritical damage develops before final failure. LEFM cannot be expected to predict this final failure load. Moreover, the problem frequently is no longer two-dimensional because of the delamination involved. Failure at the crack tip can no longer be described by a smooth fracture surface. In such a case, we are certainly not dealing with a crack "extension." We should develop a more suitable means of describing the growth of the subcritical damage until final failure.

It seems to us that a rational design criterion for filamentary composites against fracture is of the utmost importance to assure satisfactory structural performance. In order to achieve this goal, it is first of all necessary to investigate the notch sensitivity of composites as shown in this review. The mechanisms of failure and method for prediction and detection of actual flaws must be better understood than what we know today. The applicability of LEFM to this problem is not well defined at this stage. Numerous factors embedded in LEFM must be addressed, in addition to the role of laminate configuration and combined stresses, before the applicability of LEFM to composites can be accurately assessed.

## ACKNOWLEDGMENTS

The work of one of the authors (H. T. Hahn) was performed under a National Research Council Resident Research Associateship.

## REFERENCES

[1] Pagano, N. J. and Pipes, R. B., Some observations on the interlaminar strength of composite laminates, Int. J. Mech. Sci., 15, 679, 1973.

[2] Mandell, J. F., Wang, S. S. and McGarry, F. J., Fracture of graphite fiber reinforced composites, AFML-TR-73-142, Wright-Patterson AFB, July 1973.

[3] Snyder, M. O. and Cruse, T. A., Crack tip stress intensity factors in finite anisotropic plates, AFML-TR-73-209, Wright-Patterson AFB, August 1973.

[4] Paris, P. C. and Sih, G. C., Stress analysis of cracks, Fracture Toughness Testing, ASTM STP 381, 1965, 30.

[5] Harrison, N. L., Splitting of fiber-reinforced materials, Fibre Sci. and Tech., 6, 25, 1973.

[6] Erdogan, F. and Sih, G. C., On the crack extension in plates under plane loading and transverse shear, J. Basic Engng., 85, 519, 1963.

[7] Konish, H. J., Jr., Swedlow, J. L. and Cruse, T. A., Experimental investigation of fracture in an advanced fiber composite, J. Composite Materials, 6, 114, 1972.

[8] Kelly, A., Interface effects and the work of fracture of a fibrous composite, Proc. Roy. Soc. Lond., A319, 95, 1970.

[9] Beaumont, P. W. R. and Phillips, D. C., Tensile strengths of notched composites, J. Composite Materials, 6, 32, 1972.

[10] Wu, E. M., Fracture mechanics of anisotropic plates, in Composite Materials Workshop, Tsai, S. W., Halpin, J. C. and Pagano, N. J., Eds., Technomic Pub. Co., Stamford, 1968, 20.

[11] Owen, M. J. and Bishop, P. T., Critical stress intensity factors applied to glass reinforced polyester resin, J. Composite Materials, 7, 146, 1973.

[12] Guess, T. R. and Hoover, W. R., Fracture toughness of carbon-carbon composites, J. Composite Materials, 7, 2, 1973.

[13] McKinney, J. M., Mixed-mode fracture of unidirectional graphite/epoxy composites, J. Composite Materials, 6, 164, 1972.

[14] Sanford, R. L. and Stonesifer, F. R., Fracture toughness measurements in unidirectional glass-reinforced plastics, J. Composite Materials, 5, 241, 1971.

[15] Hahn, H. T. and Tsai, S. W., Nonlinear elastic behavior of unidirectional composite laminae, J. Composite Materials, 7, 102, 1973.

[16] Hutchinson, J. W., Singular behavior at the end of a tensile crack in a hardening material, J. Mech. Phys. Solids, 16, 13, 1968.

[17] McClintock, F. A. and Irwin, G. R., Plasticity aspects of fracture mechanics, Fracture Toughness Testing, ASTM STP 381, 1965, 84.

[18] Williams, J. G. and Ewing, P. D., Fracture under complex stress - the angle crack problem, Int. J. Fracture Mechanics, 8, 441, 1972.

[19] Sih, G. C., A special theory of crack propagation, in Mechanics of Fracture I, Sih, G. C., Ed., Noordhoff International Pub., Leyden, 1972.

[20] Wu, E. M., Strength and fracture of composites, in Treatise of Composite Materials, L. Broutman, Ed., Academic Press, to be published.

[21] Tsai, S. W. and Wu, E. M., A general theory of strength for anisotropic materials, J. Composite Materials, 5, 58, 1971.

[22] Phillips, D. C. and Tetelman, A. S., The fracture toughness of fiber composites, Composites, 3, 216, 1972.

[23] Zweben, C., On the strength of notched composites, J. Mech. Phys. Solids, 19, 103, 1971.

[24] Argon, A. S., Fracture of composites, in Treatise on Materials Science and Technology, Vol. 1, Herman, H., Ed., Academic Press, New York, 1972, 79.

[25] Hedgepeth, J. M., Stress concentrations in filamentary structures, NASA TND-882, May 1961.

[26] Waddoups, M. E., Jr., Eisenmann, J. R. and Kaminski, B. E., Macroscopic fracture mechanics of advanced composite materials, J. Composite Materials, 5, 446, 1971.

[27] Halpin, J. C., Jerina, K. L., and Johnson, T. A., Characterization of composites for the purpose of reliability evaluation, Analysis of the Test Methods for High Modulus Fibers and Composites, ASTM STP 521, 1973, 5.

[28] Suarez, J. A., Vulnerability of composite aircraft structures, AFFDL-TR-72-8, Wright-Patterson AFB, 1972.

[29] Experimental results by Grumman Aerospace Corp.

[30] Experimental results by Boeing Company, Vertol Division.

[31] Zweben, C., Fracture mechanics and composite materials: a critical analysis, Analysis of the Test Methods for High Modulus Fibers and Composites, ASTM STP 521, 1973, 65.

[32] Experimental results by Rockwell International, Los Angeles Aircraft Division.

[33] Durchlaub, E. and Sacco, P., Fracture mechanics and damage tolerance of S-glass composites, AFFDL-TR-72-82, Wright-Patterson AFB, August 1972.

[34] Cruse, T. A. and Stout, M. G., Fractographic study of graphite-epoxy laminated fracture specimens, J. Composite Materials, 7, 272, 1973.

# A CRITICAL REVIEW OF ANISOTROPIC FRACTURE MECHANICS

T. S. Cook and C. A. Rau, Jr.

Pratt & Whitney Aircraft, Middletown, Connecticut 06457 USA
Failure Analysis Associates, Stanford, California 94305 USA

## ABSTRACT

The fracture mechanics of anisotropic materials is discussed in two parts: (1) the calculation of the crack driving force and (2) the response of the material to the driving force. Both the crack driving force and the materials response are shown to be orientation dependent but not generally with the same dependence. Although stress intensity factors can be defined, for some geometries, equal to those for isotropic materials, the realistic crack driving force is shown to be more complex than previously realized and must be redefined for each specific mechanism of crack extension. Despite these complications, lifetime prediction of anisotropic materials can be accomplished utilizing isotropic analyses and extensive mechanical testing.

## INTRODUCTION

The deformation and failure of an anisotropic material is an area of study that has not received a great deal of attention in spite of the tremendous interest in fracture mechanics. This neglect stems partly from the difficulty involved in the analysis of anisotropic bodies and partly from the fact that within a restricted but still quite general class of problems, the stress intensity factors for isotropic and anisotropic problems can be defined to be exactly the same. However, with the increasing use of anisotropic materials in structural applications, it has become necessary to reassess these definitions and to develop methods to use them or other criteria for the prediction of fracture in an anisotropic media. The intent of this work is (1) to discuss the nature of the stress and displacement field at a crack tip in a general anisotropic material, (2) to define meaningful crack tip stress intensity factors, and (3) to review the assumptions and restrictions involved in the prediction of unstable fracture or fatigue of anisotropic materials.

From a materials usage viewpoint, the majority of work on anisotropic media has been in the area of fiber reinforced and laminated materials. These materials pose special problems for the fracture mechanist since they are both discontinuous as well as anisotropic and as such are quite difficult to treat at the gross structural level. In most instances, these materials must be treated at the "local" level where the position and orientation of the crack relative to the various phases of the composite can be accounted for. These added problems of inhomogeneity shall be excluded in this discussion; we shall instead consider the analysis of homogeneous anisotropic materials containing a crack.

## DETERMINATION OF THE STRESS INTENSITY FACTORS

When, in an isotropic material, attention is focused on the local fields of elastic stress and displacement associated with a singularity at the crack tip, it is always possible to reduce the most general situation to a sum of plane and anti-plane problems. This reduction leads to the three modes of crack tip deformation, i.e., in-plane opening (Mode 1), in-plane sliding (Mode 2), and out-of-plane sliding (Mode 3); each mode corresponds to a particular type of applied load. In the general anisotropic media, this one-to-one correspondence between load type and crack tip deformation is lost; that is, a symmetric extension load can produce both opening (Mode 1) and relative sliding (Mode 2 or Mode 3) of the crack surfaces.

Unlike the isotropic material, the crack tip deformation field in an anisotropic body is not exclusively related to the type of loading; however, it is convenient to define the three loading modes which lead to the vanishing of certain stress components on the crack plane. Specifically, the symmetrically loaded extension case is considered as the first mode ($\tau_{xy} = 0$ on $y = 0$), the skew symmetric in-plane loads are the second mode $\tau_{yy} = 0$ on $y = 0$), and the out-of-plane shear loads are the third mode ($\tau_{yy} = \tau_{xy} = 0$ on $y = 0$). Thus the problem can still be reduced to three modes, but now the mode refers to the type of loading and not the crack tip deformation as in the isotropic case. The stress intensity factors, $K_i (i = 1,2,3)$, for an infinite anisotropic body may then be defined for each loading mode such that whenever local loads on the crack surface are absent or self-equilibrating, $K_i$ are identical to those for the corresponding isotropic problem.

The formulation of the plane and anti-plane problems of anisotropic elasticity are well known, and only a minimum of expressions will be repeated here (for details see Sih and Liebowitz[1] and Leknitskii[2]). Defining

$$\frac{d\Phi_i(z)}{dz} \equiv \Phi_i'(z) \equiv \psi_i(z), \tag{1}$$

the stress distribution for the plane problem is given by

$$\begin{aligned} \tau_{xx} &= 2\,\mathrm{Re}[\mu_1{}^2\psi_1(z_1) + \mu_2{}^2\psi_2(z_2)], \\ \tau_{yy} &= 2\,\mathrm{Re}[\psi_1(z_1) + \psi_2(z_2)], \\ \tau_{xy} &= -2\,\mathrm{Re}[\mu_1\psi_1(z_1) + \mu_2\psi_2(z_2)], \end{aligned} \tag{2}$$

where the $\mu_i$ are roots of the characteristic equation

$$a_{11}\,\mu^4 - 2a_{16}\,\mu^3 + (2a_{12} + a_{66})\,\mu^2 - 2a_{26}\,\mu + a_{22} = 0, \tag{3}$$

and the $a_{ij}$ are the elastic compliances. Thus the solution to the plane problem involves the determination of two analytic functions, $\Phi_1(z_1)$ and $\Phi_2(z_2)$, of the complex variables $z_j = x_j + i\,y_j$, $j = 1, 2$.

The anti-plane problem is similar but it is somewhat simplier as it requires only the determination of a single function, $\Phi_3(z_3)$.

In order to study the fracture of anisotropic materials, we need to analyze the stress and displacement fields in the neighborhood of the crack tip. Following Muskhilishvili[3] we may express the functions $\Phi_j(z_j)$ in terms of a power series expanded about the crack tip. Further, we may separate the term containing the dominant singularity and study the singular nature of the stresses close to the crack tip. Introducing polar coordinates, $(r, \theta)$, at the crack tip, $z_o$, and letting

$$z_j - z_o = r\,e^{i\theta}\,, \tag{4}$$

we can express the stress functions near the crack tip as

$$\psi_j(z_j) = \frac{\lambda_j}{\sqrt{r\,(\cos\theta + \mu_j\,\sin\theta)}} + O(r^{1/2}),\ j = 1,2,3 \tag{5}$$

where $\lambda$ are constants dependent on the material properties and the boundary conditions of the problem.

First we observe that the stress singularity at the crack

tip, as in the isotropic case, is of the order of $r^{-1/2}$. We also see that the stress distribution near the crack tip will be dependent on the material properties (through $\mu_j$ and $\lambda_j$). Now since the $\psi_j(z_j)$ are directly related to the stresses and contain a removable singularity at the crack tip, the coefficient of $r^{-1/2}$ is a measure of the strength of the stress singularity. Any linear combination of $\psi_1$ and $\psi_2$, or equivalently $\lambda_1$ and $\lambda_2$, could be used to define a stress intensity factor. However, since the $\lambda_j$ are functions of elastic constants, the resulting stress intensity factor would be orientation dependent.

If only the stress intensity factors, rather than the entire crack tip fields, are desired, then it is most efficient to define the intensity factor in terms of the stress functions. For Mode 1, $\tau_{xy}$ vanishes on the x axis ahead of the crack, regardless of the orientation of the crack relative to the material axes. Referring to (2), the third equation is identically zero so that we may express the solution in terms of either $\psi_1$ or $\psi_2$. Eliminating either $\psi_1$ or $\psi_2$ from the second equation of (2), substituting the dominant term from (5), and requiring that, on $\theta \doteq 0$, $K_1$ is independent of the material properties, we obtain

$$\lambda_1 = \frac{\mu_2}{2\sqrt{2}\,(\mu_2 - \mu_1)}\,K_1, \quad \lambda_2 = \frac{\mu_1}{2\sqrt{2}\,(\mu_1 - \mu_2)}\,K_1. \tag{6}$$

For Mode 2 loading, $\tau_{yy}$ vanishes ahead of the crack on the plane of the crack. Following the procedure that led to (6), we obtain

$$\lambda_1 = \frac{1}{2\sqrt{2}\,(\mu_2 - \mu_1)}\,K_2, \quad \lambda_2 = \frac{1}{2\sqrt{2}\,(\mu_1 - \mu_2)}\,K_2\,. \tag{7}$$

Substituting (6) and (7) into (5), we obtain

$$\begin{aligned} K_1 &= 2\sqrt{2}\,\frac{(\mu_2-\mu_1)}{\mu_2}\,\lim_{z_1\to z_0}\,(z_1-z_0)^{1/2}\,\psi_1(z_1) \\ K_2 &= 2\sqrt{2}\,(\mu_2-\mu_1)\,\lim_{z_1\to z_0}\,(z_1-z_0)^{1/2}\,\psi_1(z_1). \end{aligned} \tag{8}$$

Alternately, we may define $K_i$ as

$$K_1 = 2\sqrt{2}\,\frac{(\mu_1-\mu_2)}{\mu_1}\lim_{z_2\to z_0}(z_2-z_0)^{1/2}\,\psi_2(z_2)$$

$$K_2 = 2\sqrt{2}\,(\mu_1-\mu_2)\lim_{z_2\to z_0}(z_2-z_0)^{1/2}\,\psi_2(z_2) \tag{9}$$

which are the same as those for a crack in the corresponding infinite isotropic plate. More importantly, they are independent of the orientation of the crack relative to the material axes as concluded previously.[4,5,6]

As an example, consider a crack in an infinite body subjected to tractions, $\sigma_o$, at infinity; the tractions make an angle $\beta$ with the crack. The stress functions are given by[7]

$$\Phi_1(z_1) = \frac{\sigma_o a^2}{4(\mu_1-\mu_2)}\left[\frac{2\mu_2 \sin^2\beta + \sin 2\beta}{z_1 + \sqrt{z_1^2 - a^2}}\right] + \Gamma_1 z_1 ,$$

$$\Phi_2(z_2) = -\frac{\sigma_o a^2}{4(\mu_1-\mu_2)}\left[\frac{2\mu_1 \sin^2\beta + \sin 2\beta}{z_2 + \sqrt{z_2^2 - a^2}}\right] + \Gamma_2 z_2 , \tag{10}$$

where $\Gamma_i$ are constants. Differentiating and substituting into either (8) or (9), the stress intensity factors are given by

$$K_1 = \sigma_o \sqrt{a}\,\sin^2\beta\ , \quad K_2 = \sigma_o \sqrt{a}\,\sin\beta\cos\beta\ . \tag{11}$$

This example shows that when the loads applied directly to the crack are zero or self-equilibrating and the crack is in an infinite body, the stress intensity factor for the anisotropic body is identical to that for the isotropic body and is independent of the material orientation. Although this definition of $K_i$ apparently affords great simplicity, it actually obscures the strong effects of anisotropy and orientation on most of the crack tip stress and displacement components. Consequently, whether or not $K_i$ are appropriate criteria for crack extension will depend on the specific mechanism of crack extension and the stress or displacement components which most strongly affect that mechanism.

When these concepts are applied to laboratory or service applications, we must deal with finite sizes. The presence of boundaries in a cracked anisotropic material introduces a material as well as a geometrical effect; hence the correspondence between isotropic and anisotropic stress intensity factors no longer holds. While the theory of anisotropic elasticity is well defined with a complex function analysis directly paralleling the isotropic analysis developed by Muskhelishvili,[8] the solution of a finite size, anisotropic cracked plate presents some difficulties. For example, the conventional analytic continuation of the stress functions from the upper half space into the lower half space cannot be done. This restriction exists because in the general finite plate $\tau_{xy}$ $(x, o)$ is non-zero for symmetric loading, although for equilibrium the integral of $\tau_{xy}$ over $y = 0$ must vanish. However, one can utilize much of the technique by applying the continuation arguments to a traction free crack; the use of such a technique has been demonstrated by Bowie.[9]

There are, in fact, no closed form solutions for finite, cracked anisotropic bodies presently available. The few solutions for cracked anisotropic panels that have been obtained numerically are for the restricted anisotropy of an orthotropic plate. For example, Bowie and Freese[10] have combined the complex analysis with a collocation technique and utilized this method to study the effect of geometry and elastic constants on the orthotropic stress intensity factors. Their study on a central crack in a sheet, and a similar one by Gandhi[11] using an inclined crack, showed that the orthotropic stress intensity factor is strongly dependent on the orientation of the material. However, the results also showed that the isotropic intensity factor provides a reasonable estimate (within 10 percent) of the orthotropic intensity factor for cases where the crack is not large in comparison to the sheet.

STRESS DISTRIBUTION

For a crack in an infinite body, the dominant term near the crack in the stress distribution is given by:[1]

$$\tau_{xx} = \frac{K_1}{\sqrt{2r}} R_e\left[\frac{\mu_1 \mu_2}{\mu_1-\mu_2}\left(\frac{\mu_2}{d_2} - \frac{\mu_1}{d_1}\right)\right] + \frac{K_2}{\sqrt{2r}} R_e\left[\frac{1}{\mu_1-\mu_2}\left(\frac{\mu_2^2}{d_2} - \frac{\mu_1^2}{d_1}\right)\right]$$

$$\tau_{xy} = \frac{K_1}{\sqrt{2r}} R_e\left[\frac{\mu_1 \mu_2}{\mu_1-\mu_2}\left(\frac{1}{d_1} - \frac{1}{d_2}\right)\right] + \frac{K_2}{\sqrt{2r}} R_e\left[\frac{1}{\mu_1-\mu_2}\left(\frac{\mu_1}{d_1} - \frac{\mu_2}{d_2}\right)\right] \qquad (12)$$

$$\tau_{yy} = \frac{K_1}{\sqrt{2r}} R_e\left[\frac{1}{\mu_1-\mu_2}\left(\frac{\mu_1}{d_2} - \frac{\mu_2}{d_1}\right)\right] + \frac{K_2}{\sqrt{2r}} R_e\left[\frac{1}{\mu_1-\mu_2}\left(\frac{1}{d_2} - \frac{1}{d_1}\right)\right]$$

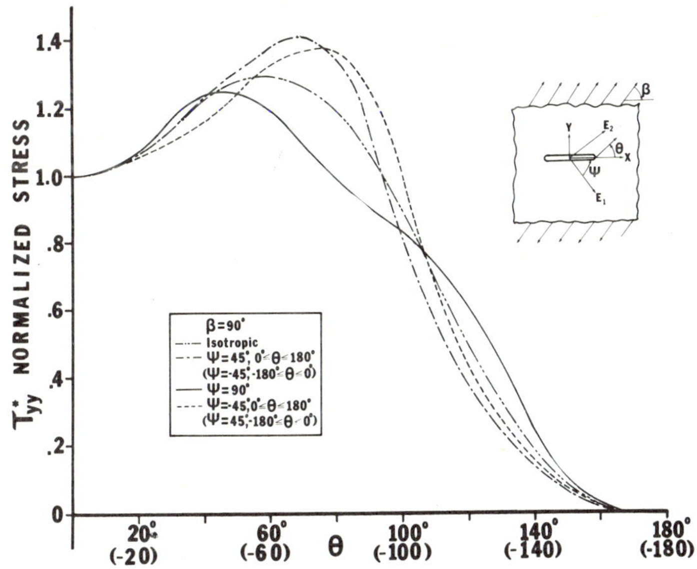


Figure 1: Normalized $\tau_{yy}$ (θ) For Mode 1 Loading

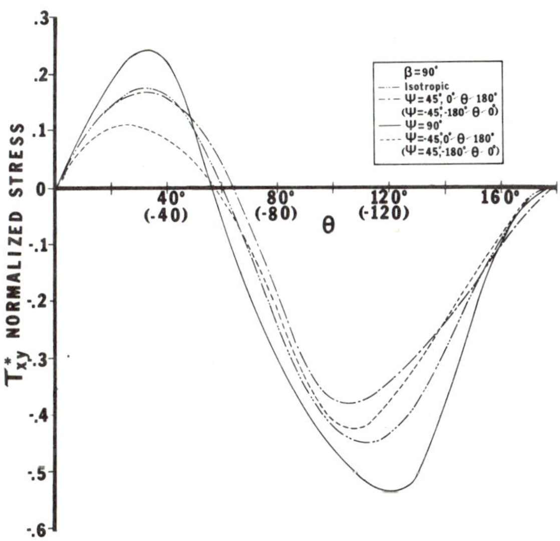


Figure 2: Normalized $\tau_{xy}$ (θ) For Mode 1 Loading

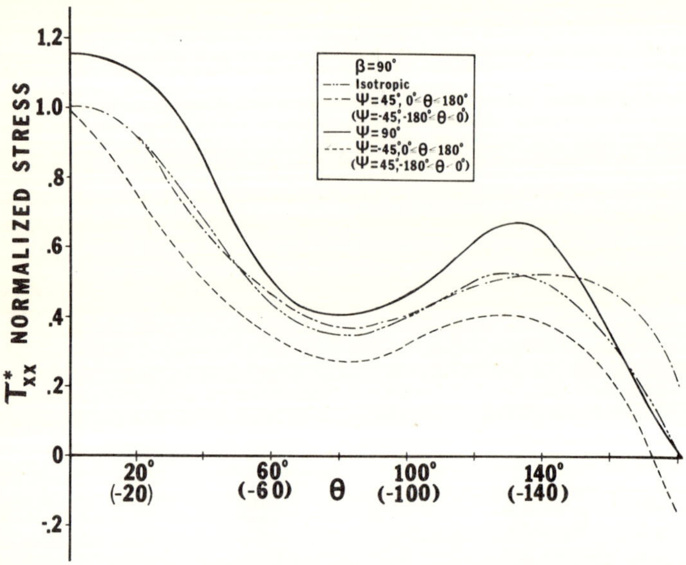


Figure 3: Normalized $\tau_{xx}$ ($\theta$) For Mode 1 Loading

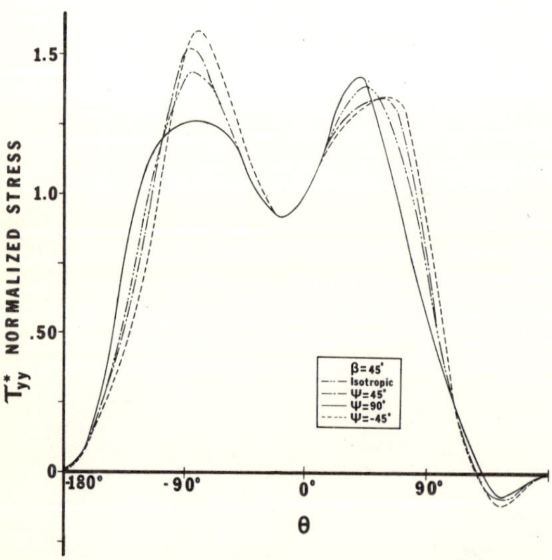


Figure 4: Normalized $\tau_{yy}$ ($\theta$) For Mixed Mode Loading

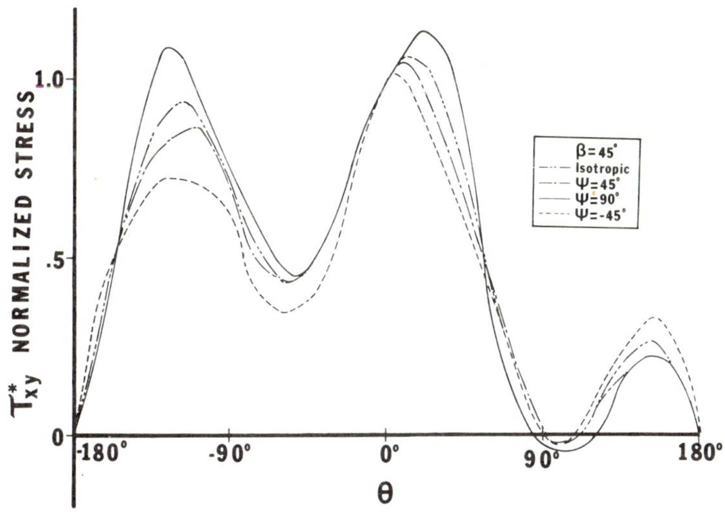


Figure 5: Normalized $\tau_{xy}$ (θ) For Mixed Mode Loading

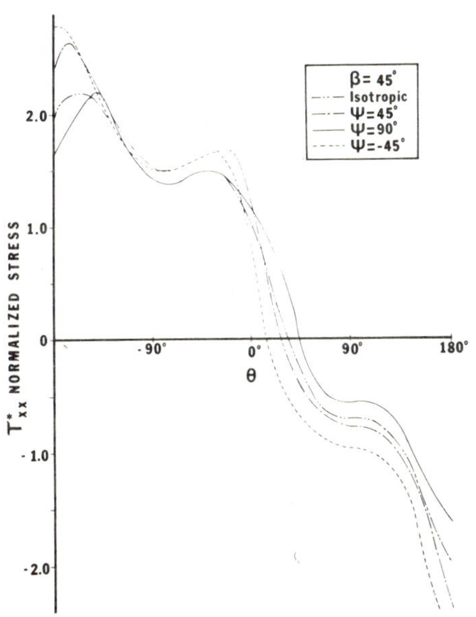


Figure 6: Normalized $\tau_{xx}$ (θ) For Mixed Mode Loading

where

$$d_j = (\cos\theta + \mu_j \sin\theta)^{1/2}$$

and the stress intensity factors have been defined in (8) and (9).

Using these expressions, the stress distribution was determined for a crack in a directionally solidified nickel-base alloy, D.S. MAR-M 200 + Hf. The elastic constants for this material are:

$$E_1 = 1.251 * 10^{12} \text{ dyne/cm}^2 \qquad \mu_{12} = 1.186 * 10^{12} \text{ dyne/cm}^2$$

$$E_2 = 1.660 * 10^{12} \text{ dyne/cm}^2 \qquad \nu_{12} = 0.391$$

Figures (1)-(3) give the stresses for remote tractions applied normally to the crack. Three orientations of the alloy, $\psi = \pm 45°$, and 90°, along with an isotropic material were used to evaluate the effect of material orientation on the stresses. With Mode 1 loading, for $\psi = 90°$ and isotropy, $\tau_{xx}$ and $\tau_{yy}$ are symmetric and $\tau_{xy}$ is anti-symmetric about the crack plane. For the $\psi = \pm 45°$ orientations, the solidification direction is 90° apart in the two cases resulting in a type of mirroring about the crack plane. That is, the stresses for the 45° orientation above the crack plane $(0 \le \theta \le 180°)$ are the same as the -45° orientation below the crack plane $(-180° \le \theta \le 0)$ except for a change in the sign of the shear stress; hence only the distribution for one side of the crack is shown in the figures. The stress components are all normalized with respect to the normal stress ahead of the crack, $\tau_{yy}(\theta = 0)$, for the particular orientation, i.e. $\tau^*_{xx}(\theta, \psi) = \tau_{xx}(\theta, \psi)/\tau_{yy}(0, \psi)$.

Figures (1)-(3) show that the orientation of the anisotropic material has a significant effect on both the magnitude and distribution of the stresses. For example, the maximum $\tau^*_{yy}$ is obtained for $\psi = \pm 45°$ while the maximum $\tau^*_{xy}$ occurs for $\psi = 90°$. Unlike $\tau^*_{xy}(0)$ and $\tau^*_{yy}(0)$, $\tau^*_{xx}(0)$ is a function of material orientation as shown in Figure 3. The figures show that the anisotropic stresses differ from the isotropic stresses by as much as 30 percent.

The maximum principal and shear stress were computed and showed strong material dependence over the entire range of $\theta$ including $\theta = 0$ which reflected the material dependence of $\tau_{xx}(0)$. The largest values for both principal and shear stresses were obtained for the $\psi = \pm 45°$ orientation, with the deviations being 12 and 10 percent higher than the isotropic principal and shear stresses, respectively. It was noted that while the extremum of the principal and shear stresses occurred 30° apart for $\psi = 90°$ and the isotropic case, the extrema for the $\psi = \pm 45°$ orientations occurred at the same angular orientation, $\theta = \pm 80°$, making this an especially critical orientation.

In order to examine the effect of Mode 2 loading, the stress distribution was computed for the tractions applied at a 45° angle to the crack; these stresses are given in Figs. (4)-(6). Comparing these components to those in Figs. (1)-(3), it is seen that the introduction of the shear load increases the deviation from the isotropic case and decreases the range over which the stresses are relatively insensitive to the material orientation. For example, all the combined mode $\tau^*_{xx}$ stresses are distinct at $\theta = 0$. The extrema of $\tau^*_{yy}$ are skewed relative to the crack plane with $\psi = -45°$ giving the maximum value below the crack while $\psi = 90°$ gives the largest value above the crack. As in the Mode 1 case, the 90° orientation gives the largest shear stress throughout most of the range of $\theta$.

In comparing the mixed mode, maximum principal and shear stresses to the pure Mode 1, the influence of the shear load on both components, but especially the principal stress, was very pronounced. The deviations of the anisotropic principal stress from the isotropic stress was much larger in the mixed mode case, the range being 23 to 30 percent higher. The maximum shear stress showed somewhat less deviation, ranging from 10 to 23 percent higher than the isotropic value. It was noted that the extrema of the two stress components occurred at about the same $\theta$ location for all three anisotropies; these locations were all near the crack faces. In both stresses, the $\psi = \pm 45°$ anisotropy produced higher stresses than the $\psi = 90°$ orientation.

## UNSTABLE CRACK PROPAGATION

The anisotropic stress and strain fields in the neighborhood of the crack tip, unlike the isotropic case, can no longer be described in terms of a single parameter, K, incorporating the geometry and the loading, but are also dependent on the material properties. This means that these fields are inadequately described by the stress intensity factor alone. We must, therefore, seek a more complex parameter than $K_i$ to describe the driving force for unstable crack extension. Unfortunately, a single parameter, regardless of complexity, may not be appropriate for all materials or even all orientations of a given material. In isotropic materials, great simplicity results because all stress and strain components are proportional to $K_i$ alone. Thus, even if the specific stress or strain component which is controlling crack extension changes, the stress intensity factors still correctly represent an appropriate driving force. However, the anisotropic components are not proportional to $K_i$ alone and the simplicity is lost.

At first glance, it appears that strain energy release rate, G, might provide a sufficiently general driving force and, along with an orientation dependent critical value, $G_c$, a suitable failure criterion. This parameter has been used by several authors, e.g., Barnett and Asaro,[12] who used a distributed dislocation approach to calculate the energy release rate in terms of the material orientation. However, it seems to have escaped these authors that the definition of G[5] requires that the crack propagate in a collinear manner. In a general anisotropic body the crack usually does not propagate in a self-similar manner and the conventional calculations of G become meaningless.

In an attempt to overcome the limitations of the energy release rate, Sih[13] has developed the strain energy density concept for isotropic materials. The theory predicts that crack propagation takes place when the strain energy density reaches a critical value and that growth takes place in a direction determined by the stationary value of the energy density. Using this theory, the tensile failure angle for a slant crack (Fig. 1) is given in Fig.7 for three orientations of D.S. MAR-M 200 + Hf. The failure angle is negative consistent with experiments with metal where cracks in a tensile field tend to propagate in such a way that they are perpendicular to the tensile field.

## FATIGUE CRACK GROWTH

While the problem of unstable fracture controls many structures, stable fatigue crack extension under cyclic loading is more

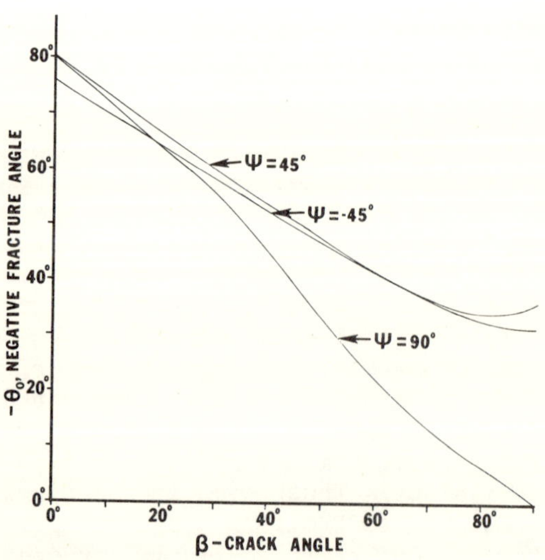


Figure 7: Crack Angle Versus Fracture Angle

often life-limiting. As noted previously, the stress and strain fields around the crack are, in general, dependent on both $K_i$ and elastic constants. Consequently, the orientation dependence of the crack driving force will depend on the specific micromechanical mechanism responsible for crack extension. For example, Mode 1 fatigue crack growth will depend upon crack opening displacements which are in turn a function of plastic deformation on numerous slip planes which intersect the plane of the crack. Although $\tau_{yy}(x, o)$ on the plane of the crack is independent of the elastic constants, all stresses above and below the crack plane are strongly dependent on orientation and elastic constants. Consequently, fatigue crack growth may be orientation dependent even when $K_i$ is not. On the other hand, unstable cleavage extension of the same crack may depend only on the tensile stress across the crack plane (which depends only on $K_1$) and thus be independent of orientation and elastic constants.

In addition to material dependence of the crack driving force, the material's response to a given driving force will, in general, be anisotropic so that crack extension need not occur in the direction predicted by maximum driving force. In view of these uncertainties, there seems to be little advantage in and justification for attempting to separate the material contribution to the crack driving force from the material's response to the driving force. Rather, the simple isotropic $K_i$ can be assumed as the driving force, and both material effects can be incorporated in the experimental measurement of the material's response to $K_i$.

## GENERAL DISCUSSION

With all the limitations that have been cited, what then is the best approach to anisotropic fracture mechanics problems? Basically, we have shown that the local stress and displacement fields around a crack are, in general, dependent on crystallographic orientation (elastic constants). Therefore, any realistic definition of crack driving force should also be orientation dependent. On the other hand, the material's response to a given driving force will itself be orientation dependent, but not necessarily with the same orientation dependence as the crack driving force. This implies that regardless of the generality of the definition of crack driving force, experiments at a single orientation cannot, in general, be used to predict response in other orientations until the mechanism controlling the material's response have been defined at different orientations. The strain energy density theory may provide a criterion for the orientation dependence of unstable fracture, but the critical value of $S = S_c$ must itself be orientation dependent. Application of strain energy density concepts to fatigue crack growth has not been investigated, but it is unlikely to correctly predict such plasticity-controlled events without extensive modifications.

Practically, it is possible to use the isotropic stress intensity factors and laboratory evaluation over a range of orientations to predict reasonably accurately the behavior of structural geometries. This approach makes no attempt to separate the distinct orientation dependencies of the driving force and the materials response. However, since the same experimental evaluation of various orientations would still be required even if the above mentioned separation could be made, this approach minimizes the analytical effort by utilizing the many known isotropic $K_i$. Inescapably, lifetime prediction of an anisotropic material will require extensive crack growth testing until research has defined some method to predict the orientation and plasticity effects on the mechanisms of crack extension.

## CONCLUSIONS

1. In the absence of nearby boundaries, certain stress components on the crack plane ahead of the crack are independent of elastic anisotropy. Hence, stress intensity factors ($K_i$) for large anisotropic plates can be defined identically equal to those for isotropic material.

2. Increasing proximity of boundaries, relative to the crack length, results in increasing differences between the anisotropic and isotropic stress intensity factors; agreement is within 10% so long as the crack occupies less than 60% of the plate area.

3. Except for those specific stress components on the plane of the crack which are dependent only on $K_i$, the stresses (and displacements) around a crack tip are strongly dependent on the crack orientation relative to the elastic anisotropy even for infinitely large plates. These orientation dependencies have been calculated for a directionally-solidified nickel-base alloy.

4. In anisotropic materials, individual stress and strain components have quite different orientation dependencies which differ from that of $K_i$.

5. Any realistic crack driving force must include both the effect of anisotropy on $K_i$ and the orientation dependencies of those crack tip stresses or strains which control the appropriate mechanism of crack extension.

6. A reasonable lifetime prediction in anisotropic materials can be obtained utilizing the isotropic $K_i$ as the driving force, and empirically lumping all the anisotropic effects into the material's response evaluations over a range of orientations.

REFERENCES

1. Sih, G. C. and Liebowitz H., Fracture: An Advanced Treatise, Vol. 2, Academic Press, New York, 1968, p 67.

2. Leknitskii, S. G., Anisotropic Plates, Gordon and Breach, New York, 1968.

3. Muskhelishvili, N. I., Singular Integral Equations, Noordhoff, Groningen, Holland, 1953.

4. Stroh, A. N., Phil. Mag., 3, 1958, p 625.

5. Paris, P. C. and Sih, G. C., Fracture Toughness Testing and its Applications, ASTM STP 381, American Society for Testing and Materials, 1965, p 30.

6. Heald, P. J., Bulletin De L'Academie Polonaise Des Sciences, Series Technique, XVIII, 6, 1970, p 211.

7. Savin, G. N., Stress Concentration Around Holes, Pergamon Press, New York, 1961.

8. Muskhelishvili, N. I., Some Basic Problems of the Mathematical Theory of Elasticity, Noordhoff, Groningen, Holland, 1953.

9. Bowie, O. L., Methods of Analysis and Solution of Crack Problems, Vol. 1, Sih, G. C., Ed., Noordhoff, Leyden, Holland, 1973, p 1.

10. Bowie, O. L. and Freese, C. E., Int. J. Fract. Mech., 8, 1972, p 49.

11. Gandhi, K. R., J. of Strain Analysis, 7, 3, 1972, p 157.

12. Barnett, D. M. and Asaro, R. J., J. Mech. Phys. Solids, 20, 1972, p 353.

13. Sih, G. C., Eng. Fract. Mech., 5, 1973, p 365.

# STRESS INTENSITY FACTORS FOR ELLIPTICAL CRACKS

A. S. Kobayashi, A. N. Enetanya and R. C. Shah

University of Washington, Seattle, Washington 98195 and
The Boeing Aerospace Company, Seattle, Washington 98124

## ABSTRACT

The combined use of a solution for an elliptical crack subjected to internal pressure, which is represented by a polynomial expression, together with a free surface solution for the alternating method is reviewed and methods for improving the efficiency of this numerical technique are proposed. Numerical differentiation for avoiding the derivation and computer programming of complex mathematical expressions is discussed. A criterion for maximum grid size for the free surface solution is proposed. Also optimum fittings of the first, second and third order polynomial distributions of internal pressure are considered.

## INTRODUCTION

Post mortem failure analysis of fractured structural components requires modeling of an embedded or surface flaw from which fracture initiated. More often, these flaws can be approximated by an ellipse and are located in regions of stress concentrations or thin cross-section, such as those shown in Fig. (1), where the stress gradient as well as the complex geometry of the structure cannot be ignored. The elliptical crack and its interaction with its adjacent or intersecting boundaries have thus been studied by various investigators during the past ten years.

The earliest paper on stress intensity factor of an elliptical crack is that by Irwin [1] who then derived the approximate expression for the stress intensity factor at the point of deepest

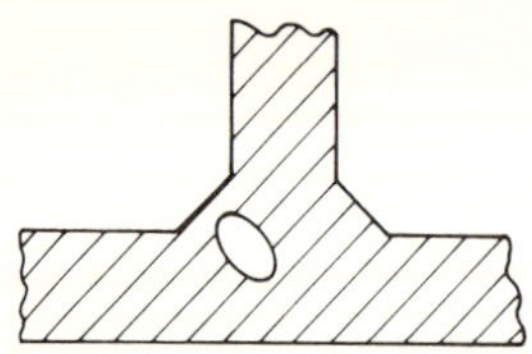

(a) EMBEDDED ELLIPTICAL WELD DEFECT IN A FLANGE - WEB CONNECTION

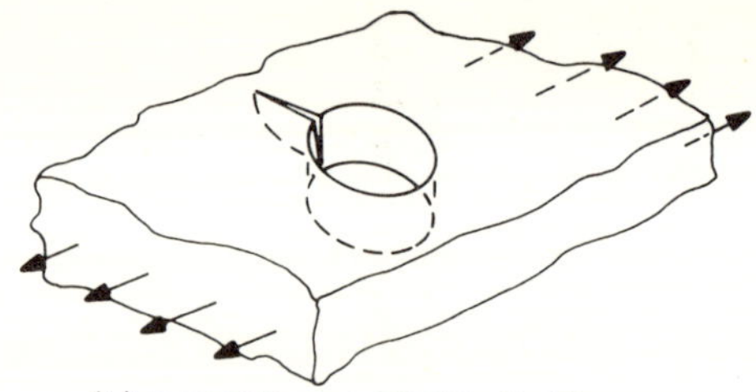

(b) QUARTER-ELLIPTICAL FLAW AT A HOLE

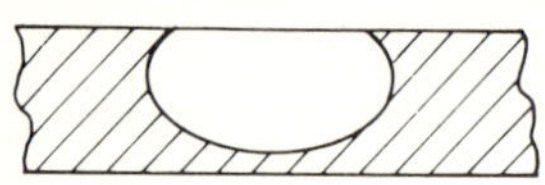

(c) DEEP PART-ELLIPTICAL SURFACE FLAW

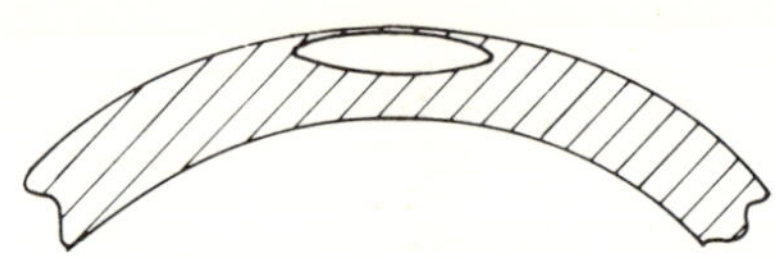

(d) EMBEDDED ELLIPTICAL FLAW IN A THIN-WALLED PRESSURE VESSEL

Fig. (1) - Typical elliptical flaws in structural components

penetration of a shallow* semi-elliptical crack in a tension plate. Among the several solutions on the "surface flaw problem" as well as embedded flaws which have been published to date, the mathematically more rigorous solutions are all based on the alternating technique suggested by Kantorovich and Krylov [2]. This procedure was first applied to crack problems by Smith [3] who determined the stress intensity factors of a semi-circular crack in a thick beam in tension, bending, and subjected to rapid heating or cooling. This method of approach was then used by Smith and his co-workers to analyze a family of problems involving circular flaws and part circular flaws [4,5,6]. Recently, Hartranft and Sih [7] reexamined one of Smith's original solutions [3,8] with an added singularity function which increased the accuracy of their solution.

The disadvantage of using the circular crack solution is its geometry which cannot be used to model oblong elliptical or semi-cracks such as those shown in Figs. (1a) and (1d)**. For better

---

* Shallow crack means that the crack depth is less than half of the plate thickness.

** Part elliptical cracks, such as those shown in Figs. (1b) and (1c), can be approximated by part circular crack as shown in [8]. Part elliptical crack with large aspect ratio, i.e., $b/a \leq 0.2$, cannot be conveniently approximated by a part circular crack.

modeling of elliptical flaws, Shah and Kobayashi [9] used the potential function of Segedin [10] to derive a polynomial representation of the internal pressure on an elliptical crack. This solution was then used in the alternating method to determine the stress intensity factors of embedded elliptical flaws under nonuniform loadings [11] as well as to estimate the stress intensity factors for semi-elliptical surface flaws [14]. Prior to the above development, Kassir and Sih [15] determined the stress intensity factor of an elliptical crack under linearly varying internal pressure and uniform shear tractions. Smith [16] extended Kassir and Sih solution [15] following the procedure described in [9] to derive a polynomial distribution of shear tractions prescribed on an elliptical crack. In addition, Smith [17] has recently used this procedure to determine the stress intensity factor of a surface flaw in a thin tension plate.

The solution procedure involving an elliptical crack is convenient for modeling actual flaw geometries. The small number of derivable terms in the polynomial representation of the internal pressure, however, limits the accuracy in which a complex residual pressure distribution on the crack surface can be fitted. Unfortunately numerical convergence of the alternating method is sensitive to this fitting and thus the accuracy of the entire solution procedures hinges on the accuracy of fitting. In particular, numerical difficulties in fitting residual stress are encountered in surface flaw problems where a portion of the ellipse protrudes through the free surface.

The purpose of this paper is to discuss numerical problems associated with the fitting process and to suggest procedures which reduce some of the immense amount of theoretical derivations and computer programming associated with this numerical analysis.

## THEORETICAL BACKGROUND

Two solutions are needed in applying the alternating method to fracture mechanics problems involving elliptical cracks. One solution involves a semi-infinite solid and the other solution involves an elliptical crack both subjected to variable surface tractions. For elliptical crack problems involving flat surfaces, Love's solution [18,19] is used for the free-surface solution of the former. Numerical procedures related to Love's solution are well documented [3,4,8,11,12,13,14] and will not be repeated here. A discussion, however, on the tolerable maximum size of a rectangle for erasing the residual surface traction on the flat surface will be presented later.

In the following, a brief review of the basic equations for

an elliptical crack with prescribed internal pressure [9], which can be represented by a polynomial in terms of Segedin's potential function [10], is presented.

Consider an elastic solid containing an elliptical crack, as shown in Fig. (2), which is located in the plane z=0, and is opened by applying an internal pressure, p(x,y), symmetrically to both surfaces of the crack. The boundary conditions for this problem are:

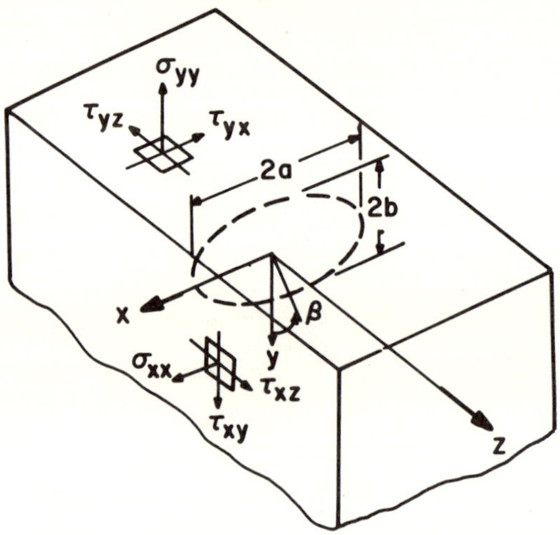


Fig. (2) - Elliptical crack in a finite solid

$$\sigma_{zz} = -\,p(x,y) \qquad \left(\frac{x^2}{a^2} + \frac{y^2}{b^2} < 1,\ z = 0\right) \tag{1a}$$

$$w = 0 \qquad \left(\frac{x^2}{a^2} + \frac{y^2}{b^2} > 1,\ z = 0\right) \tag{1b}$$

$$\tau_{xz} = \tau_{yz} = 0 \qquad (z = 0) \tag{1c}$$

$$\sigma_{xx} = \sigma_{yy} = \sigma_{zz} = \tau_{xy} = \tau_{yz} = \tau_{zx} = 0 \text{ at infinity} \tag{1d}$$

For the vanishing shear stresses on z=0 plane and in the absence of body forces, Navier's equations of equilibrium are satisfied by the harmonic function of $\phi$. The displacement perpendicular to the crack plane in terms of $\phi$ is,

$$w = z\,\frac{\partial^2\phi}{\partial z^2} - 2(1-\eta)\,\frac{\partial\phi}{\partial z} \tag{2}$$

where $\eta$ is the Poisson's ratio. The stress components necessary for solving the surface flaw problem are given by,

$$\sigma_{yy} = 2G[z \frac{\partial^3\phi}{\partial y^2 \partial z} + \frac{\partial^2\phi}{\partial y^2} + 2\eta \frac{\partial^2\phi}{\partial x^2}] \tag{3a}$$

$$\sigma_{zz} = 2G[z \frac{\partial^3\phi}{\partial z^3} - \frac{\partial^2\phi}{\partial z^2}] \tag{3b}$$

$$\tau_{xy} = 2G[z \frac{\partial^3\phi}{\partial x \partial y \partial z} + (1-2\eta) \frac{\partial^2\phi}{\partial x \partial y}] \tag{3c}$$

$$\tau_{xz} = 2G\ z \frac{\partial^3\phi}{\partial x \partial z^2} \tag{3d}$$

$$\tau_{yz} = 2G\ z \frac{\partial^3\phi}{\partial y \partial z^2} \tag{3e}$$

where G is the shear modulus.

Due to practical limitations in deriving additional terms in the polynomial representation of the pressure distribution, p(x,y), only terms up through the third powers in x and y were derived as,

$$p(x,y) = \sum_{i,j=0}^{3} A_{ij}\ x^i y^j \tag{4}$$

The harmonic function which satisfies the necessary boundary conditions was derived from Segedin's potential function. Details of the harmonic function and its derivatives are described in [9] and [14].

Finally, the stress intensity factor associated with $\phi$ can be obtained through a procedure described by Irwin [20] or Kassir and Sih [15]. Again, details of this procedure are described in [9] and [14].

## FINITE DIFFERENCE CALCULATION OF STRESSES

As mentioned earlier, the derivable terms in Eq. (4) were limited to a third order polynomial in x and y due to practical considerations. Some of this difficulty can be removed if the partial derivatives in Eq. (3) can be replaced by numerical differentiations. Analytical expressions of $\partial^2\phi/\partial x^2$ etc., which are given in the Appendix of [9], were used to obtain the third partial derivatives of $\phi$ with respect to x, y and z. Analytical expressions of $\partial^2\phi/\partial x^2$ etc., were evaluated numerically at two locations of x and x + 0.001a etc., to compute $\partial^3\phi/\partial x^3$ etc. These numerical derivatives were then combined as per Eq. (3) to determine the stresses of $\sigma_{yy}$, $\tau_{yz}$ and $\tau_{yx}$ at various values of x and

and z on various y = constant planes. These numerical results were found to agree within three significant figures with corresponding stresses obtained previously [3,9].

The above numerical differentiation procedure has been extended to compute the same third partial derivatives from analytical expressions of $\partial\phi/\partial x$ etc. Preliminary results show that the same excellent agreement between stresses, which were computed numerically and by the closed form procedure, exists except for regions of high stress gradients where the elliptical crack front intersects with the y = const. plane, e.g., y=0 plane, in Fig. (2). Effort is currently underway to determine whether the average surface traction acting on the small region surrounding such stress singularity can exhibit numerically the same effect as the residual singular surface traction as far as the free surface solution is concerned. If such replacement is possible, then the average stresses determined by the second numerical differentiations of $\partial\phi/\partial x$, etc., can also be used to compute the residual surface tractions on the bounding free surfaces.

MAXIMUM PERMISSIBLE RECTANGLE SIZE

The second step of the alternating technique is to eliminate the residual surface traction on the bounding free surfaces. A common procedure is to use Love's solution [18,19] for a half space with prescribed uniform surface tractions on a rectangle in the bounding plane. When the alternating technique was first used to analyze surface flaw problems several years ago, the total number of such rectangles was of the order of 540 [3]. Through numerical experimentation, Smith has gradually reduced the numbers of these rectangles to 180 [6] and then to 71 [17]. Hartranft and Sih [7], on the other hand, increased the number of rectangles to 1744 which together with the increased number of Fourier series used, resulted in computing time of 8 hours to solve one problem. The purpose of this section is to determine the maximum permissible rectangle size which will yield reasonably accurate results of this second step of the alternating procedure.

Since the continuously varying residual surface tractions on the bounding plane are to be replaced by uniform surface tractions over each of the rectangles forming the bounding plane, the problem reduces to the difference in the resultant stresses due to prescribed uniform or variable surface tractions on the rectangle. This difference in resultant stresses can be estimated by examining the stresses in a half space due to (1) a linearly varying stress distribution over the rectangle on the bounding plane, and (2) a uniform stress distribution over the rectangle. The uniform stress distribution is in equilibrium with the linearly varying stress distribution as shown in Fig.

(3). The stress distributions at any point in a semi-infinite plate due to a linearly varying or a uniform surface traction, $\sigma_{yy}$, over a segment 2d as shown in the legend of Fig. (3) are given by the following equations [21].

$$\sigma_{zz} = -\frac{f_o}{\pi}\left\{\frac{y}{d}\ln\frac{y^2+(z-d)^2}{y^2+(z+d)^2} + \frac{2y(z-d)}{y^2+(z-d)^2} + \frac{z+d}{d}\tan^{-1}\frac{2dy}{y^2+z^2-d^2}\right\} \tag{5a}$$

$$\tau_{zy} = -\frac{f_o}{\pi}\left\{\frac{2y^2}{y^2+(z-d)^2} - \frac{y}{d}\tan^{-1}\frac{2dy}{y^2+z^2-d^2}\right\} \tag{5b}$$

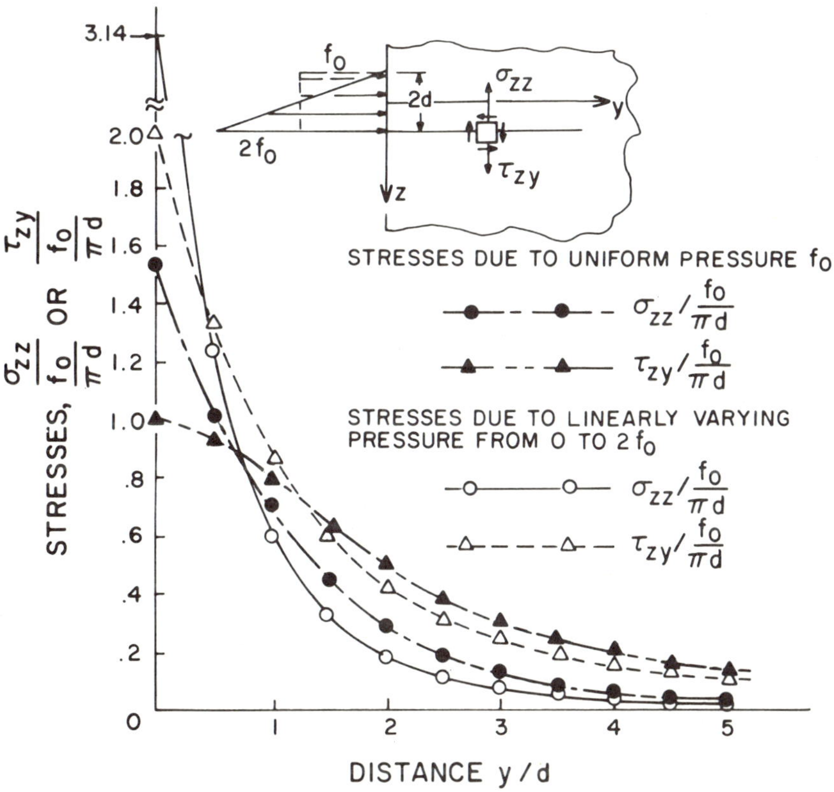


Fig. (3) - Stress distribution at z=d in a half plane due to uniform and linearly varying pressures acting on the bounding line

For a uniform surface traction prescribed over a segment of 2d, these stresses become [21]

$$\sigma_{zz} = -\frac{f_o}{\pi}\{\tan^{-1}\frac{2dy}{y^2+z^2-d^2} + \frac{2dy(-y^2+z^2-d^2)}{[y^2+(z+d)^2][y^2+(z-d)^2]}\} \quad (6a)$$

$$\tau_{zy} = -\frac{f_o}{\pi}\frac{4dyz}{[y^2+(z+d)^2][y^2+(z-d)^2]} \quad (6b)$$

Figure (3) shows the distributions of $\sigma_{zz}$ and $\tau_{zy}$ at distances of z=d from the center of the load segment for linearly varying and uniform normal tractions acting on the segment. Although the stresses due to the two loads differ by a factor of 2 at y=0 in Fig. (3), this difference when superimposed on to other stresses due to the common uniform load acting on the rectangle, reduces the overall differences to less than 5% in actual computations. Further evaluations of Eqs. (5) and (6), for most parts of the elliptical crack, show that the differences in both normal and shear stresses on the crack plane due to the above two surface tractions rapidly diminish at a distance which is half-width away from the load segment.

The above results indicate that the size of the rectangle can be equal to its closest distance to the crack plane or to the other bounding surfaces for the case of a finite thickness solid, whichever is smaller. Similar criterion was found to hold for tangential surface traction acting on the bounding free surface. The number of necessary rectangles can be reduced systematically following this criterion thus resulting in substantial reduction in computing time.

## RESIDUAL PRESSURE DISTRIBUTION ON SEMI-ELLIPTICAL CRACK

Another recognized drawback of the alternating technique based on the elliptical crack potential function is in its inability to approximate the residual pressure distribution on the crack surface with the limited third order polynomial. This drawback is for all practical purposes nonexistent for totally embedded elliptical flaw, such as those shown in Figs. (1a) and (1d), since the residual pressure distribution varies smoothly. A third degree polynomial can be fitted within 5 percent of such residual pressure distribution and thus the numerical procedure converges rapidly with typical incremental change in the stress intensity factor to be less than 0.2 percent of the original stress intensity factor after 2 to 5 iterations.

For a surface flaw where the elliptical circle penetrates the free surface, this fitting problem becomes acute since the residual pressure on the crack surface normally reaches a maximum value in the vicinity of the free bounding surface. The physical crack obviously does not continue beyond the bounding surface while the elliptical crack, used in the alternating method, does protrude into this free space. In order to represent the free bounding surface, Hartranft and Sih [7] have suggested the use of only even terms of polynomial which reduces Eq. (4) to $\sigma_{zz} = A_{00} + A_{20}x^2 + A_{02}y^2$ to which the residual pressure distribution must be fitted. The result is an incomplete erasure of the residual pressure distribution resulting in slow convergence.

In a recent paper, one of the authors [2,22] showed that the stress intensity factor for a circular crack of radius b embedded in an infinite solid and subjected to linearly varying pressure distribution of $\sigma_{zz} = p_o(1 - \frac{y}{b})$ was approximately 5 percent higher than that of a semi-circular surface crack subjected to the same pressure distribution between $0 \leq y \leq b$. This empirical finding indicated that the stress intensity magnification factor due to the free bounding surface had an effect similar to continuing the linearly varying pressure distribution in the $-b \leq y \leq 0$ region of the fictitious portion of the semi-circular crack. This approximation would obviously result in a significant error in estimation of the stress intensity magnification factor if a similar procedure were used to continue into the region of $-b \leq y \leq 0$, the nonlinear pressure distribution acting on a semi-circular crack.

The above empirical finding led to further numerical experimentation for arriving at an optimum pressure distribution on the crack surface extended into the region of $-b \leq y \leq 0$ of the elliptical crack. Three pressure distributions of linear, quadratic, and cubic with the maximum pressure at the free surface, as shown in Fig. (4) were considered. For two-dimensional problems the stress intensity factors associated with an edge crack with linear, quadratic, and cubic pressure distributions, shown in Figs. (4a), (4c), and (4e), can be computed by using the results of Stallybrass [23]. In order to obtain the same stress intensity factor at the crack tip of a totally embedded through crack, various linear distributions of pressures were prescribed on the other half of the embedded crack, i.e., $-b \leq y < 0$. It was found that pressure distributions shown in Figs. (4b), (4d), and (4f) yielded stress intensity factors which were reasonably close to those for the edge crack*. These pressure distributions were then prescribed onto the other half of the elliptical crack, i.e., $-b \leq y \leq 0$, where half of the crack was loaded by linear, quadratic, or cubic pressure distributions. Six terms in Eq. (4) or $\sigma_{zz} = A_{00} + A_{01}y + A_{20}x^2 + A_{02}y^2 + A_{21}x^2y + A_{03}y^3$

---

*Stress intensity factors at the right crack tip were computed by the procedure described in [24].

were then least square fitted to this prescribed pressure distribution on the crack surface.

The above procedure of least square fitting was also applied to the two-dimensional problem illustrated in Fig. (4). For this purpose, a pressure of $\sigma_{zz} = A_0 + A_1y + A_2y^2 + A_3y^3$ was fitted to the 3 pressure distributions illustrated in Figs. (4b), (4d), and (4f). Table I shows the fitted stress distributions, stress intensity factors at the right crack tip, and the corresponding stress intensity factors for edge cracks.

Differences between the respective stress intensity factors can obviously be reduced by further numerical experimentation

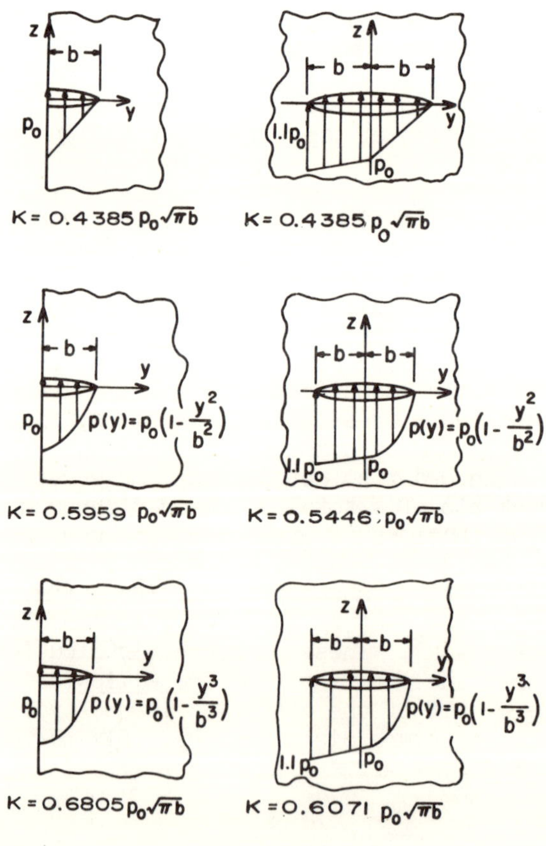


Fig. (4) - Stress intensity factor for an edge crack subjected to variable pressure loading

with various pressure distributions prescribed in the region of $-b \leq y \leq 0$. The purpose of this study, however, was to demonstrate that reasonable results can be obtained for the three cases of prescribed linear, quadratic, and cubic pressure distributions with the same additional pressure in the region of $-b \leq y \leq 0$ and thus no further attempt was made to improve the accuracy of the resultant stress intensity factors.

TABLE I

| Prescribed Pressure | Fitted Pressure | $K_I$ (Right Crack Tip) | $K_I$ (Edge Crack) |
|---|---|---|---|
| Linearly Varying Fig. (4b) | $\sigma_{zz}=0.9960-0.8714y$ $-0.4480y^2+0.3350y^3$ | $0.4636\ p_o\sqrt{\pi b}$ | $0.4385\ p_o\sqrt{\pi b}$ |
| Quadratically Varying Fig. (4d) | $\sigma_{zz}=1.0042-0.1843y$ $-0.4937y^2-0.4094y^3$ | $0.5169\ p_o\sqrt{\pi b}$ | $0.5959\ p_o\sqrt{\pi b}$ |
| Cubically Varying Fig. (4f) | $\sigma_{zz}=0.9965+0.1508y$ $-0.4480y^2-0.7276y^2$ | $0.5808\ p_o\sqrt{\pi b}$ | $0.6071\ p_o\sqrt{\pi b}$ |

Figure (5) shows the normalized stress intensity factor of a circular crack for the pressure distribution shown in Fig. (4b). Poisson's ratio for this analysis was set to $\eta = 0.3$. The polynomial pressure distribution fitted within 2% of the prescribed bilinear pressure distribution. As shown in Fig. (5), the stress intensity factor thus obtained differed as much as 9% from the corresponding stress intensity factor for a semi-circular surface crack with a prescribed linearly varying load. Part of this difference, i.e., approximately 2%, can be attributed to the differences in the Poisson's ratios used in the two analyses [8].

It is interesting to note that the residual normal surface tractions, $\sigma_{yy}$, on the y=0 plane for the problem illustrated in the legend of Fig. (4) are nearly equal to the corresponding residual normal tractions for a circular crack subjected to uniform pressure of $p_o$. It appears then that the additional residual tangential surface tractions, $\tau_{yz}|_{y=0}$, in this problem act to counteract the crack opening moments of the residual normal surface tractions thus causing the resultant stress intensity factor to approach that of a semi-circular crack in a semi-infinite solid and subjected to linearly varying pressure.

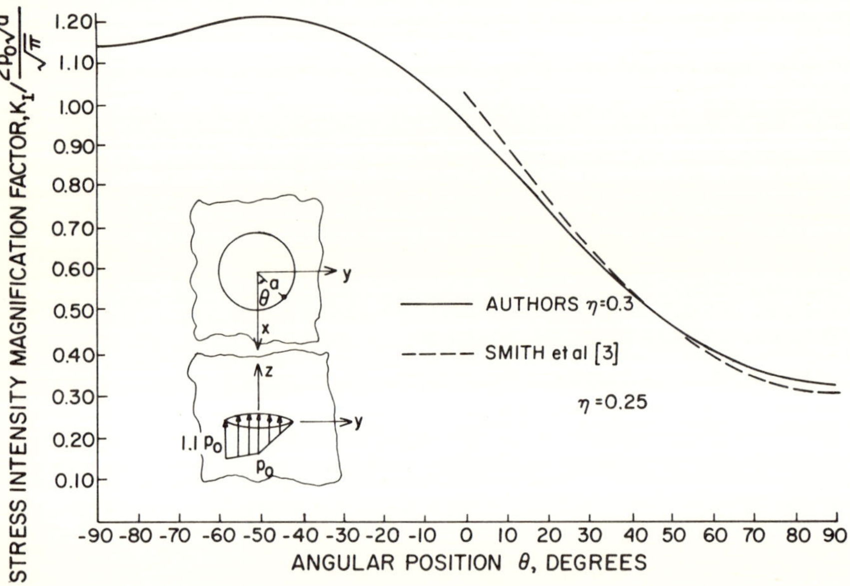


Fig. (5) - Stress intensity factors for a circular crack subjected to a combined two linear varying pressure distributions and a semi-circular surface crack with linearly varying load

The close agreement between the stress intensity factors for the two extremes of a semi-circular crack and a two-dimensional edge crack suggested that this procedure of prescribing appropriate pressure distribution in the region of $-b \leq y \leq 0$ of an elliptical crack can be used to estimate the stress intensity factors for semi-elliptical surface flaws subjected to linearly varying pressure distribution. The estimated stress intensity magnification factors thus obtained along the periphery of a semi-elliptical crack subjected to linearly varying pressure distribution are shown in Fig. (6). The stress intensity factor in the vicinity of $\theta=0$, i.e., where the semi-elliptical crack meets the free bounding surface, is not shown since the stress singularity at this point is believed to vanish following the analysis by Hartranft and Sih [7]. The error bounds of this estimation are indicated in Fig. (6) by the two known stress intensity factors for $b/a = 0$ and $1.0$. The overestimation in the stress intensity magnification factors at the deepest penetration of the ellipse or $\theta = 90°$ is consistent with the corresponding two-dimensional problem shown in Table I. For a better estimate of the stress intensity magnification factor at this position, all values could be lowered 7% following the results in Table I.

The same procedure was used to estimate the stress intensity magnification factors for semi-elliptical cracks subjected to quadratic and cubic distributions of pressure. Again, the fitted polynomial distribution of pressure fitted within 2% of the prescribed pressure. Unlike the previous case of linearly varying pressure distribution, the stress intensity factors for quadratic and cubic pressure distribution on a semi-circular surface flaw are not known. The obtained stress intensity factors, however, were consistent with known theoretical solutions at one extreme of b/a = 0, i.e., for an edge crack.

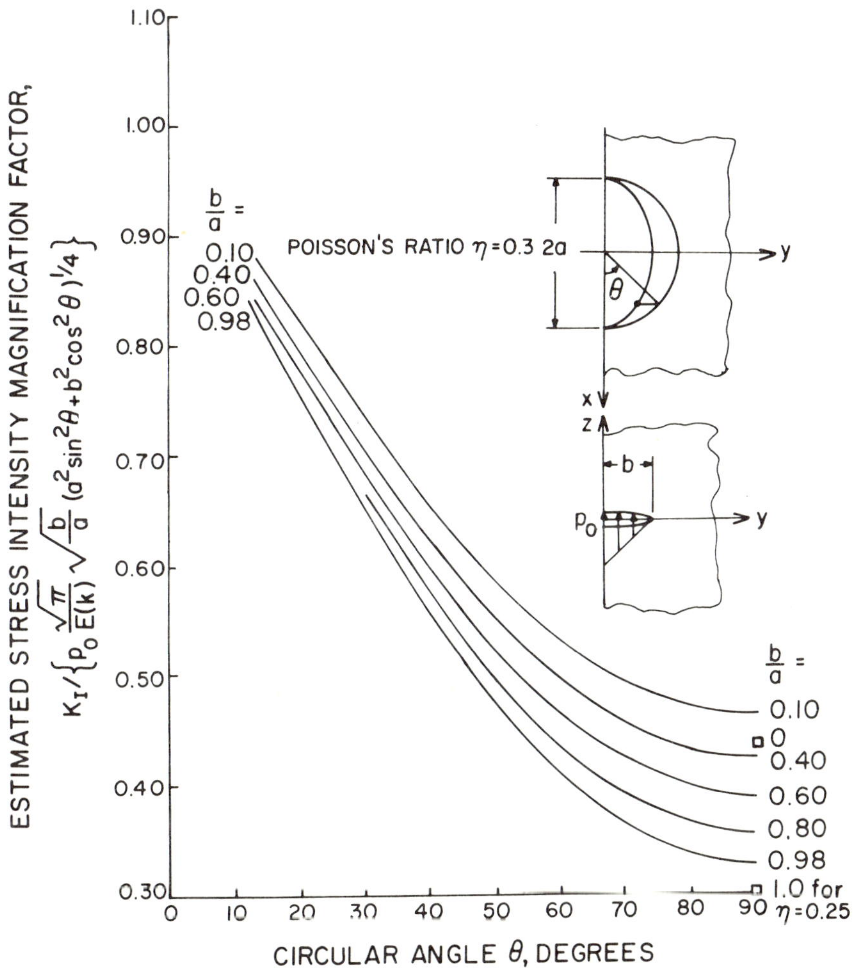


Fig. (6) - Estimated stress intensity factor of a semi-elliptical flaw in a half space subjected to linearly varying load

## DISCUSSION

The stress intensity factor of a semi-elliptical crack in a semi-infinite solid subjected to uniaxial tension can be estimated from the results of Fig. (5). Figure (7) shows the pressure distribution for erasing the residual, normal stresses on the free surface at y=0 of a semi-infinite solid after the first iteration in the alternating method. An equivalent linearly varying pressure distribution as shown in Fig. (7), which is nearly in equilibrium with the rapidly varying pressure distribution is now considered. The stress intensity factor for a semi-elliptical crack subjected to the equivalent linearly varying pressure distribution is then estimated from Fig. (6). The superposition of this stress intensity factor to the original stress intensity factor of an elliptical crack in an infinite solid and subjected to uniform pressure then yields an estimate of the front surface magnification factor for a semi-elliptical crack in a semi-infinite solid as shown in Fig. (8). It should also be noted that, as evident from the equivalent pressure distributions in Fig. (7), this procedure will yield underestimated stress intensity magnification factors close to the free surface. Nevertheless, for a semi-circular crack, the estimated value of $M_F = 1.13$ at an angular orientation of $\theta = 20°$ is sufficiently close to the original results of $M_F = 1.12$ obtained by Smith et al [3].

The estimated front surface stress intensity magnification factors together with those by Smith and Alavi [5], Kobayashi and Moss [25], and Nisitani and Murakami [26] are shown in Fig. (9). The latter results by Nisitani and Murakami, which were computed for Poisson's ratio of $\eta=0$, were corrected for $\eta = 0.25$, as shown in Fig. (9) following an investigation on the effects of Poisson's ratio by Shah and Kobayashi [27]. Also shown in Fig. (9) is the single result by Hartranft and Sih [7] for $\eta = 0.30$. Although the new estimates of the front surface magnification factor appear to be in closer agreement with the results by Smith and Alavi [5], when an adjustment is made to account for the possible overestimation discussed previously, the adjusted front surface stress intensity magnification factor will move closer to the analytically more rigorous results by Nisitani and Murakami [26].

The above mentioned approximate procedure does not eliminate the difficulties in fitting a polynomial pressure distribution. Much of these difficulties, however, can be reduced by a shrewd use of the limited terms in the polynomial using numerical procedures akin to over-and-under relaxations.

The alternating procedure with optimum rectangle spacing on the free bounding surface and the pressure distribution in the

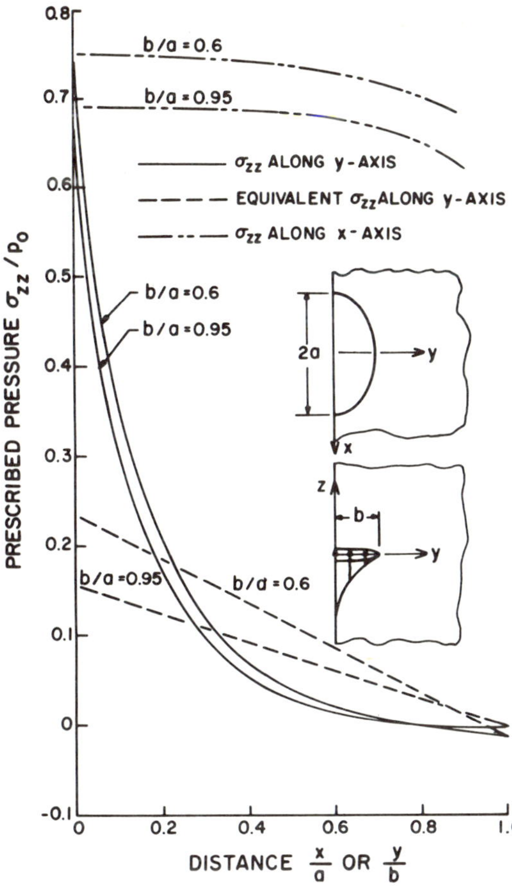


Fig. (7) - Prescribed pressure distribution on a semi-elliptical crack surface after first iteration in the alternating method, Poisson's ratio $\eta = 0.3$

region of $-b \leq y \leq 0$ of the elliptical crack are being used to solve practical problems involving semi-elliptical surface flaws. Hopefully, these results will be available for presentation at the Conference of Prospects of Fracture Mechanics.

ACKNOWLEDGEMENT

The work reported here was supported by the US Army Research Office-Durham. The authors wish to express their appreciation to Mr. J. Murray and Dr. E. A. Saibel (AROD) for their encouragement in pursuing this research. Some of the ideas for this work were conceived during the course of two of the authors' work at The Boeing Aerospace Company. The authors wish to express their

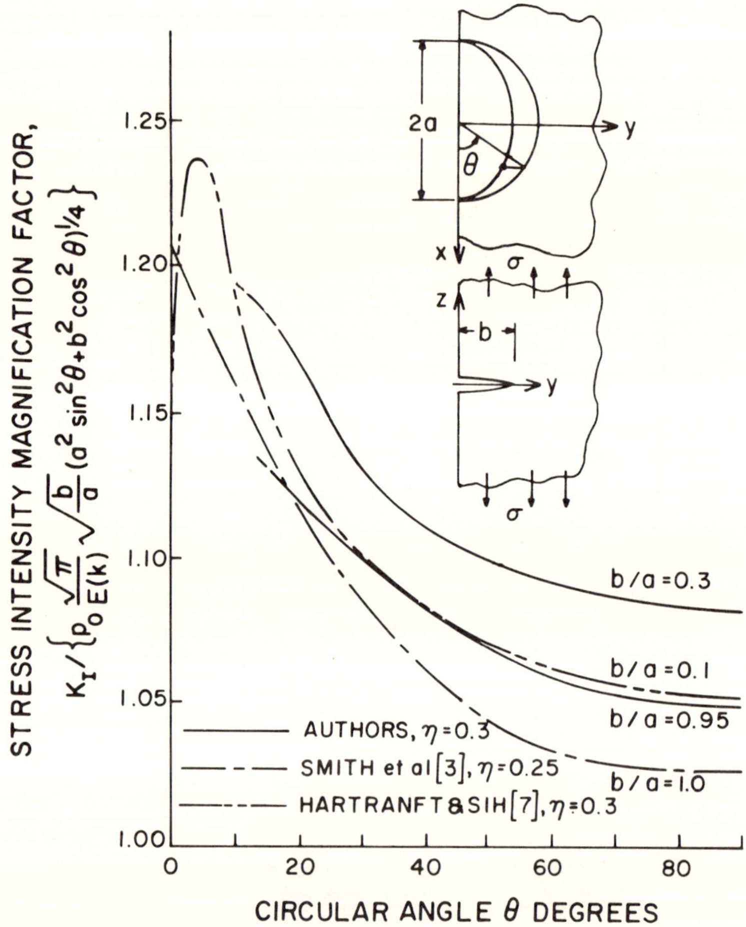


Fig. (8) - Stress intensity magnification factor along the periphery of semi-elliptical surface crack in a semi-infinite solid subjected to uniaxial tension

appreciation to Mr. J. N. Masters, The Boeing Aerospace Company, for his support.

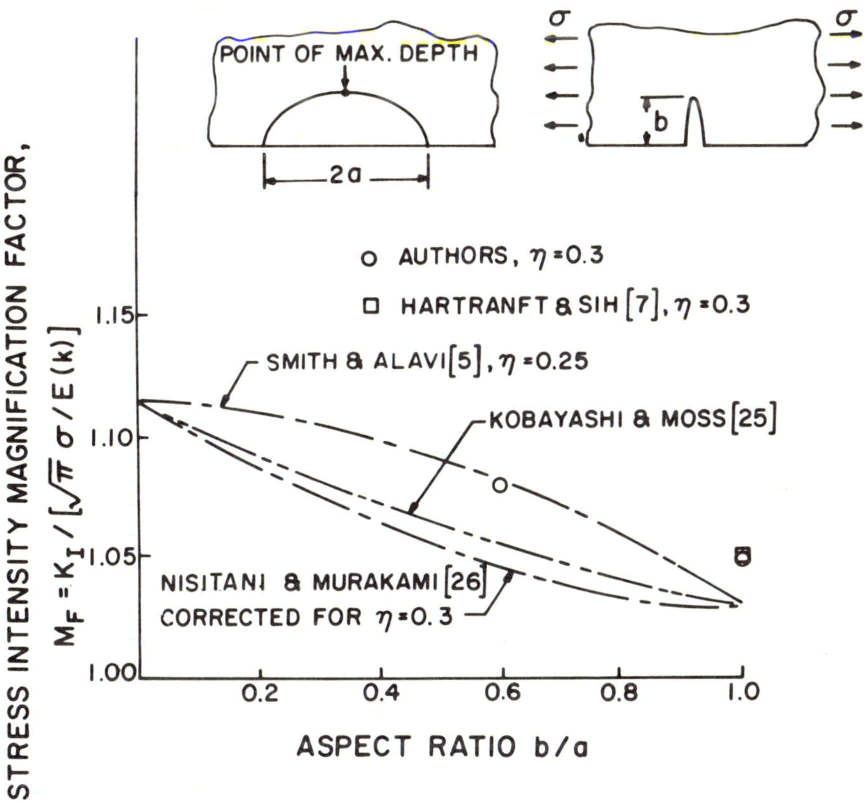


Fig. (9) - Stress intensity magnification factor at point of maximum depth of a semi-elliptical crack in a semi-infinite solid subjected to uniaxial tension

REFERENCES

[1] Irwin, G. R., "The crack extension for a part-through crack in a plate", J. of Appl. Mech., Trans. of ASME, 29, 651, 1962.

[2] Kantorovich, L. V. and Krylov, V. L., "Approximate methods of higher analysis", Interscience, New York, 1964.

[3] Smith, F. W., Emery, A. F. and Kobayashi, A. S., "Stress intensity factors for semi-circular cracks", J. of Appl. Mech., Trans. of ASME, 34, 953, 1967.

[4] Smith, F. W. and Alavi, M. J., "Stress intensity factors for a penny-shaped crack in a half space", J. of Engrg. Fracture Mechanics, 3, 241, 1971.

[5] Smith, F. W. and Alavi, M. J., "Stress intensity factors for a part-circular surface flaw", Proc. of 1st Int'l Conf. on Pressure Vessel Technology, Delft, Holland, 1969.

[6] Thresher, R. W. and Smith, F. W., "Stress intensity factors for a surface crack in a finite solid", J. of Appl. Mech., Trans. of ASME, 29, 195, 1972.

[7] Hartranft, R. J. and Sih, G. C., "Alternating method applied to edge and surface crack problems", Methods of Analysis and Solutions of Crack Problems (G. C. Sih, editor), Noordhoff Int'l., Leyden, Holland, 179, 1973.

[8] Smith, F. W., "The elastic analysis of the part-circular surface flaw problem by the alternating method", The Surface Crack: Physical Problems and Computational Solutions (J. L. Swedlow, editor), ASME, New York, 125, 1972.

[9] Shah, R. C. and Kobayashi, A. S., "Stress intensity factor for an elliptical crack under arbitrary normal loading", J. of Engrg. Fracture Mechanics, 3, 71, 1973.

[10] Segedin, C. M., "A note on Geometric Discontinuities in elastostatics", Int'l. J. of Engrg. Science, 6, 309, 1968.

[11] Shah, R. C. and Kobayashi, A. S., "Stress intensity factors for an elliptical crack approaching the surface of a semi-infinite solid", Int'l. J. of Fracture, 9, 133, 1973.

[12] Shah, R. C. and Kobayashi, A. S., "Stress intensity factors for an elliptical crack approaching the surface of a plate in bending", Stress Analysis and Growth of Cracks, ASTM STP 513, 1, 1972.

[13] Shah, R. C. and Kobayashi, A. S., "Elliptical crack in a finite thickness plate subjected to tensile and bending loading", to be published in the J. of Engrg. for Industry, Trans. of ASME, 1974.

[14] Shah, R. C. and Kobayashi, A. S., "On the surface flaw problem", The Surface Crack: Physical Problems and Computational Solutions (J. L. Swedlow, editor), ASME, 125, 1972.

[15] Kassir, M. K. and Sih, G. C., "Three-dimensional stress distribution around an elliptical crack under arbitrary loadings", J. of Appl. Mech., Trans. of ASME, 33, 601, 1966.

[16] Smith, F. W. and Sorensen, D. R., "The elliptical crack subjected to nonuniform shear loading", to be published in the J. of Appl. Mech., Trans. of ASME.

[17] Smith, F. W. and Sorensen, D. R., "The semi-elliptical surface crack - a solution by the alternating method", Colorado State Univ. Tech. Rept. 4, August 1973, NASA Grant NGL-06-002-063.

[18] Love, A. E. H., "On stress produced in a semi-infinite solid by pressure on part of the boundary", Phil. Trans. of the Roy. Soc., Series A, 228, 387, 1929.

[19] Love, A. E. H., "A treatise on the mathematical theory of elasticity", Dover Publications, New York, 241, 1944.

[20] Irwin, G. R., "Analytical aspects of crack stress field problems", Univ. of Illinois T. and A.M. Rept. No. 213, 1962.

[21] Durelli, A. J., Phillips, E. A. and Tsao, C. H., "Introduction to the theoretical and experimental analysis of stress and strain", McGraw-Hill, 225, 1958.

[22] Kobayashi, A. S., "A simple procedure for estimating the stress intensity factors in region of high stress gradient", Proc. of US-Japan Seminar 1973, Significance of Defects in Welded Structures (T. Kanazawa and A. S. Kobayashi, editors), Univ. of Tokyo Press, March 1974.

[23] Stallybrass, M. P., "A crack perpendicular to an elastic half-plane", Int'l. J. of Engrg. Sciences, 8, 351, 1970.

[24] Paris, P. C. and Sih, G. C., "Stress analysis of cracks", Fracture Toughness and Its Applications, ASTM STP 381, 30, 1965.

[25] Kobayashi, A. S. and Moss, W. L., "Stress intensity magnification factors for surface-flawed tension plate and notched round tension bar", Proc. of the 2nd Int'l. Conf. on Fracture, 31, 1969.

[26] Nisitani, H. and Murakami, Y., "Stress intensity factor of an elliptical crack or a semi-elliptical crack subjected to tension, to be published in Int'l. J. of Fracture.

[27] Shah, R. C. and Kobayashi, A. S., "Effect of Poisson's ratio on stress intensity magnification factor", Int'l. J. of Fracture, 9, 360, September 1973.

# SECTION VII

# FRACTURE EXPERIMENTS AND TESTINGS

# PRESENT STATUS ON THE EVALUATION CRITERIA OF FRACTURE TOUGHNESS FOR STRUCTURAL STEELS AND THEIR WELDED JOINTS IN JAPAN

T. Kanazawa*, S. Machida* and T. Miyata**

*University of Tokyo, Bunkyo-ku, Tokyo 113 Japan
**Nagoya University, Chikusa-ku, Nagoya 464 Japan

## INTRODUCTION

Current empirical rules, usually based on simple interpretation on industrial tests, expose their inadequacies. Advancement in technology is demanding a more fundamental and qualitative design procedure in the presence of crack like defects. The Japan Welding Engineering Society has established the WES standard, "Evaluation Criterion of Structural Steels for Low Temperature Application", based on the $K_c$-value correlations between small scale tests (V-Charpy test, Pressed Notch Charpy test) and large scale propagation arrest tests (Double Tension test). In this standard, steel plates are classified into two classes of G and A, where G is generally usable for welded structures where hazard of brittle fracture is anticipated and A is used for arresting a propagating crack.

Linear fracture mechanics concept, on which the above mentioned WES standard is based, has been very successfully applied to quantitative assessment of unstable fracture behavior under small scale yielding condition. But in most of in-service brittle failures of ordinary welded steel structures, defects that lead to catastrophic failure are located at or near a welded joint and/or structural discontinuity where high stress and strain concentrations exist. This implies that a majority of brittle fracture initiation in ordinary welded steel structures would occur under large scale yielding, i.e. a crack is accompanied by extensive plastic strain prior to unstable fracture initiation.

Consequently, G class steels in the WES standard should be revised to account for fracture under large scale yielding condition. We, thus, have been much interested in the concepts of fracture models which take more of the plasticity around crack tip. To our present knowledge, the crack opening displacement (COD) concept is considered a very promising engineering concept of brittle fracture initiation under large scale yielding.

In order to establish a revised WES standard, the TM Committee of the Japan Welding Engineering Society has conducted comprehensive cooperative research works since 1971 and the conclusions of this Committee are forthcoming soon. In this paper the present status of this cooperative research work will be briefly reported with particular emphasis on the correlations between COD bend tests and notched tension tests and related problems.

In addition to the TM Committee, the JE and the BE Committee, which intend to establish the acceptance standard for the toughness of welded joints, have been organized in the Japan Welding Engineering Society. The JE Committee carried out investigations on the notch toughness of welded joints fabricated under ordinary and actual conditions. Works in the JE Committee were succeeded to the BE Committee. The BE Committee has been conducting cooperative research works especially on the embrittlement of welded fusion line.

Using the experimental data obtained by works in the above mentioned Committees in addition to the data by the authors, the correlations between COD bend tests and Charpy impact tests on parent material and welded fusion line will be also discussed and a proposal for utilization of Charpy test will be made in the following section of this paper.

## EFFECTIVENESS OF COD BEND TEST

The bend test seems to be a practical and useful fracture test from the facts that the bend test specimen can be made smaller and fractured at considerably lower load as compared with the tension test. Many experimental investigations have been performed concerning COD bend test [1][2].

One of the troubles incidental to measurement of the critical COD on a bend specimen is that the notch tip COD is not so easily obtained from bend test as from tension test. So far COD value is estimated from the clip gauge reading using "rotational factor", on the assumption that a notched bend specimen deforms about a "pivot", the location of which depends on the applied load.

In the previous paper [2], the authors discussed the method for determining COD value at notch tip of a bend specimen of the basis of the results of experiment and calculation using the

Dugdale-Barenblatt crack model and the Finite Element Method (FEM). In this paper, the method of rotational factor and the results of calculation using FEM are combined and more effective method for determining COD is proposed. The authors show the effectiveness of COD bend test comparing the critical COD value for bend test to that for tension test.

## Experiments

### Materials

Materials tested are 21 charges of mild steel, 50kg/mm$^2$ class, 60kg/mm$^2$ class and 80kg/mm$^2$ class high tensile steel plates with thickness varying from 18mm to 50mm. The temperature dependent yield strengths and tensile strengths of test steels were obtained by using round tension bar.

### Tests and Test Procedures

In order to clarify the effects of specimen size and testing method on a critical COD measurement as well as other related problems, various tests, which included 3 point and 4 point bend tests and notched tension test, were carried out for each steel. Examples of the specimen configuration used in these tests are show in Fig. 1. COD was measured by a ring-type clip gauge where the measurement was made at the knife edges attached to the notch end for a bend specimen and at a certain distant from the notch tip for a tension specimen.

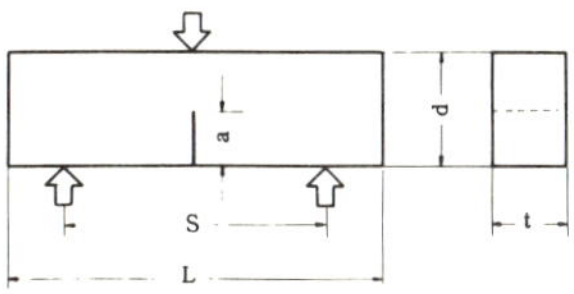


| Code | t | d | a | S | L |
|---|---|---|---|---|---|
| A | 25 | 50 | 25 | 200 | 280 |
| B | 40 | 80 | 40 | 320 | 400 |
| | 20 | 80 | 40 | 320 | 400 |

(unit in mm)

(a) 3 Point Bend Specimen

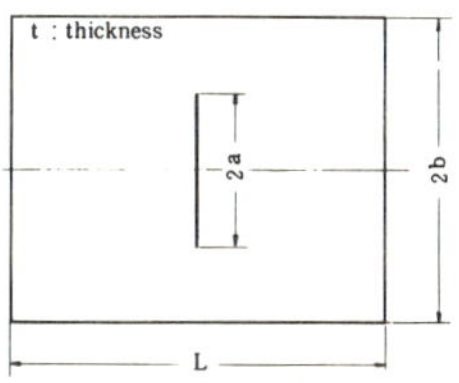


| Code | t | b | a | L |
|---|---|---|---|---|
| A | 25 | 200 | 100 | 500 |

(unit in mm)

(b) Center Notched Specimen

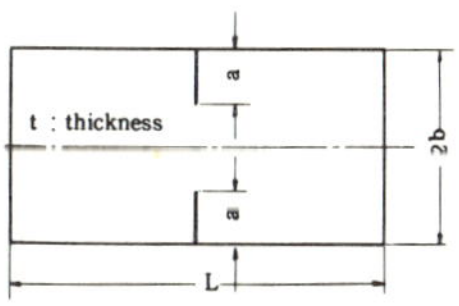


| Code | t | d | a | L |
|---|---|---|---|---|
| A | 25 | 100 | 22.5 | 500 |
| B | 40 | 100 | 60 | 500 |
| | 20 | 100 | 60 | 500 |

(unit in mm)

(c) Side Notched Specimen

Fig. 1. Examples of specimen configurations.

### Determination of Notch Tip COD

COD value is usually measured by a clip gauge for convenience of continuous and simple recording. Since the clip gauge cannot be mounted at the notch tip because of its blade thickness, procedure must be established for reducing the clip gauge reading ($V_g$) to the notch tip COD value ($\Phi$). Some of these procedures will be discussed in the following.

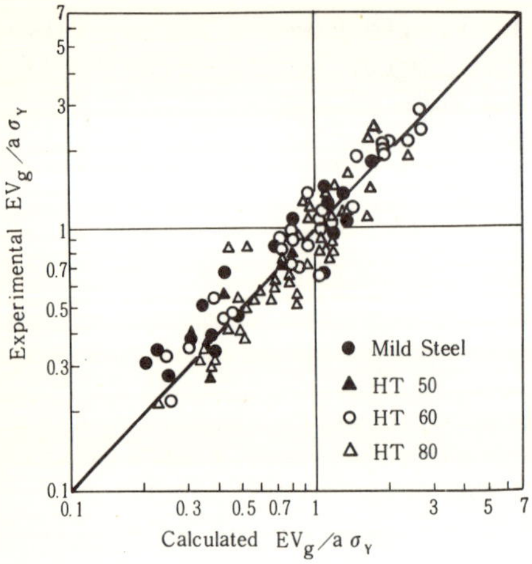


Fig. 2. Relation between measured clip gauge reading and calculated opening displacements using the D-B model for tension test.

## Tension Test

Using the Dugdale-Barenblatt crack model the opening displacement of a crack caused by the applied load can be calculated at an arbitrary point [3]. The opening displacement ($V_g$) on a crack where clip gauge is attached is calculated and compared with experimental values as shown in Fig. 2. In this figure, E is the Young's modulus, $\sigma_Y$ the yield stress of the test steel at the test temperature and 2a or a is notch length of center or side notched specimen, respectively. This figure shows that the calculated $V_g$ values compare well with the measured ones and implies that the D-B model simulates well the actual opening displacement of a crack despite of the simplicity of the model.

Thus the opening displacement, $\Phi$ , at the crack tip can be obtained from experimental $V_g$ value using the D-B model for tension test.

## Slow Bend Test

Crack opening displacements for several specimen sizes, stress-strain relations were calculated using the two dimensional, elastic-plastic computation based on the finite element method (FEM). However, as shown in Fig. 3, the relation of $E\Phi/d\sigma_Y \sim \sigma_N/\sigma_Y$ (d is the specimen depth and $\sigma_N$ is the nominal skin stress at the notched section) can approximately be represented by a single curve irrespective of specimen size, stress-strain relation for a given notch ratio, a/d, so that this curve would be applicable to the fracture data of steels and using this curve the critical COD, $\Phi_c$ , can be evaluated simply from fracture load and yield stress.

The validity of such computation by the FEM was established by comparing the experimental clip gauge reading to the calculated opening displacement at the point of clip gauge attachement. Figure 4 shows the relation between experimental and calculated critical clip gauge reading ($V_g$) in non-dimensional scale.

The $V_g$ values estimated by the FEM compare well with the experimental values at the lower load level. However, the estimated values are less than the experimental values for $EV_g/d\sigma_Y \geq 3.0$.

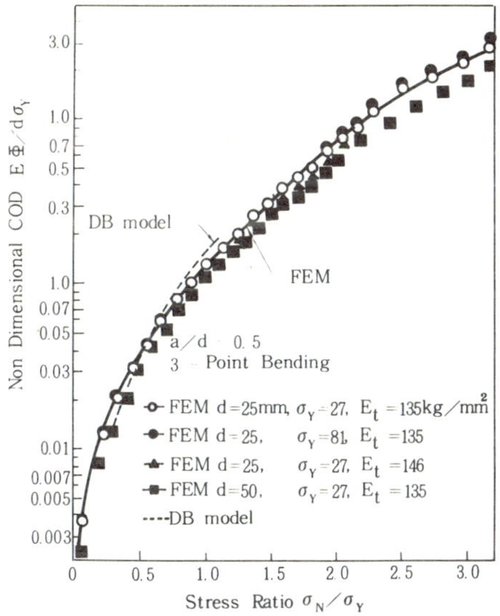


Fig. 3. Relation between non-dimensional COD and stress ratio obtained from computation by the FEM.

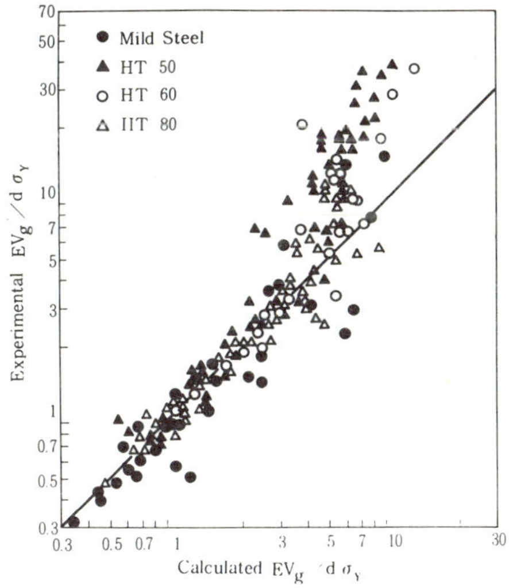


Fig. 4. Relation between measured clip gauge readings and calculated opening displacements using the FEM for bend test.

This discrepancy is probably due to the assumption of linear strain hardening in the calculation which resulted in the calculated strain being less than the actual one in the highly strained region. It is concluded from this figure that the FEM result is valid for $EV_g/d\sigma_Y \leqq 3.0$ which corresponds approximately to the condition of $\sigma_N/\sigma_Y \leqq 1.5$.

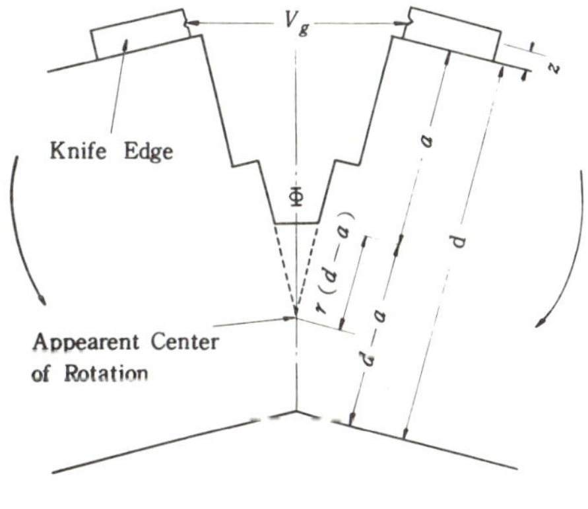


Fig. 5. Profile of notch opening associated with rotational factor.

Another method for determining COD value at crack tip of a bend specimen is the concept of rotational factor. In this concept, a notched specimen is assumed to deform about a "pivot", the location of which depends on the applied load. After the rotational factor, r, is experimentally determined, clip gauge readings can be reduced to COD values at the notch tip using a simple proportionality relation given by (see Fig. 5)

$$\Phi = V_g/[1+(a+z)/r(d-a)] \qquad (1)$$

This rotational factor was determined theoretically [4] and experimentally [5], and was found to depend on the applied load in lower load level and tend to approach a constant value as the applied load increases.

Conclusively, the concept of rotational factor can be effective taking advantage of the constancy of r at fully plastic state of the specimen or at higher value of stress ratio $\sigma_N/\sigma_Y$ ( $\sigma_N$ is nominal net skin stress). For the sake of simplicity, the value of r is taken to be 1/3 which is approximately the lower bound of the experimentally obtained value, for $\sigma_N/\sigma_Y \geqq 2.0$ in this paper. It is to be noted here that the change of crack tip COD is relatively insensitive to the large of r as is seen in the above equation.

The discussions mentioned above show that COD at a notch tip can be determined by using elastic plastic crack analysis based on FEM for stress ratio $\sigma_N/\sigma_Y \leqq 1.5$, and for $\sigma_N/\sigma_Y \geqq 2.0$ the method based on the rotational factor (r=1/3) is useful. In the region of $1.5 \leqq \sigma_N/\sigma_Y \leqq 2.0$, COD could be estimated by taking the intermediate values obtained from these two methods.

A typical result obtained by using the proposed method is shown in Fig. 6 in terms of the critical COD vs. testing temperature. The solid line represents the critical COD obtained from tension test. The dotted and broken lines represents curves obtained from the bend test using the methods of FEM and rotational factor, respectively. Temperatures which correspond to fracture loads of $\sigma_N/\sigma_Y=1.5$ and 2.0 are also indicated. As shown in this figure, $\Phi_C$ determined for $\sigma_N/\sigma_Y \leqq 1.5$ using the FEM and $\Phi_C$ determined for $\sigma_N/\sigma_Y \geq 2.0$ using the rotational factor in bend test compares well with the curve obtained from tension test. In the region of $1.5 \leqq \sigma_N/\sigma_Y \leqq 2.0$, $\Phi_C$ for the bend test may be estimated by a faired curve connecting points of $\Phi_C$ obtained by the FEM for $\sigma_N/\sigma_Y=1.5$ and $\Phi_C$ obtained by using the rotational factor for $\sigma_N/\sigma_Y=2.0$.

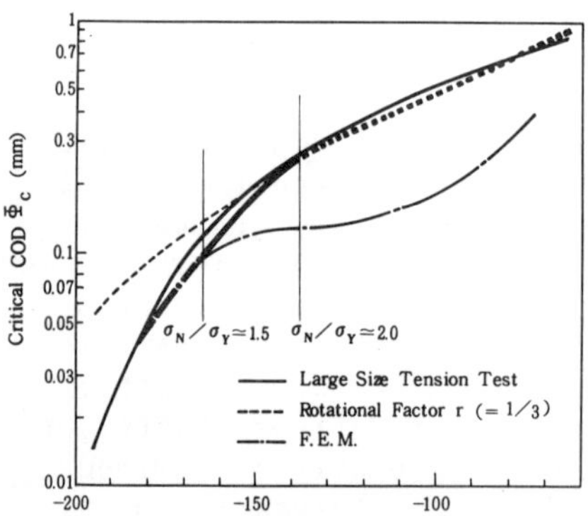


Fig. 6. Critical COD determined from bend test using the method of rotational factor and the FEM comparing with that obtained from tension test.

## Effectiveness of COD Bend Test

The relationship between the $\Phi_C$ values of the small scale yielding region and those of the general yielding region has been investigated [6]. For this purpose, notched tension tests were carried out using large specimen which has higher transition temperature in fracture stress vs. temperature relation than the small specimen. $\Phi_c$ values obtained from small size specimens compare well with those obtained from large specimens.

Figure 7 shows an example of the relation between fracture stress vs. testing temperature for large size deep notch test. The solid curve in the figure shows the predicted fracture stress of large size specimen calculated from the experimental $\Phi_c$ values of small size tension test using the D-B model. Similar results are obtained for the other steels tested. In Fig. 8 a correlation between the experimental and predicted fracture stress is shown. This figure shows that good estimation can be made for the fracture stress of large specimen on the basis of COD concept using the experimental value obtained from small specimen.

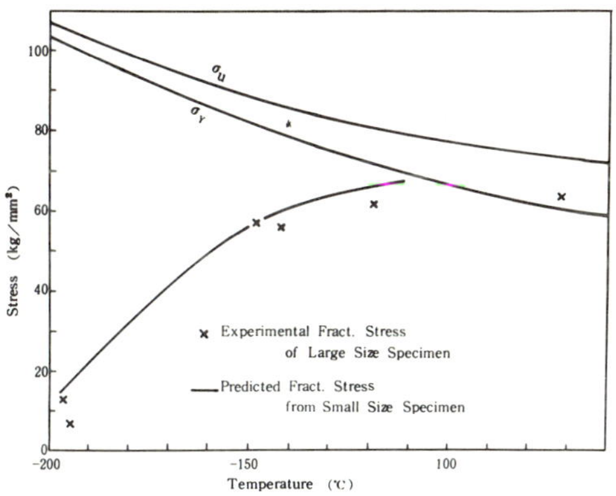


Fig. 7. Relation between measured and estimated fracture stress.

Thus we can conclude that $\Phi_c$ is qualified to be one of the most engineering fracture toughness parameters, and it is of practical importance that the fracture toughness value, $\Phi_c$,

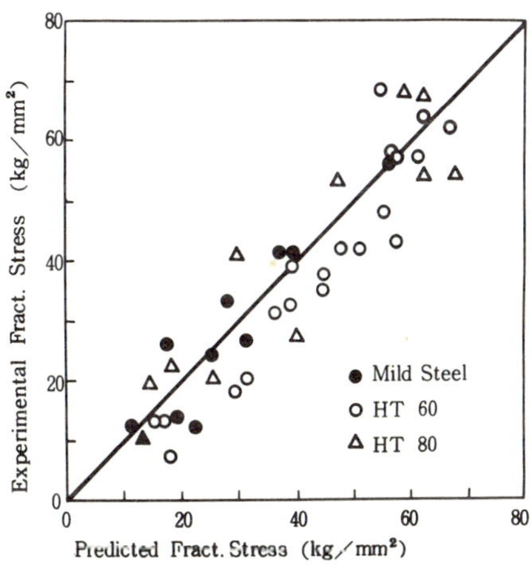


Fig. 8. Validity of estimation of fracture stress for large sized tension specimen using critical COD obtained from small tension test.

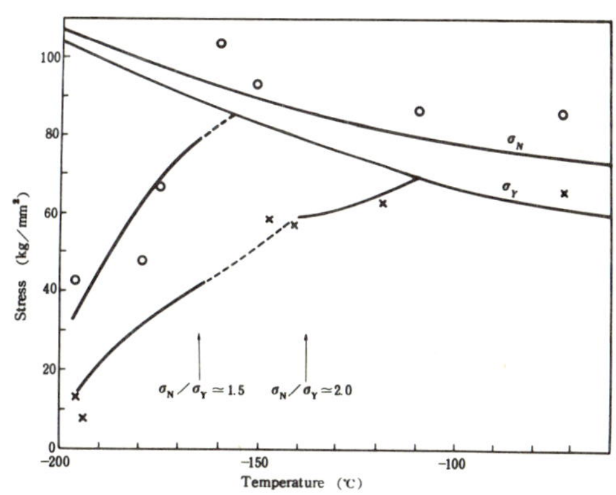


Fig. 9. Relation of measured and estimated fracture stress curve based on COD bend test.

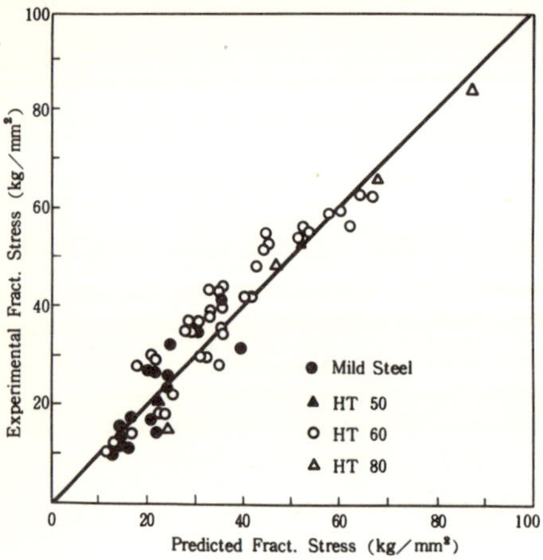


Fig. 10. Fracture stress for tension specimen using critical COD obtained from bend test.

can be measured using small sized specimen, which usually fractures under large scale yielding condition in the realistic temperature region.

The effectiveness of COD bend test was also investigated. Figure 9 shows comparison of experimental data of large and small tension tests with predicted curves from bend test. In Fig. 10, fracture stresses of large size notched tension tests are plotted against those estimated from the results of COD bend tests for various types of steel. Fracture stresses of tension specimens predicted from COD bend tests compare well with the experimental values. These results show that from practical and industrial points of view, COD bend test could be a very useful test for evaluating brittle fracture initiation characteristics of steels, since the specimens can be made smaller and can be fractured at considerably lower load than that of tension test.

## UTILIZATION OF CHARPY IMPACT TEST

Characteristics of Charpy impact test cannot be related directly to structural design but can be used as a relative comparison of the toughness of different materials for the purpose of quality control. Charpy V notch impact test, however, is one of the most widely used toughness tests and toughness requirements in current rules are established on the basis of correlation with service experience except only the WES standard.

If correlations between critical COD values and V Charpy test results could be obtained, these correlations could be used for quantitative assessment of materials because COD values are expressed in terms of the applied stress level and a given crack size. In this section the results of investigations on the correlation between slow bend COD test and V Charpy transition temperature are described.

### Correlation between COD Bend Test and V Charpy Impact Test

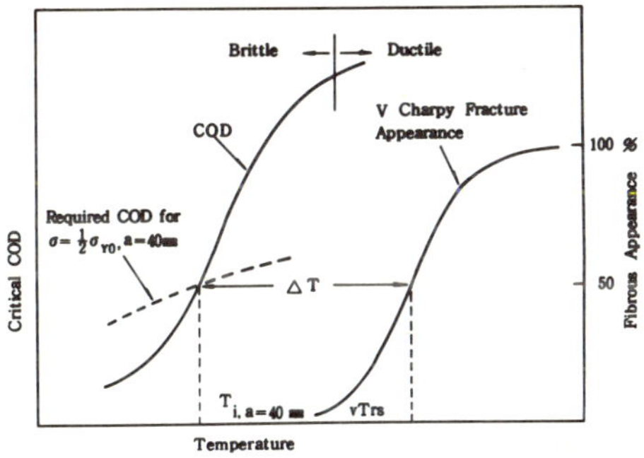


Fig. 11. Schematic representation of critical COD and V Charpy transition.

Temperature is one of the dominant factors affecting the material toughness parameters. Attempts to formulate an empirical relation for temperature dependence is important for simplifying the relation between fracture toughness parameter and more convenient industrial tests such as Charpy impact test.

Transition in the critical COD for structural steels does not occur in the same temperature range as that defined by Charpy test because of the difference of loading rate and notch acuity. As the shift in transition temperature is large, it is difficult to relate directly the critical COD to absorbed energy or some other criteria of V Charpy test at the corresponding temperature.

Then, how the shift in transition temperature would be related to any factors was investigated in this work. Fracture appearance transition temperature, $_{v}T_{rs}$, was taken for V Charpy test and as transition temperature of COD bend test the temperature, $T_i$, which possessed a certain level of the critical COD value was defined as shown schematically in Fig. 11. This level of COD was related to structural design in which a half of yield stress at room temperature was taken as the design stress, $\sigma$, and 40mm in length taken as a half crack length, a. Required COD values are obtained

$$\Phi_C = \frac{8\sigma_Y a}{\pi E} \ln\left(\sec\frac{\pi\sigma}{2\sigma_Y}\right) \tag{2}$$

where $\sigma_Y$ is yield stress at corresponding temperature.

The transition temperature, $T_{i,a=40mm}$, for $\sigma = 1/2 \cdot \sigma_{Y0}$ ( $\sigma_{Y0}$ is yield stress at room temperature) can be obtained as the temperature at which the critical COD vs. temperature curve intersects the required COD curve as shown in Fig. 11. The $\Phi_C/\sigma_Y (\approx K_C^2/E\sigma_Y^2)$ or $\Phi_C$ values at $T_{i,a=40mm}$ obtained from COD bend test are nearly constant for each material with similar level of strength. It was found from analysis for each experimental result that in case when two thirds of yield stress at room temperature, $2/3\,\sigma_{Y0}$, was taken as the design stress, the $\Phi_C/\sigma_Y$ value corresponded to the toughness value to tolerate a crack of about 20mm in half length irrespective of material strength.

Considering the fact that the critical COD should be affected by plate thickness of the material and the effects of strain rate

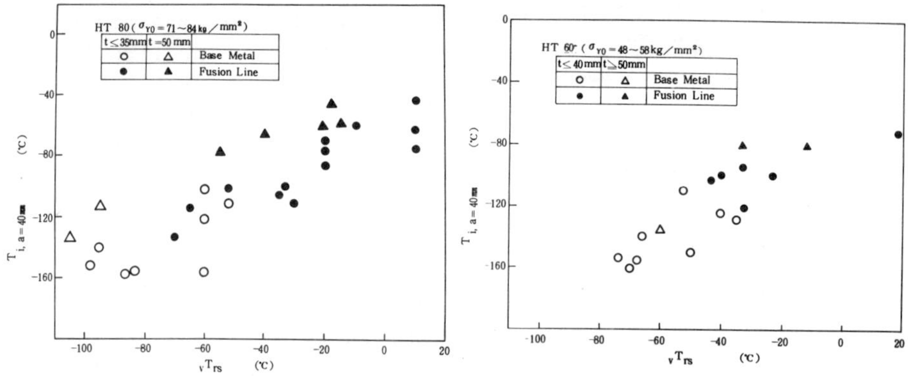


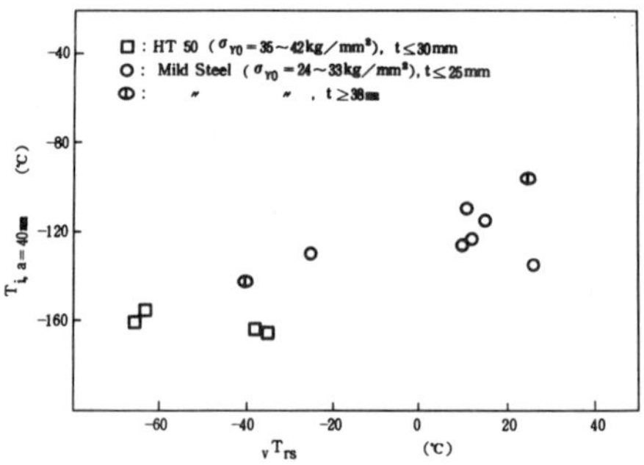


Fig. 12. Relation between $T_{i,a=40mm}$ and $_vT_{rs}$.

(a) 80kg/mm² class high tensile steel

(b) 60kg/mm² class steel

(c) Mild steel and 50kg/mm² class high tensile steel

and notch acuity differ with strength of the material, the shift in transition temperature, $\Delta T$, may be written as follows.

$$\Delta T = T_{i,a=40mm} - {}_vT_{rs} = f(\text{plate thickness, strength}) \qquad (3)$$

Figure 12 shows the relation between $T_{i,a=40mm}$ obtained from experimental results and $_vT_{rs}$. If $\Delta T$ could be expressed as the above function, slope in the relation between $T_{i,a=40mm}$ and $_vT_{rs}$ should be unity for materials with similar strength and plate thickness. The slope, however, is not necessarily unity but depends on the strength level as shown in Fig. 12. Therefore, it was assumed that the following expression for $\Delta T$ could be used instead of Eq.(3).

$$\Delta T(°K) = \left\{ R(\text{strength}) - 1 \right\} {}_vT_{rs}(°K) + \alpha\sqrt{t(\text{mm})} \qquad (4)$$

where t is plate thickness, $\alpha$ is constant irrespective of material properties, and R(strength) is a strength-dependent quantity corresponding to slope of the relationship between $T_{i,a=40mm}$ and $_vT_{rs}$.

Since the effect of plate thickness has not been studied systematically in terms of COD, the result from $K_C$ approach for Deep Notch test by Kasamatsu et al.[7] was used in the above formula. It seems rather incomprehensible that the shift in transition temperature is affected by ${}_VT_{rs}$ as expressed by Eq.(4). This may be due to the fact that $T_{i,a=40mm}$ is not a "transition" temperature which is clearly defined from phenomenological point of view such as ${}_VT_{rs}$ but a temperature corresponding to an arbitrarily prescribed level of critical COD value.

It was found empirically that the value of R could be expressed as a linear function of yield strength of the material at room temperature and was approximately equivalent to yield ratio for the case of $60kg/mm^2$ (HT60) and $80kg/mm^2$ (HT80) class high tensile steels.

Conclusively empirical formula for the correlation between COD bend test and V Charpy test could be obtained from the results on various base materials and welds (welded fusion line) as follows:

$$\Delta T(°K) = (R-1)\,{}_VT_{rs}(°K) + 11.4\sqrt{t(mm)} - 102 \qquad (5)$$

$$R = 0.0047\,\sigma_{Y0} + 0.56$$

where $\sigma_{Y0}$ =yield strength at room temperature ($kg/mm^2$), or

$$T_{i,a=40mm} = R\cdot{}_VT_{rs} + 11.4\sqrt{t} - 102 \qquad (6)$$

Comparison between $T_{i,a=40mm}$ obtained from slow bend COD tests and those predicted by Eq. (6) are shown in Fig. 13. All data fall in a scatter band irrespective of strength level and parent material or welded fusion line.

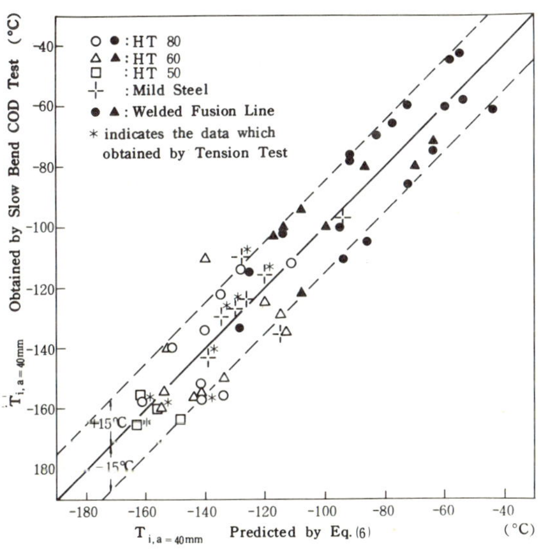


Fig. 13. Prediction of $T_{i,a=40mm}$ from Charpy impact test.

On the basis of the results of welded and notch wide plate tension tests conducted in the Welding Institute, U.K., Harrison, Burdekin and Young [8] have made a proposal on the relation between the minimum service temperature and V Charpy transition temperature for 30ft.lb or 20ft.lb absorbed energy for C-Mn steels up to 3 in. thickness. In their proposal, change in the minimum

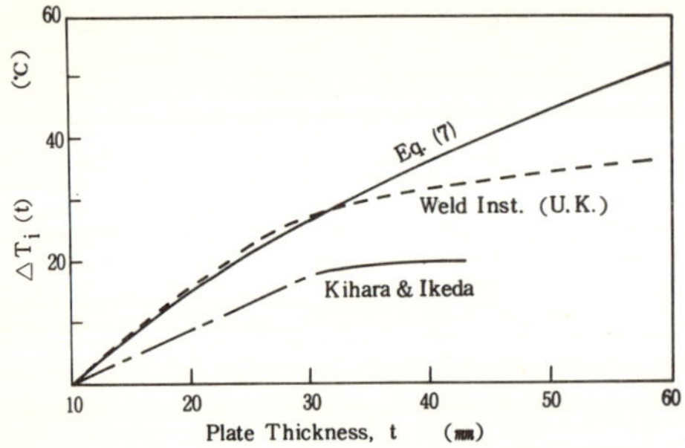


Fig. 14. Effect of plate thickness on transition temperature of fracture toughness.

service temperature due to the effect of plate thickness is independent of Charpy transition on temperature for the case of stress relieved condition, which is shown in Fig. 14 by broken line in terms of increase of the minimum service temperature from that for 10mm thick plate.

The effect of plate thickness obtained from Eq.(6) is shown by solid line in Fig. 14. The chain line shows the result by Kihara and Ikeda [9] obtained from Kc approach to their experiments for Deep Notch test.

The proposal in this paper given by Eqs.(5) or (6) agrees well with that of U.K. up to 35mm in thickness. Simple extraportation of the present proposal to a very thick material seems to result in unrealistically severe estimation. Further extensive investigations into this aspect are recommended in future works.

## Utilization of V Charpy Impact Test

When welding residual stress or stress concentrator does not exist, the transition temperature, $T_{i,a=40mm}$, is a temperature where the material is tough enough to torelate a flaw of 80mm in length at the stress level $\sigma = 1/2\sigma_{Y0}$. Such large size flaw can be detected easily in non-destructive testing. The present empirical relation between $T_{i,a=40mm}$ and $_vT_{rs}$ is valid for welded fusion line as well as base metal.

Thus, the temperature, $T_{i,a=40mm}$, can be applied to the materials including welded joints for stress relieved condition. In this case V Charpy test should be conducted after the stress relieving treatment.

As the $T_{i,a=40mm}$ predicted by Eq.(6) falls in a scatter band within $\pm$ 15°C, the minimum service temperature for design purpose may be conducted by

$$T_{min.} = T_{i,a=40mm} + 15°C \tag{7}$$

When two thirds of yield stress can be taken as the design stress through more severe defect control, the above requirement corresponds to maximum allowable flaw size of about 40mm in length.

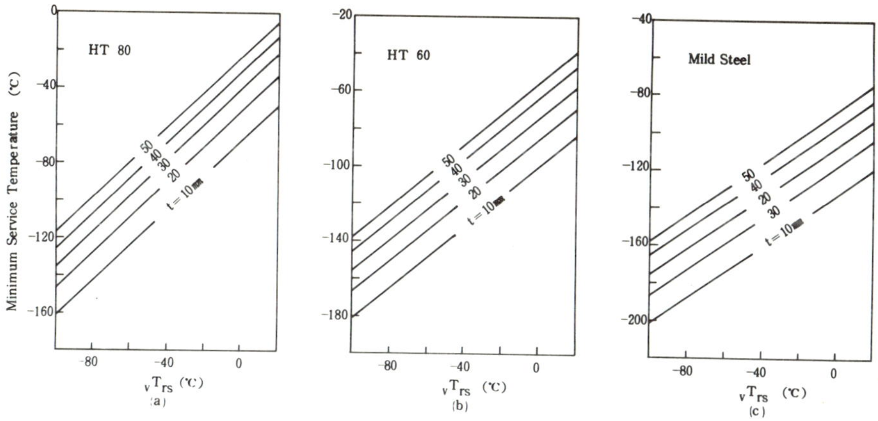


Fig. 15. Minimum service temperature and required $_{v}T_{rs}$ for stress relieved condition.

(a) 80kg/mm² class high strength steel ($\sigma_{YO} = 80\ kg/mm^2$)

(b) 60kg/mm² class high strength steel ($\sigma_{YO} = 55\ kg/mm^2$)

(c) Mild steel ($\sigma_{YO} = 30\ kg/mm^2$)

Figures 15 and 16 show examples of the relation between the minimum service temperature and the required V Charpy transition temperature calculated from Eqs.(6) and (7) with the variation in plate thickness and strength level of the materials.

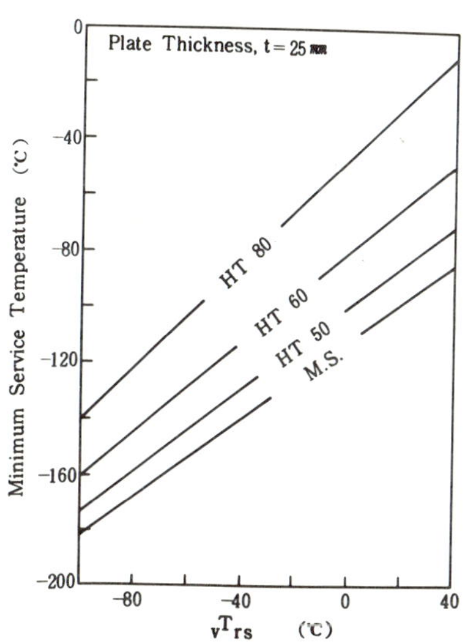


Fig. 16. Effect of strength level of materials on minimum service temperature.

## CONCLUSIONS AND RECOMMENDED FUTURE WORKS

It was confirmed that COD bend test is one of the most effective and useful test methods for the evaluation of notch toughness of structural steels since the critical COD obtained from bend test compares well with that obtained from tension test.

A proposal for toughness requirements in terms of Charpy impact test has been made with the limitation that the materials are not subjected to the effect of local stress elevation due to welding and/or structural discontinuity. This proposal is based on the correlation between transition temperature of COD bend tests and V Charpy transition temperature.

The COD concept, however, has still some difficulties in providing a useful and simple engineering design basis. One of the troubles associated with practical application of COD concept is the fact that the critical COD is to be taken as a quantity dependent not only on temperature but also on mechanical factors such as notch acuity, strain rate and stress triaxiality (plastic constraint) etc.

Another problem arises from the fact that the analytical expression of COD, as a function of flaw size and applied stress and/or strain not been obtained in the general yielding region in which the COD concept will be the one of most appropriate fracture criteria.

Despite the many problems left unsolved, the present authors believe that we should begin an attempt to draft a standard procedure for an engineering critical assessment on the significance of defects in welded steel structures by codifying vast available informations and experiences even if at the risk of oversimplifying the complex situations existing in the actual structures.

No doubt the task is not easy and the first step would include many unsatisfactory formulations, which must be improved by future researches and experiences in its application. A major step in this direction has now been taken by British research group [8]. Although the British proposal includes many controversial aspects, it provides us with many important suggestions.

## REFERENCES

[1] Practical Fracture Mechanics for Structural Steel, Proc. Symp. on Fracture Toughness Concept for Weldable Structural Steel, UKAEA, Risley, April, 1969

[2] Kanazawa, T., Machida, S. and Hagiwara, Y., Study on COD bend test as an engineering brittle fracture initiation characteristics test, J. Naval Arch. Japan, 132, 261 1972

[3] Bilby, B. A., Cottrell, A. H., Smith, E. and Swinden, K. H., Plastic yielding from sharp notches, Proc. Roy. Soc., 279 A, 1, 1964

[4] Alexander, J. M. and Komoly, T. J., On the yielding of rigid/plastic bar with an Izod notch, J. Mech. Phys. Solids., 10, 265, 1962

[5] Ingham, T., Egan, G. R., Elliot, D. and Harrison, T. C., The effect of geometry on the interpretation of COD test data, Practical Application of Fracture Mechanics to Pressure Vessel Technology, I.M.E., London, 1971, 200 (Paper C 54)

[6] Kanazawa, T., Mimura, H., Machida, S., Miyata, T. and Hagiwara, Y., ibid. 239 (Paper C 90)

[7] Kasamatsu, H., Takano, M. and Matsuoka, M., On the correlation between V Charpy impact test and Deep Notch Test, Paper presented at the Autumun Meeting of Japan Weld. Soc., 1972

[8] Harrison, J. D., Burdekin, F. M. and Young, J. G., A proposed acceptance standard for welded defects based upon suitability for service, Second. Conf. on Significance of Defects in Welded Structures, London, May, 1968

[9] Kihara, H. and Ikeda, K., On brittle fracture initiation (second report), J. Naval Arch. Japan, 118, 204, 1965

# THE USE OF SMALL, PRE-CRACKED SPECIMENS FOR QUALITY CONTROL PURPOSES

A. S. Tetelman, J. N. Robinson, I. Roman

University of California at Los Angeles, Los Angeles, California 90024 USA

## ABSTRACT

The difficulties involved in using valid ASTM $K_{Ic}$ specimens for quality control purposes are described. It is shown from tests on steels and titanium alloys that values of $K_{Ic}$ can be obtained at lower cost from small specimens. If fracture initiation occurs before general yield, $K_{Ic}$ can be determined from the load required to cause crack extension. If fracture occurs after general yield, $K_{Ic}$ can be determined from the critical plane strain crack tip opening displacement or from the notch root contraction.

The effect of a steeply rising plane strain R-curve can be corrected by using a newly developed R-extrapolation technique. Simplified versions of the methods involved are described whereby small specimens can be used to determine whether a material satisfies specifications for quality control. These simplified versions require little or no calculation for quality control-testing once master nomographs have been drawn up.

## INTRODUCTION

### Forward

The development of new high strength alloys, particularly for aerospace applications, has resulted in an increased consciousness of the need to design against catastrophic failure. One of the most useful fracture criteria from a design viewpoint is the critical plane-strain stress intensity factor, $K_{Ic}$. $K_{Ic}$ can be related to the stresses and defect sizes in a structure and is increasingly used as a parameter for characterizing material toughness, particularly in thick structures made from high strength materials. An inexpensive method is needed for determining whether a structural material has a sufficiently high value of $K_{Ic}$ to satisfy design and inspection requirements. In a production situation, this involves a program of quality control tests to check incoming material and perhaps, also, material that has gone through several fabrication processes. The objective of this report is to demonstrate that it is possible, and advantageous to use small precracked three-point bend specimens for quality control.

Problems with Large Specimens

The most obvious problem with the large specimen generally required to obtain ASTM valid $K_{Ic}$ values is the high cost. This expense is composed of the cost in material and machining necessary to produce large specimens, and the cost of the high-capacity testing machines required to conduct the tests. Furthermore, sampling of the toughness at various points in a billet is difficult because of the large quantity of material required.

Another disadvantage of using the ASTM E-399 Standard [1] stems from the crack growth resistance behavior exhibited by some materials. $K_{Ic}$ is defined by the E-399 standard as the stress intensity factor at 2% crack grwoth. The larger the specimen, the larger the absolute amount of crack growth corresponding to $K_{Ic}$. If the curve of plane strain crack growth vs. resistance (R-curve)* is similar to that shown in Fig. (1a), then increase in specimen size will not affect $K_{Ic}$, as well as a further increase in the cost of testing. These types of materials could pass a given fracture criterion if tested at one size and fail if tested at a smaller size.

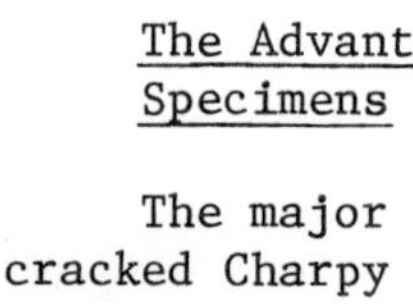


(a) Flat R-Curve

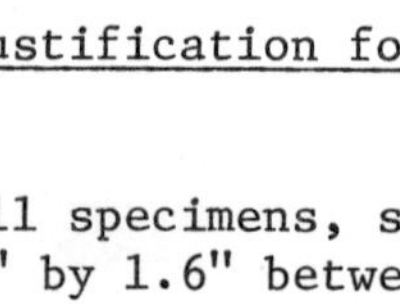


(b) Steeply Rising R-Curve

Figure 1. Schematic Representation of Extreme Types of Plane Strain R-Curve.

The Advantage of and the Justification for the Use of Small Specimens

The major advantage of small specimens, such as the fatigue cracked Charpy (0.394" by 0.394" by 1.6" between rollers), is the

---

*The "plane strain R-curve" does not result from shear lip formation and mixed mode fracture, as in the conventional definition (and cause) or R-curve phenomena. Rather, it appears a result from strain redistribution near the tip of a crack that grows stably under plane strain conditions, as described by McClintock and Irwin [25].

small cost of manufacturing, fatiguing and testing the specimens. Furthermore, it is possible to sample the toughness at various points in a billet and to obtain a measure of the statistical variation of toughness. Possible objections to small specimen testing are twofold. First, the proportion of the specimen in plain strain is smaller and a mixed-mode toughness may then result in an overly conservative (lower bound) value of $K_{Ic}$ from a small specimen.

The first objection, the smaller proportion of the specimen in plane strain, has been investigated by Robinson and Tetelman [2]. They have demonstrated that the thickness of the specimen showing transverse contraction, $B_{min}$, is given by

$$B_{min} \approx 25\ (COD) \tag{1}$$

where COD is the crack tip opening displacement in inches. Robinson and Tetelman [3,4] have experimentally shown that the COD can be converted to stress intensity factor, K, using Eq. (2)

$$K = \sqrt{\frac{E\ \sigma_Y\ COD}{1-\nu^2}} \tag{2}$$

where E is Young's modulus, $\sigma_Y$ is the yield stress and $\nu$ is Poisson's ratio. Figures (2) and (3) show the conversion between COD and $K/\sigma_Y$ for a variety of metals. For example, if $K_{Ic}/\sigma_Y = 0.5$, the plane strain on-load critical COD, $(COD)_{Ic}$, is 0.0015", and only 0.04" of the Charpy specimen is not in plane strain. Thus the central nine-tenths of such a specimen of Charph geometry (0.394" thick) is in plane strain at fracture initiation. The increase from initiation (taken as corresponding to 0.002" crack growth) stress intensity factor, $K_i$, to $K_c$ will occur as the shear lips develop after further crack growth. This was first demonstrated by Steigerwald and Hanna [5]. Their data obtained on AM355 stainless steel is shown in Fig. (4) and demonstrates the independence of $K_i$ on specimen thickness. $K_c$, however, varies considerably with change in specimen thickness.

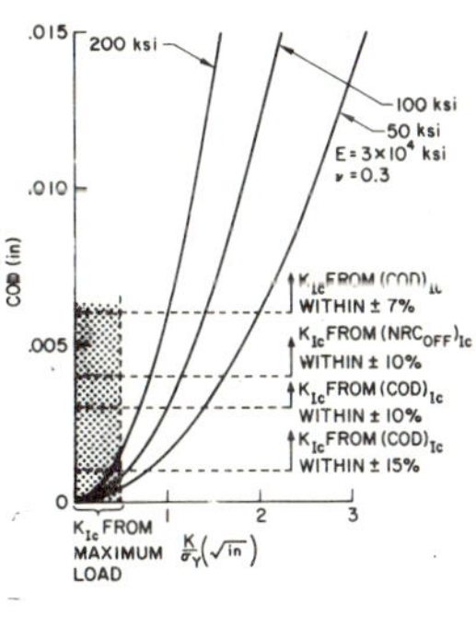


Figure 2. The Relationship Between COD and $K/\sigma_Y$ for Steel of Various Yield Strengths (Calculated from Equation (2)).

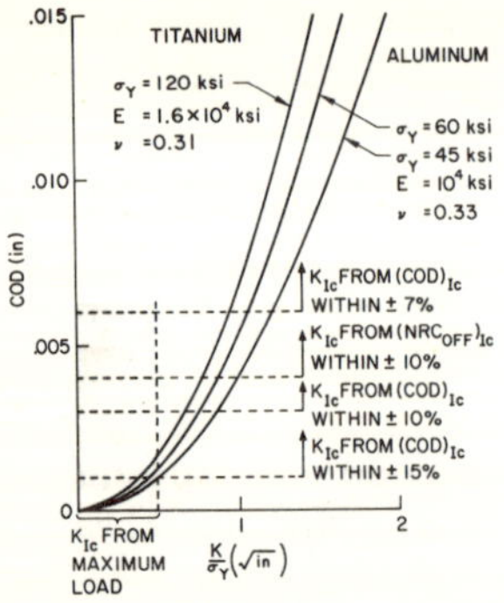


Figure 3. The Relation Between COD and $K/\sigma_Y$ for Aluminum and Titanium Alloys (Calculated from Equation (2)).

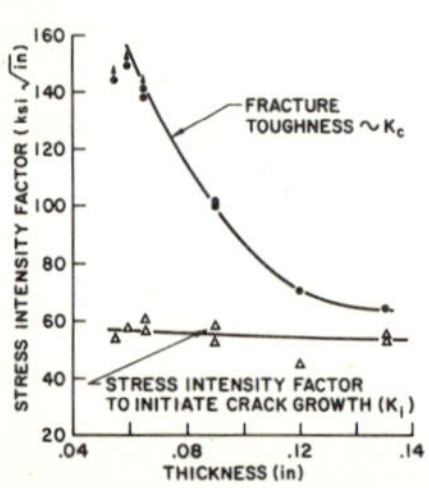


Figure 4. Influence of Specimen Thickness on Stress Intensity Parameters for AM 355 Stainless Steel (After Steigerwald and Hanna [5]).

The second objection to the use of small specimens, the small amount of crack growth prior to instability, is unimportant if the material shows an approximately level plane strain R-curve (Fig. (1a)), since toughness is not then dependent on amount of crack growth. If, however, the plane strain R-curve is similar to Fig. (1b), the toughness will depend on specimen size in the manner described in the section on large specimens. This objection can be overcome using a newly developed technique of correction for R-curve effects, a method we call "R-extrapolation". The technique is described in the following section. The method of determining $K_{Ic}$ from small bend specimens of pre-cracked Charpy geometry also depends on whether fracture initiation occurs before or after general yield. There are then four cases to consider:

(i) fracture initiation before general yield, level R-curve;
(ii) fracture initiation before general yield, rising R-curve;
(iii) fracture initiation at or beyond general yield, rising R-curve;
(iv) fracture initiation at or beyond general yield, rising R-curve;

These four cases are described in detail in the section on techniques for measuring $K_{Ic}$ on small specimens. Some practical examples are also given.

## TECHNIQUES FOR MEASURING $K_{Ic}$ ON SMALL SPECIMENS AND SOME PRACTICAL EXAMPLES

### Fracture Initiation Before General Yield

From linear elastic fracture mechanics [6], the value of the stress intensity factor for single edge cracked bend specimens is given by

$$K = \frac{6YMa^{1/2}}{BW^2} \tag{3}$$

where Y is a function only of (a/W). Values of Y are given in Ref. [6], a is the pre-crack length, M is the bending moment (calculated from the load), B is the specimen thickness and W is the specimen width.

Linear elastic fracture mechanics cannot be used when the specimen undergoes general yielding and becomes fully plastic (i.e., when $M \geqq M_{GY}$).

From slip line theory [7] the bending moment at general yield, $M_{GY}$, for a pre-cracked specimen of non-hardening material in three point bend, is given by

$$M_{GY} = 0.63\ \tau_Y\ (W-a)^2 B \tag{4}$$

where $\tau_Y$ is the shear yield stress. Combining Eqs. (3) and (4) and using the Tresca yield criterion, $\sigma_Y = 2\tau_Y$, we find that <u>at general yield</u>

$$K/\sigma_Y = \frac{1.89Ya^{1/2}\ (W-a)^2}{W^2} \tag{5}$$

For a pre-cracked specimen of Charpy geometry with a = 0.100", Eq. (5) becomes

$$K/\sigma_Y = 0.59\sqrt{\text{in.}} \tag{6}$$

Thus, if $K_i/\sigma_Y < 0.59$, fracture initiation occurs before general yield, if $K_i/\sigma_Y > 0.59$, fracture initiation occurs after general yield. The value of $K/\sigma_Y$ at general yield is not strongly dependent on crack length. For example, for a/W = 0.5, as suggested by Corten [8], Eq. (5) becomes

$$K/\sigma_Y = 0.52\sqrt{\text{in.}} \tag{7}$$

If fracture initiation occurs close to general yield, $K_i$ will be underestimated because yielding will reduce the load. A better definition of the region where a good approximation of $K_i$ can be determined from the load is probably

$$K_i/\sigma_Y \leqslant 0.5\sqrt{\text{in.}} \qquad (a/W < .5) \qquad (8)$$

For specimens of width different from the Charpy specimen, a similar relation can be obtained by combining Eqs. (5) and (8). For $K_i$ to be determined from the load

$$K_i/\sigma_Y \leqslant 0.5\sqrt{W/0.394}\ \sqrt{\text{in.}} \qquad (a/W < 0.5) \qquad (9)$$

The thickness, B, of a standard pre-cracked Charpy specimen is 0.394". Consequently, $B = 1.6\ (K_i/\sigma_Y)^2$ when Eq. (8) is satisfied at general yield. This thickness is close to the ASTM requirement $B \geq 2.5(K_I/\sigma_Y)^2$ for plane strain conditions.

Level R-Curve

Static Test. When the R-curve is level or rises only slightly with crack growth, $K_{Ic}$ for quality control purposes can be determined from the maximum load using Eq. (3). For example, ASTM standard fracture toughness tests on eutectoid rail steel showed pop-in [9] and gave a value of room temperature $K_{Ic}$ of 37 ksi$\sqrt{\text{in.}}$ The yield strength at the applicable strain rate was 86 ksi. $K_{Ic}$ determined from maximum load in tests on pre-cracked Charpy specimens was found to be 39 ksi$\sqrt{\text{in.}}$ The agreement is excellent. Similarly for 7075-T651 aluminum ($\sigma_Y$ = 60-66 ksi) a standard ASTM fracture toughness test gave a value of 28 ksi$\sqrt{\text{in.}}$ in the RW orientation and 21.6 ksi$\sqrt{\text{in.}}$ in the WR orientation [10]. $K_{Ic}$ determined from maximum load in tests on pre-cracked Charpy specimens was found to be 26 ksi$\sqrt{\text{in.}}$ in the RW orientation and 22 ksi$\sqrt{\text{in.}}$ in the WR orientation.

Dynamic Test. The fracture behavior of some materials, particularly low alloy steels, is strongly dependent on strain rate in and below the brittle/ductile transition range. In these instances, quality control tests can and should be conducted dynamically using an instrumented impact machine which records dynamic load/time curves [11,12]. The dynamic test gives a better measure of the materials' resistance to crack propagation than the static test, and thus provides a lower bound value of fracture toughness. In addition, the instrumented Charpy impact test on pre-cracked specimens is a simple extension of the standard Charpy V test presently used in quality control testing. It is therefore familiar to the technicians who conduct the tests. The dynamic test is also conducted more rapidly than the slow bend test, which is an important factor when many samples are to be run. Although most of the examples in the present paper are derived from slow bend tests, in principle there is no reason why the methods described in this paper cannot be extended to dynamic situations. Further research is, however, required on measurement techniques for dynamic COD values.

Some tests on A533B steel have been made by Server and Tetelman [11] on pre-cracked Charpy specimens. An instrumented impact hammer was employed to record dynamic load/time curves. The results are shown in Fig. (5). Figure (5) also shows $K_{Ic}$ data [13] from ASTM standard fracture toughness specimens up to 12" thick. Bearing in mind the higher strain rate used by Server and Tetelman [11] (see Fig. (5)), the agreement between their results and those from much larger fracture toughness specimens is very good.

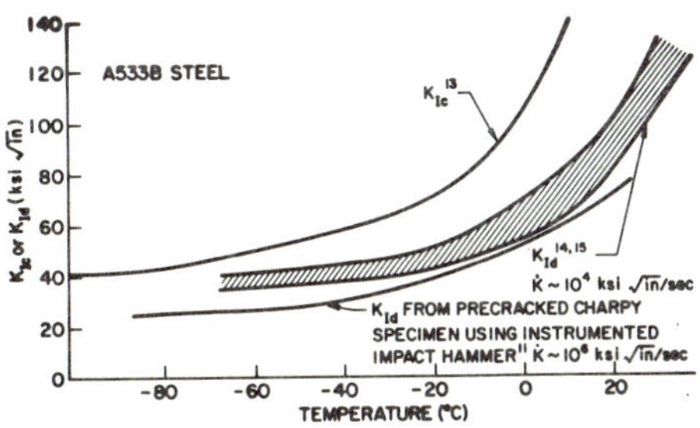


Figure 5. Comparison of Static, Dynamic, and Instrumented Pre-Cracked Impact Fracture Toughness as a Function of Temperature (After Server and Tetelman [11]).

## Rising R-Curve

If an R-curve is available for a sample of material in a class of alloys (e.g., 6Al-4V-Ti in a particular heat treatment), then it is possible to apply this curve to other samples of similar materials by using a scaling (ratio) technique. As an example, consider a material for which the R-curve is known, with a $K_{Ic}$ (for a particular size fracture toughness specimen equal to 90 ksi$\sqrt{\text{in.}}$ and a K corresponding to 0.002" crack grwoth, $K_{0.002"}$, equal to 60 ksi$\sqrt{\text{in.}}$ That is

$$K_{Ic} = 1.5\ K_{0.002"} \tag{10}$$

For a similar material where $K_{0.002"}$ (calculated from the load at 0.002" crack growth) was equal to 62 ksi$\sqrt{\text{in.}}$, then

$$K_{Ic} = 1/5 \times 62 = 93 \text{ ksi}\sqrt{\text{in.}} \tag{11}$$

If the original R-curve is approximated to a straight line, then the family of R-curves for similar materials of differing toughness will be shown in Fig. (6). This method of R-curve correction by a scaling technique (R-extrapolation) is experimentally justified below.

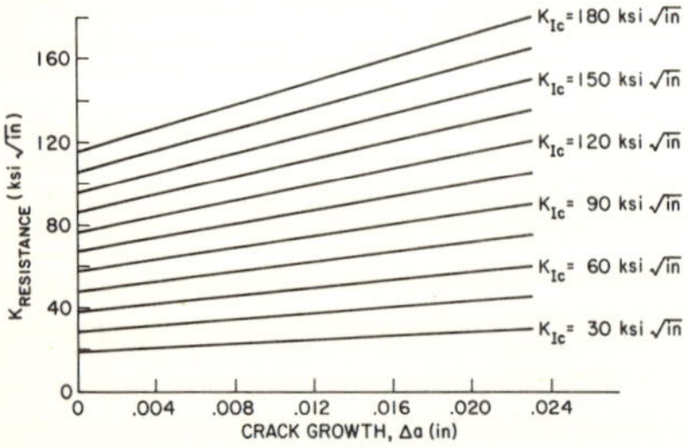


Figure 6. A Family of R-curves for use in the R-extrapolation Technique.

The material used to verify the method was 6Al-4V titanium alloy in 11 different conditions, supplied by Rockwell International. This material has been reported to show a $K_{Ic}$ which depends on the amount of crack growth and thus on the ASTM standard specimen size [16]. For all but one of these conditions (the wrought material) the condition and $K_{Ic}$ of the samples were unknown to us at the time of the tests. The purpose of the program was to determine whether $K_{Ic}$ values could be obtained from the ten small specimen tests that would agree with valid data provided by Rockwell suppliers.

It is necessary to know the stress intensity factor corresponding to a particular amount of crack growth in order to use the R-extrapolation technique. The electrical potential technique was used to detect the onset of cracking (described in the section titled Techniques for Measuring $K_{Ic}$ on Small Specimens and Some Practical Examples - Fracture Initiation at or Beyond General Yield). The specimen was then unloaded and the double bend technique used to accurately measure the amount of crack growth. The specimen was fatigued to failure. Any ductile rupture occurring in the first bending operation appears dark compared to the light appearing fatigue cracks. The extent of ductile rupture was measured using a traveling microscope calibrated in units of 0.004".

An R-curve was determined from compact tension and from precracked Charpy specimens of the wrought material. Amounts of crack growth were found using the double bend technique. K for the Charpy specimens was found from the COD (using Eq. (2)) and from the load (Eq. (4)). The $K_{Ic}$ for this material, 79 ksi$\sqrt{\text{in.}}$, determined on a 1-1/8" thick specimen and thus corresponding to approximately 0.023" crack growth, was supplied by Rockwell International. The resultant R-curve is shown in Fig. (7). There are insufficient points to determine the exact shape of the R-curve but a straight line approximation gives a good fit to the data.

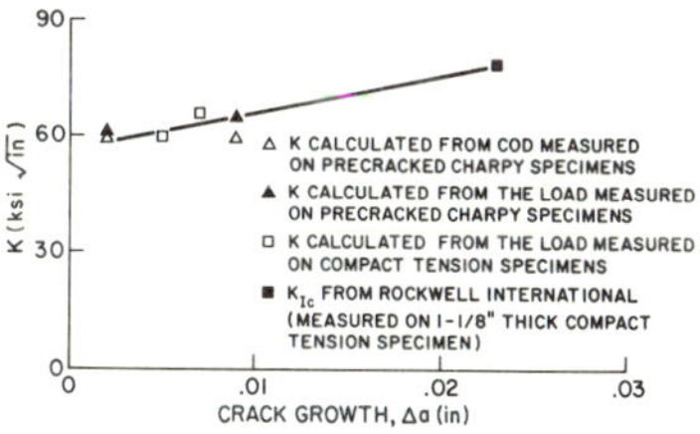


Figure 7. R-curve for 6Al-4V Wrought Titanium.

One pre-cracked Charpy specimen of each of the ten "unknown" samples was tested in three point bending to predict the $K_{Ic}$ value. With the exception of specimen 10, each specimen was bent until the deviation from the initially linear clip guage displacement vs. potential curve was from about 2 to 6% of the initial potential. This ensured that some crack growth had occurred. In specimen 10, crack growth occurred very suddenly and an approximately 10% deviation was obtained before the test was stopped. The actual amount of crack growth was determined using the double bend technique, except for specimen 10. Specimen 10, because of its very low load to failure, was inadvertently fractured completely in the first cycle of the subsequent fatigue. Specimens 1-9 were found to have undergone some crack propagation.

Three of the tests were deemed unsatisfactory at this state. Specimen 1 showed a fracture surface that was difficult to interpret. There appeared to have been two separate regions of ductile crack propagation. Specimen 9 showed a highly asymmetric fatigue pre-crack, almost nil at one surface. Specimen 10 was deemed unsatisfactory for the reasons listed above. If further specimens had been available, these tests would have been repeated. In the light of experience gained it should be possible to avoid repeating these mistakes.

All specimens showed fracture initiation prior to general yield. The K corresponding to the measured amounts of crack propagation was calculated from the load. The K values were corrected to $K_{Ic}$ by R-extrapolation. A sample of calculation is shown below.

<u>Sample 3</u>

K from load for 0.005" crack growth = 68 ksi$\sqrt{\text{in.}}$
K for wrought material with 0.005" crack growth (from Fig. (7))

$$= K_{Ic}/1.29$$

Thus, $K_{Ic}$ for Sample 3 = 68 x 1.29 ksi$\sqrt{\text{in.}}$

= 88 ksi$\sqrt{\text{in.}}$

Following these calculations, $K_{Ic}$ data, obtained on one ASTM standard specimen for each material, was furnished by Rockwell International. Rockwell had obtained this $K_{Ic}$ data from its material suppliers. The results of the experiments and calculations are shown together with this $K_{Ic}$ data, in Table 1.

For seven of the specimens, Nos. 2-5 and 7-9, the $K_{Ic}$ determined from the load is in excellent agreement with $K_{Ic}$, particularly considering that the results are based on only one specimen. For specimens 1 and 10, the need for repeat tests was indicated prior to calculating the $K_{Ic}$ values. For specimen 7, the difference is 14%. This difference is greater than normal scatter in $K_{Ic}$ tests ( ~ ± 5 - 10%) and could be improved. The valid $K_{Ic}$ for this material is based on only one specimen, however, and may be on the high side of the scatter band. It should be noted that specimen 6 would still correctly indicate that the material passed the required $K_{Ic}$ value of 75 ksi$\sqrt{\text{in.}}$, and specimen 10 would correctly indicate that the material failed the requirement. Thus, in 9 out of 10 cases, the small specimen tests were in agreement with pass/fail criteria obtained from valid ASTM specimens.

TABLE 1

Results of the $K_{Ic}$ Measurements of the Ten "Unknown Samples

| Material Number | Yield Stress ksi | Crack Growth Inches, | K for indicated amount of crack growth, ksi$\sqrt{\text{in.}}$ | $K_{Ic}$ (for 0.023" crack growth) by R-extrapolation, ksi$\sqrt{\text{in.}}$ | $K_{Ic}$ from Material Suppliers, ksi$\sqrt{\text{in.}}$ |
|---|---|---|---|---|---|
| 1 | 123 | 0.002 | 62 | 84 | 62 |
| 2 | 122 | 0.004 | 74 | 97 | 93 |
| 3 | 122 | 0.005 | 68 | 88 | 86 |
| 4 | 120 | 0.002 | 62 | 84 | 77 |
| 5 | 130 | 0.012 | 54 | 62 | 59 |
| 6 | 121 | 0.010 | 66 | 79 | 92 |
| 7 | 124 | 0.011 | 62 | 72 | 73 |
| 8 | 125 | 0.008 | 53 | 65 | 61 |
| 9 | 118 | 0.008 | 69 | 85 | 82 |
| 10 | 139 | - | 30 | ~30 | 40 |

These results indicate that pre-cracked Charpy specimens are satisfactory for $K_{Ic}$ prediction. Where fracture initiation occurs before general yield, K for a particular amount of crack growth can be calculated from the load and corrected to $K_{Ic}$ by R-extrapolation. These small specimens have the advantage of being economical on material and less costly to machine and test.

Fracture Initiation at or Beyond General Yield

Four parameters have been described in the literature as being satisfactory for the prediction of $K_{Ic}$ from small specimens fractured after general yield: the critical J integral [17,18] the critical equivalent energy, [19,20], the critical COD [4] and the critical notch root contraction (NRC) [21]. The equivalent energy approach has been shown [21] to overestimate $K_{Ic}$ by up to 40% from tests on small three point bend specimens of A533B steel. The techniques may only be applicable to large (and thus costly) compact tension specimens. This renders it unsuitable for quality control purposes. Measurements of the critical J integral using the compliance technique of Begley and Landes [17] predict $K_{Ic}$ numbers of specimens (normally at least six), is exceedingly time-consuming and tedious to use, and is therefore unsuitable for quality control. Recently, Landes and Begley [22] have described a simplified technique for measurement of the J integral requiring fewer specimens. This technique has been found, however, to underestimate $K_{Ic}$ by 16%, at least for tests on small three point bend specimens of 4340 steel [21]. The present paper describes the successful use of the COD and NRC methods for determining $K_{Ic}$ from small specimens broken beyond general yield, and the application of these methods to quality control testing.

When $K_i/\sigma_Y > 0.5\sqrt{in.}$, $K_i$ can be determined from the on-load, midsection COD at crack initiation, $(COD)_i$, using Eq. (2). $(COD)_i$ placement (CGD) using Eq. (12) [4,21] which is easily solved by standard computer programs.

$$B_0 + B_1\,COD + B_2COD^2 + B_3COD^3 + B_4COD^4 = 0$$

$$B_0 = 0.04271\ CGD,$$

$$B_1 = 0.09391\ CGD - \left(\frac{a + z}{W - a}\right) - 0.04271,$$

$$B_2 = -\ 0.009313\ CGD - 0.09391$$

$$B_3 = 0.0003678\ CGD + 0.009313 \text{ and}$$

$$B_4 = -\ 0.003678.$$

where COD and CGD are in thousandths of an inch.

In the above relations, W is the width of the band specimen (0.394" for Charpy geometry), a is the crack depth and z is the height of the knife edges above the specimen surface.

The onset of fracture initiation is determined by the double bend technique, described in detail by Robinson and Tetelman [3,4]. Briefly, the teçhnique consists of bending the specimen to a known clip guage displacement and unloading it. Any crack propagation occurring in this first bending operation can be measured by break ing the specimen in a regime where the fracture mode is different For example, if crack initiation in slow bend occurs by ductile rupture, the specimen can be fatigued to failure in the second bending operation. Any ductile rupture on the fracture surface indicates crack initiation in the first slow bend. A plot of COD (calculated form clip guage displacement) versus crack growth for three or four specimens allows initiation (defined as 0.002" crack growth) COD to be determined.

A useful adjunct to the double bend technique is the electrical potential technique. This method is based on detecting a change in cross-sectional area of the specimen due to crack growth, by measuring the electrical resistance of the specimen. The method does not work satisfactorily, however, with materials that show a very fibrous fracture surface, presumably because there is not complete separation between the crack surfaces near the crack tip. A circuit similar to that shown in Fig. (8) is set up and a curve of potential difference versus clip guage displacement is plotted. Experience with a particular class of materials indicates what derivation from the initial straight line plot of potential versus CGD corresponds to crack initiation. The actual amount of crack growth can be found using the double bend technique. The electrical potential technique is useful in studying specimens having close to the required 0.002" crack growth for measurement of $(COD)_{Ic}$ and is also useful in detecting crack initiation for the R-extrapolation method.

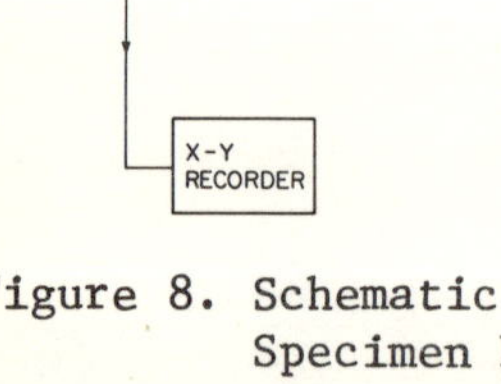


Figure 8. Schematic Representation of Circuit for Measurements of Specimen Resistance.

Another method of measuring $(COD)_i$ is the crack profile technique, described in detail in Ref. [4]. Briefly, it consists of infiltrating the propagated crack under load with a catalytically hardening silicone rubber. After the rubber has hardened into a casting of the crack, the specimen is broken. The casting is removed, sectioned at midsection and examined in the scanning electron microscope. The profile appears similar to that shown in Fig. (9). $(COD)_i$ is given by *PQ-RS*. The crack profile method has the advantage over the double bend technique that only one specimen and almost no calculation are required. The results may, however, be expected to be less accurate because the method involves an element of judgment.

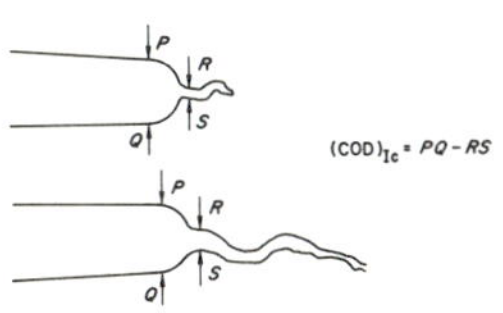


Figure 9. Schematic Representation of Castings of the Propagated Crack used for the Crack Profile Method.

A third method of measuring $(COD)_i$ is by means of the off-load notch root contraction, $NRC_{off}$ using Eq. (13) [21]

$$COD = 1.27\ NRC_{off}. \tag{13}$$

Determination of COD from $NRC_{off}$ has the advantage that $NRC_{off}$ is very easily measured with equipment no more complex than a pointed-ended micrometer. The point of initiation is determined using the double bend technique.

From Eq. (1), the $(COD)_i$ measured at the midplane of a precracked Charpy specimen (0.394" thick) is a plane strain value provided $(COD)_i \leq 0.016"$. Figs. (2) and (3) show that Charpy geometry is usable up to $K/\sigma_Y \leq 2$ for typical steel, aluminum and titanium alloys. Figures (2) and (3) also show the limits on COD and $K/\sigma_Y$ at which measurement of $K_{Ic}$ from $(COD)_{Ic}$ (from clip guage displacement) and $(NRC_{off})_{Ic}$ are accurate to within given error bands. It can be seen that the techniques are most satisfactory down to COD values of 0.003" but can be used down to COD = 0.001" with lower accuracy. If the $(COD)_i$ value is less than 0.001", then fracture will occur before general yield anyway (except for low strength steels) and $K_{Ic}$ can be obtained from the load, as described in the section entitled Techniques for Measuring $K_{Ic}$ on Small Specimens and Some Practical Examples - Fracture Initiation Before General Yield. Consequently, with the exception of low strength steel, there is no range of $(K_i/\sigma_Y)$ where the material is too tough for fracture initiation to occur before general yield, but too brittle for measurements of $(COD)_i$ to be able to predict $K_i$ to within $\pm$ 15%.

For lower strength steel it may be better to use Charpy

specimens of greater width, such that general yield occurs at higher $(K/\sigma_Y)$ values than given by Eq. (6). For example, for a steel of 50 ksi yield stress, Eq. (9) indicates that the upper limit of determination of $K_{Ic}$ from the load and the lower limit of determination of $K_{Ic}$ from the critical COD are coincident when the specimen width is approximately 1", two and one-half times the standard Charpy width.

<u>Level R-Curve</u>. When the R-curve is level or rises only slightly with crack growth, $K_{Ic}$ for quality control purposes can be determined from $(COD)_i$.

Determinations of $K_{Ic}$ from $(COD)_{Ic}$ for A533B steel at various temperatures and for 4340 steel at various yield strength levels have been reported in Ref. [4]. The results are shown, together with valid $K_{Ic}$ results, in Figs. (7) and (10). The agreement is very good and no significant R-curve effect is apparent. $(COD)_{Ic}$ values were determined both from the clip guage displacement, using Eq. (12), and the double bend technique, and from the crack profile method (see section on fracture initiation at or beyond general yield in section entitled Techniques for Measuring $K_{Ic}$ on Small Specimens and Some Practical Examples). Figures (10) and (11) show that determination of $(COD)_{Ic}$ from either the clip guage displacement or the crack profile method gives very similar results. Determination of $(COD)_{Ic}$ from $(NRC_{off})_{Ic}$ or the clip guage displacement using the double bend technique, requires three or four specimens to pinpoint initiation. The crack profile method has the advantage that only one specimen and almost no calculation are required, but the results may be expected to be less accurate because the method involves an element of judgment.

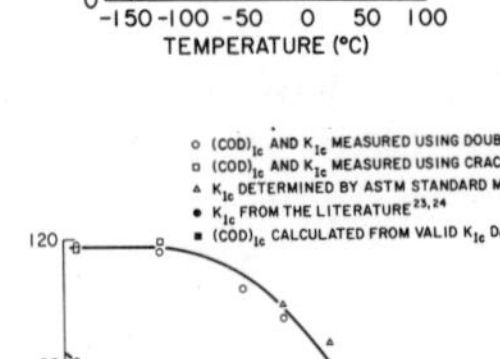


Figure 10. $K_{Ic}$ Calculated from $COD_{Ic}$ for A533B Steel Superimposed on the Scatter Band for HSST Valid $K_{Ic}$ Results (Shaded Area).

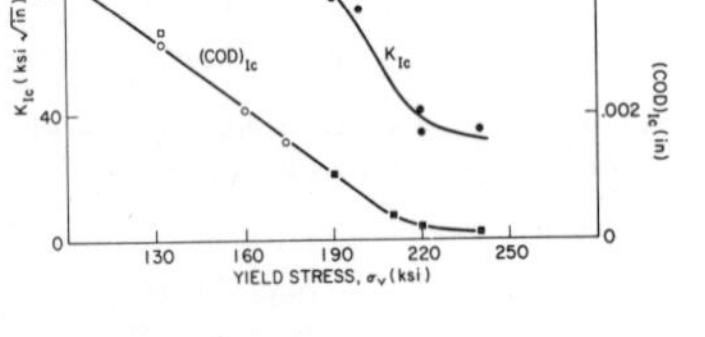


Figure 11. Variation of $(COD)_{Ic}$ and $K_{Ic}$ with Yield Stress for Quenched and Tempered 4340 Steel.

<u>Rising R-Curve</u>. $K_i$ from $(COD)_i$ can be corrected to give $K_{Ic}$ using the R-extrapolation technique described in the section entitled Techniques for Measuring $K_{Ic}$ on Small Specimens and Some Practical Examples - Fracture Initiation Before General Yield - Rising R-Curve. No experimental results are presently available for this case, (i.e., fracture occurs after general yield).

## APPLICATION TO QUALITY CONTROL

The method of testing small specimens can be greatly simplified when it is only required to determine whether or not a particular sample meets a specification of, say $K_{Ic} \geq 75$ ksi$\sqrt{in.}$ (for a particular ASTM standard specimen size). It is first necessary to run preliminary experiments on a sample of material to determine which of the four cases is applicable to the particular class of alloys. Many of the techniques given require only one specimen to characterize a material as satisfactory or unsatisfactory. It is, however, inadvisable to rely on the use of one specimen only, because of the uncertainties inherent in any test technique. At least two, or preferably three, specimens of each material should be tested before it is accepted or rejected, particularly if the value is borderline. Costly rejection of acceptable material, or acceptance of inadequate material, may otherwise result.

## ACKNOWLEDGEMENTS

We are grateful for financial support from Rockwell International under P.O. # L3P1-DM-152500-283x and from the U.S. Army Research Office-Durham under contract Number DAHC04-69-C-0008. We would also like to thank Dr. A. W. Sommer for helpful discussions and D. O. Cox for his valuable aid and advice.

## REFERENCES

[1] ASTM, Standard Method of Test for Plane-Strain Toughness of Metallic Materials, E399-72, *1972 Annual Book of ASTM Standards*, 31, ASTM, Philadelphia, Pa., 955, 1972.

[2] Robinson, J.N. and Tetelman A.S. to be submitted to *Engr. Frac. Mech.*

[3] Robinson, J.N. and Tetelman.A.S., presented at the Third Int'l. Conf. on Frac., Munich, Germany, 1973.

[4] Robinson, J.N. and Tetelman, A.S., presented at the 7th Natl. Sym. on Frac. Mech., Univ. of Md., College Park, Md. 1973.

[5] Steigerwald, E.A. and Hanna, G.L., *Proc. ASTM*, 62, 885, 1962.

[6] Brown, W.F. and Sprawley, J.E., *Plane-Strain Crack Toughness Testing of High Strength Metallic Materials, ASTM STP 410*, ASTM, Philadelphia, Pa., 1966.

[7] Green, A.P., *Quart. J. Mech. and Applied Math.*, 6, 223, 1953.

[8] Corten, H.T., Private communication.

[9] Enshah, S. and Tetelman, A.S., Report to Assn. Amer. RR, 1972.

[10] Reimer, P.G., M.Sc. Thesis, Univ. of Calif., Los Angeles, 1971.

[11] Server, W.L. and Tetelman, A.S., *Engr. Frac. Mech.*, 4, 367, 1972.

[12] Wullaert, R.A., Ireland, D.R., and Tetelman, A.S., Presented at the Frac. Prevention and Control Sym., WESTEC 72 Meeting, Los Angeles, Calif., 1972.

[13] Shabbits, W.O., Pryle, W.H., and Wessel, E.T., Westinghouse Elec. Corp. Pittsburgh, Pa., WCAP-74, 1969.

[14] Crosley, P.B. and Ripling, E.J., Mat. Research Lab., Richton Park, Ill., HSST-TR-8, 1970.

[15] Shabbits, W.O., Heavy Sec. Steel Tech. Program 5th Annual Information Meeting, 1971.

[16] Sommer, A.W., Private communication.

[17] Begley, J.A. and Landes, J.D., ASTM STP 514, Philadelphia, Pa., 1-20, 1972.

[18] Landes, J.D. and Begley, J.A., ASTM STP 514, Philadelphia, Pa., 1972.

[19] Witt, F.J. and Mager, T.R., Oak Ridge Nat'l. Lab, Oak Ridge, Tenn., ORNL-TM-3894, 1972.

[20] Witt, F.J., Conf. on Prac. Appl. of Frac. Mech. to Pressure Vessel Tech., London, 1971, Inst. Mech. Engrs., London, 163, 1971.

[21] Robinson, J.N. and Tetelman, A.S., to be submitted to *Engr. Frac. Mech.*

[22] Landes, J.D. and Begley, J.A., presented at the 7th Nat'l. Conf. on Frac. Mech., College Park, Md. 1973.

[23] Robinson, J.N. and Tuck, C.W., *Engr. Frac. Mech.*, 377, 1972.

[24] Pellissier, G.E., *Engr. Frac. Mech.*, 1, 55, 1968.

[25] McClintock, F.A. and Irwin, G.R., *Fracture Toughness Testing and its Application, ASTM STP 381*, Philadelphia, Pa., 84, 1965.

# SLOW CRACK GROWTH AND FAILURE PREDICTION OF BRITTLE MATERIALS SUBJECTED TO A TRANSIENT THERMAL STRESS

R. Badaliance, P. D. Hilton and D. P. H. Hasselman

Institute of Fracture and Solid Mechanics and Ceramics Research Laboratory, Lehigh University, Bethlehem, Pa. 18015 U.S.A.

## ABSTRACT

The effect of slow crack growth on the failure conditions of a brittle material subjected to a transient thermal stress in a stress-corrosive environment was investigated. A numerical analysis procedure was developed for calculating the instantaneous values for crack length and transient stress intensity factors of a glass rod subjected to a sudden decrease in ambient temperature under convective heat transfer conditions. The analysis demonstrated that slow crack growth significantly decreases the thermal conditions required for catastrophic failure. A form of crack growth instability is predicted at values of the stress intensity factor well below $K_{IC}$, where the rate of increase of stress intensity factor is such that catastrophic failure becomes inevitable.

## I. INTRODUCTION

The catastrophic failure of brittle materials subjected to severe thermal stress is an important technological problem. Many brittle materials in stress-corrosive environments exhibit a stage of subcritical or slow crack growth prior to catastrophic failure, which is responsible for the well-known strain rate dependence of tensile strength and the phenomenon known as static fatigue or delayed fracture. To date, in the failure prediction for brittle materials subjected to thermal stress, the stage of slow crack growth prior to fracture has been ignored [1-5]. However, the strain-rate sensitivity and delayed fracture behavior of brittle materials can be predicted successfully from independently measured rates of slow crack growth [6].

This paper reports a procedure for predicting thermally induced failure of brittle materials which exhibit slow crack growth. The approach is distinct from earlier failure predictions in that it accounts for the possibility of stable crack growth in a stress corrosive environment. It assumes a knowledge of the crack velocity as a function of the stress intensity factor (K-V curves). The transient temperature distribution and resulting stress field are calculated. From the known initial flaw size the corresponding stress intensity factor is determined. The K-V data are used to estimate the crack growth in an increment of time and the stress intensity factor is recalculated for the extended crack and the current stress distribution. This incremental procedure is continued, yielding crack size and stress intensity factor as a function of time, initial conditions, and material parameters.

## II. ANALYTICAL AND NUMERICAL PROCEDURE

The specific specimen geometry and heat transfer conditions chosen consisted of a long circular rod under conditions of generalized plane strain subjected to radial heat flow as the result of an instantaneous decrease in ambient temperature with conductive heat transfer in the solid and convective heat transfer in the surrounding medium. A decrease in ambient temperature results in a maximum tensile stress in the specimen surface and insures the maximum effect of the stress-corrosive environment on crack propagation of surface cracks. The length of the surface crack was assumed to be large in comparison with its depth, with both dimensions taken to be small relative to the cylinder radius such that the crack configuration is closely approximated by an edge-crack in a semi-infinite solid. This assumption also implies that the crack depth is not sufficient to affect the rod compliance. For simplicity it was also assumed that the crack did not affect the transient temperature field and for practical purposes it was subjected to a uniform thermal stress corresponding to those values calculated to occur at the rod surface.

For the above heat transfer condition, the heat flow equation in cylindrical coordinates is:

$$\nabla^2 T(r,t) = \frac{1}{\kappa} \frac{\partial T(r,t)}{\partial t} \tag{1}$$

The initial and boundary conditions are

$$T(r,t) = T_1 \quad \text{at} \quad t \leq 0$$

$$\frac{\partial T(r,t)}{\partial r} = 0 \quad \text{at} \quad r = 0 \text{ for } t \geq 0$$

and the convective boundary condition is

$$k\,\frac{\partial T(r,t)}{\partial r} = h[T_0 - T(R,t)], \text{ at } r = R,\ t > 0.$$

Solution of eq. (1) subjected to the above condition yields the following relationship for the transient temperature at the surface of the rod [7]:

$$T(R,t) = T_0 + (T_1 - T_0)\sum_{n=1}^{\infty} \frac{2\beta_n J_1(\beta_n)}{(m^2+\beta_n^2)J_0(\beta_n)} \exp(-\beta_n^2 \varkappa t/R^2) \qquad (2)$$

where:

$R$ is the cylinder radius

$T_0$ is the temperature of quenching medium

$T_1$ is the initial uniform temperature of the cylinder

$m$ is Biot's modulus with $m = Rh/k$

$h$ is the heat transfer coefficient

$k$ is the thermal conductivity

$\varkappa$ is the thermal diffusivity with $\varkappa = k/\rho c$

$\rho$ is the density

$c$ is the specific heat

$t$ is the time

$J_0$ and $J_1$ are the Bessel functions of the zero and first order

$\beta_n$ is the root of

$$m\,J_0(\beta_n) - \beta_n J_1(\beta_n) = 0, \quad n = 1,2,3,\ldots \qquad (2b)$$

For a cylindrical rod under conditions of generalized plane strain, the thermal stress distribution is given by [8]:

$$\sigma_r(r,t) = \frac{E\alpha}{(1-\nu)}\left[\frac{1}{R^2}\int_0^R T(r,t)r\,dr - \frac{1}{r^2}\int_0^r T(r,t)r\,dr\right]$$

$$(3)$$

$$\sigma_\theta(r,t) = \frac{E\alpha}{(1-\nu)}\left[-T(r,t) + \frac{1}{r^2}\int_0^r T(r,t)r\,dr + \frac{1}{R^2}\int_0^R T(r,t)r\,dr\right]$$

$$\sigma_z(r,t) = \sigma_r(r,t)+\sigma_\theta(r,t) = \frac{E\alpha}{(1-\nu)}\left[\frac{2}{R^2}\int_0^R T(r,t)r\,dr - T(r,t)\right]$$

where E is the Young's modulus, $\nu$ is the Poisson's ratio and $\alpha$ is the coefficient of thermal expansion.

By substitution of the transient temperature distribution in the rod into eq. (3), the transient thermal stress at the surface of the rod can be derived to be:

$$\sigma_r(R,t) = 0$$

$$\sigma_\theta(R,t) = \sigma_z(R,t) = \frac{E\alpha(T_1-T_0)}{(1-\nu)}\left\{2\sum_{n=1}^{\infty}\frac{J_1(\beta_n)}{(m^2+\beta_n^2)J_0^2(\beta_n)}\times\left[2J_1(\beta_n)-\beta_n J_0(\beta_n)\right]\exp(-\beta_n^2\,t/R^2)\right\} \quad (4)$$

Equation (4) indicates that at the surface of the cylinder the shear stress components are zero and the hoop and longitudinal stresses are equal. Consequently, within the assumption of small flaw depth the orientation of the crack is of no consequence.

The stress intensity factor $K_I$ for a surface flaw is given [9] by

$$K_I = 1.1215\,\sigma\sqrt{\pi a} \quad (5)$$

where a is the crack depth and $\sigma$ is the magnitude of uniform stress applied on the crack surface. Substitution of eq. (4) into (5) gives the time dependent stress intensity factor in terms of material properties, heat transfer conditions and crack depth. Since rates of slow crack growth generally are much lower than rates of crack propagation during catastrophic fracture, eq. (5) can be used directly without accounting for inertia effects, i.e. the problem is assumed to be a quasi-static one.

For brittle materials which exhibit slow crack growth, the crack length varies with time. The rate at which the crack length increases is dependent upon the environment, material properties, geometry, and loading conditions. A number of investigators [6,10] have suggested that for a given environment, the stress intensity factor and the temperature can be used to estimate the crack velocity for a particular material. It is assumed here that the instantaneous crack velocity is dependent on only the stress intensity factor and the transient temperature, i.e.,

$$\frac{da}{dt} = v = v(k_I, T(t)) \tag{6}$$

Substitution of eqs. (4) and (5) into this expression yields an equation for the crack velocity, v, as a function of time, t, and instantaneous crack depth a(t). Integration of this equation results in a nonlinear integral equation for crack depth as a function of time indicated symbolically as

$$a(t) = a_0 + \int_0^t v[t,a(t)]dt \tag{7}$$

An incremental computational procedure was developed to solve eq. (7) for the crack depth at any time increment and then to determine the corresponding stress intensity factor using eq. (5). For this purpose, the duration of the transient thermal stress pulse was divided into small time increments $(\Delta t)_i$ within which the stress and temperature were assumed to be constant. For a given time interval $\Delta t_i = t_{i+1}-t_i$, the temperature, thermal stress and stress intensity factor were calculated at time $t_i$ and corresponding crack depth $a_i$. Eq. (6) was then used to obtain the crack velocity at this time. The crack depth at time $t_{i+1}$ was estimated as

$$a_{i+1}^{(1)} = a_i + V(t=t_i)x(t_{i+1}-t_1)$$

where the superscript (1) refers to the first estimate for $a_{i+1}$. An iterative procedure was employed to improve this estimate, i.e., an average value for the stress intensity factor was determined at $a = [a_i + a_{i+1}^{(p)}]/2$, from which the new estimate for the crack velocity in this time interval was calculated. The new estimate for $a_{i+1}$ is:

$$a_{i+1}^{(p)} = a_i + V^{(p)}(t=t_i)x(t_{i+1}-t_i)$$

This iterative procedure was continued until the convergence condition

$$[K_{i+1}^{(p+1)}]/[K_{i+1}^{(p)}] < 1.0001$$

was satisfied. The computational procedure was then carried out for the next time interval. Failure was assumed to occur when $K_I = 0.9999K_{IC}$, at which point the calculation was terminated automatically.

The program as outlined above was used to predict the critical quenching temperature difference required for catastrophic thermal stress fracture of circular rods of soda-lime-silica glass subjected to a quench from a higher temperature into a water bath at 33°C. A soda-lime-silica glass is particularly sensitive to stress corrosion in a water environment. As shown by Wiederhorn and Bolz [10] for a number of glasses in water, the rate of slow crack growth (V) as a function of stress intensity factor can be expressed:

$$V = V_0 \exp[(-E^* + bK_I)/RT] \quad (8)$$

where $V_0$ is the pre-exponential factor, $E^*$ is the activation energy, b is a constant, R is the gas constant and T is the absolute temperature. Since for the present problem the crack size is very small compared to radius of the rod, the temperature T in eq. (8) was taken to be the instantaneous surface temperature. Further, eq. (8) was assumed valid up to the value of critical stress intensity factor, $K_{Ic}$, required for catastrophic failure. Also, as indicated by experimental data [10] for a soda-lime-silica glass no crack growth occurs when the stress intensity factor is below a value $K_0$, referred to here as the "fatigue limit". This value $K_0$ was chosen to represent the lower limit of the range of validity of eq. (8) and is shown in fig. (1). Also, since their temperature dependence is not sufficiently well established at this time, $K_0$ and $K_{Ic}$ were assumed to be constant

TABLE I

Data for Physical Properties, Crack Growth and Heat Transfer for Soda-Lime-Silica Glass Subjected to a Water Quench

|  |  |
|---|---|
| **Physical Properties:*** | |
| Coefficient of Thermal Expansion ($\alpha$) | $9.3\times10^{-6}$ °C$^{-1}$ |
| Young's Modulus of Elasticity (E) | $6.9\times10^{10}$ NM$^{-2}$ |
| Poisson's Ratio ($\nu$) | 0.25 |
| Thermal Conductivity (k) | $2.5\times10^{-3}$ cal. °C$^{-1}$cm$^{-1}$sec$^{-1}$ |
| Specific Heat (c) | 0.22 cal. gram$^{-1}$ °C$^{-1}$ |
| Density ($\rho$) | 2.53 gram cc$^{-1}$ |
| **Crack Growth Behavior:+** | |
| Fatigue-Limit ($K_0$) | $2.49\times10^5$ NM$^{-3/2}$ |
| Critical Stress Intensity Factor ($K_{Ic}$) | $7.49\times10^5$ NM$^{-3/2}$ |
| Pre-exponential Factor, $\ln V_0$ | 10.3 |
| Velocity Constant (b) | 0.11 MKS units |
| Activation Energy ($E^*$) | $1.088\times10^5$ J. mole$^{-1}$ |
| Initial Crack Length ($a_I$) | $5\times10^{-6}$ and $10\times10^{-6}$ M. |
| **Heat Transfer Data:** | |
| Temperature of Water Bath ($T_o$) | 33 °C |
| Cylinder Radius (R) | 0.236 cm. |
| Heat Transfer Coefficients** (h) | 0.13 & 1.0 cal cm$^{-2}$ °C$^{-1}$sec$^{-1}$ |

* from E. B. Shand, Glass Engineering Handbook, McGraw-Hill (1958)

\+ from reference [10]

** from references [11] and [12]

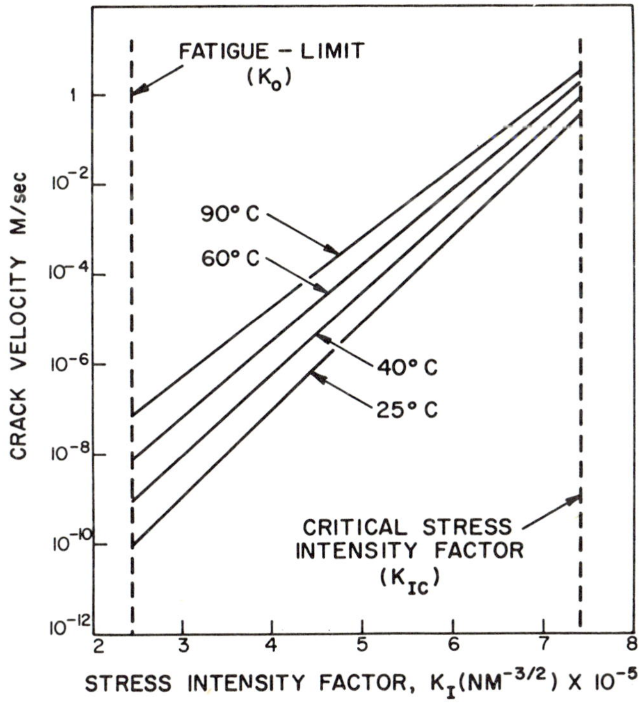


Figure 1. Crack growth behavior of soda-lime-silica glass in water as a function of temperature.

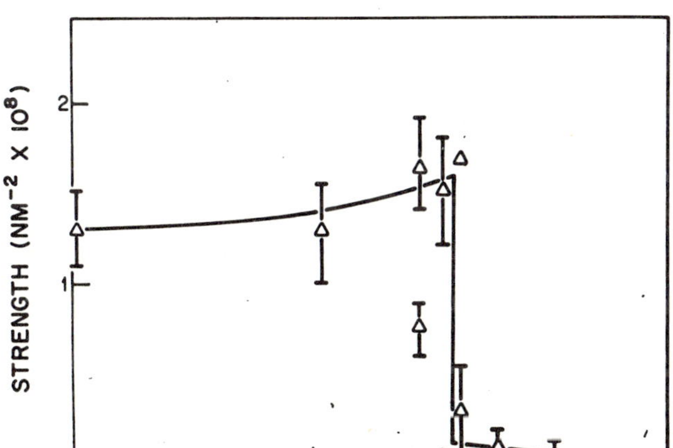


Figure 2. Strength behavior of circular rods of soda-lime-silica glass with radius of 0.236 cms subjected to a quench into a water bath at 33°C.

over the range of temperatures of the quenching experiment. The numerical values for the rod dimensions, physical properties, crack growth behavior and heat transfer conditions are listed in Table I. The value for the heat transfer coefficient h = 0.13 and 1.0 cal. $cm^{-2}$ $sec^{-1}$ $°C^{-1}$ represents the range of h representative of a water quench [11,12]. For the specific rod size and thermal conductivity these values of h give $m \approx 13$ and 100, respectively. Further, the initial flaw depth of the glass can be estimated from the ultimate strength and critical stress intensity factor as measured in a non-corrosive environment from eq. (5). A strength value of 1.94 x $10^7$ $NM^{-2}$ at liquid $N_2$ temperature [13] and fracture toughness of $K_{Ic} = 9.17 \times 10^5$ $NM^{-3/2}$ at liquid $N_2$ [14] lead to an initial flaw depth estimate of $\approx 5 \times 10^{-6}$ M. For the glass rods, fig. (2) illustrates the strength behavior as a function of quenching temperature difference. As indicated by the large loss in strength, the critical quenching temperature difference ($\Delta T_c$) required to initiate catastrophic failure under these conditions is of the order of 155-160°C. The specific objective of the present study was to determine if this value of $\Delta T_c$ could be calculated from the known information for heat transfer data, physical properties and crack propagation behavior.

## III. RESULTS AND DISCUSSION

Relative surface temperatures and thermal stresses in a circular rod of soda-lime-silica glass subjected to instantaneous decrease in ambient temperature under conditions of convective heat transfer are given in figs. (3) and (4) respectively as a function of time for various values of Biot's modulus (m) ranging from 0.1 to 100. Numerical results for stress intensity factors and crack depth as a function of time are given in figs. (5) and (6) for Biot's modulus of m = 100 and in figs. (7) and (8) for m = 13 at various quench temperatures. In figs. (5)-(8) for the curves for $\Delta T$ terminated with an arrow, calculated rates of crack growth were such that the stress intensity factor $K_I$ reached the value of $K_{Ic}$ (i.e., catastrophic fracture occurred) within the next time increment. Figure (9) gives $\Delta T_c$, the critical quench temperature difference, as a function of m for the two values of initial crack depth in the absence or presence of crack growth.

The predicted values for $\Delta T_c$ = 147 to 175°C for m = 100 and 13, respectively and $a_I = 5 \times 10^{-6}$ M, compare favorably with the observed value of 155-160°C. If slow crack growth had been ignored the corresponding calculated values for $\Delta T_c$ would have been 238 and 345°C respectively, which represent a large overestimate of the thermal stress resistance. This

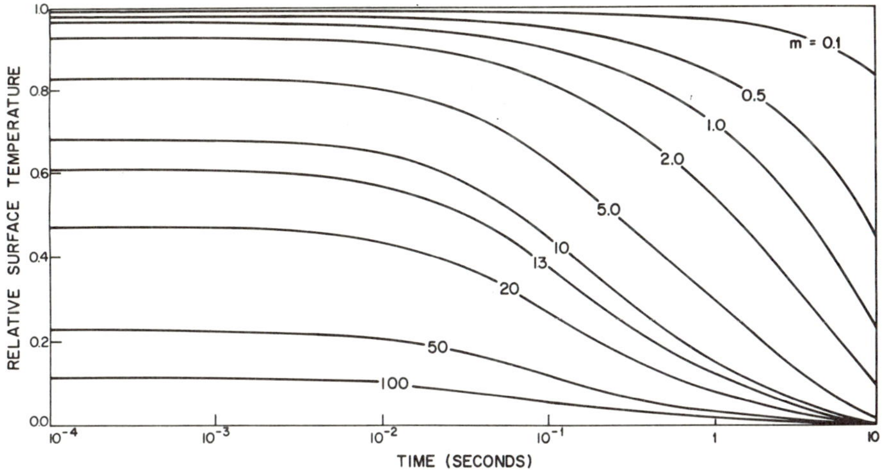


Figure 3. Relative surface temperature of circular rod of soda-lime-silica glass with radius of 0.236 cms subjected to instantaneous decrease in ambient temperature under conditions of convective heat transfer.

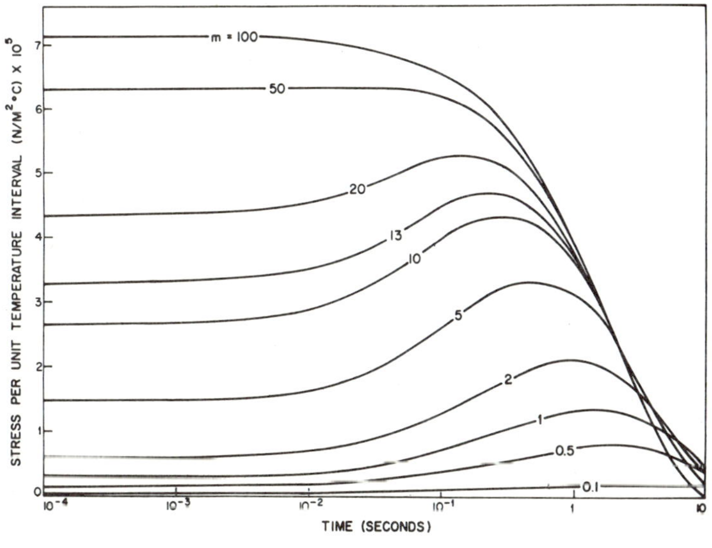


Figure 4. Tensile thermal stress in circular rod of soda-lime-silica glass with radius of 0.236 cms, per °C change in ambient temperature under conditions of convective heat transfer.

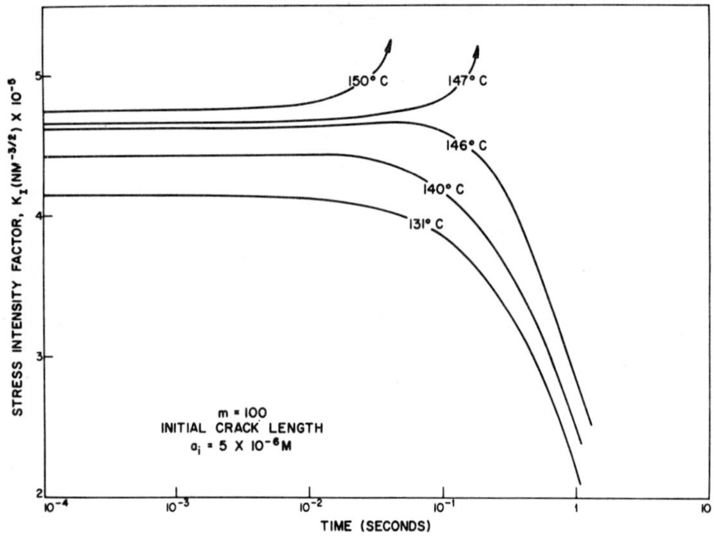


Figure 5. Thermal stress intensity factor for surface crack in circular rod of soda-lime-silica glass with radius of 0.236 cms subjected to a water quench with m = 100 and initial crack depth $a_I = 5 \times 10^{-6}$ M.

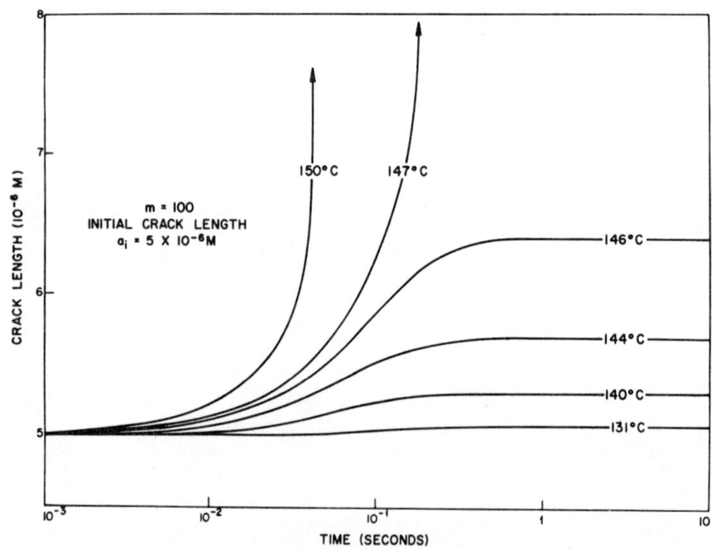


Figure 6. Growth of surface crack in circular rod of soda-lime-silica glass subjected to a water quench with m = 100 and initial crack depth of $5 \times 10^{-6}$ M.

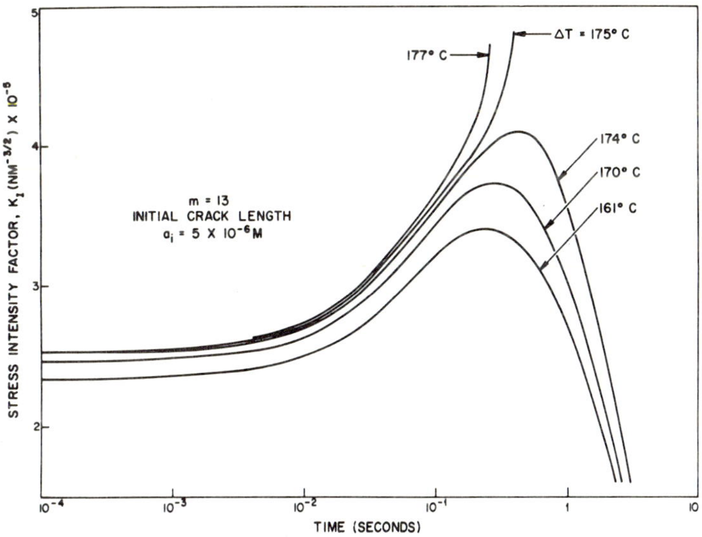


Figure 7. Thermal stress intensity factor for surface crack in circular rod of soda-lime-silica glass with radius of 0.236 cms subjected to a water quench with m = 13 and initial crack depth $a_I$ = 5x$10^{-6}$ M.

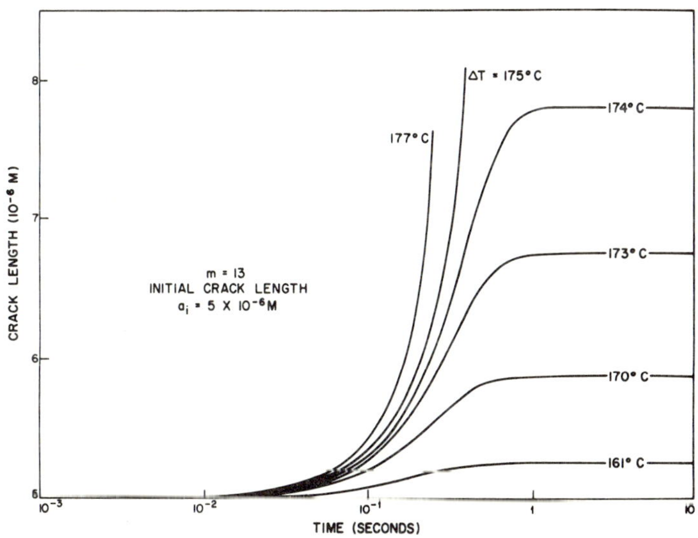


Figure 8. Growth of surface crack in circular rod of soda-lime-silica glass subjected to a water quench with m = 13 and initial crack depth of 5x$10^{-6}$ M.

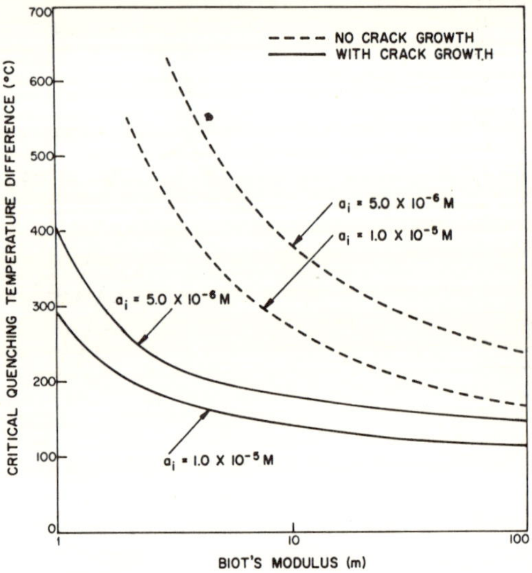


Figure 9. Comparison of calculated values of critical quenching temperature difference with and without slow crack growth for circular rods of soda-lime-silica glass subjected to a water quench.

latter conclusion is generally true for all values of m, as indicated in fig. (9).

It is of interest to examine the range of temperature over which slow crack growth occurs. For the particular quenching tests in the present study a value of $\Delta T_c$ = 147°C requires an initial specimen temperature of 180°C. This temperature is well beyond the maximum temperature of 90°C for the K-V data obtained by Wiederhorn and Bolz [10]. However, as calculated, the surface temperature even at $10^{-7}$ seconds after the initiation of the quench has reached a relative value of 0.12 of the total quenching temperature difference, which corresponds to a value of approximately 50°C, well within the range of the validity of the K-V data. This leads to the general conclusion that for transient thermal stress failure predictions unless m is very small (say $m \approx 0.1$), no need exists to generate K-V data for temperatures at or near the initial temperature of a structure subjected to sudden cooling.

It may be noted from fig. (5) that for $\Delta T < \Delta T_c$ total crack growth is significant only over a relatively narrow range of temperature. The reason behind this behavior is that as $\Delta T$ is being decreased, the thermal stress intensity factor as well as the temperature level of the specimen is being reduced. Both these effects combine to significantly reduce the total extent of crack growth with decreasing $\Delta T$. Of interest to note in fig. (4) for $\Delta T = 147°C$ is the very rapid rate of increase of stress intensity factor at values of $K_I$ well below $K_{Ic}$. This behavior is due primarily to the very rapid rate of crack growth shown in fig. (5) as the result of the given combination of stress intensity factor and temperature. This, in fact, represents a form of subcritical crack instability such that even for $K_I << K_{Ic}$ over the duration of the transient thermal stress, failure becomes inevitable. Consequently, for this particular glass in a water environment estimates of thermal stress resistance based on $K_{Ic}$ are expected to be too optimistic (see fig. (8)).

The slow crack growth also appears to have a major effect on the times-to-failure. For m = 100, the maximum thermal stress occurs at a time period even less than shown in fig. (5), i.e., at values of time $t < 10^{-4}$ seconds. At $\Delta T_c$, in the absence of crack growth, the time-to-failure should be equal to the time of maximum thermal stress, i.e., $t < 10^{-4}$ seconds. However, as indicated in figs. (4) and (5) the time-to-failure exceeds this value considerably and for m = 100 with $\Delta T_c = 147°C$ is of the order of 0.2 seconds. The reason for this is that due to the continually increasing crack depth as the result of the slow crack growth, the thermal stress intensity factor continues to increase even when the thermal stress already has passed its maximum value and is decreasing.

## SUMMARY

An analytic and numerical technique was developed for the calculation of the critical quenching temperature difference $\Delta T_c$ of circular rods subjected to a water quench, through the use of known K-V data. For soda-lime-silica glass rods of a specific radius and crack depth the critical quenching temperature difference was calculated to be 147°C which corresponded favorably with experimentally observed values of 155-160°C. In the absence of crack growth, $\Delta T_c$ was estimated to be 238°C which is well in excess of the observed value and indicative of the significant effect of slow crack growth on thermal stress resistance. It is also shown that crack growth significantly extends the time-to-failure to a value well in excess of the time of maximum thermal stress. A form of subcritical crack instability is predicted at values of $K_I << K_{Ic}$ at which catastrophic failure becomes inevitable over the duration of the transient thermal stress.

## ACKNOWLEDGMENTS

The present study was conducted under sponsorship of the Army Research Office - Durham under Grant DA-ARO-D-31-124-73-G45.

## REFERENCES

[1] Kingery, W. D., J. Am. Ceram. Soc., 38 (1), 3-15, 1955.

[2] Crandall, W. B., J. Am. Ceram. Soc., 38 (1), 44-54, 1955.

[3] Hasselman, D. P. H., J. Am. Ceram. Soc., 46 (5), 229-34, 1963.

[4] Buessum, W. R., Sprechsaal, 93 (6), 137-41, 1960.

[5] Hasselman, D. P. H., J. Am. Ceram. Soc., 50 (9), 454-57, 1967.

[6] Evans, A. G., J. of Mat. Sc., 7, 1137-46, 1972.

[7] Carslaw, H. S. and Jaeger, J. C., Conduction of Heat in Solids, 2nd ed., Clarendon Press, Oxford, 1959, 510 pp..

[8] Boley, B. A. and Weiner, G. H., Theory of Thermal Stresses, John Wiley and Sons, Inc., New York, 1960, 586 pp..

[9] Sih, G. C., Handbook of Stress Intensity Factors, Lehigh University, 1973.

[10] Wiederhorn, S. M. and Bolz, L. H., J. Am. Ceram. Soc., 53 (10), 543-48, 1970.

[11] Manson, S. S. and Smith, R. W., Trans. ASME, 78, 533-44, 1956.

[12] Kingery, W. D., Introduction to Ceramics, John Wiley, New York, 1960, 781 pp..

[13] Badaliance, R., Krohn, D. A., and Hasselman, D. P. H., J. Am. Ceram. Soc., in press.

[14] Wiederhorn, S. M., J. Am. Ceram. Soc., 53 (2), 99-105, 1969.

# EXPERIMENTAL METHODS FOR THE DETERMINATION OF STRESS INTENSITY FACTORS UNDER VARIOUS LOADING CONDITIONS

E. Sommer

Institut für Festkörpermechanik
Freiburg, W-Germany

## ABSTRACT

Two experimental methods for determining the stress intensity factor are described. Crack opening displacement measurements by an optical interference method allow the effect of residual stresses on the properties of cracks to be shown. The investigations are considered as an example of avoiding uncertainty due to residual stresses in predicting the safety of structures by fracture mechanics methods. The same optical technique is applied to investigate the crack opening in a material with time dependent behaviour. As a second method induced markings on fracture surfaces of fatigue cracks are used to determine the crack propagation rate in order to evaluate the stress intensity factors at local points of the front of part-through cracks with the aid of an empirically established calibration curve.

## INTRODUCTION

Crack-tip stress field parameters can be determined by various experimental techniques. Two of them, and some related results, are reported here. Crack opening displacement measurements using a simple optical interference method enable both the contribution of residual stresses to the fracture behaviour of an externally loaded specimen and the effect of time dependent material behaviour to be evaluated. Residual stresses induced by various processing treatments, although very often unknown in magnitude can play an important role when the methods of fracture mechanics are used to predict the safety of structures and structural components. The experimental work discussed in this paper is considered as a simple but typical example of such a prediction.

As an indirect method the measurement of the crack propagation rate da/dN in a fatigue test allows determination of the stress intensity factor at local points of the front of complicated crack configurations when the well-known empirical relation between da/dN and the stress intensity factor $\Delta K$ established for a given specimen geometry serves as a calibration curve. Experimental measurements of the extension of part-through cracks under fatigue loading are presented and compared with the analytical solutions of other authors [1,2].

## CRACK OPENING DISPLACEMENT MEASUREMENTS

### Method of Investigation

The optical interference method used to determine the opening displacement of cracks in transparent materials has been previously described [3]. The principle of the technique may be seen from Fig. (1). For incidence of parallel monochromatic light perpendicular to the two opposing crack surfaces, extinction of light by interference occurs, in the case of reflection, when

$$2v = \frac{\lambda}{2} n \quad (n = 1,2,3\ldots) \qquad (1)$$

where $2v$ = crack opening displacement,

$\lambda$ = wavelength of monochromatic light.

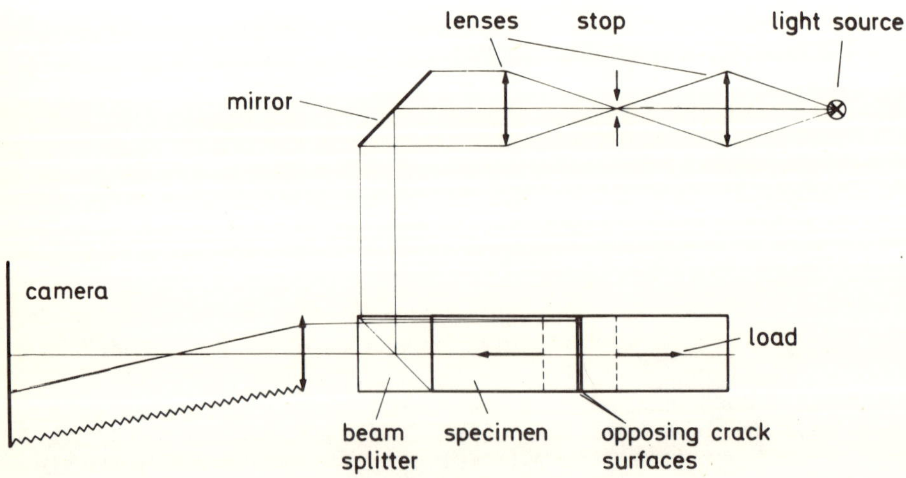


Fig. (1) - Schematic diagram of the experimental arrangement

From this equation the opening displacement at the point (r,z) as shown in Fig. (2), on the nth interference fringe can be determined. From fracture mechanics, the square of the opening displacement should be a linear function of the distance r when crack tip plasticity remains small and the condition $r \ll a$, where a is the crack length, is fulfilled (for z = constant). Assuming linear elastic material behaviour, the slope of this curve depends only upon the stress intensity factor K, i.e., for a given crack length and specimen geometry upon the load P and upon the elastic

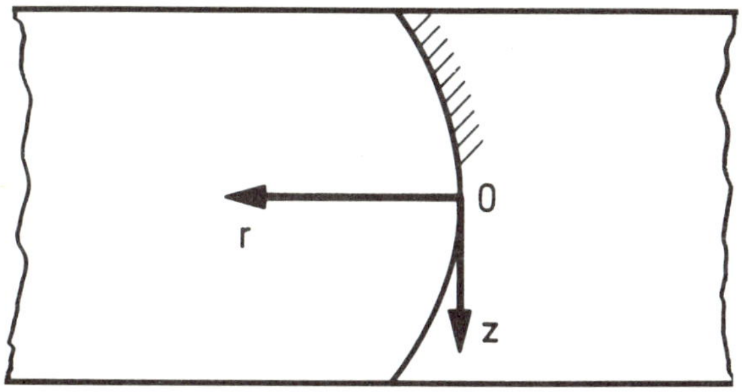


Fig. (2) - Location of the coordinate system at the fracture surface

constants E and $\nu$ of the material. It is assumed that a change of slope due to internally acting residual stresses may be expressed in terms of a modified stress intensity factor $K_{total} = K_{external} + K_{internal}$. It will be shown that this simplifying relationship holds at least for a residual stress field varying only in z-direction. If the material behaves viscoelastically the opening displacement will also vary with time. This effect can be taken into account by assuming a time dependent Young's modulus E(t). The final equation for the opening displacement which may be used to interpret the experimental findings, therefore, becomes, in the case of plane strain,

$$2v = \frac{4K_{total}}{E(t)} \left(\frac{2r}{\pi}\right)^{1/2} (1-\nu^2) + \text{higher order terms} \qquad (2)$$

where $\nu$ = Poisson's ratio. The resolution of the optical method allows the opening displacement to be evaluated in the range where this equation is valid and higher order terms may be neglected. Complex crack tip behaviour in the form of plastic or nonlinear viscoelastic deformation gives rise to disturbances only in the immediate vicinity of the tip ($r \to 0$). The factors which limit the resolution of the technique are due to:

1. the relatively poor optical quality of the investigated materials,

2. the imperfect smoothness of natural fracture surfaces,

3. the interference fringe spacing becoming too small.

Effect of Residual Stresses

To investigate the influence of residual stresses, the distribution of which will be explained in more detail in the following section, compact tension specimens (125 x 120 x 20 $mm^3$) were used. The transparent plastic material found to be suitable as a model material. Araldit B (an epoxy resin)*, shows almost no creep behaviour under load in the range considered. Its elastic constants are $E = 3240\ N/mm^2$ and $\nu = 0.3$. Care was taken in precracking specimens in order to produce smooth fracture surfaces and sharp crack tips.

A crack was propagated by means of an impact loaded wedge from a saw cut which served as a starter notch. A compressive stress acting perpendicular to the propagation direction and parallel to the crack plane helped to stop the running crack in the required range $0.45 < a/W < 0.55$ where $1.2W$ = width of the specimen.

Such a prepared specimen was adjusted in the optical device described in the previous section (Fig. (1)). In order to determine the opening displacement from the interference fringe system at different load steps photographs were taken. Two typical examples for load $P_{ext} = 0N$ where the fracture surfaces are partially opened due to residual tensile forces and for $P_{ext} = 145, 5N$ where the crack is completely open are shown in Fig. (3). In accordance with Eq. (1), the photographs can be evaluated point for point when the wave length of the monochromatic light $\lambda = 0.546\ \mu m$ and the exact fringe number n are known. A continuous but slow increase of load allows the increasing number of fringes to be followed at selected markings on the scale photographed through the same optical system.

In this way the curvature of the complete fracture surface as a function of load may be obtained. Examples are shown in the next figures. The square of the opening displacement in the center of the specimen (z=0) is plotted as a function of the distance

---

* Ciba-Geigy, Basel; plates manufactured by Tiedemann, Garmisch-Partenkirchen.

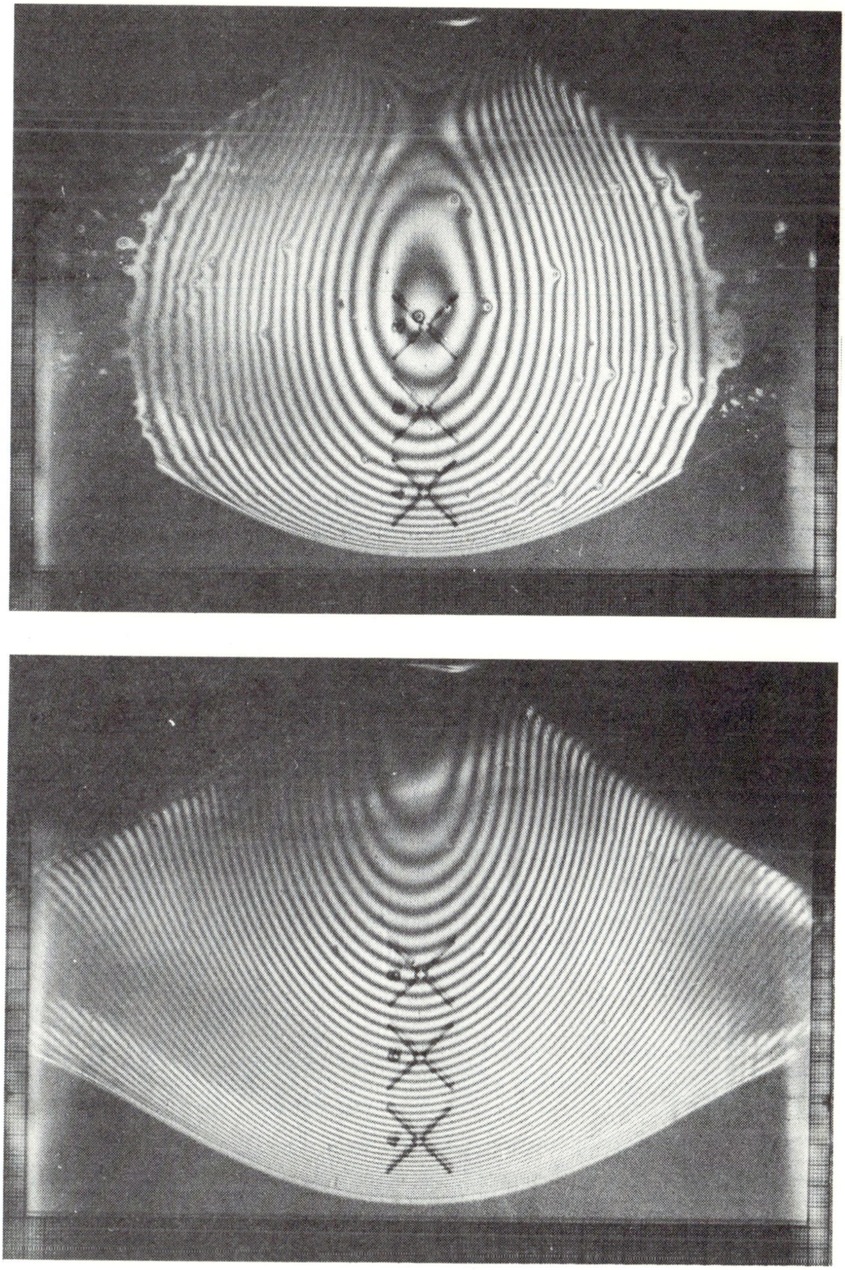

Fig. (3) - Interference pictures of a crack in a transparent specimen under two loading conditions: a) $P_{external}$ = 0N, b) $P_{external}$ = 145, 5N coordinate r for several steps of load $P_{ext}$ in Fig. (4a). A similar plot for varying z but constant load $P_{ext}$ is given in Fig. (4b). The stress intensity factor $K_{total}$ can then be determined from these measurements and Eq. (2) when the elastic con-

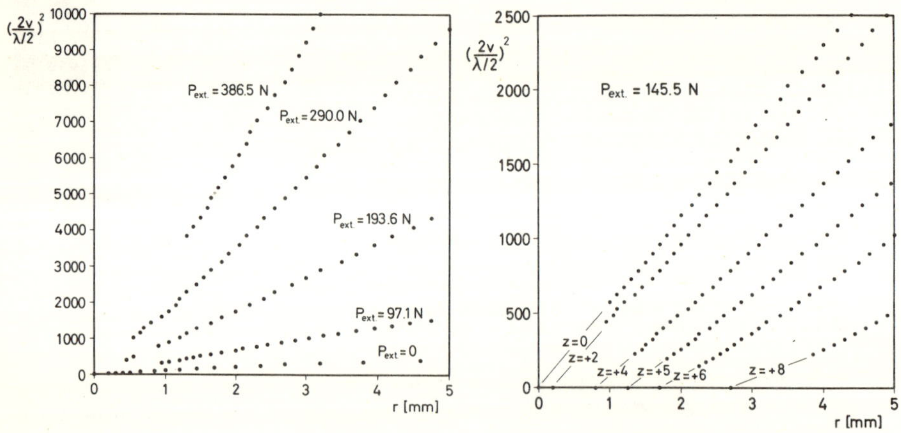


Fig. (4) - Square of normalized crack opening, $2v/(\lambda/2)^2$ versus the distance r from the crack tip.
a) for different values of $P_{ext}$, z=0
b) for different values of z, $P_{ext}$ = 145, 5N

stants E and ν are known. In accordance with Eqs. (1) and (2), the ratio

$$n/r^{1/2} \propto K_{total}(P,z) = K_{ext}(P,z) + K_{int}(z) \qquad (3)$$

is expected to be a linear function of $P_{ext}$ for constant values of the thickness parameter z. Since $K_{internal}$ is independent of the external load the slopes of the straight lines will show the variation of $K_{ext}$ with the thickness z and their intercepts with the ordinate the variation of $K_{int}$ with z. Three examples are plotted in Fig. (5).

A photoelastic examination was made of the stress distribution in the (y,z plane) on a strip of material cut out immediately ahead of the crack tip (the crack lies in the (x,z) plane). This showed that the residual stresses were distributed symmetrically about the center plane of the specimen (z=0), i.e., that they are constant for planes z = const. (neglecting effects close to the free surfaces of the specimen). From the isochromatic fringe pattern in the (y,z) plane which is shown in Fig. (6), the stresses $\sigma_{y\ res}$ (z) can be evaluated when $\sigma_z$ is assumed to be zero.

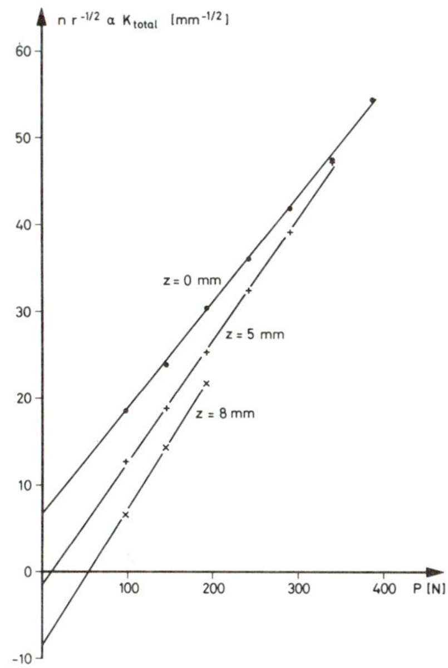


Fig. (5) - Ratio $n/r^{1/2} \propto K_{total}$ as a function of external load P.

$$\sigma_{y\ res} - \sigma_{z\ res} = \frac{NS}{D}\ \text{N/mm}^2 \tag{4}$$

where N = order of isochromatic fringes,

S = fringe value = 10.2 N/mm for Na-light [5],

D = depth of the photoelastic specimen = 39.8 mm.

To eliminate edge effects the residual stresses where photoelastically evaluated for two different values of depth D. The resulting parabolic distribution plotted in Fig. (7) shows a maximum tensile stress $\sigma_{y\ res}$ (0) = 1.2 N/mm² in the center of the plate and a maximum compressive stress $\sigma_{y\ res}$ (h/2) = 2.8 N/mm² at the plate surfaces (z = ±h/2). The neutral axes are located at z = 47 mm.

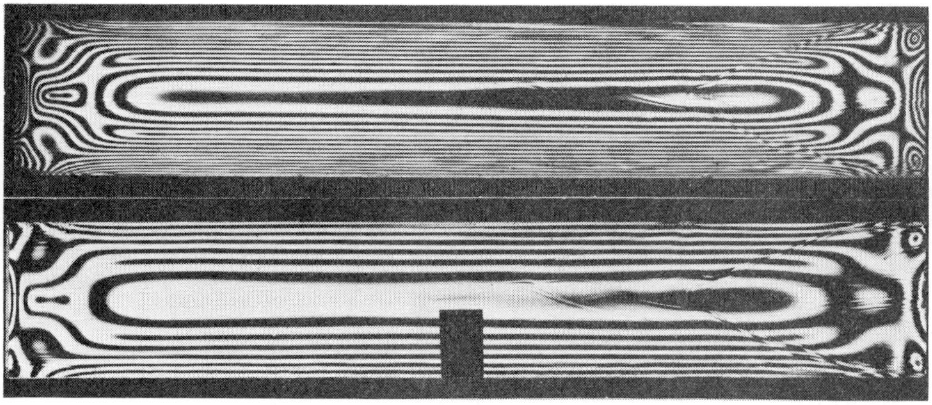

Fig. (6) - Isochromatic fringes showing the residual stress distribution in the y,z plane for depth of specimen D = 38.9 mm and 20.1 mm

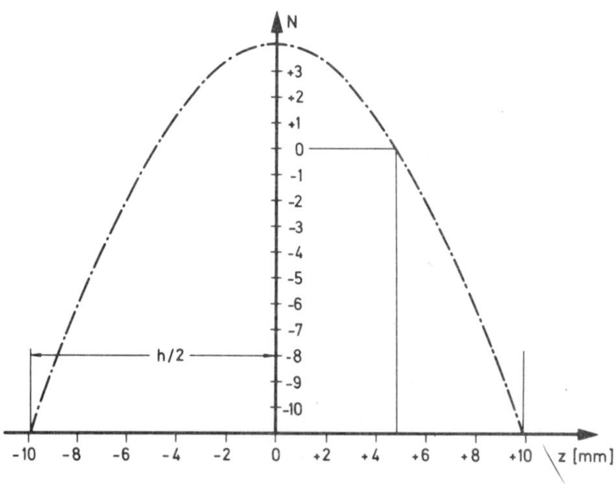


Fig. (7) - Distribution of rèsidual stresses through the plate thickness h. Fringe order N as a function of the thickness coordinate z

In Fig. (8a), a plot of the relative residual stress distribution is compared with the ratio $K_{int}(z)/K_{int}(0)$. The variation of $K_{ext}(z)/K_{ext}(0)$ with thickness shown in Fig. (8b) may be attributed to the effect of crack front curvature which can be seen in the variation of the relative crack length $a(z)/a(0)$. Both $K_{int}(z)$ and $K_{ext}(z)$ were evaluated from a least squares fit of the experimental data to Eq. (3) which is a linear function of $P_{ext}$. The error bars shown in Fig. (8) indicate the resulting standard deviations. A more detailed description will be given in [6].

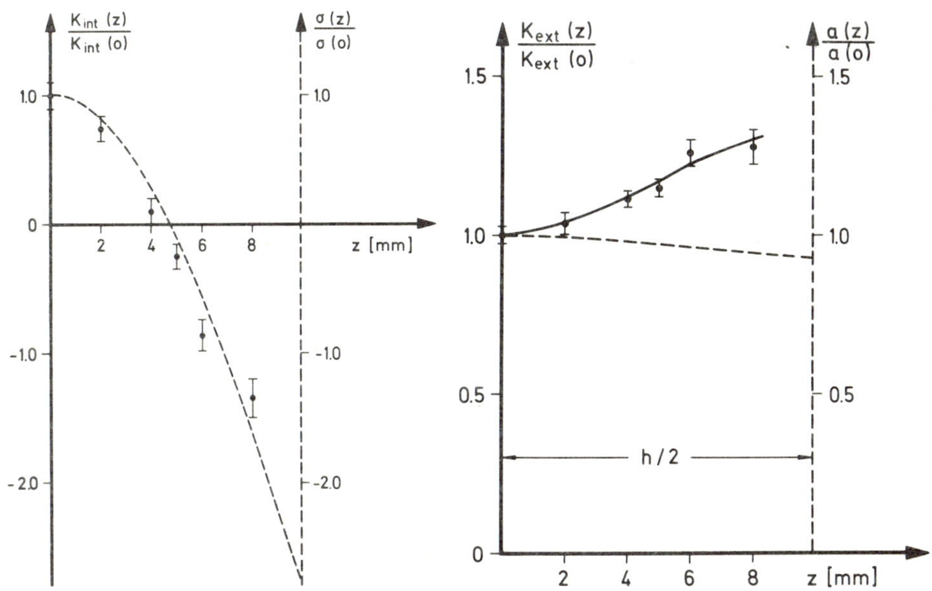


Fig. (8) - Relative stress intensity factors versus thickness coordinate z.
a) $K_{int}(z)/K_{int}(0)$ (.) in comparison to the residual stress distribution $\sigma(z)/\sigma(0)$ (---)
b) $K_{cxt}(z)/K_{ext}(0)$; relative crack length $a(z)/a(0)$ for comparison ($a(0) = 49, 9$ mm)

Effect of Time Dependent Behaviour

In materials which behave viscoelastically the opening displacement of cracks varies with time. Using the same optical arrangement, photographs of the interference patterns between fracture surfaces in a specimen of a viscoelastic material, VP 1527 (a polyester resin)*, under constant load were taken at successive time intervals starting immediately after loading. The opening displacement was determined at different times for seven values of the distance coordinate r. Since Eq. (2) yields

$$v(t)/v(0) = E(0)/E(t) \tag{5}$$

the latter ratio was plotted in Fig. (8) as a function of time.

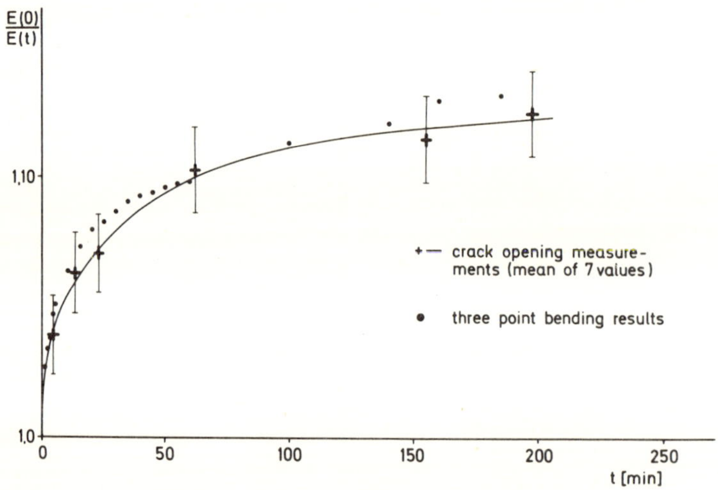


Fig. (9) - Ratio of Young's moduli E(0)/E(t) as a function of loading time

This ratio was also determined from a three point bending test using a bar measuring 120 x 10 x 10 $mm^3$ of the same material. Both sets of experimental data agree within experimental error.

## CRACK PROPAGATION RATE MEASUREMENTS

The extension of part-through cracks was investigated in a fatigue test in order to determine the stress intensity factor at local points of the crack front in non-transparent materials [7]. The experiments were performed on two different aluminium alloys,

---

* Wacker Chemie GmbH, München; plates manufactured by Tiedemann, Garmisch-Partenkirchen.

AlCuMg I and AlMg 3, in the form of plates measuring 400 x 100 x 20 $mm^3$ and semi-circular half-tubes of the same length as the plates but of thickness 16.5 mm and with two different inner radii, 60 mm and 80.5 mm. Cracks were started from notches in the form of saw cuts which were normal to the stress direction and centrally located in the specimens. The saw cuts were made with a 0.2 mm thick circular saw and were 20 mm long, 5 mm deep. To obtain the crack propagation rates of the part-through cracks, markings were induced on the fracture surface by changes in amplitude at specified numbers of fatigue load cycles. Care was taken that the effect on the propagation rate of the changes in fatigue load was kept to a minimum. After complete fracture of the specimen, the crack propagation rates da/dN were calculated by measuring the distance between the markings on the fracture surface da and knowing the corresponding number of cycles dN. A fatigue fracture surface marked in this way is shown in Fig. (10). A narrower spacing of the marking was taken for the specimens evaluated.

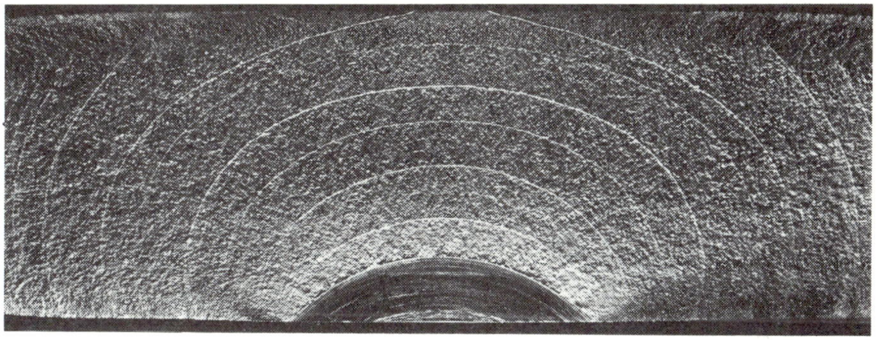

Fig. (10) - Fatigue fracture surface with markings in a plate of AlCuMg I (plate thickness 20 mm)

The relationship between propagation rate and stress intensity factor was obtained by means of a calibration experiment. Single edge notched (SEN) specimens of the same material fatigue tested under the same conditions served to establish the empirical equation

$$da/dN = C\Delta K^{n} \tag{6}$$

where C, n = constants,

$\Delta K \approx \Delta\sigma(\pi a)^{1/2} f(a/W)$,

$\Delta\sigma$ = total stress amplitude,

$f(a/W)$ = correction function for SEN-specimens.

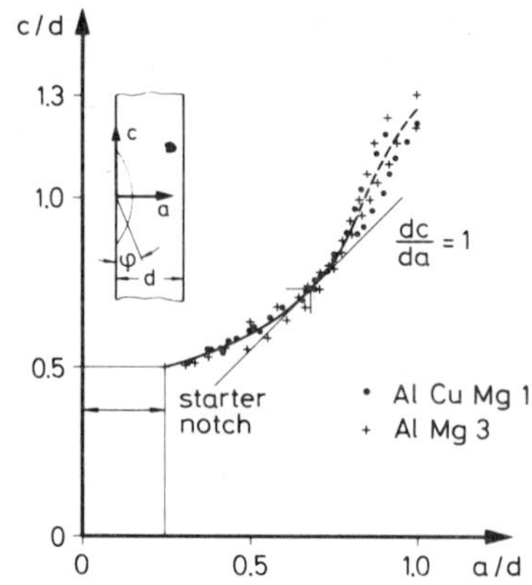


Fig. (11) - Relative crack width c/d as a function of relative crack depth a/d (d = plate thickness)

Based on this calibration curve the relative stress intensity factor $\Delta K/\Delta\sigma$ may be determined at local points of complicated crack configurations when the corresponding crack propagation rate is known and C, n are assumed to be constant in all directions of crack propagation. Similar techniques for determining the stress intensity factor have been used by other authors [8-12].

The different extensions, $c(\phi=0°)$ and $a(\phi=90°)$, of the fatigue cracks can be seen quantitatively in Fig. (11). The slope dc/da of the curve determines the ratio of the stress intensity factors $\Delta K_{0°}/\Delta K_{90°}$ in the two directions when the aforementioned assumptions hold, where

$$\frac{dc}{da} = \left(\frac{\Delta K_{0°}}{\Delta K_{90°}}\right)^{n} = \left(\frac{c}{a}\right)^{n/2} \left(\frac{f_{0°}}{f_{90°}}\right)^{n} \qquad (7)$$

where $f_{0°}$, $f_{90°}$ are functions describing the effect of crack and specimen geometry for $\phi = 0°$ and $90°$. For discussion purposes the following three cases are considered:

1. $dc/da < 1$: as the crack begins to propagate the stress intensity factor at the vertex of the part-through crack predominates and the crack extends preferentially in the direction of the plate thickness.

2. $dc/da = 1$: the equality of stress intensity factors in both directions is given for an almost semi-circular crack configuration ($a/c = 0.93$).

3. dc/da > 1: the crack extension in the direction of the plate width predominates. Since this behaviour seems to be in contradiction to analytical results, clarifying experiments still need to be carried out.

Final results obtained by the calibration method are shown in Figs. (12) and (13). The load independent geometrical part of the stress intensity factor is plotted versus the crack length. In Fig. (12), the experimental data evaluated from three

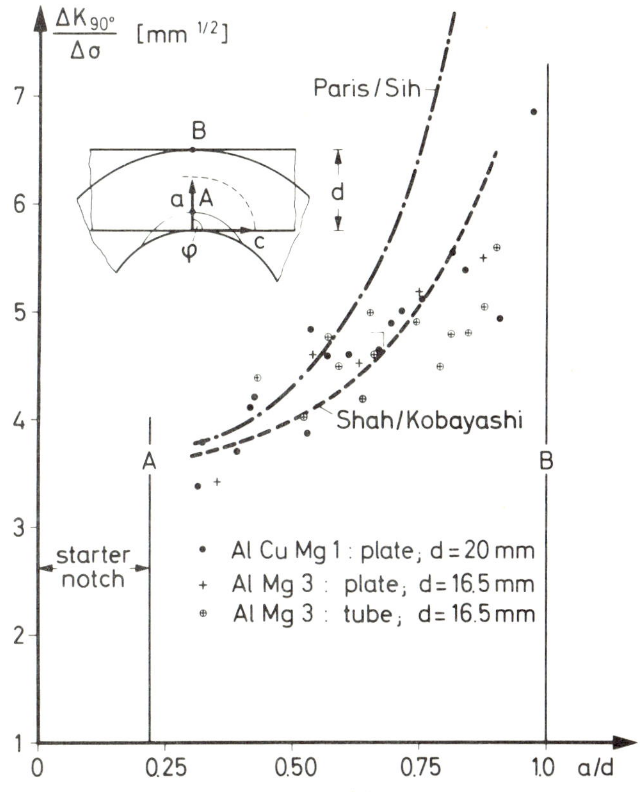


Fig. (12) - Relative stress intensity factor $\Delta K_{90^\circ}/\Delta\sigma$ versus relative crack length a/d

AlCuMg I specimens having a plate thickness of 20 mm are compared with two different analytical curves [1,2] which result for an idealized semi-elliptical crack front of the same a/c ratio. Additionally data of a plate specimen and of a tube of AlMg 3 are plotted. The high experimental scatter which is heavily influenced by the quality of the material investigated does not permit a decision on which of the curves fit best. A $\Delta K_{0^o}/\Delta\sigma$ curve is plotted in Fig. (13). Since the crack front was always found to intersect the plate surface at an angle of 90°, the microscopically measured, surface crack growth was used to determine the propagation rate dc/dN. These data show less scatter because, as the material was prepared by rolling, it would be expected to be more isotropic and homogeneous in layers parallel to the plate surface. For comparison the stress intensity factor of a through-the-thickness crack calculated for the corresponding crack length is plotted.

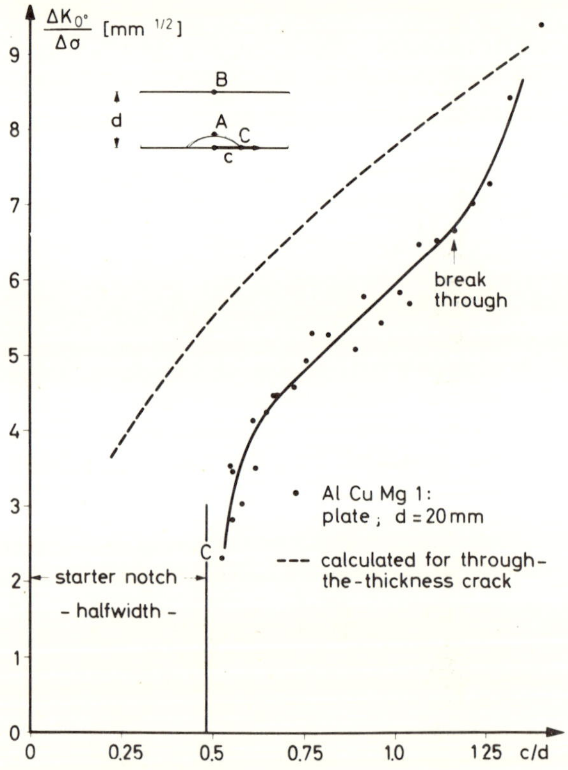


Fig. (13) - Relative stress intensity factor $\Delta K_{0^o}/\Delta\sigma$ versus relative crack length c/d

REFERENCES

[1] Paris, P. C. and Sih, G. C. M., "Stress analysis of cracks", Symposium on Fracture Toughness Testing and its Applications, ASTM STP 381, 30, 1965.

[2] Shah, R. C. and Kobayashi, A. S., "On the surface flaw problem", Proc. of ComCAM Symposium on Surface Flaw, Appl.

Mech. Div. of ASME, 1972.

[3] Sommer, E., "An optical method for determining the crack-tip stress intensity factor", Eng. Fract. Mech., 1, 4, 705, 1970.

[4] ASTM Standards, Philadelphia, 31, 1099, 1969.

[5] Soltész, U., "Experimentelle Untersuchungen zur stoßangeregten Wellenausbreitung in streifenförmigen Platten, Diss. Univers. Karlsruhe, 1973.

[6] Sommer, E. and Winkler, S., to be published.

[7] Sommer, E. and Ratwani, M., "Experimental investigations of part-through cracks in plates and thick walled shells", Second Int. Conf. on SMIRT, Berlin, 1973.

[8] James, L. A. and Anderson, W. E., "A simple experimental procedure for stress intensity factors calibration", Eng. Fract. Mech., 1, 3, 565, 1969.

[9] Mukherjee, B., Culver, L. E. and Burns, D. J., "Growth of part-through and through-thickness fatigue cracks in sheet glassy plastics", Exp. Mech. 9, 2, 90, 1969.

[10] Mukherjee, B. and Burns, D. J., "Growth of part-through thickness fatigue cracks in sheet polymethylmethacrylate", Eng. Fract. Mech. 4, 4, 675, 1972.

[11] Maddox, S. J., "Calculating the fatigue strength of a welded joint using fracture mechanics", Metal Constr. Brit. Weld. J. 49, 2, 327, 1970.

[12] Broekhoven, M. J. G., "Analysis of a nozzle-corner crack problem", Second Int. Conf. on SMIRT, Berlin, 1973.

# WRITTEN DISCUSSIONS

This section contains written discussions that were addressed to the audience at large and are not specifically related to the technical contents of any one paper.

## PLASTIC STRIP MODEL FOR THIN SHELLS

F. Erdogan
Lehigh University, Bethlehem, Pennsylvania, USA

In discussing the process of nucleation and growth of cavities and the ensuing fracture phenomenon the authors mentioned the plastic instability as the possible mechanism for the rupture of ligaments connecting the holes. It may be pointed out that a similar mechanism of plastic instability in continuum scale may be responsible for the propagation of cracks in thin plate and shell structures. Analogous to the tensile bar, in a certain portion of a given structure, the plastic instability is said to set in locally if the slope of the local strains or deformations vs. the applied load relationship reaches to a very high value. In thin plates and shells containing a crack there is usually considerable "necking" ahead of the crack tip before the actual rupture takes place. In such problems the crack opening displacement, $\delta$, obtained from a plastic strip model may be used as a quantitative measure of the local deformations.

Figure (1) shows the crack opening displacement $\delta$ in a pressurized cylindrical shell containing a through crack of length 2a obtained from a plastic strip model [1]. Here $N_0/h = pR/h = \sigma_{\theta\theta}$ is the hoop stress away from the crack region, p is the pressure, R is the mean radius, h is the thickness, $\sigma_Y$ is the yield strength used in the plastic strip model and the shell parameter $\lambda$ and the normalizing factor $d_1$ are given by

$$\lambda = [12(1-\nu^2)]^{1/4}\ a/\sqrt{Rh},\quad d_1 = 4\sigma_Y\ a/E$$

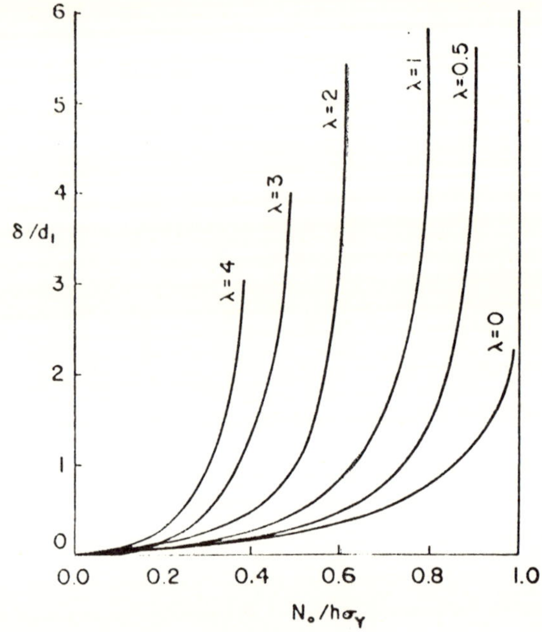


Fig. (1) - Crack opening stretch in the neutral surface and at the crack tip (x=a, z=0)

The figure shows that for each given value of $\lambda$ after the applied load reaches a certain level, a small additional increase in the load will cause a large increase in $\delta$, leading to plastic instability. The figure also shows that, particularly in shells (i.e., $\lambda>0$) this instability load may be determined by selecting an (arbitrarily high) standard slope for the $\delta$ vs. $\sigma_{\theta\theta}/\sigma_Y$ curves. A set of curves similar to those shown in Fig. (1) may be obtained for the part-through cracks in shells provided the external loads, the material, and the crack geometry are such that the net ligament along the crack border is fully yielded [2]. In this case, the maximum COD is at the center rather than at the ends of the crack.

Figure (2) shows the load-carrying capacity of the shell with a through or a part-through crack based on a slope of 1/20. Here d is the depth of the part-through crack (see insert in Fig. (4)). In this and in the subsequent figures, the yield strength $\sigma_Y$ is replaced by a "flow stress", $\bar{\sigma}$. Since the plastic strip model does not account for any strain hardenings, such a change appears to be necessary, which may be accomplished by expressing $\bar{\sigma} = \sigma_Y + \Delta\sigma$ (as in [3] where $\Delta\sigma$ was assumed to be 10,000 psi for the

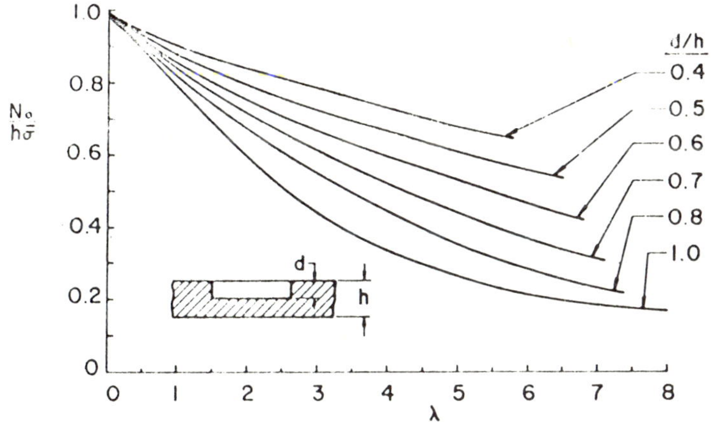


Fig. (2) - Load carrying capacity of pressurized cylindrical shells with a part-through or through crack based on the plastic instability criterion

steels tested), or $\bar{\sigma} = (\sigma_Y + \sigma_u)/2$ where $\sigma_u$ is the ultimate strength, or more generally $\bar{\sigma} = \sigma_Y(1+\alpha)$ where the factor $\alpha$ not only accounts for the strain hardening, it may also be adjusted to account for the thickness effect.

Figures (3) and (4) show the comparison of the theoretical rupture loads given by Fig. (2) with those obtained from the rupture tests on various steel pipes (X52, X60V, X65C, X60C) containing a through or a part-through crack [3]. Figure (3) shows the results for through cracks and Fig. (4) shows the results for part-through cracks where the thickness ratio d/h was approximately 0.5 ($0.492 < d/h < 0.511$). The agreement appears to be sufficiently "good" considering the nature of the problem to warrant further investigation of the concept.

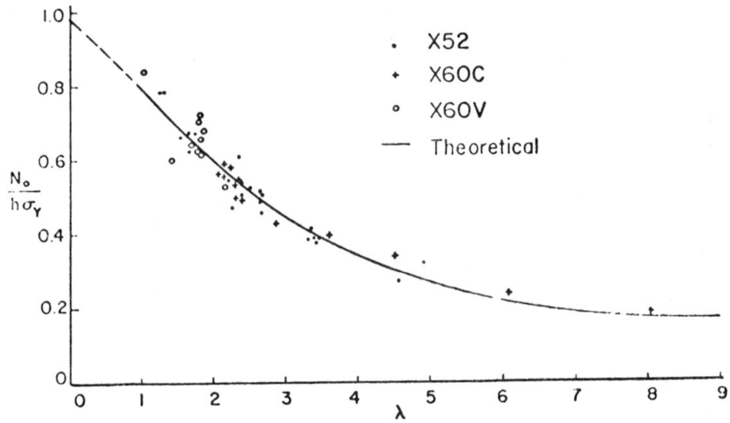


Fig. (3) - Comparison of the results of the fracture tests in steel pipes with a through crack (Kiefner, et al 1972) with that given by the plastic instability criterion

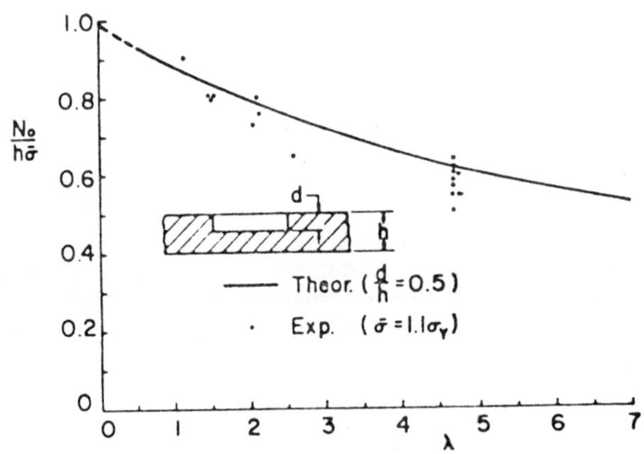


Fig. (4) - The results of the fracture tests in steel pipes with a part-through crack, $0.492 < d/h < 0.511$, $\bar{\sigma} = 1.1\sigma_Y$ (Kiefner, et al 1972). Solid line is the result predicted by fracture instability for $d/h = 0.5$

# ELASTIC-PLASTIC FRACTURE MECHANICS

G. C. Sih
Lehigh University, Bethlehem, Pennsylvania, USA

The problem of subcritical crack growth in plate specimens of polycrystalline materials under applied loads that increase slowly in time is still far from being understood. Attempts made to quantify this growth on the basis of two dimensional plasticity models are misleading and sometimes incorrect. The physical phenomenon of stable crack growth in metals is intimately associated with the effect of three dimensional necking ahead of the crack. The basic assumptions in many of the existing two dimensional models of plasticity (Fig. (1)) such as plastic strip, circular enclave* (Mode I), plastic fan, etc., are convenient for computational reasons but are unrealistic. The implication that a continuum crack propagates through a plastic zone raises several serious questions [1,2]. How does one compute the dissipated plastic energy associated with the creation of macrofracture surface? When does stable crack growth end and unstable crack extension begin? These models introduce one or more floating parameters that are often adjusted arbitrarily to fit

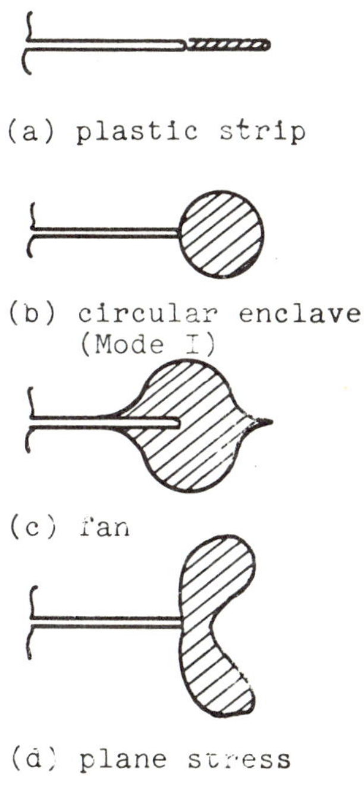


Fig. (1)-Models of plastic enclaves.

---

* This configuration holds only for Mode III loading. The material elements ahead of a crack in Mode I distort very differently. In principle, the elastic-plastic Mode III solution should not be used to explain Mode I fracture.

the experimental data*. Such an analysis is obviously load and geometry dependent. Its limitations in characterizing the fracture behavior of materials result in even less usefulness in structural design. Keep in mind that the basic aim of *fracture mechanics* is to find a parameter which is not so sensitive to changes in load, geometry, etc.

It is now well known from the plane strain** elastic-plastic analyses [3,4] that the material in the direction of crack extension does not distort appreciably and behaves elastically. The regions of high distortion, normally known as the plastic enclaves, are located away from the trajectory of the continuum crack. In addition, the crack tip radius of curvature $\rho$ affects the initial yielding*** of the material, as illustrated in Fig. (2). This is

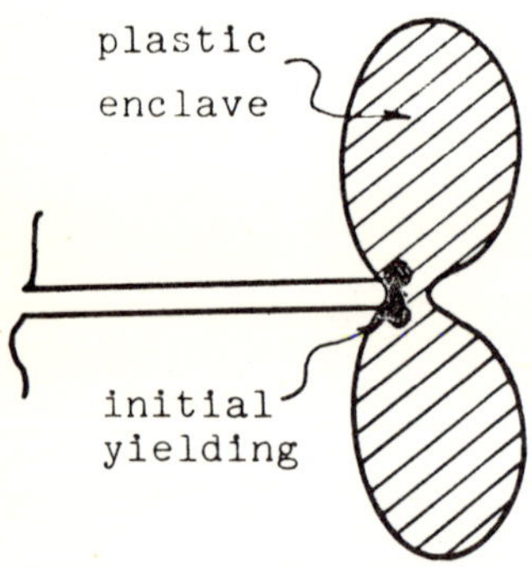


Fig. (2)-A physically blunt crack.

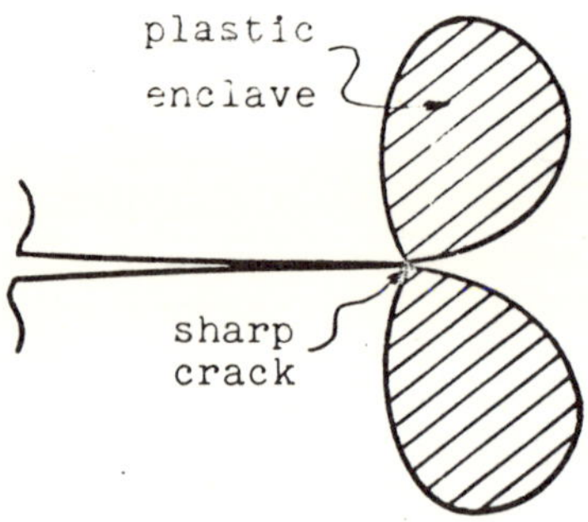


Fig. (3)-A mathematically sharp crack.

normally neglected in the numerical analysis because of the additional complexity. The plastic enclaves for a mathematically

---

* A classical example, as pointed out in [1], is the plastic strip model accounting for yielding in the thickness direction of thin metal sheets. However, the analysis considers no such effect and is strictly a two dimensional one. Agreement between theory and experiment can only be considered as fortuitous.

** Elastic-plastic plane stress solutions should not be used to analyze the state of affairs near the crack front.

*** The shape of the initial plastic zone is not a circle.

sharp crack, $\rho=0$, are shown in Fig. (3). Note that the material along the line of expected crack extension is elastic. However, the analytical solution of a finite crack in an infinite solid cannot adequately account for the influence of plasticity in a real specimen. The size of the plastic enclave can be as large as the crack* and will not affect the failure load appreciably [3]. Plasticity has a negligible influence on the available energy stored in the large elastic zone ahead of the crack. Physically, permanent deformation becomes significant only when the crack tip approaches a free surface as in the necking of a tensile bar or when two free surfaces approach each other such as in necking in the thickness direction of a plate. In other words, the plane strain elastic-plastic solution of a finite crack in an infinite solid does not contain sufficient length parameters to describe permanent deformation and has the gross feature of being an elastic problem. Aside from the evidence in [3], further conclusions can also be drawn from the results of [4] which deals with a crack in an elastic-plastic material subjected to off-axis tensile load, Fig. (4). The positions of the plastic enclaves for different angles between the load and line crack clearly indicate that there exists a path of elastic material leading to

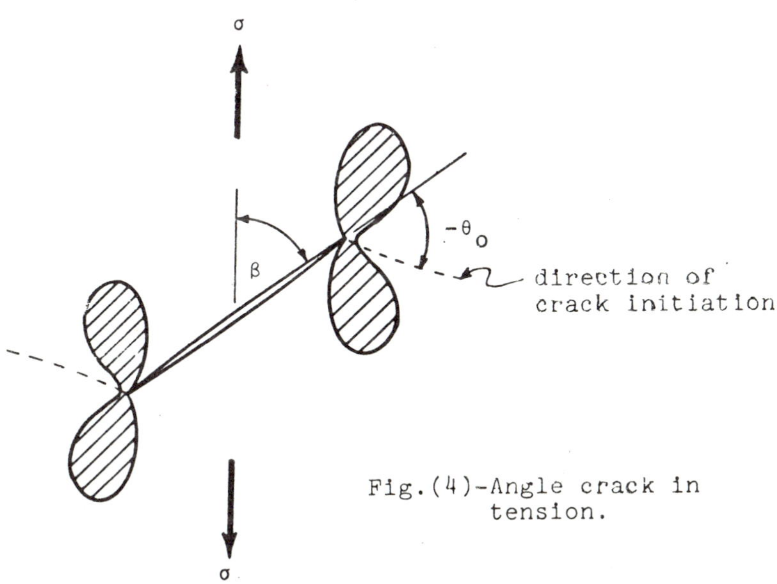


Fig.(4)-Angle crack in tension.

*The commonly used terminology of "small scale" yielding is not sufficient to describe the limitation of linear elastic fracture mechanics. This is clearly indicated by the results in [3].

the crack tip*. This path coincides within a few degrees to the direction, $-\theta_o$, of crack initiation predicted from the strain energy density criterion [5] using an elastic stress analysis. These observations reinforce the urgent need for some realistic elastic-plastic crack solutions which contain the essential physical parameters for describing permanent deformation.

The elastic-plastic aspects of a crack passing transversely through a metal plate of finite thickness have only been studied phenomenologically but not analyzed. This is primarily because of the lack of an understanding of the problem and of an effective criterion of fracture. Some of the most important features of thickness fracture associated with subcritical crack growth depend on the interaction of fracture surfaces that are normal or approximately 45° to the direction of loading. Figures (5) to (7) schematically display the various fracture surface appearances of a given metal. These different patterns may be obtained by either changing the temperature or the specimen thickness relative to the crack length. Figure (5) shows an initially straight crack front that grows forward and attains a thumbnail shape prior to unstable fracture. Under plane strain, the fracture surface is predominantly flat and normal to the applied load. As the thickness and/or temperature change, the fracture growth patterns in Figs. (6) and (7) are developed. Interior to the plate, there prevails a nearly hydrostatic tension which is most likely to lead to a flat fracture dissipating energy by volume change. In this interior region, the crack front is ob-

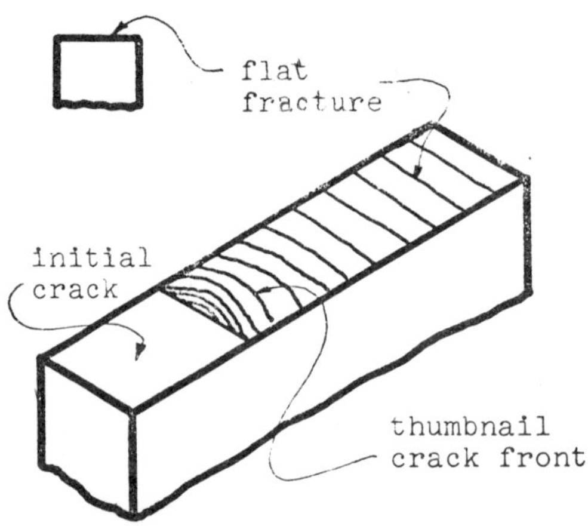


Fig.(5)-Predominantly flat fracture.

---

*There is always a small region [6] around the crack tip within which the mathematical results and experimental data become uncertain.

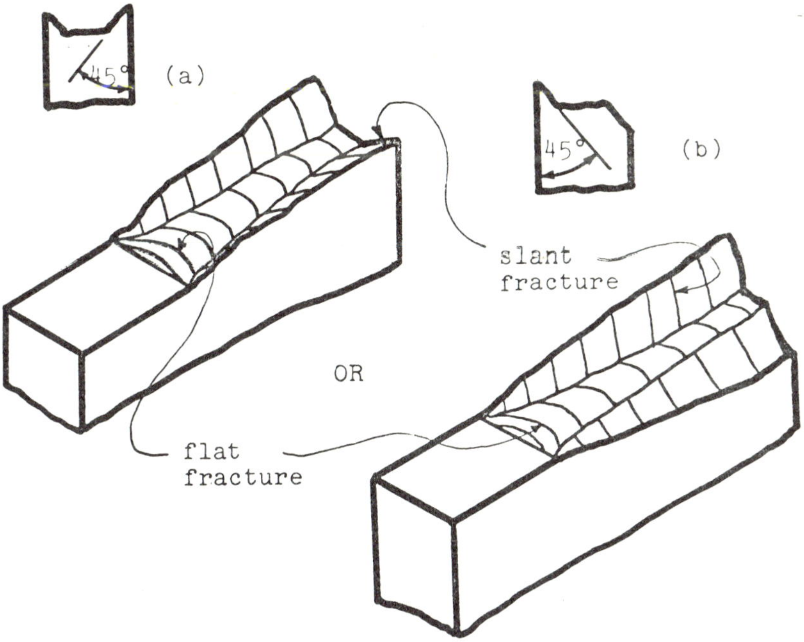


Fig.(6)-Flat and slant fracture.

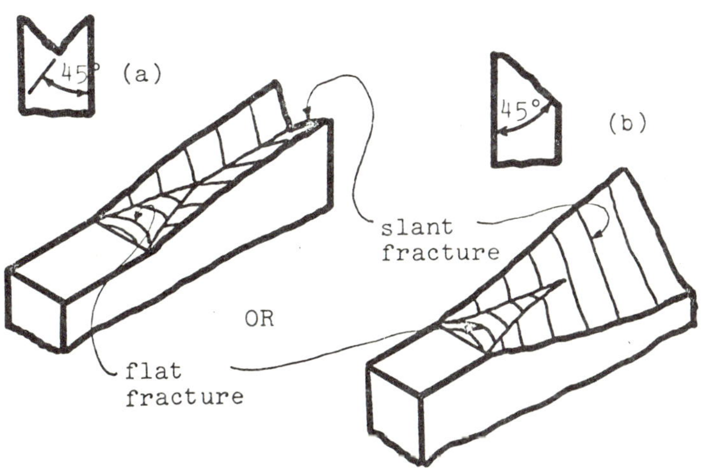


Fig.(7)-Predominantly slant fracture.

served to grow forward along the plane of the initial crack. The material near the surfaces of the plate tends to change its shape more readily and to dissipate more energy in distortion. The crack surface is observed to rotate in this region so as to make an angle of approximately 45° with the direction of tension, Figs. (6) and (7). These inclined portions of the crack surface near the plate surfaces are generally known as "shear lips". The proportions of each type of crack surface depend on material properties (yield strength, strain hardening exponent, etc.), specimen geometry (crack length, plate thickness, etc.), and loading. The proportion of the fracture surface associated with slant fracture or shear lips increases with decreasing plate thickness. Figure (7) shows that the slant fracture dominates across the plate leaving only a small portion of the flat fracture surface. This occurs in very thin plates in which the shear lips merge into each other and is evidenced by the localized dimpling or necking of the material in the region around the crack tip.

In order to predict the fracture growth patterns shown in Figs. (5) to (7), it is necessary to have a fracture criterion that accounts simultaneously for the volume change (dilatation) and shape change (distortion) experienced by each of the material elements. The strain energy density criterion [7] is particularly well suited for this purpose since it incorporates both of these effects automatically* and predicts that fracture is most likely to follow a path along which the material elements experience more volume change than shape change. This implies that macrofracture takes place in the elastic portion of the material and releases only elastic energy. The plastic portion of the material is relatively soft and can only contain damage or fracture at the microscale level. Hence, for moderately thick plates with a fracture pattern as shown in Fig. (6), the continuum crack tends to spread around the shear lips in the form of crack branching, Fig. (8). In reality, the crack never branches symmetrically, always leaving a shear lip on only one half of the separated pieces. The size of the thumbnail crack front can be predicted according to the strain energy density criterion by locating points with the same strain energy density, $dW/dV$.

A series of possible new crack front shapes as shown in Fig. (9) can be determined for various values of the parameters involved, say $\sigma$ (applied stress), $\sigma_{ys}$ (yield strength), $n$ (strain hardening exponent), $a$ (half crack length), $h$ (plate thickness), and $\rho$ (crack tip radius of curvature). For a particular set of parameters, the

---

*For details, refer to [7] in which the fracture surfaces following positions of minimum strain energy density around penny-shaped and elliptical cracks are displayed.

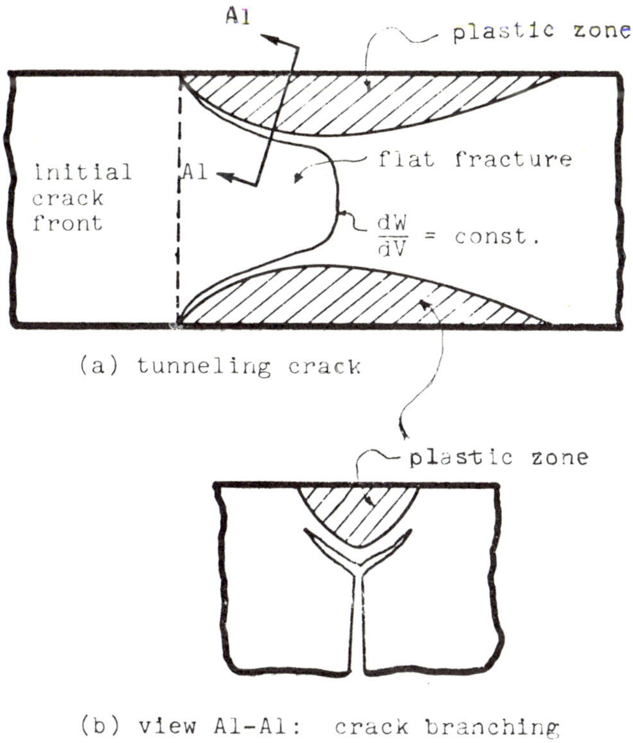


Fig.(8)-Shear lip formation in a moderately thick metal plate.

new crack front shape as predicted by the strain energy density criterion will be the basis of a second set of stress calculations. If the analysis for the extended crack indicates that an increased load is required to obtain further crack extension using the same growth criterion, then the crack extension will be considered to have been stable and the analysis will be repeated to study further growth. The compliance values for the various stages of crack growth can then be determined as the crack front tunnels through the mid-section of the plate and possibly branches near the plate surface (see Fig. (8)). The results may be summarized in the following mathematical statement

$$S\left(\frac{\sigma}{\sigma_{yd}}, n, \frac{\rho}{a}, \frac{h}{a}\right) = S_{cr}$$

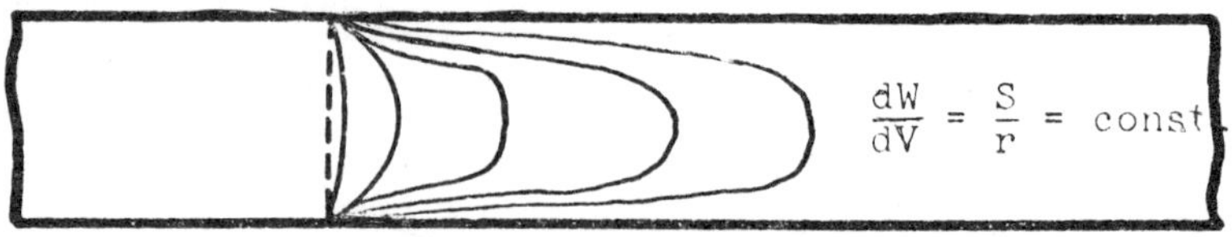


Fig.(9)-Crack growth patterns.

where the critical value of some strain energy density factor, $S_{cr}$, will be used as a measure of the fracture toughness of the material at the point of global instability or unstable crack propagation. For values of $S < S_{cr}$, the crack will tunnel and may even branch but travels in a stable manner as the load is gradually increased. A study of this problem is being carried out in the Institute of Fracture and Solid Mechanics, Lehigh University.

In conclusion, it should be emphasized that there remains a great deal of work to be done in the area of elastic-plastic fracture mechanics. The classical continuum plasticity approach is fundamentally inadequate for understanding the problem of elastic-plastic fracture. It is necessary to include a measure of the amount of physical damage to the material within the plastic zone into the analysis. Otherwise, it would not be possible to determine the energy, dissipated in plastic flow, which is not available to drive the continuum, or macro, crack. Such an approach will be further discussed in [8].

## REFERENCES

[1] Sih, G. C., Discussion of "Some observations on Sih's strain energy density approach for fracture prediction", by I. Finnie and H. O. Weiss, Int. J. of Fracture, 10, 279, 1974.

[2] Sih, G. C., "Application of strain energy density theory to fundamental fracture problems", Proc. of 10th Anniversary Meeting of Soc. of Engin. Sci., (in press).

[3] Hilton, P. D. and Hutchinson, J. W., "Plastic intensity factors for cracked plates", J. Engin. Frac. Mech., 3, 435, 1971.

[4] Shih, C. F., "Small scale yielding analysis of mixed mode plane strain crack problems", AFOSR Tech. Report 73-1692, 1973.

[5] Sih, G. C., Methods of analysis and solutions to crack problems, Mechanics of Fracture, Vol. I, Ed. G. C. Sih, Noordhoff International Publishing, Leyden, 21, 1973.

[6] Sih, G. C., Handbook of stress intensity factors, Institute of Fracture and Solid Mechanics, Lehigh University, Bethlehem, Pa., 1973.

[7] Sih, G. C., Three dimensional crack problems, Mechanics of Fracture, Vol. II, Ed. G. C. Sih, Noordhoff International Publishing, Leyden, (in press).

[8] Sih, G. C., "What are linear and nonlinear fracture mechanics?", J. Engin. Frac. Mech., (in press).

## STRESS INTENSITY FACTOR DETERMINATION FOR POLYMERS

J. C. Radon
Imperial College, London, United Kingdom

A method of K-determination is discussed. Special care is called for if the investigated material is strain rate sensitive, for instance, mild steel and certain polymers. Moreover, special attention should also be given to the range of temperature under which measurements are made.

$K_{1c}$ was investigated over a wide range of temperature and testing speeds on specimens of various geometries (SEN, DCB, 3-point bend tests) of polymethylmethacrylate. The graphs of $K_{1c}$ vs. temperature or strain rate showed distinct toughness peaks and in some regions the base line toughness changed abruptly. Such results obtained in impact tests are shown in Fig. (1). The pronounced features of the toughness curves were found to be subject to time-temperature relationships of the Boltzmann type and were explained in terms of relaxation motions of parts of the polymer molecule. For this purpose the "time to fracture" was regarded as an experimental macro-relaxation time and was directly measured or calculated from crack speed. It was observed that the well-known fracture mode transition in the glassy state was a consequence of the $\beta$ relaxation and a sharp toughness peak associated with the minor $\gamma$ transition was also established, Fig. (1). The apparent activation energies for the two processes obtained here correspond well with the published data.

In the field of plastics the use of dielectric data in fracture studies is limited to polar or polarisable polymers due to the very low values of $\varepsilon''$ and the nearly constant value of $\varepsilon'$

found in nonpolar polymers. Also, many relaxations have not yet been identified with particular molecular motions. However, evidence linking the fracture behaviour of polymers with changes in other physical properties occurring at second-order transitions is forthcoming. A coincidence of fracture mode transition with the abrupt modulus change and step in heat capacity accompanying the principal second-order transition Tg in PVC-ABS copolymer has been reported in [2]. Methods used on polymers may also serve as a useful model for other materials. Here, changes in optical properties, mechanical energy losses in vibration experiments and nuclear magnetic resonance results may be investigated. Preliminary results obtained in such tests on metals are discussed in [3].

## REFERENCES

[1] Johnson, F. A. and Radon, J. C., Eng. Fr. Mech., 4, 555, 1972.

[2] Radon, J. C., Materialprufung, 13, 253, 1971.

[3] Radon, J. C., "Fracture process at critical temperatures", 5th Conf. on Dimensioning and Strength Calculations, Budapest, October 1974.

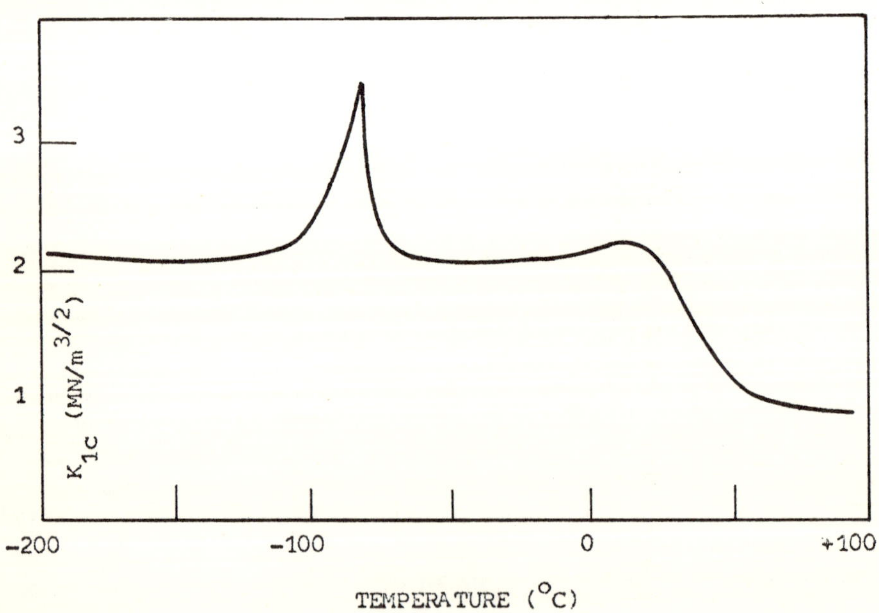


Fig. (1) - PMMA. $K_{1c}$ vs temperature in impact tests

# ANTIPLANE STEADY STATE CRACK MOTION IN A PERFECTLY PLASTIC MATERIAL

H. Andersson
Lund Institute of Technology, Lund, Sweden

The problem of steady-state motion of a crack-tip in a perfectly plastic material in the antiplane strain mode has been solved assuming a line discontinuity ahead of the crack tip, as proposed by Broberg [1]. The equation of equilibrium and the yield condition makes the stress vector $\bar{\tau} = (\tau_x, \tau_y)$ constant along straight lines in the plastic zone, Fig. (1). Instead of forming a centered fan, the straight lines are assumed to start from the zone of discontinuous slip with angles $\psi$ determined as a function $a(\psi)$ of the distance a from the crack tip.

A set of orthogonal coordinates is constructed from the stress characteristics in analogy with the cylindrical coordinates used with a centered fan field of characteristics. The equation of compatibility and the flow rule at steady-state conditions then give equations for the strains $\gamma_t$, $\gamma_\psi$ in these coordinates.

Inside the plastic zone these equations are solved numerically by a finite element method developed by Lynn and Arya [2], where the least squares criterion is used to minimize the error when fitting the element shape functions to the differential equations. A restricted solution is also performed by analytical methods.

For different assumptions about the function $a(\psi)$ different plastic zones are found. Standard finite element methods, Andersson [3], are used to optimize the choice of $a(\psi)$ by insertion of the boundary values of deformation and stress of the plastic zone into computer runs for the surrounding elastic area, assuming small-scale yielding conditions.

The deformations in the plastic zone obtained in this way are shown in Fig. (2). For comparison the corresponding deformations in the static case are shown as well.

It is found that the energy let through to the zone of discontinuous slip is about 22 percent of the total energy flow into the plastic region. This result supports the theory presented by Professor Broberg in his paper. Further, I think that this type of calculations for the in-plane modes should be of interest for discussions of micromechanisms at catastrophic failure after some stable crack growth.

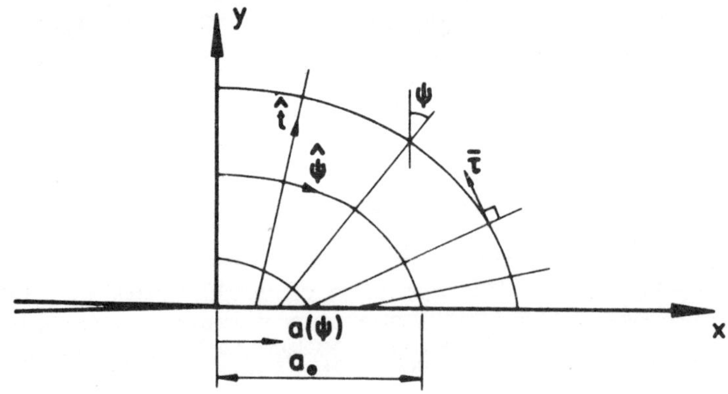
y
ψ
$\hat{t}$
$\hat{\psi}$
$\bar{\tau}$
a(ψ)
$a_o$
x

Fig. (1)

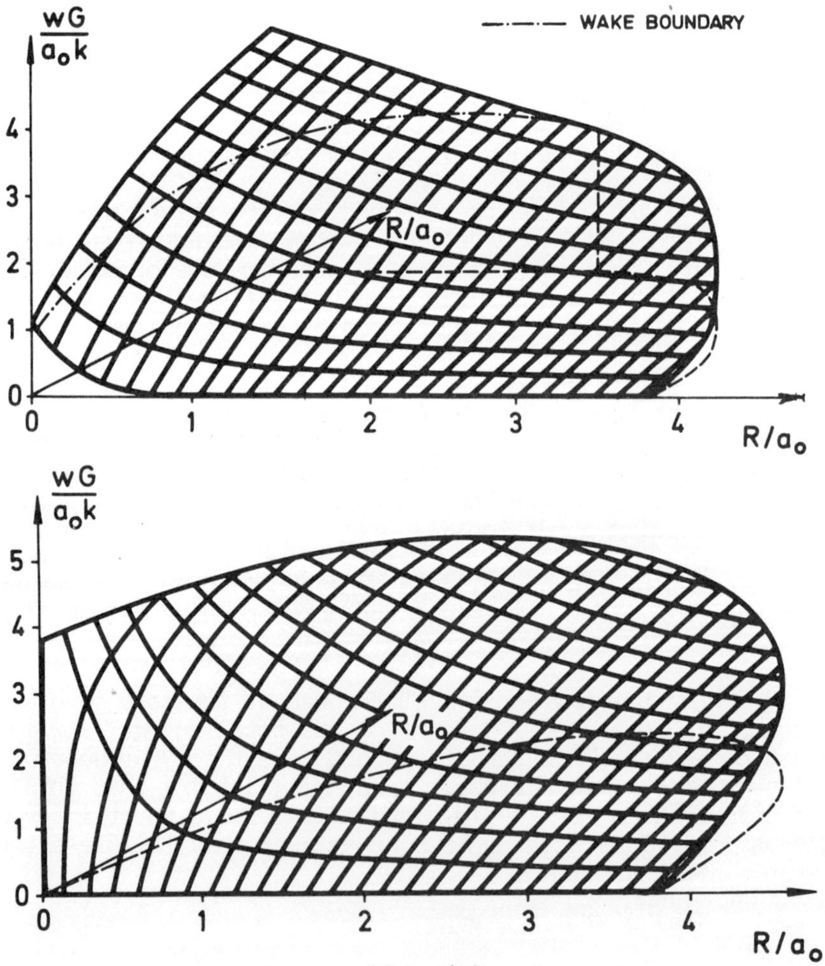
WAKE BOUNDARY
$\frac{wG}{a_o k}$
R/$a_o$
0
1
2
3
4
5

Fig. (2)

## REFERENCES

[1] Broberg, K. B., "On stable crack extension", Report from Div. of Solid Mechanics, Lund Institute of Technology, Lund, Sweden, 1973:4.

[2] Lynn, P. P. and Arya, S. K., Int. J. Num. Methods in Engng., 6, 75.

[3] Andersson, H., "Finite element treatment of a uniformly growing elastic-plastic crack tip", Report from Div. of Solid Mechanics, Lund Institute of Technology, Lund, Sweden, 1973:1.

## ON BRITTLE CRACK PROPAGATION AND ARREST

H. Bergkvist
Lund Institute of Technology, Lund, Sweden

In the last few years we have studied brittle crack propagation and arrest from a theoretical as well as an experimental point of view at Lund Institute of Technology.

The basis is a balance between the release of energy and the energy consumption in the dissipative processes. In the quasi-dynamic approach, the release of energy G from the stress field in the dynamic elastic case is assumed to be dependent on crack length a and on crack tip velocity c but not on higher time-derivatives of crack length. The dissipation is expressed in terms of a specific fracture energy $\gamma_F$ which is assumed to be dependent on crack tip velocity only. Thus the motion of the crack is governed by

$$G(a,c) = 2\gamma_F(c) \tag{1}$$

The function $\gamma_F = \gamma_F(c)$ for the material of interest can be determined from experiments on the long strip configuration as described by Paxson and Lucas at the Lehigh conference in 1972.

Since G is known analytically only for a very few configurations and since it is also desirable to handle sheet configurations where Young's modulus or thickness varies in the thickness direction, an expression for the energy flow to the crack tip is proposed:

$$G(a,c) = J(a)\ w(c) \tag{2}$$

Equation (2) is the product of the static J-integral value times some dynamic function depending only on the crack tip velocity.

It is easily seen that in the infinite sheet configuration, the function w(c) corresponds to the Broberg dynamic function.

J(a) can be obtained from a finite element calculation. The data from Paxson and Lucas for PMMA equations can be used to give $\gamma_F(c)$ and Broberg's dynamic function, w(c). The combination of Eqs. (1) and (2) gives

$$J(a) = \frac{2\gamma_F(c)}{w(c)} \tag{3}$$

which yields a relationship between crack length and crack tip velocity.

## DESIGN CONSIDERATIONS IN FRACTURE MECHANICS

T. Varga
Sulzer Brothers Limited, Winterthur, Switzerland

In evaluating the merits and limitations of different fracture mechanic methods, one should bear in mind that all of them may potentially contribute to solving practical problems. The emphasis should be on their proper use, and their limits of applicability; sources of error should be taken into consideration rather than condemning one or the other.

A "systems approach" to establish brittle fracture safety of structural steel constructions gives as a "user's guide" some considerations for a stepwise evaluation [1]. With plastic and brittle fracture behaviour there is little doubt: a non-ductile fracture is excluded in the former, linear elastic methods are given for the latter. In the intermediate, transitional range (or, if fracture behaviour is not known) first screening may be done by Charpy-V-testing. If the method does not seem to be adequate or criteria are not met, the next step, limit value procedures, may be used. Pellini drop-weight specimen, Schnadt pressed sharp notch, fatigued Charpy, Battelle drop weight and Robertson tests offer more stringent testing conditions as they can be found in many structures.

With a further increase in the testing effort, more exact information is obtainable by using "applied (or simulated) methods". Wells and Kihara wide plate tests are best examples.

Finally, if quantitative defect size or load stress information is desired, fracture mechanics methods apply. This is the case also, when criteria are not met in foregoing steps.

A fracture mechanics evaluation is usually done at the time being only on notched, fatigue pre-cracked specimen, provided a calculated or experimental calibration is available.

The vast majority of structures will always be only Charpy-tested. Limit value procedures as a next are considered to be safe, but necessitate often higher material quality than anticipated. Applied methods which need much work and cost, give a good but not quantitative result.

Fracture mechanics gives the most accurate information. Again we begin with the simplest test: fatigue pre-cracked instrumented Charpy to determine $K_{Id}$. If this test is applicable, we have some quantitative information. $K_{Ic}$ determination is a greater effort (except that the transformation $K_{Id}$ to $K_{Ic}$ can be used). Unless thickness independence becomes proven, specimens up to near the full wall thickness have to be tested in this case.

By using the special fracture mechanics specimens, their geometrical and testing condition adequacy in respect of the structure has to be verified.

For fracture mechanics evaluation, one needs maximal stresses, defect sizes, shapes and location. More exact results are obtainable for the stress (or strain), distribution over the whole volume and if defects mapped according to size, shape and location are disposable. So the effort can be adapted also here to the necessary amount.

Similarly, systematic considerations have been set-up and are in use for fatigue, environmental and temperature loadings.

## REFERENCE

[1] Varga, T., "A systems approach to assess brittle fracture safety", Proc. 2nd Int. ASME Conf. on Pressure Vessel Technology, Part II, pp. 1059-1070.

# LIST OF PARTICIPANTS

Achenbach, J. D.
Northwestern University
Evanston, Illinois

Adamis, P.
Institut fur Maschinenwesen
Clausthal, W. Germany

Andersson, H.
Lund Institute of Technology
Lund, Sweden

Angelino, G.
C.I.S.E.
Milan, Italy

Anstee, R. F.
Royal Aircraft Establishment
Farnborough, Hants, England

Atkinson, C.
Imperial College of Science
and Technology
London, England

Bergkvist, H.
Lund Institute of Technology
Lund, Sweden

Bradshaw, F. J.
Royal Aircraft Establishment
Farnborough, Hants, England

Bristoll, P.
Royal Shell Laboratory
Amsterdam, The Netherlands

Broberg, K. B.
Lund Institute of Technology
Lund, Sweden

Broek, D.
Institute of Technology
Delft, The Netherlands

Broekhoven, M. J. G.
Delft University
Delft, The Netherlands

Bui, H. D.
Ecole Polytechnique
Paris, France

Burgers, A.
Delft University
Delft, The Netherlands

Caglioti, G.
Institute Ingegneria
Milan, Italy

Cannon, D. F.
British Railways Board
Derby, England

Carlsson, A. J.
Royal Institute of Technology
Stockholm, Sweden

Cook, T. S.
Pratt and Whitney
Middletown, Connecticut

Cowan, A.
U.K.A.E.A.
Lancs, England

de Boer
The Netherlands

De Pater, C.
Technical University
Twente, The Netherlands

Diez, J. M.
C.R.M.
Liege, Belgium

Doker, H.
Institut fur Werkstoff Forschung
Porz-Wahn, W. Germany

Dvorak, G. J.
Duke University
Durham, North Carolina

Eccles, D.
Hawker Siddeley Aviation
Surrey, England

Elst, H. C. van
Metal Research Institute
Apeldoorn, The Netherlands

Erdogan, F.
Lehigh University
Bethlehem, Pennsylvania

Eshelby, J. D.
University of Sheffield
Sheffield, England

Ewald, T.
Delft University
Delft, The Netherlands

Folias, E. S.
University of Utah
Slat Lake City, Utah

Geier, W.
Messerschmidt-Bolkow-Blohm
Munchen, W. Germany

Gestel, G. F. J. A. van
Delft University
Delft, The Netherlands

Grange, D. H.
Hawker Siddeley Aviation
Woodford, England

Hartranft, R. J.
Lehigh University
Bethlehem, Pennsylvania

Hasselman, D. P. H.
Lehigh University
Bethlehem, Pennsylvania

Hayes, D. J.
Royal Shell Laboratory
Amsterdam, The Netherlands

Huth, H.
Lab. fur Betriebsfestigkeit
Darmstadt, W. Germany

Ingham, T.
U.K.A.E.A.
Lancs, England

Irvine, W. H.
U.K.A.E.A.
Lancs, England

Janssen, G. T. M.
Iweco TNO
Delft, The Netherlands

Jones, D. L.
George Washington University
Washington, D. C.

Jonker, G. J.
N.V. Nederlandse Gasunie
Groningen, The Netherlands

Kan, J. L. van
Central Laboratory DSM
Geleen, The Netherlands

Kanazawa, T.
University of Tokyo
Tokyo, Japan

Kanninen, M. F.
Battelle Laboratories
Columbus, Ohio

Kerkhof, F.
Institut fur Festkorpermechanik
Freiburg, W. Germany

Kilpatrick, I. M.
Naval Construction Establishment
Dunfermline, Scotland

Kirkby, W. T.
Royal Aircraft Establishment
Farnborough, Hants, England

Knott, J. F.
University of Cambridge
Cambridge, England

Kobayashi, A. S.
University of Washington
Seattle, Washington

Koning, A. V. de
National Aerospace Laboratory
Emmeloord, The Netherlands

Kretzschmannn, W. B.
Technische Hochschule
Aachen, W. Germany

Labbens, R. C.
Creusot-Loire
Paris, France

Lachat
CETIM
Senlis, France

Lambert, J. A. B.
Hawker Siddeley Aviation
Hatfield, Herts, England

Liebowitz, H.
George Washington University
Washington, D. C.

Lipzig, H. T. M. von
Institut fur Werkstoff
Forschung
Porz-Wahn, W. Germany

Los, A. J.
Delft University
Delft, The Netherlands

Louwaard, E. P.
Delft University
Delft, The Netherlands

Low, J. R., Jr.
Carnegie-Mellon University
Pittsburgh, Pennsylvania

Maarse, J.
Twente University
Enschede, The Netherlands

Maddox, S. T.
The Welding Institute
Cambridge, England

Marcus, H. L.
Rockwell International
Thousand Oaks, California

McCartney, L. N.
National Physical Laboratory
Teddington, Middlesex,
England

McEvily, A. J.
University of Connecticut
Storrs, Connecticut

Meijers, P.
Iweco TNO
Delft, The Netherlands

Miyata, T.
Nagoya University
Nagoya, Japan

Nibbering, J. J. W.
Delft and Ghent University
Delft, The Netherlands

Nicholson, C. E.
Safety in Mines Establishment
Sheffield, England

Oberparleiter, W.
Industrieanlagen-Betriebs GmbH
Ottobrunn, W. Germany

Ostensson, B.
AB Atomenergi
Nykoping, Sweden

Overbeeke, J. L.
Technical Highschool
Eindhoven, The Netherlands

Pelissier-Tanon
Creusot-Loire
Paris, France

Pelloux, R. M. N.
Massachusetts Institute of Technology
Cambridge, Massachusetts

Pineau, A. G.
Ecole des Mines de Paris
Corbeil-Essonnes, France

Pook, L. P.
National Engineering Laboratory
East Kilbride, Glasgow, Scotland

Pratt, P. L.
Imperial College
London, England

Quirk, A.
U.K.A.E.A.
Risley, Warrington, England

Radenkovic, Dr.
Ecole Polytechnique
Paris, France

Radon, J. C.
Imperial College
London, England

Reynen, J. W.
Euratom CCR
Ispra (Varese), Italy

Riedel, H.
MDI Eisenforschung
Dusseldorf, W. Germany

Rietjens, P. A. H.
N. V. Nederlandse Gasunie
Groningen, The Netherlands

Robinson, J. N.
Imperial College
London, England

Rongen, H. J. M. van
Metal Research Institute
Apeldoorn, The Netherlands

Rooke, D. P.
Royal Aircraft Establishment
Farnborough, Hants, England

Rosenfield, A. R.
Battelle Laboratories
Columbus, Ohio

Rousselier, G.
Electricite de France
Chatou, France

Schijve, J.
National Aerospace Laboratories
Amsterdam, The Netherlands

Schlonski, A.
Institut fur Maschinenwesen
Clausthal, W. Germany

Sih, G. C.
Lehigh University
Bethlehem, Pennsylvania

Simpson, A.
Hawker Siddeley Aviation
Woodford, Cheshire, England

Smith, E.
University of Manchester
Manchester, England

Smith, S. H.
Battelle Laboratories
Columbus, Ohio

Sommer, E.
Institut fur Festkorpermechanik
Freiburg, W. Germany

Staskewitsch, E.
IFAM Institut fur angewandte
Materialforschung
Bremen, W. Germany

Street, K. N.
Ecole des Mines de Paris
Corbeil, France

Swift, T.
Douglas Aircraft Company
Long Beach, California

Tetelman, A. S.
University of California
Los Angeles, California

Thomason, P. F.
University of Salford
Lancashire, England

Trense, R. W.
Technological Laboratory
Rijswijk, The Netherlands

Tsai, S.
Wright Patterson Air Force Base
Dayton, Ohio

Turner, C. E.
Imperial College
London, England

Varga, T.
Sulzer Brothers Limited
Winterthur, Switzerland

Vigier, P. J.
Aluminium Pechiney
Voreppe, France

Vlieger, H.
National Aerospace Laboratory
Amsterdam, The Netherlands

Vogelensang, L. B.
Technical University
Delft, The Netherlands

Vrtel, J.
Research Institute for
Materials
Prague, Czechoslovakia

Wanhill, R. J. H.
National Aerospace Laboratory
Emmeloord, The Netherlands

Wei, R. P.
Lehigh University
Bethlehem, Pennsylvania

Wildschut, H. H.
Metal Research Institute
Apeldoorn, The Netherlands

Wilhem, D. P.
Northrop Corporation
Hawthorne, California

Williams, T. S.
Hawker Siddeley Aviation
Surrey, England

Wnuk, M.
South Dakota State University
Brookings, South Dakota

Wullaert, R. A.
Dynatup Operations
Santa Barbara, California

# AUTHOR'S INDEX

# SUBJECT INDEX